# UNDERSTANDING LIFE

## Perspectives from Physical, Organic, and Biological Chemistry

Revised First Edition

By Ronald J. Duchovic
*Indiana University—Purdue University Fort Wayne*

Bassim Hamadeh, CEO and Publisher
Michael Simpson, Vice President of Acquisitions
Jamie Giganti, Senior Managing Editor
Jess Busch, Senior Graphic Designer
Angela Schultz, Senior Field Acquisitions Editor
Natalie Lakosil, Licensing Manager
Mandy Licata, Interior Designer



First published in the United States of America in 2016 by Cognella, Inc.



Printed in the United States of America

ISBN: 978-1-63487-138-9 (pbk)/ 978-1-63487-139-6 (br)

www.cognella.com 800-200-3908

# Contents

# CHAPTER ZERO
## A Conceptual Orientation and Invitation

### How Did Chemistry Begin?

The field of chemistry represents one of the earliest fields of study pursued by humankind with a systematic and methodical persistence. Long before the rise of the modern scientific method (traceable to the early conceptualizations of Francis Bacon in the sixteenth century and the magnificent steps taken by Galileo Galilei and Isaac Newton in the seventeenth century), only a handful of human investigations—astronomy, geometry, medicine, and chemistry—focused on the world in which the human family found itself. Of these, chemistry uniquely stood at the fateful intersection of careful observation and innovative application. On the one hand, it was realized by the earliest practitioners of the chemical sciences (the metallurgists who focused on the properties of ores and the extraction of metals and the seventh-century Chinese discovers of gunpowder) that the key to their success lay in paying meticulous attention to the behaviors of substances. Unlike the Aristotelian perspective, which relegated inductive empiricism to a secondary role in scientific reasoning, these earliest chemists (and, later, the alchemists) amassed a rich catalog of **observable** properties that characterized the many diverse substances that make up our physical world. But the real heart of these early chemical investigations lay elsewhere; it was the innovative **application** of these observations to create new materials and new phenomena (the development of new alloys such as brass and bronze well before the Christian era and the application of colorant materials to clothing throughout the Middle Ages) that distinguished chemistry from other human intellectual pursuits.

As a consequence of the impact of Galileo and Newton, by the end of the seventeenth and the beginning of the eighteenth centuries, the science of chemistry had become well-recognized as a major contributor to the wide-ranging endeavor known as "natural philosophy." In fact, with the quantitative description of gaseous behavior, chemistry had begun its fateful trek down the road pioneered by physics with its use of powerful mathematical techniques. Further, because chemistry had long stood at the crucial intersection of observation and application, it was not surprising, by the mid-nineteenth century, to see the impact of the worldwide dye industry on the science. With the rising dominance of the textile industry (stimulated by the introduction of technologies from the Industrial Revolution) and the importance of colorants to that industry, chemistry became a respectable profession; that is, one could earn a living as a professional chemist. The dye industry throughout Europe, but particularly in Germany, relied on the use of both synthetic and newly synthesized natural-product dyes. By the end of the century, advances in the dyestuff industry were the result of forging intimate links between basic chemical research and the technology of dye manufacturing.

Concurrent with this marriage between chemical research and industrial manufacturing technology, a significant conceptual paradigm was introduced into the chemical sciences. By the late 1850s Kekulé had proposed his ring structure for benzene and, along with Couper (interestingly, the son of a Scottish textile mill owner), had articulated the concept of the tetravalent carbon atom. These ideas placed structure at the heart of chemical thinking, transforming the study of chemistry from the dual focus on observation and application that had characterized chemistry throughout most of its long history into a discipline that increasingly attributed observed properties to putative structures. This connection, in turn, became an avenue for the rational introduction of novel materials and processes. No longer was chemistry simply limited to making observations, and then, to utilizing those observations in specific and often limited applications. As structure began to assume a central position in the chemical science, there was an almost inexorable realization that chemical reactivity is intimately and inseparably linked to chemical structure.

As a result, at the end of the nineteenth century chemistry was rapidly becoming a mature intellectual discipline that began to exhibit the diversity and sophistication that had formerly been attributed to only a select few human activities. The field of study that had simply been called "chemistry" was now identified as "organic chemistry." In the opening decades of the twentieth century, chemistry began to be differentiated into a number of distinct subdisciplines (again, a sign of growing intellectual maturity): organic chemistry, physical chemistry, inorganic chemistry. With the growing presence of sophisticated instrumentation, analytical chemistry was added to this constellation of subspecialties. The dramatic developments in the life sciences during the twentieth and twenty-first centuries—the germ theory, antibacterials, antivirals, genomics—propelled biochemistry to a position of prominence; by the end of the twentieth century many departments of chemistry had been renamed departments of chemistry and biochemistry. Finally, we would be remiss not to note again the significant impact on the chemical sciences of the tremendous growth of the pharmaceutical, petrochemical, and petroleum industries throughout the past century. Like the dye industry of the nineteenth century, these industries have had a transformative effect on the science of chemistry.

There is a long-standing tradition that organizes the study of chemistry around its major subdisciplines: analytical chemistry, biochemistry, inorganic chemistry, organic chemistry, and physical chemistry. For nearly a century both departments of chemistry and standard chemistry textbooks have followed this "subdisciplinary" approach, which is a highly effective organizational tool. However, it has the unfortunate potential to fragment the science by implying that there are well-defined boundaries separating and, in fact, isolating all the subdisciplines of chemistry. The reality is that the science of chemistry is a highly integrated discipline, being both a **body of knowledge** and a **process of discovery**, in which common fundamental principles form the essential structure of each area of chemistry. The existence of these common principles means that the phenomena and processes on which each subdiscipline focuses are not in some sense "special," thereby requiring "special" explanations or interpretations that apply only to a specific subdiscipline. On the contrary, the knowledge and processes characteristic of each area of chemistry all rest on a set of fundamental physical principles that belong not just to a single subdiscipline but to the science as a whole. In fact, from the broader perspective of all the physical sciences, one of the most profound lessons of the twentieth century has been the realization the individual physical sciences are all deeply interrelated by and dependent on the same fundamental physical principles.

## Key Concepts

In recognition of the importance of the common fundamental principles that lie at the foundation of all the areas of chemistry, this text is organized around central themes that will recur repeatedly throughout the text. Rather than focusing on an "organic" idea or a "biochemistry" idea, we will highlight the themes of chemical **structures**, chemical **reactivity**, and chemical **energetics** as they recur throughout our story of chemistry. In each chapter or topical section, the student is strongly encouraged to identify these recurring themes:

1. Structure
2. Reactivity
3. Energetics

Because chemistry is a science with a very long history, chemistry texts, sometimes attempting to be comprehensive, sometimes responding to respected pedagogical traditions, and sometimes simply choosing an appealing but less accurate description of a physical phenomenon, often present explanations that do not incorporate the best and most consistent contemporary understanding of the physical universe. In many instances, this choice seems to be the result of an assumption that a partial explanation, easily described and quickly grasped by the student, is better than a discussion of the subtle complexities that are characteristic of the world in which we live. Further, because the physical sciences over the past three hundred years have utilized an increasingly mathematically sophisticated set of theories to understand our universe, texts, at the introductory level, are reluctant to embrace this mathematical sophistication. There is good reason for this. The beginning student frequently does not command the mathematical tools needed to address these contemporary explanations. However, this does not mean that the conceptual framework of a modern theory that is grounded in sophisticated mathematics should be ignored.

## Models and Mathematics

This text will focus on the best set of consistent modern paradigms in order to understand the rich diversity of chemical knowledge (both the collections of observed facts and the process of "doing chemistry") that has been assembled over the course of human history. This approach does not mean that we will ignore previous chemical models in favor of models or theories that are in some sense "fashionable." Rather, for those cases in which multiple models are capable of providing both insight into the operation of chemical principles and powerful predictions verified by experiment, the text will highlight and examine a range of models in an effort to identify a consistent set of modern paradigms. In some instances, this analysis will transform earlier chemical models into footnotes to a modern paradigm of greater explanatory and predictive power. Consequently, the student will not simply be presented with and asked to memorize a single model or algorithm; the focus will be on understanding the best explanation that is consistent with a global set of modern paradigms. In fact, the process of "memorizing without understanding" will not work.

The role of mathematics in modern chemistry, particularly at the introductory level, is much more problematic. Because our contemporary understanding of chemical phenomena relies heavily on sophisticated mathematics (multivariable calculus, the mathematics of quantum field theory, abstract algebra, and group theory, to name just a few), the best modern chemical paradigms are replete with powerful mathematics. To the introductory student, this becomes an almost insurmountable barrier to understanding the essential chemical concepts. However, "almost insurmountable" is not the same thing as "impossible." This text will not shy away from the mathematics, but the goal of our presentation is not to achieve mathematical proficiency in topics well beyond the introductory level. Consequently, the mathematics used in modern chemistry will be discussed, often at the conceptual level, so that the student is introduced to the underlying assumptions of modern chemical models and understands their connections to other branches of modern science.

However, one of the key progressions that have characterized Western science since the Renaissance has been the movement from a qualitative understanding to a quantitative understanding. Beginning most clearly with physics, and continuing in every other science, there has been an unrelenting effort to quantify, to progress from a generally descriptive presentation toward quantitative analysis and prediction. Chemistry is no exception to this general trend characterizing all areas of science. This means, inevitably, that this text will use mathematics. Consequently, while there are no expectations that a student must bring sophisticated mathematical tools to the study of this text, students will use and be expected to both know and understand elementary mathematical relationships. Equations, calculations, and fundamental problem-solving techniques will form an integral part of the text. The student will find that the key to mastering chemistry lies in solving problems; as a wise chemistry instructor once noted, "Chemistry is not a spectator sport."

## What Is Science?

It has been a mere seven hundred years since the intellectual awakening called the Renaissance began in the city of Florence and only four hundred years since the dawn of modern Western science with the work of Galileo and Newton in the seventeenth century. Even more amazingly, the critically important disciplines of thermodynamics (eighteenth century), statistical mechanics (nineteenth century), and quantum mechanics (twentieth century) are mere infants in comparison; yet the technology of the twenty-first century is the direct consequence of the rapid development and ubiquitous application of these (and many other) disciplines. How did this happen? More importantly, at the beginning of the twenty-first century, what is science?

The Aristotelian model of knowing (in conjunction with Platonism, named for Plato, the teacher of Aristotle) emerged in Greece during the fourth century BCE and, after its transmission to Western Europe via Arab scholars as the first millennium CE came to an end, became the dominant framework for understanding the physical universe. Fundamentally for Aristotle, the highest form of knowledge consisted of understanding the cause of a phenomenon. In short, he asked the question, "Why?" Importantly, he required the answer to this question to fall into one of four categories, Aristotle's four causes. The most important method to answer the question was the series of deductive steps from premises to conclusion, the Aristotelian syllogism. As we noted earlier, the process of inductive empiricism, the reasoning from many individual instances to a general statement, was considered to be an inferior route to knowledge.

As we have already seen, the work of both Galileo and Newton dramatically changed the entire focus of humankind's attempt to understand the physical universe. The question changed from "Why?" to "How?" In fact, Newton, in his monumental work, *Philosophiae Naturalis Principia Mathematica* (*Mathematical Principles of Natural Philosophy*), explicitly refuses to make hypotheses, that is, to suggest causes. He writes, "I have not as yet been able to deduce from phenomena the reason for these properties of gravity, and I do not feign hypotheses. For whatever is not deduced from the phenomena must be called a hypothesis; and hypotheses, whether metaphysical or physical, or based on occult qualities, or mechanical, have no place in experimental philosophy. In this experimental philosophy, propositions are deduced from the phenomena and are made general by induction."[1] His focus is to demonstrate how the universe behaves, based on empirical observations. There is no attempt by Newton to explain (in the sense of answering the question of "why") the phenomena that are observed. Following the revolution in thinking made by Galileo and Newton, empirical observations of the physical universe became the final arbiters of all scientific knowledge.

Throughout the eighteenth and nineteenth centuries a growing cadre of "natural philosophers" widened and deepened the body of knowledge that was gleaned from the surrounding world through ever-more-sophisticated experimental techniques and persistent observations. The invention of the microscope (very late sixteenth century) and the telescope (first decade of the seventeenth century) extended the ability of humankind to observe directly both the very small and the very distant. It was, in fact, the telescope that enabled Galileo to confirm visually the Copernican ideas that forever changed humankind's position in the universe. In addition to the growing body of observational data, natural philosophers began the arduous effort of constructing generalizations (what may be called "theories") that both summarized the observational data and seemed to offer encompassing explanations with

1 Isaac Newton, *Philosophiae Naturalis Principia Mathematica* (London: 1687); Isaac Newton, *The Principia: Mathematical Principles of Natural Philosophy: A New Translation*, trans. I. B. Cohen and Anne Whitman, preceded by I. B. Cohen, "A Guide to Newton's *Principia*" (Berkeley: University of California Press, 1999), 943.

varying degrees of predictive power. In effect, it seemed that science (the term now commonly used in place of "natural philosophy," although the first recorded use of this term dates from the fourteenth century) was approaching a "true" understanding of the physical universe. But even further, buoyed by these successes, the scientific enterprise became identified as a foolproof process by which humankind could discover every possible "truth" about our universe. Just as Euclidean geometry was accepted as the sole "true" (some may have said "unique") geometry describing the concept of "space" in our universe, the scientific method was humankind's unerring method of discovering that which is "true" about the physical universe.

However, beginning in the late eighteenth century and clearly by the middle of the nineteenth century, Euclidean geometry's distinction as the "true" geometry correctly describing the concept of space in the universe was challenged by the creation of the so-called non-Euclidean geometries. (Curiously, while spherical geometry and trigonometry had been studied two millennia earlier by the Greeks, no one seemed bothered that the *surface* of a sphere is a non-Euclidean geometry. Perhaps, there was a simple acceptance of balls and spheres as *surfaces* in the three-dimensional space described so successfully by Euclid.) With the dawn of the twentieth century and Einstein's presentation of his general theory of relativity, suddenly the Newtonian description of gravity that was fundamental to humankind's conception of the universe was replaced by Einstein's elegant, highly mathematical, and most importantly, fundamentally different theory of gravitation. This immediately raised a serious question: Whose description is "true," Newton's or Einstein's? Ironically, the pursuit of a scientific theory by Einstein had called into to question the assumption that science is a discovery process that invariably yields a "true" answer. How could two so totally different theories be simultaneously "true"?

With the advent of the quantum theory during the first quarter of the twentieth century, another challenge to the scientific enterprise emerged. Since the time of Galileo and Newton, the question of "how" had supplanted the question of "why" that was asked by the Greek philosophers. The structure of the quantum theory forced humankind to begin asking a new question: "What?" While the quantum theory that burst upon the world at the beginning of the twentieth century would correctly describe every set of experimental data, it could no longer address the question of "how." The initial theory and its successors (developed painstakingly throughout the twentieth century) correctly accounted for what happened in each experiment (the outcome of an experiment) but could not provide a description of a physical process at the microscopic level that matched humankind's macroscopic experience of the physical universe. As a matter of fact, the quantum field theories that emerged as the century ended represent the most successful scientific theories ever created by the human mind; the disagreement between the experimental measurements and the theoretical predictions is on the order of 1 part in $10^{11}$, an astounding achievement.

Where do these developments leave science? Rather than understanding science as a discovery process, with each new discovery increasingly approaching a "true" description of our universe, we will suggest that in light of the developments that span the past 125 years, science is better understood as a "model-building process." Consequently, science is not focused on "truth"; this is the purview of logic and mathematics. The so-called "scientific method" (which we will examine in some detail in chapter 1) is **not** attempting to discover the "truth" about our universe. Science is a model-building process based on observations of the universe; its models (theories) are both **explanatory** and **predictive**. Note that it is not sufficient to be explanatory; both mythology and magic are explanatory, but they are not predictive.

Science possesses both characteristics. Further, scientific theories are **never proven**; scientific theories are only falsified. That is, there is no definitive experiment that provides a final proof that a theory is correct; each experimental result that is consistent with a theory's explanation and predictions certainly verifies the theory, but does not prove it. A single validated experiment, on the other hand, can falsify a theory. The reader should observe an important corollary of this understanding of science: if a theory or model is **not** falsifiable, it is unscientific.

As the twenty-first century unfolds, science has taken on a guise distinctly different from the activity known to the natural philosophers of the seventeenth and eighteenth centuries. The contemporary scientist is a builder of falsifiable models that possess both explanatory and predictive power. The elegantly simple image of the scientist as a "seeker of truth" has been replaced by the more complex and subtle image of the scientist as a builder of models. The touchstone of these models is the consistency between model and experiment. It is not that the science is "true"; rather, we all share a single common truth: the universe itself. Further, the scientist of the twenty-first century no longer asks "Why?" or "How?" but only "What?" It can be said that modern science creates models so that humankind is more easily able to remember what the universe does, to remember the marvelous majesty and diversity of activity found in our world.

This changed perspective of what constitutes modern science in no way diminishes the demands that the scientific disciplines place on their practitioners. In fact, the demands may be even greater in the sense that one of the great lessons of the previous two hundred years, in addition to the changed understanding of science itself, is the further understanding of the interconnectedness of the individual scientific subspecialties. While each scientific discipline, particularly during the last century, has been repeatedly subdivided into more and more specific domains, the stunning and truly awesome realization has dawned that common and consistent principles underpin all the subdisciplines. Each individual discipline is not an island unto itself; there is a deep connectivity among all the scientific fields.

This connectivity demands the use and clear understanding of a basic terminology, a set of fundamental concepts and terms on which the sciences are built. While there are many terms that do not seem to be "chemical" in the sense that they do not belong to chemistry alone, it is critical that we take time to demarcate unambiguously the meaning of these primitive (i.e., fundamental) terms. We will devote a part of chapter 1 to this task. Without a successful appreciation of these fundamental terms, we cannot appreciate the deep interconnectedness of modern science.

## A Map or Outline of Concepts

As this text unfolds, the reader will see a progression that begins (chapter 2) with a discussions of **origins**. That is, because chemistry is focused in a particular way on the properties of matter, on all the transformations it undergoes, and on the energy changes associated with those transformations, the most basic question is "From where did our universe come?" We will briefly explore humankind's best current understanding of the origins of the physical universe, the very beginnings of the atoms and energy, and of the space and time that constitute the fabric of our physical existence. We will articulate a fundamental question, the answer to which will consume a large part of this text: "Why is there **something** and not **nothing**?"

In fact, the framing of this question will initiate the first great theme that will recur throughout this text: **structure**. What is the structure of atoms? What is the structure of molecules? Why do atoms and molecules exist in our world? But as the reader may suspect, the answers to these questions will inevitably involve the second and third central themes of this text: **reactivity** and **energetics**. The three themes, while apparently separate and distinct concepts, in actuality are tightly linked in a complex ballet that makes chemistry not simply a collection of disconnected facts but rather a rich process of interacting components. We shall see that the themes of structure, reactivity, and energetics both enable as well as circumscribe one another, producing a multifaceted richness that goes beyond the apparently simple characteristics of each individual theme.

The reader will find that the introduction of the idea of **chemical structure** marked a watershed in chemical thinking. The identification of specific properties with putative atomic arrangements introduced an organizing principle of great power and, as the nineteenth century ended, transformed the science of chemistry. The suggestion that a specific grouping of atoms correlated with distinct properties brought order to the multitude of observations that had been cataloged over the centuries. But, most importantly, these groupings of atoms came to be called **functional groups** precisely because the very arrangement of the atoms in the group determined the **reactivity**, the *function*, of the structure. Consequently, from the point of view of chemistry, the most important property of a substance is not simply a static characteristic but rather the dynamic reaction(s) in which it participates. This dynamism has achieved a level of immense complexity in the **structures** and **reactivity** that demarcate biochemistry, the chemistry of life as we know it on this planet. The biochemical structures of life forms, the ability to replicate, and the symmetric processes of anabolism (building up) and catabolism (breaking down) that together define metabolism represent an exquisite interplay of structure and reactivity.

But it will come as no surprise to the reader that the idea of reactivity extends far beyond the elementary picture of something either happening or not happening. Each transformation that occurs in every part of the physical universe is characterized and determined by a concomitant energy change. The energy determinants, that is, the **energetics**, of a transformation are central to our understanding of all the dimensions of the transformation because the transformations we observe are not simply on-or-off or all-or-nothing processes. On the contrary, there are issues of energy differences (intimately related to the concept of chemical equilibrium), energy conservation (a thermodynamic concept, with the word "thermodynamics" coming from two Greek words: *therme*, meaning "heat," and *dynamis*, meaning "power"), and, finally, rates of reactions (chemical kinetics, from the Greek "*kinesis*," meaning "to move"). Even more subtly, there are questions of directionality, which can be addressed in terms of spontaneity or nonspontaneity (another thermodynamic concept). Finally, all of these discussions raise a question about the meaning of the word "energy" itself. What is energy? What does it mean to say, "I am conserving energy"?

All of these intriguing questions lie before us as we begin to examine the science of chemistry. We have sketched here a brief road map of a fascinating world. It is now time to take our first steps into a realm of great beauty that will both challenge us and reward us as we uncover its amazing dimensions.

# Chapter 0 Exercises

1. As chemistry began, the focus of the science was on two activities. What were those activities?
2. Who were the two people whose work in the seventeenth century and at the beginning of the eighteenth century changed science?
3. As chemistry became a highly integrated science, what were the two characteristics that defined the science?
4. List the five subdisciplines of chemistry.
5. What are the three central themes that will be emphasized in this text?
6. In the nineteenth century one of the important themes discussed in this text became central to chemistry. What is that theme?
7. Why is mathematics important to the study of chemistry?
8. Does a scientist stop her investigations when she has achieved a qualitative explanation of a physical phenomenon?
9. If science is not the search for "truth," what is science?
10. Which question did the science of Aristotle try to answer?
11. Galileo and Newton also asked a question. What is that question?
12. What did Newton refuse to do?
13. Modern quantum mechanics also asks a characteristic question. What is it?
14. What are the two characteristics of modern science?
15. Can a scientific theory be proven?
16. Chemistry's emphasis on *structure* led to the identification of groupings of atoms. What is the name given to these groups of atoms?

# Chapter 0 Exercises

1. [illegible] the focus of the [illegible] on two activities. What were those activities?
2. [illegible] in the [illegible] century and at the beginning of the [illegible]?
3. [illegible] became a highly [illegible] science, what were the two characteristics that defined [illegible]?
4. [illegible]
5. [illegible] in the text.
6. [illegible] in this text are central to chemistry. What are these?
7. [illegible]
8. [illegible] explanation of a [illegible]
9. [illegible]
10. [illegible]
11. [illegible]
12. [illegible]
13. [illegible]
14. [illegible]
15. [illegible]
16. [illegible] of [illegible]. What is the [illegible]

# CHAPTER ONE
## Tools of Chemistry

We begin our study of the chemical sciences by equipping ourselves with a collection of basic tools that we will use repeatedly in the coming chapters. These tools include

1. Some very general principles, often collectively called "the Scientific Method"
2. A collection of fundamental terms and concepts that underpin all of modern science in the twenty-first century
3. A set of mathematical tools, data analysis methods, and problem-solving skills

All of the tools introduced in this chapter are essential to the study of a quantitative science. Because these tools are so critical, it is imperative that the reader pay close attention to this chapter and master each topic; they are, in very many ways, foundational to our entire study of chemistry. Consequently, mastery of these topics cannot be postponed, or, perhaps better stated, postpone them at your peril! The time to achieve mastery of these topics is **now**, during chapter 1.

### Nature of Science

We noted in chapter 0 that science is a model-building process based on observations of the universe; its models (often called "theories") are both **explanatory** and **predictive**. Our first goal is to turn this general understanding of science into an effective instrument with which can begin to interrogate the universe around us. As a fundamentally empirical science, chemistry begins with **observations**; that is, it is rooted in the physical universe. Chemistry does not exist apart from the universe nor in isolation of the universe. But crucially, it does not stop with asking questions of the universe. Once a question is framed and an answer identified, the science attempts to **represent** the answer in some form, often in the language of mathematics. The representation has the powerful effect of leading humankind to **interpretations** of the answers. The process of interpreting our observations and representations is characterized by a transition from initial **qualitative** statements or analyses to increasingly **quantitative** statements. This dynamic immediately implies that mathematics, systems of units, and the critical mathematical analysis of observed data all play a central role in the "doing" of science as a model-building exercise. Thus, contemporary physical science is a model-building process that requires **observation**, **representation**, and **interpretation**, three steps that inevitably move from the **qualitative** level to the **quantitative** level. The consequence of this model-building process is the creation of an integrated conceptual framework that is both explanatory and predictive.

## The Scientific Method

Fundamentally, the scientific method is an organized procedure designed to enable humankind to understand an external reality. Because of its focus on an external reality, it does make a fundamental assumption: all of humankind shares a common external reality that is equally accessible to each individual. Anyone can investigate the **same commonly shared** phenomenon in an attempt to understand it. Pay close attention to the fact that the reality is both **external** and **shared**. Note that without this basic assumption (and, make no mistake, it is an assumption), humankind is reduced to the state of a disembodied "brain in a vat" being stimulated by an alien super intellect (or supercomputer). In this scenario, our "external reality" is nothing more than a sequence of neural stimulations. However, keeping the perspective offered by this assumption in mind, we find that the scientific method has been effectively used in both the physical and the biological sciences, and, over the past century, in the social sciences.

There is a second key point (in addition to the assumption made above) associated with the scientific method. While it is **a highly effective** way to understand our external world, it is **not the only** way to understand the external world. Clearly, art, dance, literature, music, poetry, and song (to list only a few) are legitimate approaches to articulating and expressing an understanding of humankind's external reality.

Because the scientific method is an organized and systematic approach to understanding, it consists of a number of well-defined steps. We begin by asking a question. Fundamentally, this means clearly identifying a part of the external universe to study, picking a topic to investigate, and then simply formulating a question (or, more daringly, a series of questions). Once a question (or questions) has been articulated, the investigator simply **observes** the selected phenomenon. The observations may or may not involve experiments which focus on a single aspect (i.e., a variable) of the chosen phenomenon. At this stage, the scientist is simply playing with a part of the world, observing and recording what she observes. If possible, there is an attempt to quantify what may have been an initial qualitative observation. (Remember that a key characteristic of science is the movement from a qualitative observation to a quantitative statement about the phenomenon.)

At this point, after making initial observations, an investigator can formulate a **hypothesis**. What is this? A hypothesis is a **tentative explanation** (here the emphasis is on the word *tentative*) for a collection of observations. It is not a definitive answer, but merely a suggestion. What next? Once a hypothesis has been articulated, the first wave of hard work begins. It is now time to design controlled experiments, isolating variables, collecting data, and inventing new experiments based on the results of the initial experiments. The center of attention is testing the hypothesis by interrogating the universe. The process can continue for extended periods of time: days, months, or years. The hypothesis testing is, in fact, even more complicated than I have suggested. The most obvious step to testing a hypothesis is a physical experiment. However, hypothesis testing does not and cannot stop at this level. A complementary process is the completion of a theoretical analysis, which could be a very detailed mathematical analysis of accumulating data or a simple "back-of-the-envelope" calculation. As a result, science is really neither a purely experimental activity nor a purely theoretical one; rather, there is a complex dance in which theory and experiment are intertwined.

However, as the twentieth century ended, a new dimension entered the hypothesis-testing phase of the scientific method: computer simulation. The advent of powerful computer hardware and innovative software near the end of the previous century ushered in the ability to simulate complex physical processes. Problems in structural analysis, weather and climate change, fluid dynamics, aircraft design, and chemical processes (to mention only a few examples) could now be modeled and tested via the modern supercomputer. It became possible to make direct comparisons with collected experimental data as well as realistic extrapolations than went far beyond our experimental capabilities. With the aid of this sophisticated computational technology, science now had three windows through which to view the universe: experiment, analytical analysis, and computer simulation.

Having amassed a variety of data testing the initial hypothesis, the next step for the investigator is to formulate of a **law**: a concise verbal or mathematical statement of a relationship among phenomena that is always identical when observed under the same conditions. It is important to notice that the term "law" as used here is very different from the social or political concept of "law." Whereas in the political context, a law represents the act of a legislative body, requiring compromises and agreements or societal approval, a scientific law rests solely on the outcome of the hypothesis testing that has taken place. It is a consequence of interrogating and simulating the physical universe and **not** the result of reaching a compromise agreement. Hence, scientific law is rooted in the commonly shared external reality and is not simply a reflection of humankind's desires, decisions, or compromises.

But science is much more than the careful collection of data, the analysis of that data, or a simulation of a portion of the physical universe. These activities have indeed produced a representation of the physical universe. But what is missing? Remember that science is a model-building process that is both explanatory and predictive. Consequently, the tentative explanation suggested by a hypothesis and the meticulous testing of the hypothesis is certainly necessary but **not sufficient**. The next step is the articulation of a **theory**: a unifying principle that explains a body of observations and the "laws" that are based on the observations. It is important to realize that theories are continuously tested; they are **not** final explanations. Further, a scientific theory is formulated to be falsifiable; it must make predictions that either are repeatedly verified by interrogating the universe or predictions that can be shown to be invalid.

It is essential to recognize that the word "theory" as it is used in science has a distinctly different meaning than the colloquial use of this term. Commonly, we hear the statements "It's only theoretical," or "Theoretically speaking …," which connote that the idea being discussed is somewhat uncertain, ill-defined, or able to be accepted or rejected with equanimity. However, we must recognize that, at any given point in time, a scientific theory represents the **best** efforts of humankind to produce a model of some portion of the physical universe that is both explanatory and predictive. It is a model resulting from the dynamic processes of observation, representation, and interpretation that often involves groups of investigators distributed in both space and time. It does not matter if the theory comes from physics, such as quantum mechanics, from cosmology, such as the general theory of relativity, or from biology, such as the theory of Darwinian evolution. As scientific theories, they all represent humankind's **best effort** at a particular point in time. Are they "true" like a mathematical proposition (e.g., in Euclidean geometry, the ratio of the circumference of a circle to the circle's diameter defines the transcendental and irrational number $\pi = 3.1415926535\ldots$)? No, they are not "true." Science is not focused on "truth"; it is a model-building process. What is "true" is the universe we share; our models are either more accurate or less accurate models of that truth.

Consider the history of astronomy. In the second century CE, Ptolemy of Alexandria proposed a geocentric model of the universe that depended on circular motion and extended the ideas of the Greek astronomer Hipparchus. Ptolemy provided tables in his work, the *Almagest*, which predicted planetary positions accurately enough to coincide with naked-eye observations. The model was consistent with the observations of its time. In the sixteenth century, the heliocentric proposal of Copernicus, strongly supported by the observational efforts of Kepler and Galileo, and the concept of universal gravitation proposed by Newton replaced Ptolemy's geocentric model. Again, this new Newtonian model was consistent with observations. In 1915, Albert Einstein proposed the general theory of relativity, which was able to account for observational anomalies (first noted in the nineteenth century) not explained by the Newtonian model of the universe. Again, the work of Einstein was consistent with the larger set of observations. Are any of the three models, Ptolemaic, Newtonian, or Einsteinian, "true"? No, they are all models of increasing explanatory and predictive power that describe our shared truth, the universe itself.

## Comments on Scientific Method

The outline of the scientific method discussed here leaves the distinct impression that the doing of science is a very sequential, almost rigid, step-by-step process. It appears to be almost automatic, proceeding from one clearly understood activity to the next, culminating in a powerful explanatory and predictive model. In reality, this impression is highly misleading; the doing of science follows a highly nonlinear and circuitous path, often characterized by repetition, dead ends, and simply incorrect analyses and conclusions. Kepler, for example, spent on the order of two decades working toward the formulation of his three laws, much of the time exploring nonproductive investigations. Scientific understanding often advances serendipitously. There are numerous examples (e.g., the identification of the antibiotic properties of penicillin and the recognition of naturally occurring radioactivity, to name only two) in the long history of science in which an individual, simply being curious and (we might imagine) muttering the phrase "I wonder," has made stunning observations or important breakthroughs in understanding our world. In 1854 Louis Pasteur noted, "*Dans les champs de l'observation le hasard ne favorise que les esprits préparés,*" which can be translated as "Where observation is concerned, chance favors only the prepared mind."[1] Finally, it is important to recognize that many investigations in science are not the work of a single individual, doggedly following a step-by-step process, but rather are the culmination of work completed by many teams of individuals widely separated both temporally and physically. A more effective way to understand the scientific method is to view it as a cyclical process in which each activity in the method is revisited multiple times and in no specified or determined order. In a sense, the comment of Wernher von Braun is a simple encapsulation of this cyclical understanding of science. He noted, "Research is what I'm doing when I don't know what I'm doing."[2]

## Fundamental Concepts

At the very heart of the scientific process is the need for one human being to communicate clearly and unambiguously with another human being. In order for this communication to be successful, the speakers must share a set of commonly understood concepts and the terms associated with these concepts. While it is often assumed that this language is well-known and universally accepted (texts rarely, if ever, discuss these terms), this is far from being clear. We pause now to face the most basic of terms, asking, "What do they mean?"

### Space

We assume the existence of the concept of **extension**. That is, our physical universe is not a mathematical point having a zero dimension. This assumption is consistent with our **experience** (recall that the first step of the scientific method is observation). Further, our experience is consistent with three **mutually perpendicular** (i.e., orthogonal) extensions, which we call **directions** or **dimensions**. The sum total of these directions we call **space** (sometimes labeled as three-space). In the classical world, we assume that this space is **continuous** and **homogeneous**. Consequently, at each point, we can define three mutually perpendicular axes, a coordinate system, and we can define distance along each of these axes. With such a coordinate system we can define positions in space, which we call **points**. We are then able to measure the distance between points.

### Time

We **experience** (i.e., observe) events **sequentially** and call this physical but nonspatial separation between events our psychological experience of **time**. We measure such nonspatial separations with devices called **clocks**. (There are both manmade and naturally occurring clocks.) In the classical world we assume that time is both **continuous** and **homogeneous**.

1 Louis Pasteur, Lecture, University of Lille, December 7, 1854.

2 Wernher von Braun, Interview, *New York Times*, December 16, 1957.

As a result of these foundational definitions of **space** and **time**, we can make several further specifications, which provide additional tools to describe our universe:

1. Each event in the universe can be uniquely specified by four coordinates: $x$, $y$, $z$, and $t$. This defines a **position vector r** $= (x, y, z, t)$, a mathematical object that possesses both **magnitude** and **direction.**
2. We can now use our concepts of space and time to define the **velocity**, **v**, which is also a **vector** (indicated by the use of boldfaced type) because it possesses both magnitude and direction. The units of velocity are

   (change in distance)/(change in time).

   Symbolically, we write

   $$\mathbf{v} = \frac{\Delta \mathbf{r}}{\Delta t}. \tag{1.1}$$

3. Using the idea of velocity, we can now define the concept of **acceleration**, **a**, also a **vector** quantity. The units of acceleration are

   (change in velocity)/(change in time).

   Symbolically, we write

   $$\mathbf{a} = \frac{\Delta \mathbf{v}}{\Delta t}, \tag{1.2}$$

   which is equivalent to

   (change in distance)/(change in [time × time]).

   Symbolically, we write

   $$\mathbf{a} = \frac{\Delta \mathbf{r}}{\Delta t^2}. \tag{1.3}$$

## Mass

Conventionally, because the terms "matter" and "mass" are too often used in an *almost* interchangeable manner, the distinction between these terms, and, more importantly, between their fundamental meanings, remains cloudy and obscure. As a matter of fact, it is very likely that the reader has only a foggy idea that the terms "matter" and "mass" are in some way related to "stuff" or, maybe more quantitatively, "weight." Commonly, definitions similar to the following are provided: "Matter is anything in the physical universe that has mass and occupies space." "Mass is the measure of matter in an object." While the definitions *seem* quite reasonable when taken separately, their juxtaposition makes it painfully clear that they are **circular**. Notice that each definition depends on the other one to specify a critical term (in one case, *mass*; in the other case, *matter*). This interdependence means that the definitions **do not** successfully define anything.

In order to avoid the failure of these ordinary definitions, we will take another approach. First, let's examine the term **mass**. We begin with the **observation** that, for historical reasons, is called Newton's third law of motion. Consider two isolated bodies that **only** interact with one another. If we compare their mutually induced (through this interaction) accelerations, we observe that

$$m_1 \times \mathbf{a}_1 = -(m_2 \times \mathbf{a}_2). \tag{1.4}$$

Their accelerations are **oppositely** directed, but **not** necessarily equal. Newton found that it was necessary to include the two factors, $m_1$ and $m_2$, in order to write an **equation**. We can rewrite the equation to show that the accelerations are proportional:

$$\mathbf{a}_1 = -(m_2/m_1) \times \mathbf{a}_2. \tag{1.5}$$

Equation 1.5 simply states that, in an isolated two-body system, the accelerations are always proportional to one another, whereas the quantity, $(m_2/m_1)$, is a scalar, a **constant** that is independent of the bodies' positions, velocities, and internal states. We have used the variables $m_1$ and $m_2$ because we will eventually call these scalar quantities the **masses** of the two bodies.

Now, suppose we define the first body to be our **standard** body, that is, the body that will be compared to every other body in the universe. (This is what we mean by the word *standard*.) Because it will be our standard, we are free to assign to it **unit mass** (i.e., $m_1 = 1$). Then,

$$\mathbf{a}_1 = -(m_2 \times \mathbf{a}_2). \tag{1.6}$$

Using our **standard** body and measuring the accelerations $\mathbf{a}_1$ and $\mathbf{a}_2$, we can measure the **mass** of any other body in the universe:

$$m = |\,\mathbf{a}_1\,| / |\,\mathbf{a}_2\,|. \tag{1.7}$$

Notice that we have used the **absolute values** of the two acceleration vectors, and we have ignored the minus sign because we are not concerned with the direction, only a **magnitude** of the scalar $m$. More importantly, note that equation 1.7 makes sense because we have already defined the terms **space**, **time**, **velocity**, and **acceleration**.

So, what have we accomplished by this long discussion? Based on the fundamental observation of how two material bodies influence one another, that is, how their accelerations behave, we have assigned a **scalar** property called the mass to the objects of our universe. Further, by selecting a standard mass, we can assign a **numerical** value to every mass. It is important to note that the concept of mass is a **numerical abstraction**; that is, it is defined as the ratio of two numbers, $|\mathbf{a}_1|$ and $|\mathbf{a}_2|$, which describe our **observations** of the universe.

## Matter

We can now define **matter** as anything in the physical universe that has mass and occupies some volume of space. Unlike the earlier definition, our understanding of matter is no longer circular. Its definition does depend on the word *mass*, but our definition of mass is no longer connected to the word *matter*.

## Force

The quantity ($m \times \mathbf{a}$) plays a fundamental role in our description of the universe. Hence, we give it a special name, defining it as follows:

$$\mathbf{f} = m \times \mathbf{a}, \tag{1.8}$$

where **f** is a **vector** quantity (possessing both magnitude and direction) and is called the **force**. This was Isaac Newton's definition of force.

## Momentum

Another fundamental physical quantity is ($m \times \mathbf{v}$). Again, we give it a special name, defining it as follows:

$$\mathbf{p} = m \times \mathbf{v}, \tag{1.9}$$

where **p** is a **vector** quantity (possessing both magnitude and direction) and is called **momentum** (more properly, **linear momentum**). Now, the reader may wonder how force and momentum are related. Writing equation 1.2 using the notation from calculus, we have

$$\mathbf{a} = \frac{d\mathbf{v}}{dt}. \tag{1.10}$$

Hence, we can write the force as

$$\mathbf{f} = m \times \frac{d\mathbf{v}}{dt}. \tag{1.11}$$

We will now make the assumption that the mass is **constant** (in light of the developments in twentieth-century physics, this assumption will need further discussion; see below), allowing us to write

$$\mathbf{f} = \frac{d\,(m \times \mathbf{v})}{dt}. \tag{1.12}$$

However, the reader should note that the quantity $(m \times \mathbf{v})$ appears in equation 1.12, and from equation 1.9 we have $\mathbf{p} = m \times \mathbf{v}$. Combining equations 1.9 and 1.12, we can now write

$$\mathbf{f} = \frac{d\mathbf{p}}{dt}. \tag{1.13}$$

This expression for the force makes the momentum the **more fundamental** concept, and this expression is consistent with Einstein's theory of special relativity, first published in 1905. Einstein's work fundamentally changed humankind's understanding of the terms **space** and **time.** The definitions we gave to these terms (see above) are now called the classical or Newtonian definitions to distinguish them from the interpretations given by Einstein. Because Einstein's definitions become significant only when extremely large mass concentrations or velocities comparable to that of the speed of light are encountered, we will continue to use the **classical** definitions of space, time, and force as we build our model understanding of chemistry. This will cause no difficulty because the description of the universe in chemical terms that we will build here rarely depends on velocities approaching that of light. Consequently, the model we will build here is a very effective model of the chemistry of our universe. (One should note, however, that sophisticated mathematical calculations of certain metal properties do exhibit critical relativistic effects. For example, the color of gold is a relativistic effect. If we ignore Einstein, the calculations would predict that the metal gold is white in color, not yellow!)

## Work

We again begin with an observation. Suppose we have an object that is at rest and wish to move it from point A in space to point B in space (where A and B are not coincident). To accomplish this we must put the object in motion for some time, thereby changing its velocity from zero to some nonzero value. That is, we have **accelerated** the object. The operation just described involves three fundamental quantities:

1. Mass
2. Acceleration
3. Distance

From equation 1.8 we already know that force = mass × acceleration.

We now define work as follows: **work** = force × distance. Hence, **work is force acting through or over a distance.**

Mathematically (in one dimension),

$$w = \sum_{i=1}^{n} \mathbf{f}_i \times \Delta x_i. \tag{1.14}$$

Using calculus (also one dimension),

$$w = \int_0^x \mathbf{f}(x)\,dx. \tag{1.15}$$

Note that, again, we now have a precise definition of the term *work*, which uses our previously defined fundamental ideas of **distance, acceleration**, and **mass.**

## Energy

The term **energy** is one of the most central ideas in our modern model of the physical universe. So, we ask, "What is energy?" One possible answer is "Energy is the capacity to do work." While this is a commonly used expression, it does not reflect the most fundamental characteristic or descriptive feature of the term energy.

A simple story will illustrate this important point. This is a paraphrase of a description given by the American theoretical physicist Richard Feynman, in chapter 4, volume 1 of *The Feynman Lectures on Physics.*[3] A mother places a young child in the playroom with his twenty-six alphabet blocks. At the end of the day, she counts the blocks: twenty-six. There are twenty-six blocks at the beginning of the day and twenty-six at the end of the day. On the second morning the child and his blocks are in the playroom. As the day ends, the mother sees only twenty-four blocks. But then she looks in the toy box to find two more blocks: a total of twenty-six blocks. There are twenty-six blocks at the beginning of the day and twenty-six at the end of the day. As the third day ends, the mother sees only twenty-two blocks in the room; "Aha," she says to herself, "check the toy box." But the toy box is empty. She then notices the open window, looks out, and sees four blocks on the front lawn: again, a total of twenty-six blocks. There are twenty-six blocks at the beginning of the day and twenty-six at the end of the day. Finally, as the week comes to an end and the boy's playmate has left for her home, the mother notices twenty-eight blocks. But there are two "A" blocks and two "Q" blocks. The friend didn't take all of her blocks home. Deleting the two duplicates, the mother now has twenty-six unique blocks. There are twenty-six blocks at the beginning of the day and twenty-six at the end of the day.

The key thing to notice is that no matter what has gone on, however involved the child's day of play has been, there are twenty-six blocks at the beginning of the day and twenty-six at the end of the day.

3 Richard P. Feynman, Robert B. Layton, and Matthew Sands, *The Feynman Lectures on Physics*, 3 volumes (Reading, MA: Addison-Wesley, 1963–1965).

That is, the number of blocks is **constant**. In fact, the blocks themselves are not as important as the constancy of the **number of blocks**. When a quantity remains constant, like the number of blocks, a scientist will say that it is a **conserved quantity**.

How, you might ask, is the story of the child and the blocks connected to energy? After all, the term energy is the focus of our attention. To formulate an answer, we begin with an **observation**: we observe a universe that is characterized by **transformations**, that is, activities, changes, and rearrangements, much like the intricate daily play of the child. Over the course of the past five hundred years, humankind has discovered that we can compute a number, a **numerical invariant**, which is associated with every observed transformation. (The details of the actual calculation are not important at this point; we'll learn how to do these calculations as the course progresses.) We call this **number** the energy. If the number is calculated at the **beginning** of a transformation and at the **end** of a transformation (and we take into account "open windows" and "duplicate blocks," as in the story of the child and his blocks), the two calculated numbers are the **same**. We say that energy is a **conserved quantity**. Because this **numerical invariant** is a number, the energy is also an **abstraction** (like mass) that is used to characterize every transformation observed in the physical universe. The energy is like a tag, identifying every particular transformation.

We have defined energy as an abstract quantity that we can compute mathematically. As a result, we do **not** understand energy as "little blobs of stuff" that can be put into a bag. Energy is an **idea**, a **concept** that allows us to understand in a **very general way** all the transformations that occur in our universe. Rather than viewing energy as a material substance that you can hold, you should see it as an idea that ties together all the various processes of the universe. It describes and characterizes all the changes, all the transformations that we **observe** in the physical world. A useful analogy for understanding energy as an idea is to compare it to the concept of **color**. You cannot hold "blueness" or "greenness" in your hand, but color is certainly a part of the world we observe.

Depending on the context in which a transformation occurs, we say that energy has different **forms**:

**Heat** energy
**Electrical** energy
**Mechanical** energy
**Radiant** energy

We do distinguish two major **types** or **kinds** of energy:

**Kinetic Energy:** This is the energy associated with a mass in motion. The rule for calculating the kinetic energy is given by the following equation

$$E_k = \tfrac{1}{2}\, mv^2. \quad (1.16)$$

In this equation, $m$ is the mass of an object and $v$ is the velocity of the object whose mass is $m$.

**Potential Energy:** This is the energy associated with a mass as a result of the mass's position in space and the action of a conservative force. Defining the technical meaning of the concept of a conservative force is not crucial to our current discussion. The central idea is that the potential energy does **not** depend on the **path** over which one moves the object. The energy is determined solely by the **position** of the mass. **Gravity** is the most commonly encountered conservative force. One simple mathematical expression for potential energy (here the potential energy caused by the force of gravity) is

$$E_p = mgh, \tag{1.17}$$

where $m$ is the mass of an object, $h$ is its height (i.e., position) above a reference point, and $g$ is the acceleration caused by gravity.

You may wonder, why is energy such an important concept? In the above discussion, we've already indicated that every change (transformation) in our universe is characterized by the concept of energy. A brief look at history underscores the importance of the concept of energy. While studying mechanical systems, in particular planetary motion, Isaac Newton observed that the quantity that we calculate and call energy is a **constant of the motion** of the planets (i.e., it does **not** vary with time). Although the velocity of a planet **changes** along its orbital path so that the kinetic energy is **not** constant, and although the gravitational potential energy is **not** constant (because the distance of the planet from the sun varies along the elliptical path of its orbit), the **sum** of the kinetic energy and the potential energy (i.e., **total energy) does not change. It is a constant.**

Prior to the beginning of the twentieth century, chemistry paired the conservation of energy with another critical conservation principle that was based on the careful **observation** of chemical processes: the **conservation of mass** in a chemical reaction. The fundamental observation that **all** chemical transformations caused neither the disappearance of mass nor the sudden appearance of new mass suggested that mass itself is conserved in the course of a chemical transformation. Until Einstein's remarkable insight, $E = mc^2$, demonstrated that mass and energy are **equivalent**, the two conservation principles were viewed as separate and distinct. The modern understanding is that there is only a single conservation principle in our universe: the conservation of mass-energy. However, because the processes that are described by chemistry occur in a very low-energy regime (yes, chemical explosions, like those associated with TNT, are low-energy!), we can continue to treat very effectively the conservation of mass-energy as if there are two separate principles at work. In fact, when we **observe** chemical processes in the laboratory, at the limit of our ability to measure, **no** change in mass is observed during the course of a chemical reaction. Consequently, the conservation of mass is a highly effective principle on which to base our model of chemical behavior.

Before leaving this brief introduction to the concept of energy, we pause to note a highly significant and possibly surprising relationship between two of the fundamental terms we have discussed. In equations 1.14 and 1.15, the mathematical statements defining **work** are given, while in equations 1.16 and 1.17, the mathematical definitions of kinetic energy and gravitational potential energy are given. The crucial point is that **work** and **energy** have the **same** units. (Shortly, we will learn about the SI system, the International System of Units, which is the collection of fundamental quantities that have been adopted, by international agreement, as the standards of measurement. With the SI system, we can measure our world, that is, provide a quantitative description of the universe.) In the SI system, the unit of energy is named the **joule (J)**, and $1\ J = 1\ \text{kg} \times (\text{m} \times \text{s}^{-1})^2$.

We begin with the definition of **force** as **mass** × **acceleration** (see equation 1.8; in terms of the units, this is $\text{kg} \times \text{m} \times \text{s}^{-2}$). The unit of force is called the **newton** (*N*). But **work** is defined as **force** × **distance** (see equation 1.14); in symbols this is $\boldsymbol{N \times m}$ (where m is the symbol for the "meter," an SI unit of length that measures distance). But in terms of the SI units we have $(\text{kg} \times \text{m} \times \text{s}^{-2}) \times (\text{m}) = \text{kg} \times \text{m}^2 \times \text{s}^{-2} = \text{kg} \times (\text{m} \times \text{s}^{-1})^2$. But this final collection of units is energy (see equation 1.16, for the case of kinetic energy).

Consequently, the terms work and energy are intimately related in our physical universe. Their joint study, after the beginning of the Industrial Revolution circa 1750, spurred the development of the science of thermodynamics (see chapter 10). We will discover that one of the central themes of this book, energetics, will draw on central concepts from this science and will provide the critical link between the two themes of structure and reactivity that will organize our study of chemistry.

### Temperature

Intuitively, the term **temperature** identifies the "hotness" or "coldness" of some region of the physical world. This **qualitative** statement represents our most elementary approach to comprehending the many and varied dimensions of this term. However, the scientific enterprise has taken great strides to convert this **qualitative** understanding into a **quantitative** description through the construction of effective temperature scales, measuring instruments, and powerful theoretical constructs. The kinetic molecular theory of gases (one of the most significant successes of nineteenth-century science) identifies temperature as a quantitative measure of the average kinetic energy of molecules. Further, thermodynamics has employed temperature (i.e., **absolute temperature**; this meaning will become clear when we study temperature scales) as an index of the *random* motion of molecules, a phenomenon now called **thermal motion**. The perceptive reader will note the use of the word "molecule"; we will see that this term plays a central role in the modern chemical paradigm. Hence, the term temperature, which started as a very **qualitative** description of humankind's observations, is now linked to the most central ideas of modern chemistry.

## Mathematics

As we have already noted, one of the major themes that has characterized science since the Renaissance is the progression from **qualitative** observations to **quantitative** explanations and predictions. This move in the direction of the quantitative has meant that scientific disciplines have increasingly used mathematics as a central tool. This certainly is the case for chemical sciences, and the variety of mathematical concepts used by chemistry is very great, ranging from the most elementary arithmetic to powerful tools from multivariable calculus, differential equations, and abstract algebra. Does this mean that you, as a student, must be a master mathematician to understand the chemistry that we will study here? Certainly not! But it is the case that mathematics will play a key role in what we do, and, consequently, there is a collection of basic mathematical tools that you will need to master. It's important to note that these tools are **not optional**; you will need them. By the same token, they are **not** impossible to master. Let's begin.

## Percent

A commonly used measure of a part of a whole is called the **percent** or **percentage**. Its definition uses two of the basic arithmetic operations, division and multiplication. Percent is defined as follows:

$$\text{Percent} = \frac{\text{Part}}{\text{Whole}} \times 100. \tag{1.18}$$

Example 1.1: What percent of 325 is 65?

Answer:

Identify the whole: 325

Identify the part: 65

What am I looking for? Percent

Apply the formula:

$$\text{Percent} = \frac{65}{325} \times 100$$

$$\text{Percent} = 20\%$$

Note that after the arithmetic operations are complete, we add "%" to the answer. This is called the percent sign.

Example 1.2: Suppose that 85.0 is 62.5% of some number. What is that number?

Answer:

Identify the part: 85.0

Identify the whole: This is unknown; the percent is given as 62.5%

What am I looking for? The whole

Apply the formula: $62.5 = \frac{85.0}{\text{Whole}} \times 100.$

Notice that in this problem we must rearrange the formula using the basic **rules of algebra**:

$$\text{Whole} = \frac{85.0}{62.5} \times 100.$$

$$\text{Whole} = 136$$

We will find that the idea of percentage has many applications throughout chemistry, particularly when we discuss the idea of a *solution*. As we will see in greater detail later, a solution is a mixture of two or more substances. Consequently, it is a perfect place to apply the concepts of "part" and "whole." We will see much more of this as we proceed further in our study of chemistry.

## Equations and Algebra

In defining the idea of **percent** we used one of the most fundamental mathematical relations: **the equation**. You will recall that an equation is like a finely balanced teeter-totter; both sides are exactly equal. Elementary algebra is a branch of mathematics focused on the manipulation of equations and the mathematical expressions that make up equations. (The word "elementary" is used here because the word "algebra" has many specific and sometimes highly specialized meanings in mathematics. We are focusing **only** on the section of mathematics that gives us powerful and effective tools to work with equations.) Fundamentally, elementary algebra tells us how to work with equations so that both sides of an equation remain **equal**. There are only six mathematical operations in elementary algebra: addition, subtraction, multiplication, division, exponentiation ("raising a number to a power"), and extracting roots (square roots, for example). The operations are related in that subtraction is the inverse of addition and division is the inverse of multiplication. For the two pairs of operations, the word "inverse" simply means "undoing" the operation. In contrast, extracting roots is not exactly the inverse of exponentiation, but is very close to "undoing" the effects of exponentiation.

Example 1.3: Addition: $8 + 2 = 10$
Subtraction: $10 - 2 = 8$
Notice that in the "addition step," the number 2 was added to 8. By subtracting 2 from 10, we are back at the starting value of 8.

Example 1.4: Multiplication: $5 \times 7 = 35$
Division: $35 \div 7 = 5$
Notice that in the "multiplication step," the number 5 was multiplied by 7. By dividing 35 by 7, we are back at the starting value of 5.

Example 1.5: Exponentiation $2 \times 2 \times 2 = 2^3$
Exponentiation is repeated multiplication. The phrase that is used is "raising a number to a power." In this example, 2 is raised to the power of 3.
Note that it is possible to have **negative** exponents:

$$2^{-3} = 1 \div 2^3.$$

Similarly,

$$1 \div 2^{-5} = 2^5.$$

The last two expressions demonstrate that an exponent's sign changes whenever the quantity is moved **either** from the numerator to the denominator **or** from the denominator to the numerator.

Example 1.6: Extracting Roots

Square Roots: The number $49 = 7^2$. Consequently, the **square root** of 49 is "almost" the number 7 (remember, extracting roots is not exactly the inverse of exponentiation). The reason for the word "almost" is seen in the following:

$$\sqrt{49} = \pm 7,$$
$$(49)^{1/2} = \pm 7.$$

Similarly, for **cube roots**, we write

$$(27)^{1/3} = 3,$$

and

$$(-27)^{1/3} = -3\ .$$

Note that these last two equations make sense because

$$27 = 3^3,$$

and

$$-27 = (-3)^3.$$

The fact that care must be taken with the "signs" when extracting roots means that this process is only "almost" the inverse of exponentiation.

The previous six examples are intended to be reminders of the basic mathematical operations in algebra. They do **not** encompass all the subtleties and complexities of the subject; the reader is **strongly encouraged** to review these basic algebraic operations because they will be necessary skills to approach chemistry as a quantitative science.

## Logarithms

The logarithm is a mathematical tool that was invented in the seventeenth century to simplify calculations that involve very large numbers by turning complex multiplications into additions. More importantly, it was also understood that the logarithm is the "true" inverse operation of exponentiation. It has applications in numerous areas and will be important to our study of several areas of chemistry. Let's examine some of the important logarithmic properties we will need in the future.

Because the logarithm is the "true" inverse operation of exponentiation, it's not unexpected that properties of exponentiation will play a significant role in understanding the logarithm. Recall that

$$z^a \times z^b = z^{a+b}. \tag{1.19}$$

In equation 1.19 $z$ is called the **base**, and $a$ and $b$ are the **exponents**. Notice that in order to add the exponents, the bases **must** be the same. We now define two **different** logarithms, the first called the **natural** logarithm and the second called the **base-10 logarithm**. We will use both later in our study of specific chemical topics. For the natural logarithm we use a very special base, $e$, which is called Euler's number, and we write an exponential expression:

$$e^{a} = b. \tag{1.20}$$

We now define the **natural logarithm**, symbolized by ln, as

$$\ln(b) = a. \tag{1.21}$$

Note that this definition simply uses Euler's number, $e$, as the **base** in equation 1.20.

For the case of the **base-10 logarithm**, we begin with

$$10^{c} = d. \tag{1.22}$$

We now define the **base-10 logarithm**, symbolized by log, as

$$\log(d) = c. \tag{1.23}$$

We now observe a very remarkable property of logarithms that made them powerful calculating tools in the seventeenth century (long before the invention of either electromechanical or digital calculators). Suppose we have the following:

$$10^{a} = A \text{ and } 10^{b} = B.$$

From equation 1.19, we have

$$10^{a} \times 10^{b} = 10^{a+b}.$$

or

$$A \times B = 10^{a+b}.$$

Now applying equations 1.22 and 1.23, we have

$$\log(A \times B) = a + b.$$

Similarly,

$$10^{a} \div 10^{b} = 10^{a-b}.$$

or

$$A \div B = 10^{a-b}.$$

Now, applying equations 1.22 and 1.23, we have

$$\log(A \div B) = a - b.$$

Example 1.7: What is the base-10 logarithm of 13.598?
Answer: Use equation 1.23: $\log(13.598) = x$
$x = 1.13348$

Example 1.8: Suppose $\log(y) = 0.789$. What is $y$?
Answer: Use equations 1.22 and 1.23: $\log(y) = 0.789$
means
$y = 10^{0.789}$
$y = 6.15$

That is, via the definition of the logarithm, we have converted a potentially difficult multiplication (because $A$ and $B$ could be very large numbers) into a simple addition. With the preparation of tables that give the logarithm of a number (and, implicitly, the value of "10 raised to a power," where the power is greater than or equal to zero and strictly less than 10), very tedious multiplications were reduced to simple additions. Exactly the same property applies to the **natural logarithm**. We will find that both the **natural logarithm** and the **base-10 logarithm** are used in our later study of various topics in chemistry.

## Equations and Problem Solving

The last several sections have focused our attention on the abstract operations that play a central role in algebra and on the definition of the logarithm. However, we are not primarily interested in these abstract concepts for their own sake. This is the role and the primary focus of mathematics. The real goal is to apply these abstractions to our observations of the physical universe from the perspective of the chemical sciences. The application of these abstractions means developing the ability to use various chemical paradigms to analyze, evaluate, interpret, and draw inferences about the physical universe. It means understanding and knowing the limitations of modern chemical paradigms and being capable of communicating chemical concepts to experts and nonexperts alike. Succinctly, the reader is challenged to become a critical thinker capable of using these mathematical abstractions. The key to this step is becoming an effective problem solver. How do we do this?

Problem solving involves a minimum of **four** steps. These steps require the problem solver to **organize** information, to **identify** an unknown, to **translate** ideas from a primarily verbal or graphical form to a mathematical equation, and, finally, to **solve** the equation using the tools of algebra. Let's examine each step and then look at examples of good problem-solving techniques.

1. **Organizing Information**
   When a problem is presented, the first step is to organize **all** the given information. In its simplest form, this means **writing** out in tabular form every piece of datum provided by the problem description. It is not sufficient to simply "read the problem"; you must actively evaluate and list every piece of information provided by the problem statement.
2. **Identify the Unknown**
   What quantity must you find? In effect, what is the question that the problem is asking? **Write** it down.
3. **Translate Ideas**
   The statement of a problem or question invariably begins in verbal form (after all, language is one of the most fundamental identifying characteristics of human beings) often supported by graphical information. The problem solver's most fundamental task is to translate this information into an **algebraic** statement (or logical statement) that connects the given information with the unknown. It is only when a relationship between the given information and the unknown is clearly created that a solution is possible.
4. **Solve the Equation**
   The final step involves using the tools of algebra (or, more generally, the tools of logic) to find the unknown. In more colloquial terms, this means "solve the equation."

Let's now look at some examples and apply this process. We'll start simply and then increase the complexity of the problem we are attempting to solve.

Example 1.9: A rectangular plot of land is 85 meters wide and 115 meters long. What is the area of the plot in square meters?

Answer:

**Organize:** Width = 85 meters
Length = 115 meters
Note that the given units are included in the organization.

**Unknown:** Area

**Translate:** Area = length × width
This statement makes use of a geometric idea.

**Solve:** Area = (115 meters) × (85 meters)
= 9,775 square meters
(In **scientific notation** and as a result of the rules for **significant figures**, the answer is limited to $9.8 \times 10^3$ square meters. We will study these concepts shortly.)

Example 1.10: Suppose that your friend Carrie is 13 years old. While Carrie's father is three times as old as Carrie, he is only one-half the age of Carrie's grandmother. How old is Carrie's grandmother (in years)?

Answer:

**Organize:** Carrie is 13 years old
Carrie's father's age is 3 times Carrie's age

| | |
|---|---|
| | Carrie's father's age is ½ Carrie's grandmother's age. |
| **Unknown:** | The grandmother's age. |
| **Translate:** | Let *GA* be the symbol for the grandmother's age. |
| | Let *FA* be the symbol for the father's age. |
| | Let *CA* be the symbol for Carrie's age. |
| | $FA = ½ \times GA$ |
| | $(3 \times CA) = ½ \times GA$ |
| | $(3 \times 13 \text{ years}) = ½ \times GA$ |
| | (This final equation is the one that must be solved.) |
| **Solve:** | $GA = 2 \times (3 \times 13 \text{ years})$ |
| | $GA = 78$ years |
| | Final answer: Carrie's grandmother is 78 years old. |

## Mensuration

As we have already noted, a key theme that has characterized modern science since its inception in the Renaissance has been the inevitable progression from a **qualitative** description of physical phenomena to a **quantitative** assessment that provides both explanation and prediction. Such a transition requires two fundamental components: (1) the increased use of mathematics and (2) the definition of a rational measurement system, a system of units that is used to quantify the physical universe. Each civilization in human history has devised a system of weights and measures. There are numerous examples, ranging from the early Egyptian "rope-stretchers" through the development of "English" or "standard" units based on a combination of Anglo-Saxon and Roman systems and the introduction of the metric system at the time of the French Revolution. Today, the system of measurement used in all branches of science is the International System of Units, often shortened to the SI system (from the French *Système International d'Unités*). While all parts of the world have made a commitment to use the SI system (the United States Congress authorized the use of the metric system in 1866, and the United States signed the Treaty of the Metre in 1875), the implementation of the SI in the daily culture is a slow process.

The SI system is an expanded version of the metric system of units and is built from seven **base** units. The base units are listed in Table 1.1.

Table 1.1

| | Physical Property | Unit | Abbreviation |
|---|---|---|---|
| 1. | Length | meter | m |
| 2. | Mass | kilogram | kg |
| 3. | Time | second | s |
| 4. | Temperature | kelvin | K |
| 5. | Amount of substance | mole | mol |
| 6. | Electric current | ampere | A |
| 7. | Luminous intensity | candela | cd |

Historically, each of the seven base units was defined by a human artifact, an actual object or device created by humankind. For example, the unit of length, the meter, was represented by a platinum-iridium bar, located physically in Paris, on which there were inscribed two lines defining the distance of exactly one meter. In recent years, there has been a systematic effort to define the base units of the SI system in terms of characteristics of the physical universe. For example, since 1967, the second has been defined to be the duration of 9,192,631,770 periods of the radiation corresponding to the transition between the two spectroscopic levels of the cesium atom. The SI unit of mass is the kilogram and is (as of 2014) the only SI base unit that is still directly defined by a human artifact (a block platinum-iridium alloy, the International Prototype Kilogram, *Le Grand K*) rather than a fundamental physical characteristic that can be independently reproduced in a laboratory. Because four of the seven base units in the SI system are defined relative to the kilogram, its stability is critical to the entire SI system, yet the mass of *Le Grand K* varies with time as surface atoms evaporate or contaminants are deposited. There is now a concerted effort to replace this last remaining human artifact with a new definition of the kilogram. A decision could come as early as 2015.

Given the seven base units, the remainder of the SI is built successively by defining the **derived** units. Each derived unit is based on two or more base units. Table 1.2 summarizes several key examples.

Table 1.2: Volume Units

| Derived Unit | Definition |
|---|---|
| Cubic meter | $m^3 = m \times m \times m$ |
| Cubic centimeter | $1\ cm^3 = (1.0 \times 10^{-2}\ m)^3 = 1.0 \times 10^{-6}\ m^3$ |
| Cubic decimeter | $1\ dm^3 = (1.0 \times 10^{-1}\ m)^3 = 1.0 \times 10^{-3}\ m^3$ |

The definition of one particular unit plays a central role in chemistry but is **not** part of the SI system. This is the definition of the **liter**, which became universally used in chemistry before the adoption of the SI system. Table 1.3 summarizes key connections to the liter.

Table 1.3: The Liter (L)

| |
|---|
| $1\ L = 1\ dm^3$ |
| $1\ mL = 1.0 \times 10^{-3}\ L$ |
| $1\ dm^3 = 1.0 \times 10^3\ cm^3$ |
| 1 mL of water at 3.98°C = $1\ cm^3$ |
| 1 mL of water at 3.98°C = 1 gm |

The last two entries of the above table will prove to be extremely useful. They establish a connection between **mass** and **volume** using water as the reference point. This connection will be incredibly valuable in our effort to understand all water-based chemistry. As we discuss a system of measurement, it is worthwhile to pause and emphasize the difference between the terms **mass** and **weight**. These terms are often confused; it is important to distinguish them. Earlier in this chapter we defined the term **mass** using Newton's third law of motion and measuring relative accelerations. In contrast, **weight** is a **force** (see equation 1.8). In is very important to note that the terms mass and weight refer to two distinct and different concepts.

In order to represent easily either very large or very small numerical quantities, a notational system called **scientific notation** was devised and is now universally used in the sciences. We shall also discover shortly that scientific notation offers another critical benefit by allowing scientists to convey explicitly the quality or, better, the **significance** of reported measurements. This second feature will follow from understanding a set of universally agreed-upon conventions called **significant figures**. Let's start by exploring the details of scientific notation. A number written in scientific notation has the following general form:

$$N \times 10^{m}.$$

The number $N$ is called the **mantissa** (sometimes, the significand) and can take on the following values:

$$1 \leq N < 10.$$

The **exponent**, $m$, is an integer that may be either positive, negative, or zero (recall that $10^0 = 1$). The introduction of scientific notation now allows us to take a look at one other part of the SI system of units: the SI prefixes (see table 1.4).

Table 1.4: SI Prefixes

| Factor | Name | Symbol | Factor | Name | Symbol |
|---|---|---|---|---|---|
| $10^{24}$ | yotta | Y | $10^{-1}$ | deci | d |
| $10^{21}$ | zetta | Z | $10^{-2}$ | centi | c |
| $10^{18}$ | exa | E | $10^{-3}$ | milli | m |
| $10^{15}$ | peta | P | $10^{-6}$ | micro | m |
| $10^{12}$ | tera | T | $10^{-9}$ | nano | n |
| $10^{9}$ | giga | G | $10^{-12}$ | pico | p |
| $10^{6}$ | mega | M | $10^{-15}$ | femto | f |
| $10^{3}$ | kilo | k | $10^{-18}$ | atto | a |
| $10^{2}$ | hecto | h | $10^{-21}$ | zepto | z |
| $10^{1}$ | deka* | da | $10^{-24}$ | yocto | y |

* The SI spelling of this prefix is deca-, but the US National Institute of Standards and Technology (NIST) recommends deka-. This is consistent with other spelling variations permitted in the SI system, for example, *meter* in the United States and *metre* in the United Kingdom. Note that the kilogram is the only SI unit having a prefix as part of its name and symbol. **Multiple** prefixes may not be used in the SI system. Consequently, in the case of the kilogram, the prefix names are used with the unit name *gram*,

and the prefix symbols are used with the unit symbol g. Except for this case, any SI prefix may be used with any other SI unit.

## Factor Label Method or Dimensional Analysis

Once a system of units has agreed upon, the question facing every user of the system is "How do I convert from one unit to another unit?" This is actually two questions, because the conversion can take place between units **within the same system of units** or between units **of two different systems of units**. There is a direct and clear approach called the **factor label method** (sometimes called **dimensional analysis**) that removes all of the apparent difficulty associated with these questions. It is an approach that requires virtually no memorization and avoids the often confusing phrases "going from a smaller unit to a larger unit" and "going from a larger unit to a smaller unit." Central to the factor label method is the idea that both the **units** themselves and their **associated numerical values** are treated **algebraically**. Let's look at an example of the factor label method at work.

Example 1.11: We begin with an equation (this is information taken from a reference book or information you have memorized; this is the "memory" part of the method) that relates two different units, in this example, the "dollar" unit and the "penny" unit:

**1 dollar = 100 pennies.**

This equation allows us to write two different factors. Dividing both sides of the equation by 100 pennies gives the factor as follows:

$$\mathbf{1 = \frac{1\ dollar}{100\ pennies}}.$$

Dividing both sides of the equation by 1 dollar gives the factor as follows:

$$\mathbf{1 = \frac{100\ pennies}{1\ dollar}}.$$

**The most important idea to understand is that *both* factors are equal to 1.** Consequently, if we multiply by these factors, because they are equal to 1, **the multiplication *cannot* change the value of any quantity.**

**Problem:** Convert 2.46 dollars to pennies.

$$(2.46\ \text{dollars}) \times \frac{100\ \text{pennies}}{1\ \text{dollar}} = 2.46 \times 10^2\ \text{pennies}$$

Pay close attention to the fact that the multiplication by 1 does **not** change the value ($2.46); only the units are changed (from dollars to pennies).

**Another Problem:** Convert 57.8 m to cm.

Here we need a starting equation:

$1\ \text{cm} = 1 \times 10^{-2}\ \text{m}$

Using this equation, we again create two factors, both of which are exactly equal to 1:

$$1 = \frac{1\text{cm}}{1 \times 10^{-2}\,\text{m}},$$

and

$$1 = \frac{1 \times 10^{-2}\,\text{m}}{1\text{cm}}.$$

We can now convert the units by multiplying by 1:

$$(57.8\text{m}) \times \frac{1\text{cm}}{1 \times 10^{-2}\,\text{m}} = 5.78 \times 10^{3}\,\text{cm}$$

## Significant Figures

Having agreed on both a system of measurement (the SI system) and a convenient and a powerful notation with which to represent numerical data (scientific notation), we now face a significant challenge: How do we communicate the **quality**, that is, the **significance** of the data we collect about the observed universe? Because every measurement made by humankind is uncertain to some extent (limited by the instruments, the techniques, or, as we will learn shortly, the universe itself), measurements and their numerical representations are important **only** if they communicate a clear understanding of this inherent uncertainty. To do this, humankind uses a set of rules, **the rules of significant figures**, which reflect the manner in which a measurement was made. There are five fundamental rules:

1. Any **nonzero** digit is a **significant** figure.

   845 cm → 3 significant figures
   1.234 kg → 4 significant figures

2. Zeros **between** significant figures are also **significant** figures.

   606 m → 3 significant figures
   40501 kg → 5 significant figures

3. Zeros to the **left** of the **first nonzero** digit in a number are **not** significant figures. They are called "placeholders" and simply specify the location of the decimal point.

   0.08 L → 1 significant figure
   0.0000349 g → 3 significant figures

4. Numbers **with** decimal points:
   a. If a number is greater than or equal to 1, **all** the zeros written to the **right** of the decimal point are **significant** figures.

   | | | |
   |---|---|---|
   | 2.0 mg | → | 2 significant figures |
   | 40.062 mL | → | 5 significant figures |
   | 340.0 | → | 4 significant figures |

   b. If a number is less than 1, **only** zeros that occur at the end of the number **or** between nonzero digits of the number are **significant** figures.

   | | | |
   |---|---|---|
   | 0.090 kg | → | 2 significant figures |
   | 0.004020 min | → | 4 significant figures |

5. Numbers **without** decimal points:
   Trailing zeros (i.e., zeros occurring **after** the last nonzero digit in the number) **may** or **may not** be **significant** figures.

   | | | | |
   |---|---|---|---|
   | 400 | → | **May** have | 1 significant figure |
   | | | | 2 significant figures |
   | | | | 3 significant figures |

   We cannot know which case applies without further information. The number 400, in the example above, is **ambiguous**. If no additional information is provided, we assume that there is only **one significant figure** (the nonzero digit).

   Use **scientific notation** to avoid ambiguity:

   | | | |
   |---|---|---|
   | $4. \times 10^2$ | → | 1 significant figure |
   | $4.0 \times 10^2$ | → | 2 significant figures |
   | $4.00 \times 10^2$ | → | 3 significant figures |

The rules of significant figures communicate to everyone who observes the universe information about the **quality** or **significance** of the numbers that constitute a measurement. They inform every reader and user of the data about the inherent limitations of that data, because the **number** of significant figures immediately identifies the level of uncertainty in a measurement. If four significant figures are reported, the uncertainty is 1 part in $10^3$; if 7 significant figures are reported, the uncertainty is 1 part in $10^6$. For example, if a measurement is reported as $3.000 \times 10^3$ m, the number has 4 significant figures; the uncertainty is ±1 meter, meaning that the measurement lies between $3.001 \times 10^3$ m and $2.999 \times 10^3$ m. In contrast, a measurement of $3.00 \times 10^3$ m (3 significant figures) yields an uncertainty of ±10 meters, meaning that the measurement lies between $3.01 \times 10^3$ m and $2.99 \times 10^3$ m.

Once we know how to represent measurements, including the uncertainty inherent in each measurement, by using significant figures, the next task is to define a consistent procedure for doing calculations with significant figures. There are only **two** rules, one governing the operations of addition and subtraction, and a second one governing the operations of multiplication and division:

**Rule 1:** In **addition** and **subtraction**, the number of significant figures to the right of the decimal point in the final answer is determined by the **fewest number of digits** to the right of the decimal point in **any** of the **original numbers**.

Example 1.12: 89.332

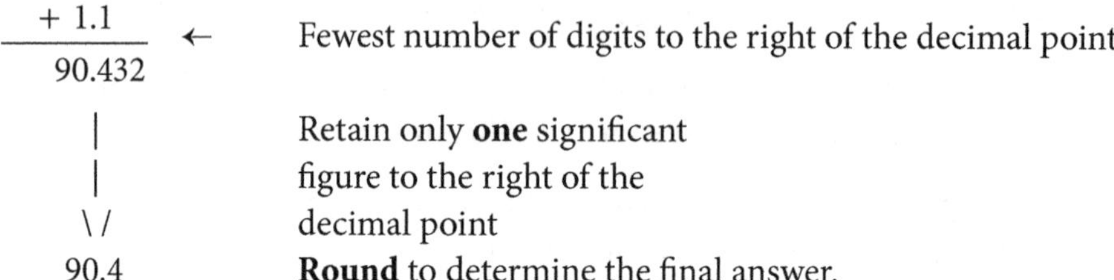


A procedure called "**rounding**" provides a systematic approach to limiting the number of digits reported in the final answer. The first step is to identify the "rounding digit," which identifies the location (decimal place) of the last significant digit that will remain in the final answer (i.e., after applying the rounding procedure). Once the rounding digit has been identified, examine the digit immediately to the **right** of the rounding digit. Three possibilities exist:

1. The digit **immediately to the right** of the rounding digit is **strictly less than five**:
   Then, **drop all digits to the right** of the rounding digit and **retain** the rounding digit.

Examples 1.13: Suppose that in the number 8.724, the rounding digit is 2.
Then, 8.724 → 8.72 .
Suppose that in the number 85.397, the rounding digit is 5.
Then, 85.397 → 85.

2. The digit **immediately to the right** of the rounding digit is **strictly greater than five**: The phrase "strictly greater than five" means that the digit immediately to the right of the rounding digit is either greater than 5 or is a 5 followed by **at least one other nonzero digit**.
   Then, **drop all digits to the right** of the rounding digit and **increase** the rounding digit by one unit.

Examples 1.14: Suppose that in the number 8.724, the rounding digit is 2.
Then, 8.724 → 8.72.
Suppose that in the number 85.397, the rounding digit is 3.
Then, 85.397 → 85.4.
Suppose that in the number 85.350001, the rounding digit is 3.
Then, 85.350001 → 85.4.

3. The digit **immediately to the right** of the rounding digit is **exactly five**: The phrase "exactly five" means that the digit immediately to the right of the rounding digit is a 5 followed either by no other digits or followed by **only** zeroes.

Then, there are two possibilities:

a. If the rounding digit is **even**, drop the 5 and **retain** the rounding digit.

Example 1.15: Suppose that in the number 0.425, the rounding digit is 2.
Then, 0.425 → 0.42.

Example 1.16: Suppose in the number 0.725000, the rounding digit is 2.
Then, 0.725000 → 0.72.

b. If the rounding digit is **odd**, drop the 5 and **increase** the rounding digit by one unit.

Example 1.17: Suppose that in the number 0.535, the rounding digit is 3.
Then, 0.535 → 0.54.

Example 1.18: Suppose that in the number 0.775000, the rounding digit is 2.
Then, 0.775000 → 0.78.

While the above rounding procedure may seem to complicate unnecessarily the "exactly five" case, note that the rounding digit is **increased** only when it is **odd**, rather than being **increased every time** the digit to the right of the rounding digit is "exactly five." This is a statistically more evenhanded approach to approximating measured data.

What happens when we multiply or divide significant figures? The second rule governs these two operations:

**Rule 2:** In **multiplication** and **division**, the number of significant figures in the final answer is equal to the number of **significant figures** in the original operand having the **fewest number of significant figures.**

Example 1.19: Multiply 2.8 by 4.5039.
The numbers 2.8 and 4.5039 are called the **operands** of a binary operation.

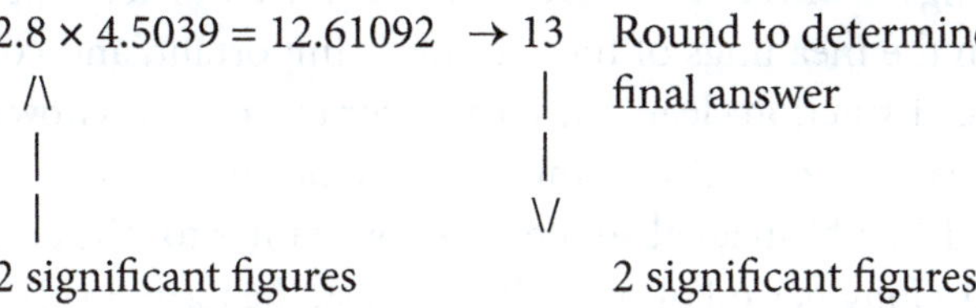


Example 1.20: Divide 6.85 by 112.04.
The numbers 6.85 and 112.04 are the operands.

6.85 / 112.04 = 0.0611388789

0.0611 Round to determine this final answer.

3 significant figures (6.85) — 3 significant figures (0.0611)

The two rules listed above (along with the rounding procedure) provide the basic tools for manipulating all measured data. Further, because the numerical square root algorithm only requires successive multiplications and subtractions, the rules allow us to extract square roots (and, in fact, cube roots) and to maintain an appropriate number of significant figures in our answers.

There are two remaining situations that occur as measured data is analyzed: First, how do we treat **exact** numbers that arise from either a **direct count** or from a **definition**? The answer to this question is straightforward: **exact numbers that are obtained by a direct count or from a definition possess an *infinite* number of significant figures.**

Example 1.22: Suppose that you require the total mass of 8 objects, each having a mass, $m = 0.2786$ g.
Total mass = 8 × (0.2786 g) = 2.229 g
**Here we treat the number 8 as if it possesses an infinite number of significant figures.**

Example 1.22: Calculate the **average** of two measurements:
(6.64 cm + 6.68 cm + 6.67 cm)/3 = 6.66 cm
**Because the number 3 arises as a result of the definition of an "average," it is treated as if it possesses an infinite number of significant figures.**

Second, the analysis of measured data often requires more than one calculation; very commonly a **chain** of calculations is necessary. As in the first situation, the answer is straightforward: **retain in the answer of all intermediate calculations *one more* significant figure than the rules for significant figures allow, then round *only* the final answer.** This means that, at each step of a chain of calculations, care must be exercised to determine the number of significant figures by **each individual calculation.** The step that allows the minimum number of significant figures will determine the number of significant figures that are permitted in the **rounding step.** It is important to note that you may round **only once** in a series of calculations.

The two terms **accuracy** and **precision** are used repeatedly to discuss a measurement or series of measurements, but the meanings of these terms are either confused or left imprecisely defined. However, the distinction between the meanings of these terms is important, and colloquial usage of these terms often fails to make the distinction clear. The term accuracy indicates how close a measurement is to the **true value** of the measured quantity. In this context, the phrase *true value* does not refer to a mathematical or metaphysical "truth" in an absolute sense but rather to "the accepted value after repeated measurements spanning a long period of time." In contrast, the term precision indicates how closely two or more measurements of the **same** quantity agree with one another.

Figure 1.1 displays graphically the difference between accuracy and precision. Let's assume that the center of each image in the figure represents the **true value.** Then, panel a shows measurements that are **neither** accurate nor precise; panel b displays the case in which the measurements are **both** accurate and precise; panel c shows the case in which the measurements are **not** accurate but they **are** precise.

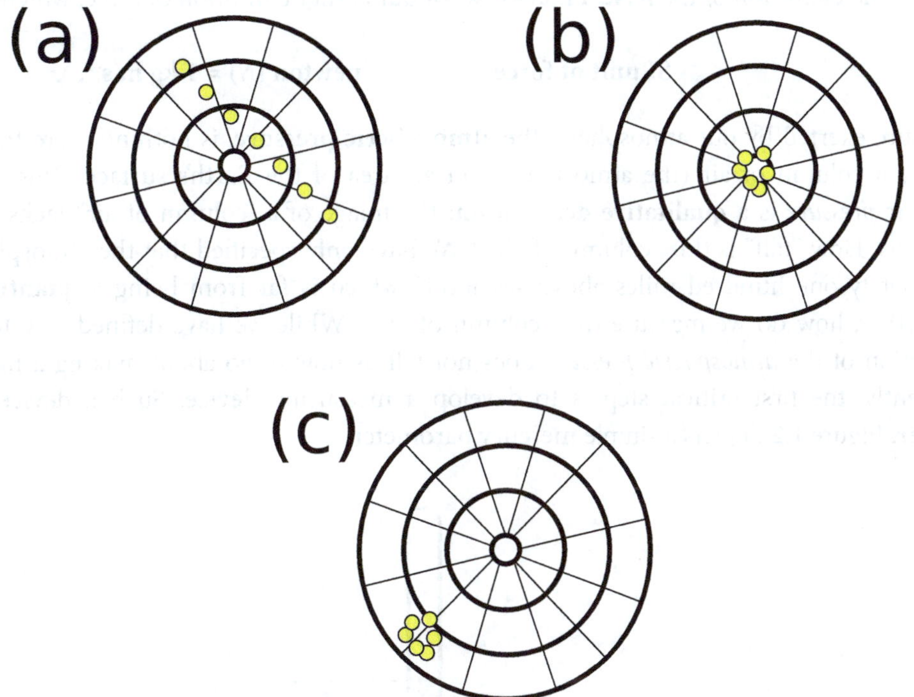


Figure 1.1: Accuracy and Precision

## The Gaseous State: An Early Measurement of the Way Matter Behaves

Our planet is encased in a blanket of matter called the **atmosphere**. Extending approximately one hundred miles above sea level, this **state of matter** is called a **gas**. (Shortly, we shall meet several other *states of matter*.) The atmosphere is responsible for the great diversity of weather conditions we experience throughout the course of a year—pleasant spring days, violent storms (thunder, tornados, and hurricanes), and the bitterness of a midwinter blizzard. However, the most significant property of this gaseous layer to the development of chemistry is its immediate accessibility. Beginning in the late sixteenth century, humankind began a serious, quantitative study of the gaseous state of matter. We will now retrace the steps taken more than three hundred years ago and use the tools we have developed in this chapter to become acquainted with some of the most significant characteristics of the gaseous state.

We will begin by making some definitions (starting from the fundamental concepts we defined earlier). The word **pressure** is defined as follows:

**Pressure = Force ÷ Area.**

That is, pressure is the force exerted per unit area. Using the SI system of units, we use the following units as we measure pressures:

**SI unit of pressure:** **pascal ($Pa$) = 1 $N\,m^{-2}$.**

Notice that the definition of the **pascal** makes use of our earlier definition of force, which is called the **newton**:

**SI unit of force:** **newton ($N$) = 1 kg m s$^{-2}$.**

The pressure exerted by our atmosphere, the **atmospheric pressure**, is nothing more than the force exerted by a column of air (the atmosphere) on an area of the Earth's surface. This definition of *atmospheric pressure* is a **qualitative description**; the image of a "column of air" lacks precision in several ways. How "tall" is this "column of air"? We have only specified that the atmosphere extends "approximately one hundred miles above sea level," which is far from being a **quantitative** statement. Further, how do we measure this "column of air"? While we have defined a system of units, the definition of the *atmospheric pressure* does not tell us how to go about making a measurement. Consequently, the first critical step is to develop a measuring device. Such a device is called a **barometer.** Figure 1.2 depicts a simple mercury barometer.

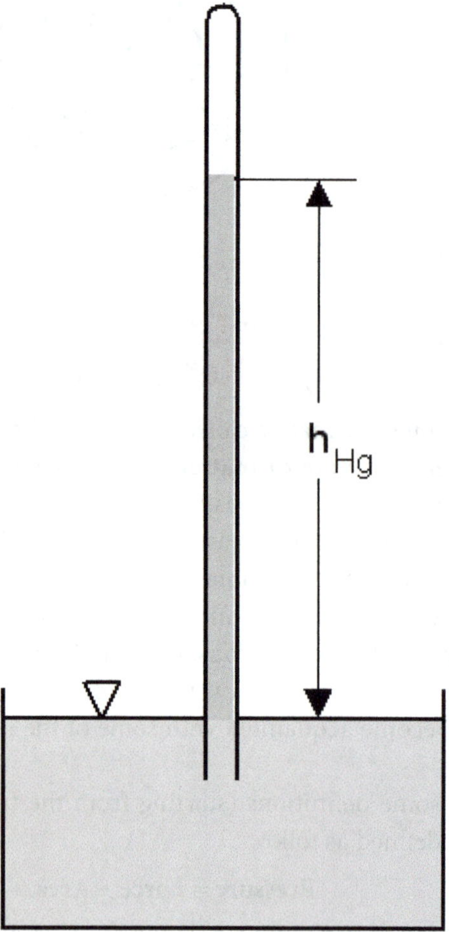


Figure 1.2: Mercury Barometer

The barometer was invented in the mid-seventeenth century by Evangelista Torricelli and utilizes Newton's third law of motion (used earlier to define *mass*) to convert the pressure exerted by the atmosphere into a **distance**, the height of the column of mercury metal ($h_{Hg}$). The column of mercury is stationary because the force exerted on each unit of area by the atmosphere (the atmospheric pressure) is exactly equal to the pressure exerted on each unit of area by the mercury column (the mercury pressure). If the atmospheric pressure increases, the column of mercury **increases** in height (more mass is needed to increase the force exerted by the mercury, balancing the force exerted by the atmosphere); if the atmospheric pressure decreases, the column of mercury **decreases** (less mass is needed to decease the force exerted by the mercury, again balancing the force exerted by the atmosphere). Consequently, the height of the mercury column responds directly to changes in atmospheric pressure, and we report atmospheric pressure in units of **length** rather than units of **pressure**.

You will encounter a variety of units. They are summarized in table 1.5.

Table 1.5: Pressure Units

| | | |
|---|---|---|
| 1 atm = 760. mm Hg | | |
| 1 torr = 1 mm Hg | | |
| 1 atm = 760. torr | | |
| 1 atm = 1.01325 × $10^5$ Pa → SI unit | | EXACT |
| 1 bar = 1. × 105 Pa | → non-SI Unit | EXACT |

In 1981 a decision was made to change the **reference pressure** under which thermodynamic data (we will study aspects of the *science of thermodynamics* throughout this course) were measured from **1 atm** to **1 bar**. This change made minor, **but very important, changes** to the very large body of thermodynamic measurements contained in reference tables. Consequently, when using data from **any** thermodynamic table, it is critically important to identity the reference pressure for the data in that table.

Beginning in the mid-seventeenth century several different individuals investigated the behavior of the pressure and volume of a constant amount of gas held at a constant temperature. In 1662 Robert Boyle published the results of his experiments, and, consequently, this **pressure-volume** relationship is known as **Boyle's law**. Boyle made two important assumptions:

1. The amount of gas in the sample is kept **constant**.
2. The temperature of the sample of gas is kept **constant**.

If these two assumptions are made, Boyle's law can be formally stated as follows:

**Boyle's law:** **The volume of a constant amount of gas maintained at constant temperature is *inversely proportional* to the gas pressure.**

**Mathematically:** $$P \times V = k, \tag{1.24}$$

**where $k$ is a nonzero, positive constant.**

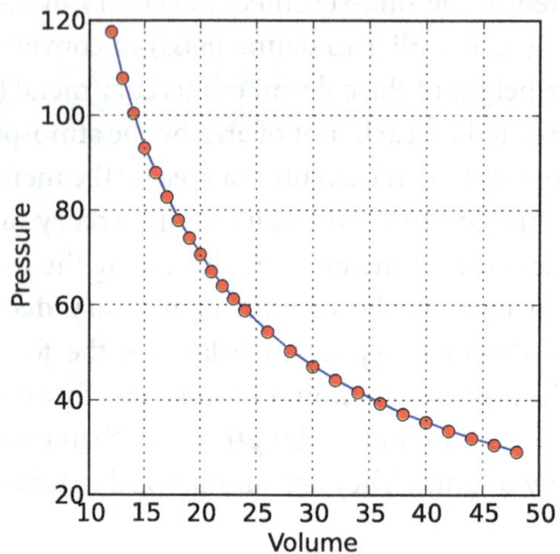


Figure 1.3: Boyle's Law

Note from figure 1.3 that as the pressure **increases**, the volume **decreases**; similarly, as the volume **increases**, the pressure **decreases**. That is, the behavior of one variable (pressure or volume) is the **exact opposite** of the behavior of the other variable. This is the meaning of the phrase *inversely proportional.*

Let's now ask what happens if we maintain the two assumptions (**constant** temperature and **constant** amount of gas), but consider **two** different pressures, $P_1$ and $P_2$. Boyle's law then states that there are **two** different volumes, $V_1$ and $V_2$ such that

$$P_1 \times V_1 = k,$$

and

$$P_2 \times V_2 = k,$$

where $k$ is the **same number** because the temperature is held **constant** and the amount of gas is **constant**. But these two equations can be combined into a single equation:

$$P_1 \times V_1 = P_2 \times V_2. \tag{1.25}$$

This last expression means that given **any** three of the four quantities, you can always find the fourth. Let's look at an example:

Example 1.23: Suppose that a sample of gas at a temperature of 273.0 K occupies a volume of 7.3 L and exerts a pressure of 3.5 atm. Assuming that the temperature and the amount of gas are both kept constant, suppose that the volume of the container is reduced to 2.5 L. What pressure does the gas now exert?

Answer: **Organize:** $P_1 = 3.5$ atm
$V_1 = 7.3$ L
$V_2 = 2.5$ L
Amount of gas is kept constant

Temperature of gas is kept constant at 273.0 K

**Unknown:** $P_2$

**Translate:** $P_1 \times V_1 = P_2 \times V_2$

This is the mathematical statement of Boyle's law. Note that the temperature value does **not** appear in the equation. Hence we **do not use** the number 273.0 K to solve this problem. We **only** require that the temperature is **constant.**

**Solve:** $P_2 = (P_1 \times V_1) \div V_2$

Now, substitute the actual numbers into the equation. Be sure to substitute **both** the value **and** the unit.

$P_2 = (3.5 \text{ atm} \times 7.3 \text{ L}) \div 2.5 \text{ L}$

**Stop and check units.** Please note that the L unit cancels, leaving only the atm units in the answer. But this is **correct** because the problem asks you to calculate the new pressure.

$P_2 = 10. \text{ atm}$ (or $1.0 \times 10.^1$ atm)

(Note: Only two significant figures.)

Because one of the two critically important assumptions requires that the temperature of the gas sample be kept **constant**, we must now refine our **qualitative** understanding of the term **temperature** (defined earlier) and arrive at a **quantitative** understanding of this concept. To accomplish this, we need to define **temperature scales**. In the course of our study we will encounter three different temperature scales:

1. Fahrenheit scale: $^{o}$F
2. Celsius scale: $^{o}$C
3. Kelvin scale: K SI temperature scale

While only the third scale is part of the SI system of units, the first two scales are commonly used. As noted earlier, the conversion to the SI system of units is a slow process, and, at many points in time, multiple scales of measurement coexist. This is certainly true in the case of temperature.

The left panel of figure 1.4 displays the relationship between the Celsius and Fahrenheit scales; the right panel shows the relationship between the Celsius and Kelvin scales.

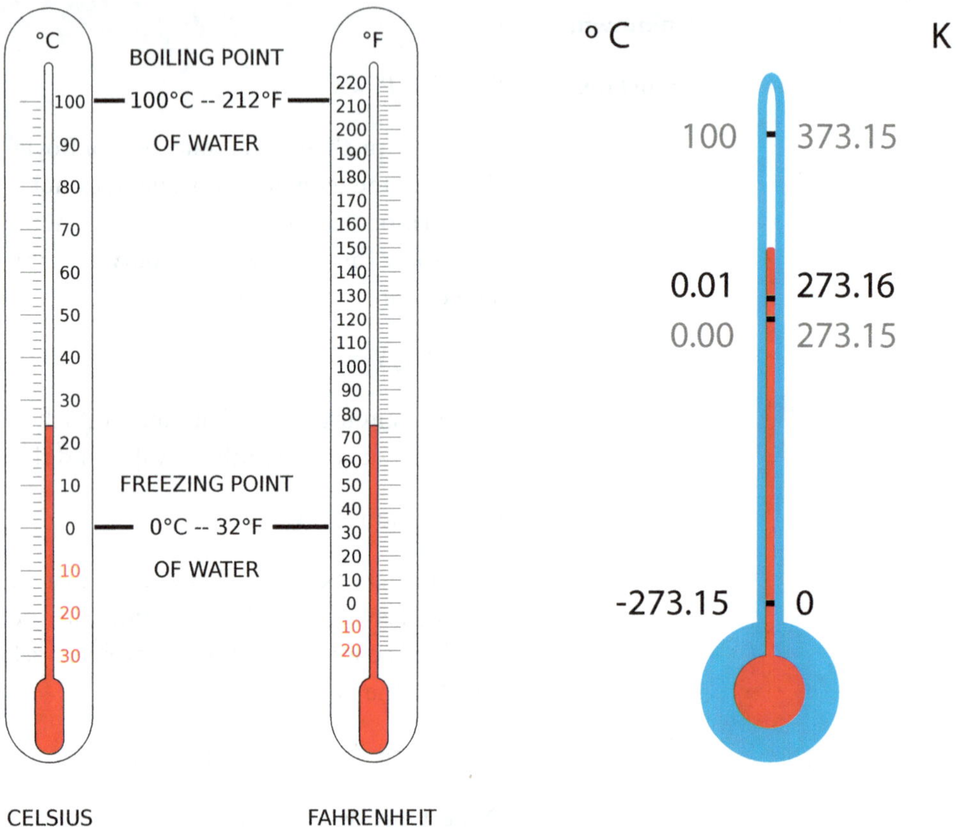


Figure 1.4: Fahrenheit, Celsius, and Kelvin Temperature Scales

As you might expect, there are exact mathematical relationships that connect the three temperature scales. These are given as follows:

$$T_C = (°F - 32\ °F) \times (5\ °C \div 9\ °F), \tag{1.26}$$

$$T_F = °C \times (9\ °F \div 5\ °C) + 32\ °F, \tag{1.27}$$

$$T_K = (°C + 273.15\ °C) \times (1\ K \div 1\ °C). \tag{1.28}$$

In these three equations, **all** the numerical values are assumed to be followed by an **infinite** number of zeroes; consequently, the addition/subtraction or multiplication/division by these constants does **not** limit the number of significant figures in the final answer. The number of significant figures in the final answer is solely determined by the **measured** temperature data.

Example 1.24: Suppose that a sample of gas has a temperature of 298.15 K. What is its temperature on the Fahrenheit scale?

Answer:

**Organize:** $T_K = 298.15$ K

**Unknown:** The value of $T$ on the Fahrenheit scale

**Translate:** $T_K = (°C + 273.15\ °C) \times (1\ K \div 1\ °C)$

This equation relates the Kelvin scale to the Celsius scale.

$T_F = °C \times (9\ °F \div 5\ °C) + 32\ °F$

This equation relates the Celsius scale to the Fahrenheit scale.

**Solve:** Two-step process:

1.Convert the Kelvin temperature to Celsius temperature.

$T_C = K \times (1\ °C \div 1\ K) - 273.15\ °C$

2.Convert the Celsius temperature to Fahrenheit temperature.

Substitute from step 1 to get

$$T_F = [K \times (1\ °C \div 1\ K) - 273.15\ °C] \times (9\ °F \div 5\ °C) + 32\ °F$$

Now, substitute the given value:

$$T_F = [298.15\ K \times (1\ °C \div 1\ K) - 273.15\ °C] \times (9\ °F \div 5\ °C) + 32\ °F$$

$$T_F = 77.000\ °F$$

Because we now have **quantitative** temperature scales, the next question that we can ask is, "What is the effect of temperature **changes** on a gas?" Remember, Boyle's law makes the strict assumptions that the gas sample's temperature and the amount of gas are kept **constant**, allowing **only** the pressure and volume to change. While a relationship between pressure and temperature was recognized in the early eighteenth century, it was not until the early nineteenth century that Joseph Louis Gay-Lussac (citing the unpublished work of Jacques Charles dating from the late eighteenth century) articulated the relationship between the **volume** of a gas and its **temperature** as measured on the **Kelvin** scale. Known today as Gay-Lussac's law (or, sometimes, as Charles's law), this **volume-temperature** relationship, like Boyle's law, makes two important assumptions:

1. The amount of gas in the sample is kept **constant**.
2. The pressure of the sample of gas is kept **constant**.

If these two assumptions are made, Gay-Lussac's law can be formally stated as follows:

**Gay-Lussac's law: The volume of a constant amount of gas maintained at constant pressure is *directly proportional* to the gas temperature when *measured on the Kelvin scale.***

**Mathematically,** $V = k \times T,$ (1.29)

**where $k$ is a nonzero, positive constant.**

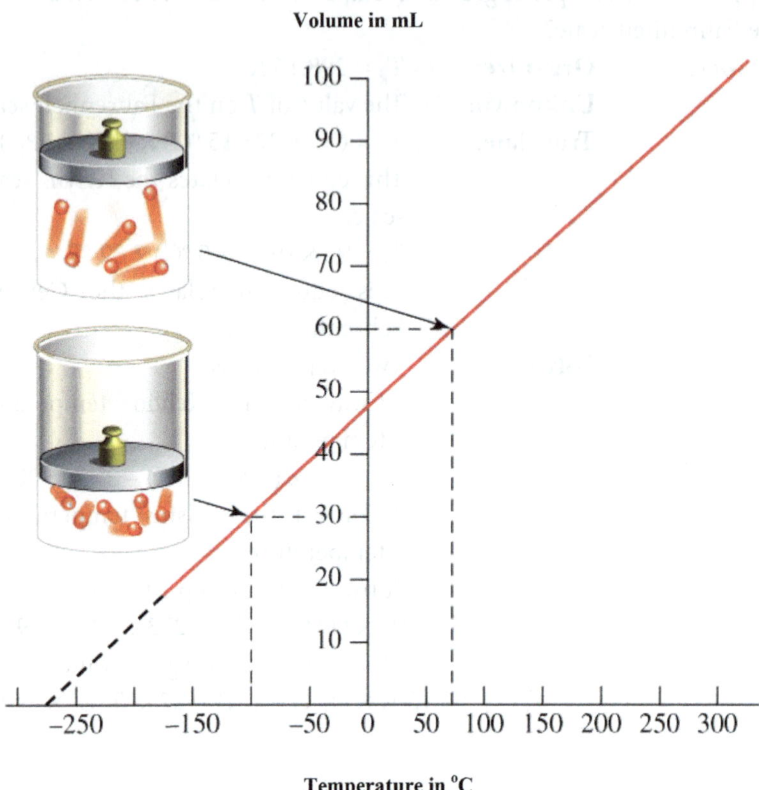


Figure 1.5: Gay-Lussac's law

Figure 1.5 provides a graphical display of the relationship between volume (plotted on the vertical axis in mL) and temperature (plotted on the horizontal axis in °C). Note from the figure that as the temperature **increases**, the volume **increases**; similarly, as the volume **increases**, the temperature **increases**. That is, the behavior of one variable (temperature or volume) is **identical to** the behavior of the other variable; either **both** increase or **both** decrease. This is the meaning of the phrase *directly proportional.*

Let's now ask what happens if we maintain the two assumptions (**constant** pressure and **constant** amount of gas), but consider **two** different temperatures, $T_1$ and $T_2$. Gay-Lussac's law then states that there are **two** different volumes, $V_1$ and $V_2$, such that

$$V_1 = k \times T_1,$$

$$V_2 = k \times T_2,$$

where $k$ is the **same number** because the pressure is held **constant** and the amount of gas is **constant.**

But these two equations can be combined into a single equation:

$$V_1 \div T_1 = V_2 \div T_2. \tag{1.30}$$

This last expression means that given **any** three of the four quantities, you can always find the fourth. Let's look at an example:

Example 1.24: Suppose that a sample of gas at a temperature of 273.0 K occupies a volume of 7.3 L. Maintaining a constant pressure of 3.5 atm and keeping the amount of gas constant, suppose that the volume of the container is reduced to 5.5 L. What is the new Kelvin temperature of the gas?

Answer:

**Organize:**
$P = 3.5$ atm
$V_1 = 7.3$ L
$T_1 = 273.0$ K
$V_2 = 5.5$ L
Amount of gas is kept constant
Pressure of gas is kept constant at 3.5 atm

**Unknown:** $T_2$

**Translate:** $V_1 \div T_1 = V_2 \div T_2$
This is the mathematical statement of Gay-Lussac's law.
Note that the pressure value does **not** appear in the equation. Hence we **do not use** the number 3.5 atm to solve this problem. We **only** require that the pressure is **constant**.

**Solve:**
$T_2 = V_2 \div (V_1 \div T_1)$
$= V_2 \times T_1 \div V_1$
Now, substitute the actual numbers into the equation. Be sure to substitute **both** the value **and** the unit.
$T_2 = 5.5 \text{ L} \times 273.0 \text{ K} \div 7.3 \text{ L}$
**Stop and check units.** Please note that the L unit cancels, leaving only the K units in the answer. But this is **correct** because the problem asks you to calculate the new temperature.
$T_2 = 2.1 \times 10^2$ K
(Note: Only two significant figures)

The reader should pay close attention to the fact that the above example using Gay-Lussac's law **requires** the temperatures to be measured on the **Kelvin** scale. In fact, because Gay-Lussac's law describes a fundamental behavior of gases in our universe, the mathematical relationship

$$V = k \times T$$

also identifies a fundamental characteristic of temperature (when measured on the Kelvin scale). Because the smallest volume that has physical meaning is a **zero** volume, Gay-Lussac's law also implies that the **lowest temperature attainable in our universe (again measured on the Kelvin scale) is *zero*.** As a result, the zero on the Kelvin scale represents not simply a conventional choice but rather a fundamental characteristic of the universe. The zero on the Kelvin scale, for this reason, is often called **absolute zero.**

The two relationships that we have encountered involve four distinct variables: pressure, temperature, volume, and the amount of the gas sample (recall that the SI unit of the amount of substance is called the "mole"; see table 1). As the reader may surmise, there is a connection among **all four** variables that can be summarized succinctly in a single mathematical statement:

$$P \times V = n \times R \times T, \tag{1.31}$$

where $n$ is the amount of the gas sample measured in **moles** and $R$ is a constant of our universe called the **universal gas constant.** The value of the universal gas constant can appear in a variety of units. The two most commonly used values are as follows:

$$R = 8.206 \times 10^{-2}\ \text{L atm K}^{-1}\ \text{mol}^{-1},$$

and

$$R = 8.314\ \text{J K}^{-1}\ \text{mol}^{-1}.$$

Equation 1.31 is called the **ideal gas equation of state** because it completely characterizes the manner in which gases behave; that is, it provides a complete description of the **state** (this is a term from *thermodynamics*, which we will explain in greater detail as we need it) of a gas. The term "ideal gas" has a specific meaning in the *science of thermodynamics* and suggests that this is a model of gas behavior that is in some way "special" (i.e., "idealized"). The gases of our atmosphere at room temperature and up to pressures of approximately 5.0 atm are very well-described by this equation. Consequently, we can use equation 1.31 to examine the realistic behavior of a constant amount of gas but allowing the pressure, volume, and temperature to vary. We write

$$P_1 \times V_1 = n \times R \times T_1,$$

and

$$P_2 \times V_2 = n \times R \times T_2,$$

where $P_1$, $V_1$, and $T_1$ are an initial set of pressure, volume, and temperature values, while $P_2$, $V_2$, and $T_2$ are a second set of pressure, volume, and temperature values. Notice that the $n$ is the **same** in both equations because the amount of gas (the gas sample) is **constant** and $R$ is the universal gas constant.

Because both $n$ and $R$ are constants, their product is a constant. We rewrite the above two equations as follows:

$$(P_1 \times V_1) \div T_1 = n \times R,$$

and

$$(P_2 \times V_2) \div T_2 = n \times R.$$

But, because $n \times R$ is a constant, we simply have

$$(P_1 \times V_1) \div T_1 = (P_2 \times V_2) \div T_2. \tag{1.32}$$

Equation 1.32 is significant because, if you are given information for **any** five of the six variables in the equation, you can **always** find the sixth value. This means that if you know a set of three initial values, $P_1$, $V_1$, and $T_1$, and two of the three values in a second set, you can determine the third value of the second set. This also works in reverse; knowing the three values, $P_2$, $V_2$, and $T_2$, of a second set and two of the three values in the initial set, you can determine the third value of the initial set. Let's look at an example.

Example 1.25: Suppose that a sample of gas at a temperature of 298.0 K occupies a volume of 8.7 L and exerts a pressure of 3.2 atm. Keeping the amount of gas constant, suppose that the volume of the container is reduced to 6.5 L and the temperature of the gas is increased to 75.0°C. What is the pressure that the gas now exerts? (Assume ideal gas behavior in solving this example.)

Answer:

**Organize:**
$T_1 = 298.0$ K
$V_1 = 8.7$ L
$P_1 = 3.2$ atm
$V_2 = 6.5$ L
$T_2 = 75.0$°C
Amount of gas is kept constant

**Unknown:** $P_2$

**Translate:** $(P_1 \times V_1) \div T_1 = (P_2 \times V_2) \div T_2$
This is the mathematical statement that comes from the ideal gas equation of state.

**Solve:** $P_2 = (P_1 \times V_1) \div (T_1 \times V_2) \times T_2$
Before we can substitute the actual numbers into this equation, **we must convert the** $\mathrm{T}_2$ **temperature to Kelvin.**
From equation 1.28, we have the following:
$T_K = (°C + 273.15\ °C) \times (1\ K \div 1\ °C)$
Substituting the value for $T_2$ into equation 1.29 gives the following:

$T_K = (75\ °C + 273.15\ °C) \times (1\ K \div 1\ °C)$

$T_K = 348.15\ K$ **(Don't round.)**

Now we can substitute all the values in order to calculate $P_2$.

$P_2 = (3.2\ atm \times 8.7\ L)$
$\div (298.0\ K \times 6.5\ L)$
$\times 348.15\ K$

**Stop and check units.** Please note that the L units and K units cancel, leaving only the atm unit in the answer. But this is **correct** because the problem asks you to calculate the new pressure.

$P_2 = 5.0\ atm$

(Note: Only two significant figures)

# Chapter 1 Exercises

1. After making observations of physical phenomena, a scientist performs two additional steps. What are these two steps?
2. The conceptual framework of science has two important characteristics. What are these two characteristics?
3. What basic assumption does science make?
4. What is the first step in the scientific method?
5. What is a hypothesis?
6. How are experimental observations and theoretical analyses related?
7. What activity now complements experiments and theoretical analyses?
8. What is a law?
9. What is a scientific theory?
10. What does it mean to say that a scientific theory is "falsifiable?"
11. Is the scientific method simply a sequential, step-by-step process?
12. Define the terms "space", "time", "velocity", and "acceleration."
13. Do the terms "mass" and "matter" have identical definitions?
14. What does the word "momentum" mean?
15. What is the key characteristic of energy?
16. List examples of different forms of energy.
17. What are the two types or kinds of energy?
18. Suppose that a mass of 18.5 kg is moving with a velocity of 35 m $s^{-2}$. What is the kinetic energy associated with the moving mass?
19. Suppose that a kinetic energy of 6375 J is associated with a mass of 155 kg. What is the velocity of the mass?
20. Suppose that a student has a collection of twenty identical bowling balls. Three of the balls are red, five of the balls are blue, and the remainder are black. What percent of the balls are red?
21. Suppose that 65% of coins in a collection are gold and the remainder are silver. There are 325 gold coins in the collection. How many silver coins are in the collection?
22. Calculate the square roots of the following: (a) 625; (b) 81; (c) 144; (d) 265.
23. Calculate the cube root of the following: (a) 27; (b) 8; (c) –729; (d) 1728.
24. Use a calculator to evaluate the following logarithms: (a) ln(2.874); (b) log(3.589).
25. Suppose that $A$ = 4.798 and $B$ = 9.367. Use the properties of logarithms to calculate log(A × B); (b) log(A ÷ B).
26. What are the seven base units of the SI system?
27. Identify the number of significant figures in each of the following: (a) 1.1190 m; (b) 1300. kg; (c) 57.00091 m; (d) 0.97801 mL.

28. Perform each of the following calculations. Be sure that your answer is reported with the correct number of significant figures:
    a. $2006 \times 375 =$
    b. $1.567 \div 3.0675 =$
    c. $0.617 + 67.3 =$
    d. $89.03 + 43 =$
29. Perform each of the following calculations. Be sure that your answer is reported with the correct number of significant figures:
    a. $2206 \times 375 \div 32.1 =$
    b. $(1.867 \div 5.0675) + 15.1 =$
    c. $(0.697 + 17.3) \times 2.5 =$
    d. $(89.451 + 43) \times 1.478 =$
30. Round each of the following values, such that there are only two digits to the right of the decimal point in the reported answer:
    a. 34.67901
    b. 5.343
    c. 78.9950001
    d. 13.625
31. Convert the following:
    a. 567.3 m to kilometers
    b. 6.987 kg to grams
    c. 13.907 cm to meters
    d. 1.6793 ml to liters
    e. 4.65 L to cubic decimeters
32. Convert the following:
    a. 45.50°F to the Celsius scale
    b. 37.00°C to the Fahrenheit scale
    c. 25°C to the Kelvin scale
33. Suppose that a balloon in a room has a volume of 12.0 L when the temperature of the room is 77°F. What is the volume of the balloon when after the air conditioning has cooled the room to 65°F?
34. Suppose that the cylinder of a tire pump has an initial volume of 1.5 L, and the pressure of the gas in the cylinder is 1.0 atm. A student pushes on the pump's plunger, reducing the volume of the cylinder to 0.65 L. Assuming that the temperature of the gas does not change, what is the pressure of the gas in the pump's cylinder?
35. Suppose that a constant amount of gas is contained in a vessel with an initial pressure of 2.0 atm, a volume of 3.5 L, and a temperature of 38.5 °C. Now suppose that the temperature of the gas is increased to 50°C and its volume is reduced to 1.25 L. What is the new pressure of the gas?

# CHAPTER TWO
## Origins

At the end of chapter 1, we briefly investigated the several characteristics of gaseous behavior. These early investigations, beginning in the late seventeenth century, represent some of the first successes of converting chemistry from a qualitative science to a quantitative science in which basic principles can be formulated mathematically. As we'll see during the course of our study of chemistry, this success at making chemistry a quantitative science possessing both explanatory and predictive capabilities has continued spectacularly over the subsequent three hundred years. Underlying this remarkable development lurks a fundamental question: Why is there something and not nothing? Can we actually offer a plausible understanding of the sequence of events that brought the universe to its current state? This question is actually multiple questions because it can be asked again and again as we explore our universe at different levels. This question will be the focus of many of our discussions over the next several chapters.

At the most naïve level we can begin by asking, why is there a universe? That is, can we understand a causally linked sequence of events that brought the universe to the state we observe today? While this question may seem to lie beyond the reach of science, if we understand science to be a model-building process rooted in observation, remarkably, this is not the case. Our observations of the current universe allow us to propose a model of the events that may have brought the universe to its current state of existence. Remember, however, that this model is not the final answer, but it the best consensus model that seems to reflect the observational data.

As the nineteenth century drew to a close and the twentieth century dawned, humankind's understanding of the universe underwent a dramatic transformation. Through the work of a group of astronomers—Heber, Shapley, Slipher, Leavitt, Hubble, and Humason—our understanding expanded from viewing the Milky Way as the totality of our universe to the recognition that the Milky Way is only one of billions of galaxies rapidly receding from one another. The correlation between distance and recessional velocity (Hubble's law) introduced the idea that the universe is a dynamic environment, a concept diametrically opposed to the old notions of static space and time. By the early 1930s, in an effort to model the expanding universe, cosmologists, beginning with Georges Lemaître (whose work appeared after earlier suggestions by Alexander Friedmann), proposed a big bang model, which suggests that the universe began to expand rapidly approximately 13.8 billion years ago (while figure 2.1 indicates that the age of the universe is approximately 13.7 billion years, data newly released in 2013 suggest that the age of the universe is closer to 13.8 billion years) from an extremely dense and extremely hot (millions of kelvin!) initial state (Lemaître's *primeval atom*). Based on Einstein's general theory of general relativity (1915) and some basic uniformity assumptions, this model explains that the universe has not simply been expanding into "space" during this "time," but, rather, has been creating its own **space-time**, the very fabric of our existence. The discovery of the microwave background radiation (*radiation* is the transfer of energy either as mass or as electromagnetic disturbances through space-time) in 1964 by Penzias and Wilson (Nobel Prize, 1978) provided convincing evidence supporting the big bang model, making it the currently best model of the origin of our universe.

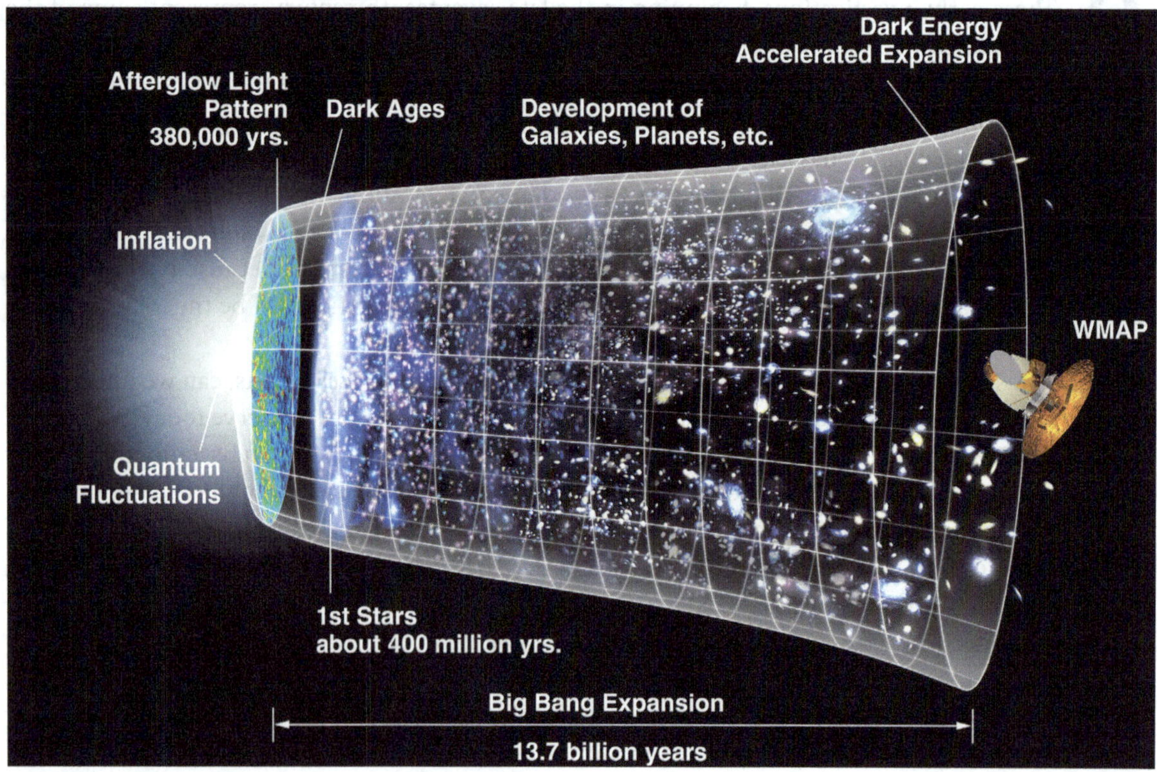


Figure 2.1: Big Bang Diagram

As the universe cooled, only four **elements**, hydrogen (~75%), helium (~25%), and very small amounts (less than ~1%) of lithium and beryllium, were formed. On the other hand, by 2013 humankind had identified a total of 118 distinct elements; living organisms on this planet depend on a variety of elements: hydrogen, carbon, oxygen, nitrogen, phosphorus, and iron, to name only the most prominent ones. An **element** is matter that both possesses well-defined chemical properties and **cannot** be changed by chemical processes into matter possessing different, well-defined chemical properties. The concept of an element is a very old idea; it originated at the beginning of humankind's observation of the way matter behaves in our world. We will shortly connect the idea of an element to more sophisticated models of matter. But, for now, we want to ask, from where did the elements come? In fact, they are the result of a synthesis process that took place in the first generation of stars. (Yes, there have been multiple generations!) The key elements, on which life as we know it depends, were synthesized in ancient stars. In effect, every living organism on this planet, in colloquial terms, is composed of "star stuff."

One of the central points of contention in the big bang model has centered on the extreme uniformity of the cosmic background radiation. In the early 1970s Alan Guth, in an attempt to explain this uniformity, proposed an inflationary big bang model, which posited an extremely rapid expansion of space-time during the first fleeting moments after the initial big bang. This proposal remains controversial. A series of high-altitude balloon and satellite experiments beginning in 1989—BOOMERanG (Balloon Observations of Millimetric Extragalactic Radiation and Geophysics), the COBE (Cosmic Background Explorer) satellite, the WMAP (Wilkinson Microwave Anisotropy Probe) satellite, and the Planck satellite—has both provided support for the inflationary hypothesis and raised questions about its validity. However, in early 2014, cosmologists announced the apparent observation of the residual effects of gravitational radiation (predicted by Einstein's general theory of relativity) in the cosmic microwave background radiation that pervades our universe, providing strong support of Guth's inflationary hypothesis.

But the inflationary proposal is not the only the mystery lurking in the halls of astronomy. Beginning in the 1930s (see the work of Zwicky, 1933) astronomers noted that there appeared to be insufficient mass to explain the dynamics of the newly recognized galaxies. The data collected by Zwicky were either dismissed as incorrect or simply ignored until astronomers attempted to determine the shape of the universe in the 1990s. Consequently, astronomers have suggested that there is another type of undiscovered mater, known as **dark matter**, which affects the dynamics of the galaxies scattered throughout the universe. However, astronomers were at this same time pursuing another question: What is the geometry of our universe? Convincing evidence from the COBE and WMAP satellites suggested that the universe is **flat** (in the sense that Euclid would understand). But, again, there was insufficient **visible** matter in the universe to enforce this geometry. In fact, the dark matter that had been proposed to account for the motions of the galaxies was, even when combined with all the visible matter in the universe, unable to account for the apparently flat geometry of our universe.

An interesting intersection of studies now occurred. In light of the contentious nature of the inflationary hypothesis at the end of the twentieth century, astronomers had continued to study the apparent expansion of the universe. In 1998, a startling discovery was made. Conventional wisdom expected that the expansion (first seen in the early years of the twentieth century) would gradually slow. However, Perlmutter, Schmidt, and Riess observed that the expansion rate is **accelerating**! There appears to be a component of our universe that is acting to counter the effects of the gravitational attraction between any two masses in the universe; rather than slowing, the expansion rate is increasing. It has been suggested that there is a component of our universe called **dark energy** that is responsible for the observed increase in the expansion rate. In recognition of the significance of the discovery, Perlmutter, Schmidt, and Riess were award the 2011 Nobel Prize in Physics.

Consequently, as the twenty-first century began, it now appeared that our universe was made up of three distinct components: **ordinary mass-energy** (using Einstein's 1905 statement of the equivalence of energy and mass, $E = mc^2$), **dark matter**, and **dark energy**. The most surprising characteristic of these three components is the proportion each contributes to the total composition of the universe. Table 2.1 summarizes our current understanding of the composition of the universe.

While the two questions "What is dark matter?" and "What is dark energy?" remain unanswered (as of 2014), table 2.1 has serious implications for the study of chemistry. The data in the table indicate that **everything** that has been studied by the science of chemistry from the earliest times constitutes **less than** 5% of the total universe in which we live. This is important because it puts our study of chemistry into perspective; it is awesome that such a small fraction of the total universe contributes to the tremendous richness and diversity of phenomena we will encounter. At the same time, it is important to remember that the story of chemistry addresses only a very small portion of our total universe.

Table 2.1: Composition of the Universe

| | |
|---|---|
| ~ 0.01% | Cosmic microwave background |
| ~ 0.1 % | Neutrinos |
| ~ 4.9 % | Normal matter (mass-energy) |
| ~ 26.8 % | Dark matter |
| ~ 68.3 % | Dark energy |

## Early Atomic Theory

We again ask, why is there something and not nothing? Now, rather than asking about the universe as a whole, we focus our attention on "normal matter," remembering that "normal matter" constitutes only approximately 5% of the entire universe. The earliest effort to describe the underlying structure of this portion of the universe dates from approximately the fifth century BCE. The Greek philosopher Democritus proposed that all matter (and, here, we are speaking only about "normal matter") is composed of small, indivisible particles. The English word "atom" is derived from the Greek *atomos*, meaning "indivisible" or "that which cannot be cut." This proposal represents an early attempt to explain macroscopic observations by a microscopic interpretation. Further, this proposal represents an implementation of both a materialist and a mechanistic viewpoint, attributing observed phenomena to the existence of natural laws that organize the material universe. However, this conceptual framework was rejected by both Plato and Aristotle and their successors. For nearly the next 2,300 years the concept of the atom played no significant role in Western thought.

Near the end of the first decade of the nineteenth century, the Englishman John Dalton resurrected the atomic concept pioneered by Democritus many centuries earlier. However, the atomic hypothesis was now viewed within the context of a materialist and mechanistic perspective that had come to characterize Western science since the work of Isaac Newton. Dalton articulated four hypotheses that influenced chemical thinking throughout the nineteenth century:

1. All matter consists of indivisible atoms.
2. The atoms of any single element are **identical** in mass and other properties. They possess the same **size**, the same **mass**, and the same **chemical properties**. The atoms of a single element are **different** from the atoms of any other element. (Note: This statement indicates that Dalton had no knowledge of *chemical isotopes*, which we will soon meet.)
3. A **chemical** reaction involves **only** the separation, combination, or rearrangement of atoms. It does **not** result in either the **creation** or **destruction** of atoms. This is a statement of the observed conservation of mass. In particular, atoms of one element **cannot** be converted into atoms of another element by a **chemical reaction.**
4. Compounds are composed of **numbers of atoms of more than one element**. The numbers of atoms of each element present in a compound occur in a **fixed, specific ratio**. These **ratios** are either **integers** or **simple fractions** (i.e., ratios of small whole numbers).

Dalton's hypotheses provided a strong theoretical framework within which to understand many of the observations made by earlier chemists (and alchemists). In particular, the theoretical framework proposed by Dalton in 1808 was consistent with the law of definite proportions, enunciated by the Frenchman Joseph Proust between 1789 and 1804. This "law" simply asserted that compounds always contained the same proportions of their constituent substances by mass. (These substances later came to be called "elements"; we will shortly provide a more formal definition.) This means that the mass ratios of the elements making up a specific compound are **always** the same. For example, in a sample of water (composed of the elements hydrogen and oxygen), it is found experimentally (to the limit of the number of significant figures in the measurement) that one-ninth of the compound's mass is the element hydrogen and eight-ninths of the compound's mass is the element oxygen. This experimental observation is consistent with Dalton's hypotheses: hypothesis 1, matter is composed of **atoms**; hypothesis 2, all the atoms of a single element are **identical**; hypothesis 3, conservation of mass; and hypothesis 4, compounds are made of atoms of different elements in **fixed** and **specific** proportions.

However, the theoretical framework envisioned by Dalton went beyond simply being consistent with earlier chemical observations. The four atom-centered hypotheses formed the foundation of a new generalization called the law of multiple proportions, which provided a theoretical basis for understanding the formation of two **different** compounds by the same elements. Articulated by Dalton in 1808, the law of multiple proportions provides an explanation for the formation of two different compounds by two or more elements. Consider the simplest case, in which two distinct compounds are formed by the reaction of two different elements. Choose a constant mass of one of the elements and allow it to form the two compounds by reacting it with two **different** masses of the second element.

The **ratios** of the masses of the second element that completely react with the constant mass of the first element to form two distinct compounds will be **ratios of small whole numbers** (again, to the limit of significant figures in the respective measurements). As we shall see, carbon monoxide and carbon dioxide are two different compounds formed by the elements carbon and oxygen. For a fixed (i.e., constant) amount of carbon, the ratios of the mass of oxygen to the mass of carbon in each of the compounds are ratios of **small whole numbers**.

While Dalton's revival of the atomic hypothesis and his insightful explanations of the myriad of chemical reactions that had been cataloged by humankind are often recognized as the beginning of chemistry as a modern science, the preoccupation of the nineteenth century with electrical phenomena would ultimately lead, by the end of the century, to an entirely new conception of the atom. The identification of two distinct electrical charges, *positive* and *negative*; the recognition that the mixing of water with a variety of chemical compounds exhibited *electrical* properties consistent with the distinct *positive* and *negative* charges; the monumental theoretical synthesis of electricity and magnetism (now called electromagnetic phenomena) by James Clerk Maxwell; and the general growth of the electrical industry led to widespread speculation about the nature of the so-called "electrical fluid." By the middle of the century it was suggested that perhaps an atom was not "indivisible"; rather, it was a core of matter surrounded by concentric spheres of subatomic "electrical" atoms or particles. In 1874, George Johnstone Stoney, an Irish physicist, speculated that a single quantity of electricity existed but that this quantity of electricity was permanently bound to the atom. The German physicist Herman von Helmholtz proposed in 1881 that the *positive* and *negative* charges were divided into elementary parts and that these parts behaved like electrical atoms. In 1891, Stoney used the name "electron" for his fundamental electrical unit, although the relationship between this fundamental unit of electricity and matter was unclear.

It is remarkable that by the middle of the century, the concept of the "indivisible" atom, so recently reintroduced by Dalton, was being reconsidered. Further, the tremendous explosion in the study of electrical phenomena had raised the possibility of "electrical atoms" or "fundamental units of electricity," which seemed to be separate and distinct from the atoms of matter that Dalton had used so effectively and so recently to provide a theoretical framework for chemistry. However, unlike the experimental work of a number of chemists and physicists during the nineteenth century, at least some of these speculations lacked the full force of rigorous experimental observation; there would soon be a shift from qualitative speculations to quantitative experimentation that would dramatically change humankind's understanding of the microscopic.

Beginning in the late 1890s a number of experimental observations were made that provided a puzzling array of empirical data challenging Dalton's idea of the atom as well as the overall picture of the physical universe created by Isaac Newton and his successors. In 1895, Wilhelm Röntgen observed a new, highly penetrating form of radiation, soon to be named X-rays. Henri Becquerel, working in Paris in 1896, discovered the amazing phenomenon of *radioactivity* (a term coined by Pierre and Marie Curie in 1898) in which atoms **spontaneously** emit radiation. Becquerel's observation immediately suggested that if matter is indeed composed of atoms, these atoms are neither indivisible nor eternal. There appeared to be constituent parts in atoms that were emitted as an atom changed from an initial, higher energy state to a state of lower energy. The mystery and surprise deepened in 1897 when J. J. Thomson and his students at the Cavendish Laboratory in Cambridge,

England, studied cathode "rays" and identified them as minute, *negatively* charged pieces of matter. They were deflected by electric and magnetic fields and appeared to be at least one thousand times less massive than the lightest element, hydrogen. Thomson was able to determine the charge-to-mass ratio of the cathode rays, but could provide only very imprecise independent estimates of the charge and the mass. It was not until 1907 that an American physicist, Robert Millikan, was able to make more accurate estimates of both the charge and the mass of theses rays. While Thomson gave the name "corpuscles" to cathode rays, the name "electron" became the accepted term for these *negatively charged* fragments of atomic matter. Thomson continued to use the term "corpuscle" until 1913. It is particularly important to notice that Thomson's critical insight was the recognition that the cathode rays were, in fact, subatomic pieces of matter. This insight led to the recognition of the connection between matter and electricity.

The experimental work of Thomson determined the charge-to-mass ratio of an electron to be approximately $-1.76 \times 10^{8}$ $Cg^{-1}$ (where "C" is the symbol for the *coulomb*, the SI unit of electric charge). After a careful measurement of the electron's charge as $-1.6022 \times 10^{-19}$ C, Millikan was able to estimate the electron's mass as:

$$m_e = \text{(charge)/(ratio)} = (-1.60 \times 10^{-19}\,\text{C}) \div (-1.76 \times 10^{8}\,\text{C/g})$$
$$= \sim 9.09 \times 10^{-28}\ \text{g}.$$

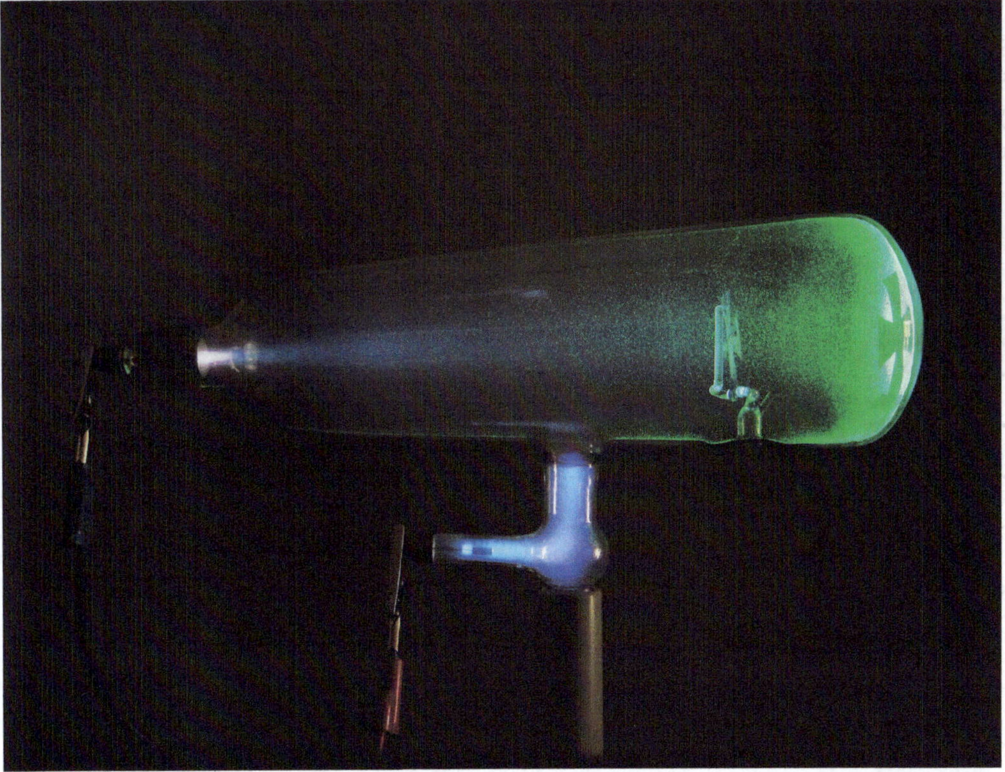

Figure 2.2: Cathode Ray (Crookes) Tube

The modern value is $9.10938291 \times 10^{-28}$ g. The important point of this calculation is that the mass of the electron is an **extremely small** number, confirming Thomson's original judgment that the electron is a subatomic species.

During the next twenty years the world of experimental physics was dominated by the imposing presence of New Zealander Ernest Rutherford. Beginning with his work at McGill University in Montreal, continuing at the Victoria University of Manchester (commonly known as the University of Manchester) in Great Britain, and finally, succeeding Thomson as director of Cambridge's Cavendish Laboratory, Rutherford became the world's leading expert on radioactivity. In 1899, he demonstrated that the uranium radioactivity observed by Becquerel was not homogeneous but rather was composed of several distinct types of radiation. Rutherford identified two types: **α radiation**, later shown by Rutherford to be helium atoms that had lost two negative charges (electrons), and **β radiation**, later shown to be identical to the electrons observed by Thomson in cathode rays. Both α radiation and β radiation are associated with nonzero mass. In 1900, Paul Villard identified a highly penetrating third type of radiation emanating from uranium, **γ radiation**, that is a purely electromagnetic phenomenon (like light) and with which there is **no** associated mass.

In 1909, Rutherford and his students, Hans Geiger and Ernest Marsden, performed the famous gold-leaf experiment, in which α particles were used to bombard an extremely thin sheet of gold foil. A zinc sulfide screen surrounded the source of α particles and the thin gold foil; small flashes would appear on the screen each time an α particle hit the screen. The prevailing atomic theory of the time suggested that the α particles should easily pass directly through the thin foil experiencing, at most, only a slight deflection from their initial path. However, contrary to their expectations, the experimenters observed that a few α particles reflected at very **large** angles, with some reflected α particles moving directly back toward their source! The two panels of figure 2.3 show the expected behavior (left panel) and the results of the actual experiment (right panel).

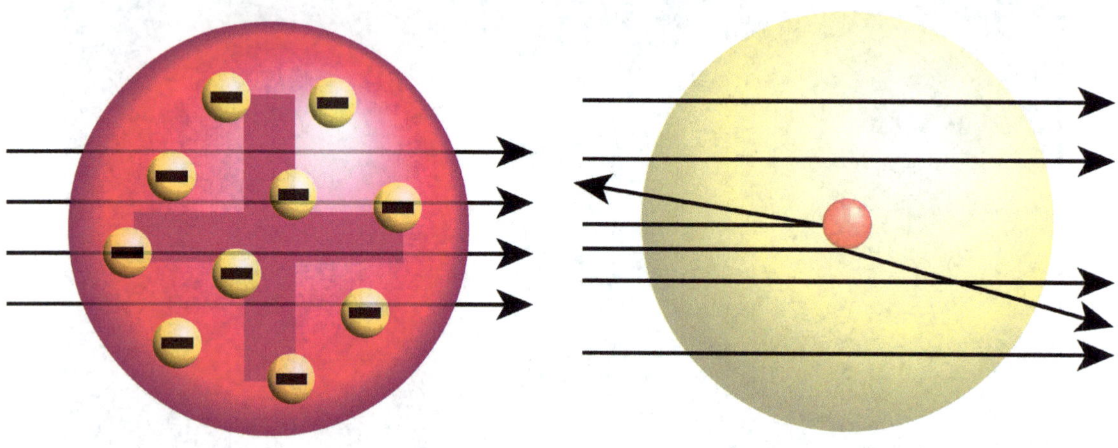


Figure 2.3: Gold-Leaf Scattering

(It should be noted that the diagram is **not** drawn to scale. The diameter of the scattering center in the right panel is actually one thousand times smaller than the diameter of the spherical shape representing the entire atom.) For Rutherford, this was an astounding observation, the equivalent of a large naval gun shell bouncing off a piece of tissue paper. It suggested that in the gold foil there were regions of extremely dense mass interspersed among regions containing almost **no** mass. In fact, to explain the nearly backward scattering of some of the α particles required that almost **all** of the gold mass must be concentrated in extremely small volumes of space. In effect, the gold foil, despite its solid appearance, must be mostly empty space.

Rutherford also experimented by passing α particles through chambers filled with various gases and observed the scintillations (the bright flashes appearing on a zinc sulfide screen) produced by the *positively* charged radiation stream generated by the interaction of the gases with the α particles. This type of *positively* charged radiation had been observed as early as 1886 by Eugen Goldstein in early cathode ray tubes and was called at that time "canal rays" or "anode rays." In the case of hydrogen, Rutherford observed a particularly penetrating, *positively* charged type of radiation that produced characteristic flashes in his scintillation counter.

In 1917, Rutherford bombarded a virtually pure sample of nitrogen gas (nitrogen gas composes approximately 78% of the Earth's atmosphere today) with α particles and observed the same type of *positively* charged radiation (based on the characteristic scintillations) that he had observed when hydrogen gas had been bombarded with α particles. Because there was no hydrogen present in his test sample of gas, Rutherford (in his 1919 report of these experiments) proposed that he had achieved the first artificial **transmutation** (the transformation of one element into another by bombarding the dense central core, later called the *nucleus,* of the first element with nuclear particles or other nuclei):

$$^{4}_{2}\mathrm{He} + {}^{14}_{7}\mathrm{N} \rightarrow {}^{13}_{6}\mathrm{C} + {}^{4}_{2}\mathrm{He} + {}^{1}_{1}\mathrm{H}. \tag{2.1}$$

While Rutherford was correct in suggesting that an atomic transmutation had taken place, it was not until 1924–1925 that P. M. Blackett showed the process Rutherford had observed was actually as follows:

$$^{4}_{2}\mathrm{He} + {}^{14}_{7}\mathrm{N} \rightarrow {}^{17}_{8}\mathrm{O} + {}^{1}_{1}\mathrm{H}. \tag{2.2}$$

Rutherford used the symbol $^{1}_{1}\mathrm{H}$ to represent a hydrogen atom that has **lost** a single negative charge (electron); it is a species with a single *positive* charge, and in 1920, he gave the name **proton** to $^{1}_{1}\mathrm{H}$ symbol. (Early experimental work by Wilhelm Wien in 1898 had provided the first evidence for a positively charged particle with the mass of a hydrogen atom. However, the name "proton" was not suggested until 1920 by Rutherford.)

## Three-Particle Model of the Atom

By the mid-1920s, thirty years of intense experimental effort had provided the evidence that would force humankind to change dramatically the framework used to understand the world. Even the meaning of the symbols used in equations 2.1 and 2.2 by Rutherford and Blackett to describe their experimental results was undergoing repeated modification as new theoretical models were constructed to describe the rapidly growing body of empirical results. Let's now examine the development of these models and their relation to the current best understanding of atomic structure.

Following his experiments in 1897, Thomson in 1904 proposed what came to be known as the "plum pudding" model of the atom in which the electrons (Thomson's "corpuscles") were free to move in a uniformly distributed region of positive charge. While Thomson initially thought that there were many thousands of electrons in the hydrogen atom moving in a series of concentric shells, he was later forced to accept that this picture could not be correct. Even though Thomson's model included some sophisticated ideas, it could not successfully explain other experimental data and was inconsistent with Maxwell's electromagnetic theory. Further, it did not account for the positively charged α particles that had been identified by Rutherford in his 1899 analysis of Becquerel's radiation and the fact that α particles were associated with a definite mass. Thomson's model proposed only positive charge, with no a definite mass associated with that charge. In fact, Rutherford showed in 1908 that α particles are the nuclei of helium atoms.

The gold-leaf experiments in 1909–1911 led Rutherford to suggest a model of the atom in which **virtually all** the atom's mass was concentrated in a very tiny region of space. Further, the positive charge was now identified with the highly concentrated mass of the atom. (Interestingly, it was a prominent British astrophysicist, John Nicholson, who coined the term **nucleus** in 1911. This term became associated with Rutherford's model, identifying the central mass concentration of an atom, even though he had not used that term. Today, we continue to refer to this dense central core of an atom as the nucleus.) The experimental data collected by Rutherford also suggested that the number of units of positive charge (each equal in magnitude to the negative charge associated with the electron) was approximately one-half the mass of an atom (where an atom's mass is measured in units of the mass of a hydrogen atom). This relationship was not made quantitative until Henry Moseley's work in 1913, identifying a connection between X-rays and the number of positive charges (later called the **atomic number**) in the nucleus. Rutherford's model was entirely consistent with his scattering observations that atoms (or at least the gold atoms he studied) were virtually empty space. It was only on those rare occasions when an α particle encountered the nucleus that it was significantly deflected from its path. However, Rutherford's new model of the atom still left many questions unresolved. In fact, Rutherford's preoccupation with the phenomenon of radioactivity meant that he was not focused on the question of atomic structure. His model of the atom failed in several ways: it was inconsistent with Maxwell's electromagnetic theory, and consequently, was unstable; it did not provide any understanding of the structure of the nucleus; it was unable to explain the origin of the various types of radiation associated natural radioactivity; and it could not explain the spectral data (think of the variety of colors observed in a fireworks display) that had been collected during the nineteenth century.

In 1913 a young Danish physicist working with Rutherford in England, Niels Bohr, proposed a more sophisticated version of **Rutherford's nuclear atom** (called the **Rutherford-Bohr atom**, or simply, the **Bohr atom**), which was both electromagnetically stable and successfully explained a wide range of spectral data. Further, Bohr built his model of the atom using a new idea, **quantum theory** (which we will study very shortly), that was to revolutionize humankind's conception of the microscopic. While Bohr's ideas won wide acclaim (and eventually a Nobel Prize) and provided an initial understanding of the structure of electrons in the atom, they were generally silent about the structure of the nucleus.

During the period 1914–1920 there was much speculation about the nucleus and its composition. While it had been well-established that α, β, and γ radiations were produced spontaneously by a number of different elements, the relationship of these radiations to the atomic nucleus was unclear. We now return to equations 2.1 and 2.2 and ask more carefully about the meaning attached to the symbols by Rutherford (and Blackett) and the connection between the meaning attached to the symbols and contemporary models of the atom:

$$ {}^{4}_{2}\text{He} + {}^{14}_{7}\text{N} \rightarrow {}^{13}_{6}\text{C} + {}^{4}_{2}\text{He} + {}^{1}_{1}\text{H}. \tag{2.1} $$

$$ {}^{4}_{2}\text{He} + {}^{14}_{7}\text{N} \rightarrow {}^{17}_{8}\text{O} + {}^{1}_{1}\text{H}. \tag{2.2} $$

Note that the relative size of the superscripts and subscripts associated with each nucleus is consistent with Rutherford's earlier supposition that the number of units of positive charge (subscript) is approximately one-half the mass number (superscript). Further, by writing equation 2.1, Rutherford argued that the experimental data confirmed that his **protons** (the symbols in the equations) must reside in the nucleus. However, two possible explanations were offered to explain the structure of the nucleus. In the first model, **both** electrons and protons reside in the nucleus and the **number of protons is equal to the atomic mass number**. To reduce the number of units of positive charge to that which is experimentally observed, it was proposed that sufficient electrons were also present in the nucleus to partially cancel the positive charge of the protons. For example, in the case of nitrogen, this proposal suggested that there are fourteen protons (superscript) along with seven electrons in the nucleus, leaving a net positive charge of seven (subscript). The second model postulated the presence of an entirely new particle, which Rutherford called the **neutron**. The neutron was nearly equal in mass to the proton, but carried **no** electrical charge. In this model, the nucleus of the nitrogen atom with a mass of fourteen units (superscript) contains seven protons (subscript) and seven neutrons. The matter was not settled until James Chadwick's experimental discovery of the neutron in 1932.

Consequently, after nearly forty years of experimental and theoretical effort, the atom was visualized as being composed of **three** elementary particle types, two of which, the protons and the neutrons, reside in the nucleus. The electrons, which make up the third particle type, are arranged around the nucleus. Because the protons and the electrons carry **equal but opposite** electrical charges, the numbers of electrons and protons in a **neutral** atom are **equal**. Table 2.2 summarizes the properties of these three elementary particles. There are several key characteristics to notice in the table. First, the masses of all three particles are extremely small. Second, the masses of the proton and the neutron are **nearly equal**, but both are **much more massive** than the mass of the electron, the proton by a factor of more than 1,836 and the neutron by a factor of more than 1,838. Third, the electrical charge of the electron is **negative**, while the electrical charge of the proton is **positive, equal in magnitude** but **opposite** that of the electron; the neutron, as its name implies, is **neutral**, that is, **uncharged**.

As a result of the extensive effort of the first third of the twentieth century, a number of basic terms are now routinely used to describe the defining characteristics of the atom. These terms will provide us with a working language to study many ideas in chemistry.

Table 2.2: Three-Particle Model of the Atom (Modern Values)

| Particle | Mass | Charge |
|---|---|---|
| Electron | $9.10938291 \times 10^{-28}$g | $-1.6022 \times 10^{-19}$ C |
| Proton | $1.672621777 \times 10^{-24}$ g | $+1.6022 \times 10^{-19}$ C |
| Neutron | $1.674927351 \times 10^{-24}$g | 0 C |

## Atomic Number

The **atomic number** of an element is the **number of protons** in the nucleus of each atom of an element. The letter Z is used as the symbol for the atomic number. In a neutral atom (using our three-particle model) the **number of protons** is *equal* to the **number of electrons.** Consequently, in a neutral atom, Z also indicates the number of electrons. As we study chemistry more thoroughly, we'll discover that the chemical behavior of an element depends **primarily** on two characteristics of every atom of that element: (1) the arrangement of the electrons in each atom (called the **electronic configuration**) and (2) the number of protons in the nucleus of each atom. Consequently, the chemical identity of all neutral atoms is primarily determined by the value of Z.

## Atomic Mass Number

The **atomic mass number** of an atom of a given element is the **total** number of **neutrons** and **protons** in the nucleus of an atom of that element. The symbol for the atomic mass number is the letter A. Consequently, the number of neutrons in the nucleus of any atom is simply the difference of A–Z. It is important to note that **both** the atomic number and the atomic mass number are **integers.** This also means that the number of neutrons is also an **integer.** This again makes sense in terms of the three-particle model of the atom because protons and neutrons are distinct units in this model. The atomic number and atomic mass number simply count these distinct units.

## Isotopes

We have already noted that chemical behavior of an element depends primarily on the arrangement of the electrons (**electronic configuration**) in an atom of the element and on the number of protons in the nucleus of each atom (**atomic number**). There means that the number of neutrons does **not** affect the chemical behavior of an element and raises the possibility that **all** the nuclei of atoms one element are not required to have the identical number of neutrons. In fact, this occurs in our universe. Atoms that have the **same atomic number** (meaning all the same element because the number of protons is the same) but **different mass numbers** (differing numbers of neutrons) are called **isotopes** (from two Greek words: *iso*, meaning "equal," and *topos*, meaning "place").

By the final third of the nineteenth century, chemists had agreed on the outlines of a simple system of symbols to represent all the known elements that was based on each element's name. While this symbolic system was routinely used by chemists, its most profound application came with the creation, shortly after midcentury, of the periodic table of elements (by Dmitri Mendeleev and Julius Lothar Meyer, in 1869–1870). The periodic table has become an iconic symbol of chemistry and summarizes in a very compact form a large amount of chemical information. For the moment, let's concentrate on the symbols. The chemical symbols for hydrogen, carbon, and oxygen are H, C, and O, respectively. If we add the additional information of atomic mass number and atomic number to these symbols, we have

$$^{1}_{1}\mathrm{H} \qquad ^{12}_{6}\mathrm{C} \qquad ^{16}_{8}\mathrm{O}.$$

In each symbol, the superscript represents the atomic mass number, and the subscript is the atomic number. By specifying the atomic mass number for each element, we have been able to specify an exact isotope of that element. In fact, three distinct isotopes exist for all three of the elements, and they are represented as follows:

$$^{1}_{1}\mathrm{H} \qquad ^{2}_{1}\mathrm{H} \qquad ^{3}_{1}\mathrm{H}$$

$$^{12}_{6}\mathrm{C} \qquad ^{13}_{6}\mathrm{C} \qquad ^{14}_{6}\mathrm{C}$$

$$^{16}_{8}\mathrm{O} \qquad ^{17}_{8}\mathrm{O} \qquad ^{18}_{8}\mathrm{O}.$$

The three isotopes of hydrogen are given the names protium, deuterium, and tritium. The isotopes of carbon are referred to as carbon-12, carbon-13, and carbon-14; similarly, the isotopes of oxygen are called oxygen-16, oxygen-17, and oxygen-18.

Example 2.1: For each of the following neutral elements, determine the number of protons, neutrons, and electrons: $^{14}_{7}\mathrm{N}$, $^{13}_{6}\mathrm{C}$, $^{238}_{92}\mathrm{U}$, $^{207}_{82}\mathrm{Pb}$.

Give the name of each element.

Answer:

**Organize:** Four neutral elements given
Superscripts and subscripts along with the chemical symbol are the information needed to answer the questions.

**Unknown:** Each element's name
The number of electrons
The number of protons
The number of neutrons

**Translate:** Superscript: atomic mass number
Subscript: atomic number

**Solve:** $^{14}_{7}\mathrm{N}$ nitrogen
7 neutrons

7 protons
7 electrons

$^{13}_{6}C$ carbon
7 neutrons
6 protons
6 electrons

$^{238}_{92}U$ uranium
146 neutrons
92 protons
92 electrons

$^{207}_{82}Pb$ lead
125 neutrons
82 protons
82 electrons

## Atomic Mass and Average Atomic Mass

Because the three-particle model of the atom is the result of the experimental and theoretical work completed during the first third of the twentieth century, the terms atomic number and atomic mass number were not part of the vocabulary of chemistry prior to the last century. As a result, chemists have been focused on measuring the masses of samples, on understanding the mass ratios of substances participating in chemical reactions, and on quantifying the conservation of mass as a central chemical tenet. In fact, we will see later in our story that the effort to organize the chemical elements in the periodic table utilized mass as one of the organizing principles. But, you might object, atoms are so tiny! (Review the masses of the electron, proton, and neutron in table 2.2.) How is it possible to measure such small masses?

The key to answering this question lies in the very clever approach adopted by chemists in the early nineteenth century to the problem of measuring masses. Rather than using a **macroscopic** quantity (the kilogram, for example) as the standard of comparison, chemists chose to construct a system of **relative mass**, in which an atom of a single element was chosen, **by agreement**, to be to the standard to which atoms of every other element are compared. That is, the atomic mass of every element is determined **relative to the atomic mass of a single element**. Historically, the first element used for this purpose was the element oxygen. While physicists used a specific isotope of $^{16}_{8}O$, to define the relative mass standard, chemists based their relative mass scale on a weighted average of the naturally occurring oxygen isotopes, oxygen-16, oxygen-17, and oxygen-18. Since 1961, a single isotope of carbon has been used to define the **unified atomic mass unit**, which is symbolized by u. We define the following:

$$\mathbf{12\ u = Mass\ of\ one\ atom\ of\ ^{12}_{6}C} \quad (2.3a)$$

Another way stating **exactly the same idea** is

$$\mathbf{1u = One\text{-}twelfth\ the\ mass\ of\ one\ atom\ of\ {}^{12}_{6}C} \tag{2.3b}$$

It is important to note that the definition of the unified atomic mass unit specifies a single isotope of carbon, the carbon-12 isotope. The unified atomic mass unit is **not** part of the SI system of units that we defined earlier but is used ubiquitously in chemistry. There are two other terms that often occur in discussions relative atomic masses. First, one often encounters the symbol **amu** for the atomic mass unit. This was the official designation prior to the adoption of the unified atomic mass unit standard in 1961. It is no longer appropriate to use the symbol **amu**; however, old traditions are not easily replaced. The second term is the **dalton**, symbolized by Da, which is also a non-SI unit used frequently in biochemistry and molecular biology, although it has not been approved by the Conférence Général des Poids et Mesures (CGPM, the General Conference on Weights and Measures). The dalton is simply an alternate name for the unified atomic mass unit; it currently has exactly the same definition as the unified atomic mass unit. However, as we noted in our discussion of the SI system, there is currently (as of 2014) an effort to define the kilogram in terms of fundamental constants rather than the currently created human artifact *Le Grand K*. Such a redefinition of the kilogram will require a complementary reexamination of the definitions of the unified atomic mass unit and the dalton.

The definition of the unified atomic mass unit establishes a **relative mass scale**, which can now be used to determine the **atomic mass** of an atom of any element by comparing it directly to the mass of a carbon-12 atom. Indeed, the mass of an atom of every isotope of an element can be determined. But this raises a practical question: If an element, say, oxygen, for example, exists in several isotopic forms (oxygen-16, oxygen-17, and oxygen-18), which mass number should represent an atom of the element oxygen? The answer to this question is found in the definition of the **average atomic mass**, which is based on the relative abundance of **all** the naturally occurring isotopes of that element. The **average atomic mass of** an atom of any element is computed as the weighted average of the atomic masses of the naturally occurring isotopes. As a result of this definition, a single number, the average atomic mass of an element's atom, is assigned to each of the 118 elements of the periodic table. In the case of the synthetic elements (elements that are human artifacts), the total nucleon count (total number of protons and neutrons in the nucleus) of the most stable isotope is adopted as the atomic mass of an atom of the element. In order to acknowledge that the relative isotopic abundances do vary in terrestrial samples for number of reasons, the International Union of Pure and Applied Chemistry (IUPAC) in 2011 recorded the average atomic masses of ten elements as an interval rather than as a single number.[1] The most recently published versions of the periodic table will reflect this change by including a range of numbers (i.e., [a,b], with "a" and "b" being lower and upper bounds of the range) rather than a single value as the atomic mass of an element's atom.

1 See Michael E. Wieser and Tyler B. Coplen, "Atomic Weights of the Elements 2009 (IUPAC Technical Report)," *Pure Appl. Chem.* 83 (2011): 359–396.

Example 2.2: The element lithium occurs as two isotopes: lithium-6, with an atomic mass of 6.015 u and a relative abundance of 7.420%, and lithium-7, with an atomic mass of 7.016 u and a relative abundance of 92.58%. Calculate the average atomic mass of the element lithium.

Answer:

**Organize:** Lithium-6: Atomic mass: 6.015 u
Abundance: 7.420%
Lithium-7: Atomic mass: 7.016 u
Abundance: 92.58%

**Unknown:** Average atomic mass of lithium

**Translate:** Calculate a weighted average by converting the relative abundance percentages to decimals.

**Solve:** **Average atomic mass** of Li
= (0.07420) × (6.015 u)
+ (0.9258) × (7.016 u)
**Average atomic mass** of Li = 6.942 u
(Note: Four significant figures)

Example 2.3: The element boron has an average atomic mass of 10.81 u and occurs as only two isotopes: boron-10 and boron-11. The boron-10 isotope has an atomic mass of 10.013 u and a relative abundance of 19.80%. Calculate the atomic mass of the boron-11 isotope.

Answer:

**Organize:** Boron-10: Atomic mass: 10.013 u
Abundance: 19.80%
Average atomic mass of boron: 10.81 u

**Unknown:** Atomic mass of boron-11.

**Translate:** Because there are **only** two isotopes of boron, the problem **implicitly** gives the relative abundance of boron-11.
Calculate the relative abundance of boron-11:
1.0000 – 0.1980 = 0.8020

**Solve:** 10.81 u = (0.1980) × (10.013 u)
+ (0.8020) × (**Unknown** u)
**Unknown** u = [10.81 – (0.1980) × (10.013 u)] ÷ (0.8020)
**Unknown** u = 11.01 u
(Note: Four significant figures)

## The Mole

Up to this point, our discussion has focused on single atoms, their atomic number, atomic mass number, atomic mass, and average atomic mass. But, you object, how is a chemist able to work with a single atom? Atoms, by our very measurements, are extremely small bits of matter (review table 2.2 for the masses of the electron, proton, and neutron). In order to work with amounts of mass that are measurable at the macroscopic level of human beings, chemists have made use of the clever relative atomic mass scale that we introduced above to define a quantity of matter called the **mole.** A **mole** is defined as **the amount of substance of a system that contains as many distinct elementary objects as there are atoms in exactly 0.012 kilogram of carbon-12.** There are several important characteristics of this definition. First, a mole is an **amount of substance**; it is **not** a number. Recall that the mole is one of the base units of the SI system (see table 1.1) and is a measure of the amount of substance. Second, note that the mass used to define the mole is **exactly** 12 grams (0.012 kg) of a **single** isotope of carbon, specifically the carbon-12 isotope. Third, the actual number of distinct elementary objects in a mole is an **experimentally determined number** because the actual number of atoms in exactly 12 grams of carbon-12 is also an **experimentally determined number.** This number depends on the ability of humankind, at any point in time, to count individual atoms; this ability to count changes over time. The 2010 recommendation from the Committee on Data for Science and Technology (CODATA) for the value of this number is $6.022\ 141\ 29 \times 10^{23}\ mol^{-1}$. It has been given the name **Avogadro's number,** in honor of the nineteenth-century Italian scientist Amedeo Avogadro, who first proposed (in 1811) that the pressure exerted by a sample of gas, whose pressure and temperature are kept **constant,** is directly proportional to the number of atoms or molecules and is **not** dependent on the identity of the gas. We will use the symbol $\mathbf{N_A}$ to represent Avogadro's number. It is important to note that the number quoted earlier **will change;** it is **not** a constant because, as noted earlier, the number of distinct elementary objects in a mole is an **experimentally determined number.** In this text we will use $6.022 \times 10^{23}\ mol^{-1}$ as our working value of Avogadro's number.

Example 2.4: Using the working value of Avogadro's number, how many atoms are there in 3.275 moles of the element iron?

Answer:

**Organize:** 3.275 moles of Fe
Avogadro's number: $6.022 \times 10^{23}\ mol^{-1}$
In this problem, the "elementary objects" are the atoms of Fe

**Unknown:** Number of Fe atoms.

**Translate:** Write information in concise form, including all the units:
3.275 mol
$6.022 \times 10^{23}$ atoms Fe $mol^{-1}$

**Solve:** Number of Fe atoms =
(3.275 mol) × ($6.022 \times 10^{23}$ atoms Fe $mol^{-1}$)
Number of Fe atoms = $1.972 \times 10^{24}$ atoms Fe

The decision to define the mole using the same carbon-12 isotope that was used to define the unified atomic mass unit means that the mole and the unified atomic mass unit are directly and simply related to one another. Focusing for a moment on an atom of the carbon-12 isotope, its mass was defined to be exactly 12 u. On the other hand, one mole of the carbon-12 isotope was defined to be a mass of exactly 12 grams. But this means that these definitions yield the following relationship, using the symbol $\mathbf{N_A}$ for Avogadro's number:

$$(12\ u) \times (\mathbf{N_A}) = 12\ grams/mole. \quad (2.4)$$

This is exactly equivalent (dividing both sides by 12 and cancelling units, remembering that $N_A$ has units of $mol^{-1}$) to

$$(1\,u \times N_A) = \text{gram.} \tag{2.5}$$

But what does this mean? This equation states that Avogadro's number ($N_A$, where the actual numerical value will change, depending on how well we are able to count at a particular point in time) of unified mass units is equal to one gram. Suppose we select 39.10 grams of potassium (symbol K on the periodic table), then, using the unified atomic mass unit scale, we have a mass of 39.10 ($u \times N_A$). But because the **average** mass of the potassium atoms in our sample is 39.10 u (on the unified atomic mass scale), the 39.10-gram sample contains exactly $N_A$ atoms. Hence, the 39.10-gram sample is exactly one mole! This is the case for every element on the periodic table: Identify the average atomic mass of the element. Select a sample in which the number of grams in the sample is **numerically** equal to the element's average atomic mass. The sample constitutes one mole of the element. This mass is called the **molar mass** of the element.

Example 2.5: How many grams of Fe (iron) with an average atomic mass of 55.85 u must be present in a sample to have one mole of Fe?

Answer:

| | |
|---|---|
| **Organize:** | Element Fe<br>Average atomic mass: 55.85 u |
| **Unknown:** | How many grams are required to equal one mole? |
| **Translate:** | The number of grams must be **numerically equal** to the average atomic mass of the element. |
| **Solve:** | Sample size = 55.85 g Fe |

Example 2.6: Suppose that a sample contains 287.00 grams of chromium. How many moles of chromium are in the sample?

Answer:

| | |
|---|---|
| **Organize:** | Element Cr<br>287.00 g of Cr<br>Average atomic mass: 52.00 u |
| **Unknown:** | How many moles of Cr are in the sample? |
| **Translate:** | The molar mass of Cr is **numerically equal** to the average atomic mass of Cr. Hence, the molar mass of Cr is 52.00 g $mol^{-1}$.<br>Moles of Cr = Mass ÷ Molar mass |
| **Solve:** | Moles of Cr = (287.00 g) ÷ (52.00 g $mol^{-1}$)<br>Moles of Cr = 5.519 mol<br>(Note: Four significant figures) |

Example 2.7: Suppose that a sample contains 325.00 grams of iron. How many atoms (average number) are present in the sample?

Use $N_A = 6.022 \times 10^{23}\ mol^{-1}$

Answer:

| | |
|---|---|
| **Organize:** | Element Fe<br>325.00 g of Fe<br>Average atomic mass: 55.85 u |

| | |
|---|---|
| **Unknown:** | How many atoms of Fe are in the sample? |
| **Translate:** | The molar mass of Fe is **numerically equal** to the average atomic mass of Fe. Hence, the molar mass of Fe is 55.85 g $mol^{-1}$.<br>Moles of Fe = Mass ÷ Molar mass<br>Number Fe atoms = Moles of Fe × $N_A$ |
| **Solve:** | Number of Fe atoms = [(325.00 g) ÷ (55.85 g $mol^{-1}$)] × 6.022 × $10^{23}$ $mol^{-1}$<br>Number of Fe atoms = 3.504 × $10^{24}$ atoms<br>(Note: Four significant figures) |

## Modern Atomic Theory

The three-particle model of the atom was completed by the early 1930s with the experimental identification of the neutron. However, the early 1930s are now eighty or more years in the past, and the reader might rightly wonder if there have been further developments in our understanding of atomic structure. The answer to this question is a resounding "Yes!" As we shall shortly learn, the development of a model of the atom consistent with the accumulating experimental data was accompanied by additional developments that have dramatically altered humankind's understanding of the universe. Although the three-particle model of the atom will prove highly effective in our attempt to understand a vast array of chemical phenomena as we progress through this text, the reader should clearly understand that the contemporary model of the microscopic is very different from the picture created by the early atomic pioneers. While the sophisticated mathematical details of what is called the **standard model** lie well beyond the scope of this presentation, it is possible to approach a limited, qualitative understanding of this modern model of the microscopic.

The standard model, first proposed in 1964, was completed in the early 1970s as the premier theory of particle physics and is a comprehensive description of the submicroscopic domain, representing almost a half century of theoretical and experimental study. The standard model identifies two classes of particles, the **leptons** and the **hadrons.** The leptons are a group of six particles, arranged in three generations of two particles each: the first generation, made up of the **electron** and the **electron neutrino**; the second generation, made up of the **muon** and the **muon neutrino**, and the third generation, made up of the **tau** (sometimes called a **tauon**) and the **tau neutrino.** The six particles in the lepton class are often identified as six different lepton flavors. Three of the leptons (electron, muon, and tau) carry a unit negative charge ($-1e$, where the symbol $e$ is the charge of an electron), and the other three leptons (the neutrinos) are uncharged (i.e., neutral). Each of these six particles has an associated antiparticle (When particles and antiparticles encounter one another, the masses of the two particles are converted completely into energy, obeying Einstein's mass-energy relationship, $E = mc^2$) making a total of twelve particles in the lepton class.

The hadron class is also a group of six particles called **quarks** and, like the leptons, is arranged in three generations of two particles each: the first generation, made up of the **down quark** and the **up quark**; the second generation, made up of the **strange quark** and the **charm quark**; and the third generation, made up of the **bottom quark** and the **top quark**. Just like the leptons, the six particles in the hadron class are often identified as six different hadron flavors. Surprisingly, the six hadrons carry **fractional electric charges**: $+\frac{2}{3}e$ (up, charm, top) and $-\frac{1}{3}e$ (down, strange, bottom). Each of these six particles also has an associated antiparticle, making a total of twelve particles in the hadron class.

The reader has no doubt noticed that we have not mentioned the proton and the neutron, two of the three particles that played central roles in our earlier model of the atom. Only the electron remains. In fact, the standard model views protons and neutrons as **composite** particles; they are no longer considered to be *elementary*. In the least complex view, a proton is composed of three quarks, two up quarks and one down quark, and the proton carries a **positive** charge (notice that the charges still add to $+1e$, as in the earlier model). Similarly, the neutron is also understood to be a combination of three quarks, two down quarks and one up quark. As before, the neutron is **uncharged**. (The reader should again add the charges of the neutron's constituent quarks.) Note that the standard model only applies to normal matter, that is, less than 5% of our universe (see table 2.1).

In the framework of the standard model, there are **four forces** that determine all the transformations in the universe: the **electromagnetic force (electromagnetism)**, the **weak nuclear force**, the **strong nuclear force**, and **gravity**. Two of these forces, electromagnetism and gravity, are part of our daily experiences: gravity is responsible for the acceleration of mass that we call "weight"; light, television and radio transmissions, microwaves, and electricity are only a few examples of the electromagnetic force. The remaining two forces, as their names imply (strong and weak *nuclear* forces), are forces that **only** operate on the shortest distance scales, at the subatomic level of leptons and hadrons. They are extremely short-range forces, unlike gravity and electromagnetism, which are effective over an infinite range. As a result of their nonzero masses (even the neutrinos appear to have mass), all the leptons and all the hadrons experience the force of gravity, even though it is the weakest of the four forces. The **charged** leptons and hadrons respond to the electromagnetic force, while **all** the leptons and hadrons are affected by the weak nuclear force. Finally, only the hadrons interact via the strong nuclear force; consequently, only the hadrons experience all four fundamental forces.

In order to describe the **interactions** of the leptons and hadrons, the standard model represents the fundamental forces in terms of an exchange of particles. (The standard model is one example of a **quantum field theory**, the successor to the quantum theory used by Bohr in 1913 to construct his model of the atom. All modern quantum field theories employ this same approach.) As a result, the standard model includes one further class of particles, the **bosons**. The members of this class of particles are the **photon** (electromagnetic force); the three **vector bosons**, ($W^+$, $W^-$, $Z^0$; weak nuclear force); and eight **gluons** (strong nuclear force). Because the hadrons are the only class of particles that undergo transformations caused by the strong nuclear force, they carry a new type of charge (similar to the electromagnetic charge with its values of *positive* and *negative*) that is called the **color charge**. Unlike electromagnetism's two values, the color charge has **three** distinct values: red, blue, and green. Because the proton and the neutron are now understood to be composite particles made up of hadrons interacting through the strong nuclear force, the gluons literally hold the hadrons together to form protons and neutrons. (Think here of glue holding two objects together.) There are now a total of forty-eight "elementary" particles that are the building blocks of matter as we experience it, twelve leptons, twelve hadrons, and twenty-four colored gluons. (From the point of view of the strong nuclear force, there is a total of thirty-six colored hadrons.) We should not think of the values of the color charge as we think of the macroscopic colors "red," "blue," and "green" that are used to describe a blue sky or green grass. The color charge is a characteristic of the hadrons required by quantum field theory, just like the negative electromagnetic charge is a characteristic of the electron. The three names, red, blue, and green, reflect a whimsical choice made by physicists to simply identify the three distinct varieties of the color charge. These whimsical names gave rise to the name for the portion of the standard model that describes the strong nuclear force: **quantum chromodynamics** (the Greek word *chromos* means "colored").

The reader will note that we have said nothing about gravity, the fourth fundamental force. While gravity has been the force most commonly experienced by humankind, it is also the most problematic from the viewpoint of contemporary science. Since the advent of quantum theory at the beginning of the twentieth century, the most powerful and successful descriptions of the microscopic behavior of the universe have been in terms of the quantum theory and its successors (quantum field theories). In contrast, the most successful gravitational theory (Einstein's general theory of relativity) is **not** a quantum-based theory. It is called a **classical** theory and is rooted in a pre-twentieth century scientific paradigm. Surprisingly, Einstein's theory has been consistent with every experimental test that has been performed since 1919. In fact, today (2014), it is, without question, the best model of gravity the human mind has produced; there is no completely successful quantum theory of gravity. **If** a quantum field theory of gravity **were to exist**, it would require the existence of another boson, called the **graviton**, to explain gravitational interactions in terms of an exchange of particles. No such particle has been observed, and no successful quantum theory of gravity has been constructed; consequently, the study of gravity remains a very active area of current (2014) research.

Finally, there is one last critical concept about which the standard model, with its complement of leptons, hadrons, and bosons, was initially silent: Why do all (with the exception of the photon and the eight gluons) the particles of the model possess nonzero mass? **Experimentally**, the leptons, hadrons, and all the bosons (again, the photon and the gluons are the exceptions) possess a mass that is measurable, yet the original **theoretical** prediction of the standard model indicated that all the particles are massless. To remedy this stark disagreement between experimental measurements and theoretical predictions, several groups of physicists independently proposed (initially in 1962 and with continuing development through the early 1970s) that all the massive particles of the standard model acquire their mass by interacting with a quantum field through the exchange of a new particle, which, for historical reasons, is called the **Higgs boson**. This proposal launched a nearly forty-year search, which culminated with an announcement in July 2012 at the Large Hadron Collider (LHC) that the Higgs particle had been experimentally identified.

While all the theoretical predictions of the standard model have now been verified experimentally, the quest to understand the elementary structure of the universe is far from complete. The standard model, our current best model of the microscopic organization of the universe, still requires that the experimentally measured values of approximately nineteen parameters (more may be needed to explain the nonzero masses of the three neutrinos) be added to the model. The model provides an understanding of neither the relative magnitudes of the parameters nor the relationships among these parameters. There are a number of additional technical questions that the standard model does

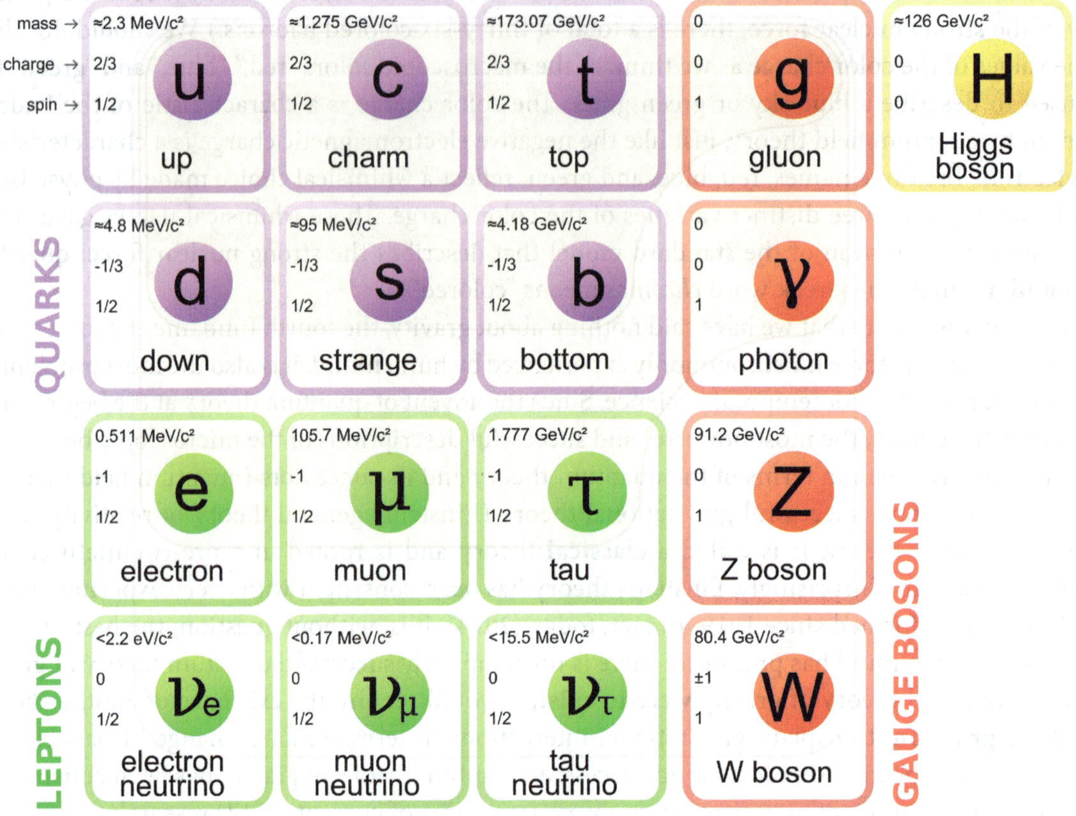


Figure 2.4: Diagrammatic Summary of the Standard Model

not address. Finally, the questions surrounding the inclusion of gravity in the model (a so-called quantum theory of gravity) remain unresolved. While the theoretical framework of the standard model is the product of more than a century of concerted effort, and the model's predictions exhibit spectacular agreement with a vast collection of experimental measurements, it is very far away from being a "theory of everything."

## Radioactivity and Radioisotopes

As we have already observed, by the early twentieth century the phenomenon of **radiation** (recall that *radiation* is the transfer of energy either as mass or as electromagnetic disturbances through space-time) had become a central focus of scientific research. Cathode rays (and their positive counterparts, *canal rays*) had been intensely studied by a number of investigators throughout the last third of the nineteenth century (leading to Thomson's crucial recognition of the connection between electricity and matter), and the enigmatic X-rays were first observed by Röntgen as the century ended. However, all of these types of radiation were intentionally generated by the experimenters. In contrast, **radioactivity**, a process by which atoms **spontaneously** emit radiation (first observed by Becquerel in 1896 and named by the Curies two years later) is **not** initiated by an investigator. This process of spontaneously emitting radiation is called **disintegration** because the radiation is emitted from the atomic nucleus and the nucleus, in turn, is changed. That is, the **unstable** nucleus of the original atom is said to "disintegrate." It is a process that occurs without the human intervention. By 1900, Rutherford and Villard had shown that radioactive materials emitted three distinct types of radiation:

1. **α radiation** carries a +2*e* charge and was later shown to be made up of helium atoms that had lost two negative charges (associated with the electrons of the three-particle model of the atom). In effect, α radiation possesses nonzero mass and is composed of helium nuclei.
2. **β radiation** carries a −1*e* charge and was later shown to be identical to the corpuscles observed by Thomson in his cathode ray experiments (ultimately associated with the electron of the three-particle model of the atom). β radiation also possesses nonzero mass.
3. **γ radiation** carries no charge, is unaffected by electric and magnetic fields, and is a purely electromagnetic phenomenon (like light but possessing much higher energies, similar to the energies associated with X-rays) with which there is **no** associated mass.

As radioactivity continued to be investigated and the understanding of the three-particle model of the atom matured during the first third of the twentieth century, two additional types of radiation emanating from radioactive materials were identified:

4. **Positrons** carry a +1*e* charge and have the **same** mass as the particles of β **radiation.**
5. **Neutrons** are nearly equal in mass to protons, but carry **no** electrical charge.

Consequently, by the early 1930s radioactive materials were shown to emit five distinct types of radiation. Table 2.3 contains a summary of their properties. As table 2.3 indicates, the various types of radiation associated with radioactive materials have differing abilities to penetrate matter. This ability to penetrate matter is determined by a complex interplay of the radiation's energy, characteristics, charge, and mass. Figure 2.5 displays qualitatively the effectiveness of various types of radiation to penetrate matter:

Table 2.3: Types of Radioactivity

| Radiation | Symbol | Characteristics | Charge | Velocity | Penetrating Power |
|---|---|---|---|---|---|
| α radiation | α | Helium nuclei | +2*e* | 5–10% that of light | Low |
| β radiation | β | Electrons | −1*e* | Up to 90% that of light | Moderate |
| γ radiation | γ | Electromagnetic | 0 | Light Speed | High |
| Positron | $\beta^+$ | Positive electron | +1*e* | Up to 90% that of light | Moderate |
| Neutron | n | Free neutron | 0 | Up to 20% that of light | High |

Note: The speed of light = $c = 2.99792458 \times 10^8$ m $s^{-1}$

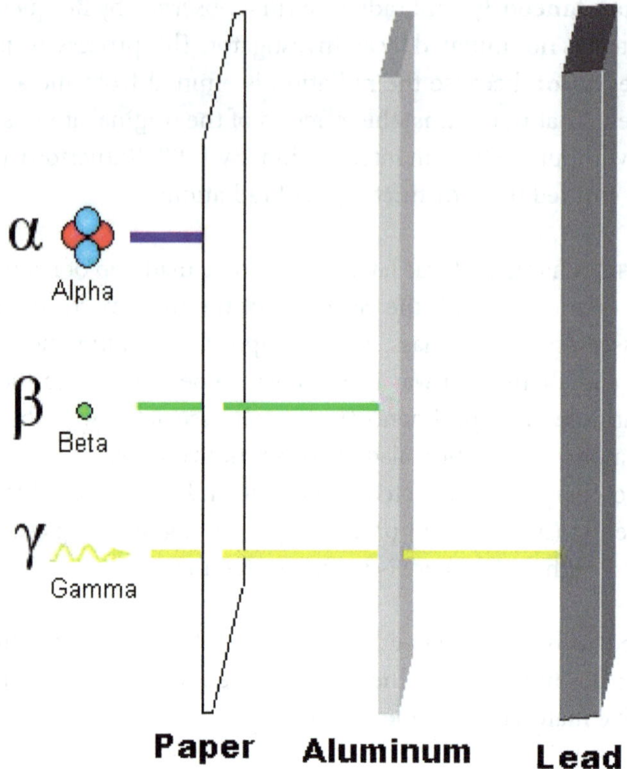


Figure 2.5: Penetrating Ability of Radiation

A variety of units to measure radiation and the effects of radiation on human beings has been developed over time. The most direct measure of radioactivity is to specify the number of disintegrations per unit time. Two units have been defined to identify this quantity; only one of them is an SI unit. The **curie** (Ci) is a non-SI unit originally defined in 1910 using a specific mixture of the radioactive elements radon (a radioactive gas) and radium in which thirty-seven billion atoms disintegrate per second. However, in 1964, the twelfth CGPM agreed to accept the curie as a measure of **activity** and defined the curie as thirty-seven billion disintegrations per second. The SI unit of radioactivity is called the **becquerel** (Bq) and is defined as one disintegration per second. The fifteenth CGPM in 1975 formally adopted the **becquerel** for the SI unit of activity. Consequently, 1 Ci = $3.7 \times 10^{10}$ Bq.

However, to quantify the effects of radiation on both inanimate materials and living organisms, two additional categories of units are necessary; one is used to measure the energy transferred per unit mass, the second to measure the average radiation absorbed by biological tissue, taking into account the fact that the different types of radiation cause differing amounts of damage to biological systems. As before, in each category there is a SI unit and a non-SI unit. The first of these categories is called the **absorbed dose** (also known as the **total ionizing dose**). The SI unit of the absorbed dose is the **gray** (Gy), which is defined as 1 Gy = 1 J $kg^{-1}$. The older, non-SI unit is the **rad**, which is defined as 1 rad = $10^{-2}$ J $kg^{-1}$. Hence, 1 Gy = 100 rad. Note that this category of units simply indicates how much energy (here, measured in joules) is deposited in every kilogram. The second of the two categories is called the **equivalent absorbed radiation dose** (often shortened to either **equivalent dose** or **dose equivalent**). The SI unit of equivalent absorbed radiation dose is the **sievert** (Sv). For alpha (α) radiation, 1 Sv = $5.0 \times 10^{-2}$ Gy, while for beta (β) and gamma (γ) radiation, 1 Sv = 1 Gy. The older, non-SI unit of equivalent absorbed radiation dose is called the **rem.** Measuring the effects of radiation with the non-SI units, we find that for alpha (α) radiation, 1 rem = $5.0 \times 10^{-2}$ rad, while for beta (β) and gamma (γ) radiation, 1 rem = 1 rad. Consequently, 1 Sv = 100 rem.

## Half-Life

Radioactive substances exhibit a disintegration rate (the disintegration of a nucleus per unit time, sometimes called the **decay rate**) that is virtually constant; to a very high degree of precision, measurements (with only a few apparent exceptions) indicate that conditions outside the atomic nucleus have no effect on the rate at which an unstable nucleus spontaneously emits radiation. Not only does the rate of decay appear to be insensitive to external conditions; the decay rate also exhibits a simple exponential behavior, plotted in figure 2.6.

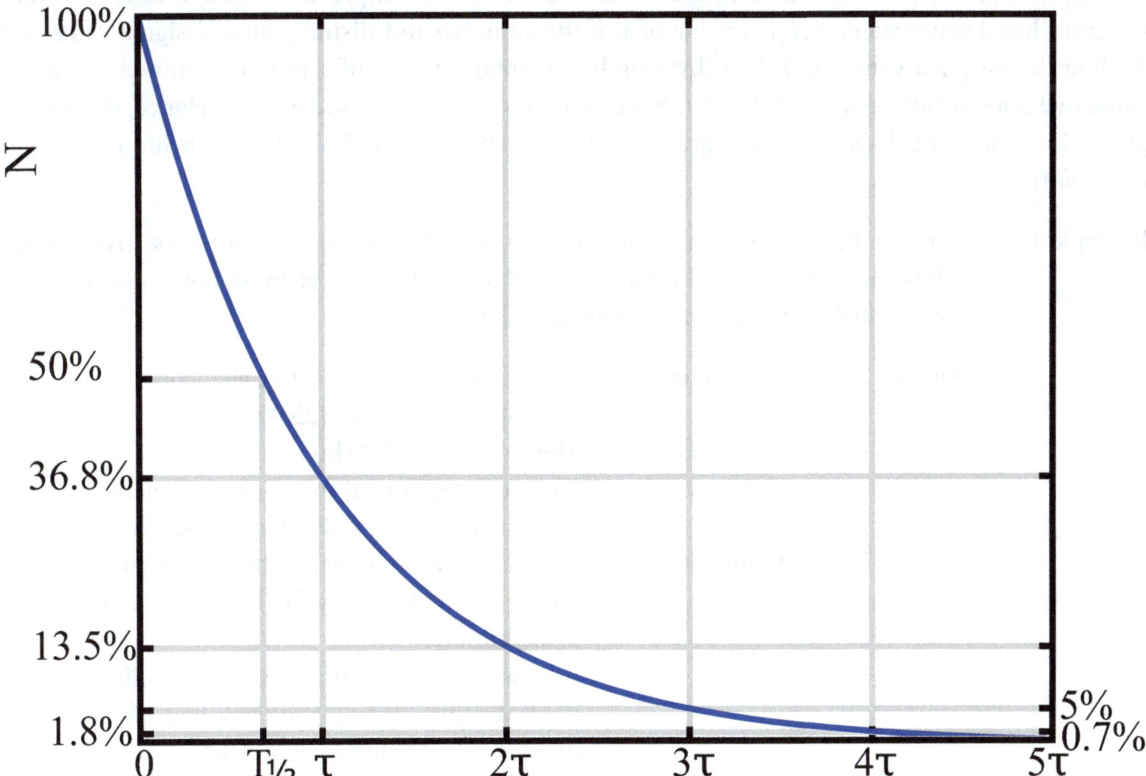


Figure 2.6: Graph of Exponential Decay

The decay rate of a radioactive nucleus is measured by its **half-life**. The half-life of a radioactive substance is most simply defined as the amount of time required for one-half of the nuclei in a specified sample of the material to disintegrate. (In figure 2.6, the half-life is represented by the symbol $T_{½}$.) However, this conceptually simple definition becomes meaningless if the sample has only one atom. Clearly, at the end of one half-life, there is either a single atom that **has disintegrated** or a single atom that **has not disintegrated**; it is impossible to have a single atom that has half-disintegrated! Consequently, a better definition of a radioactive substance's half-life can be given in terms of *probability*. We define a radioactive substance's half-life as the amount of time needed to guarantee that the average probability that a nucleus has disintegrated is 0.5. Why is this definition consistent with the simplest definition given above? Consider the following process: Suppose that you have a very large number of samples, each containing 1000 atoms. For each sample, you wait one half-life and then count the number of atoms that have disintegrated. The **exact** number in each sample will be **different**, but the **average** number will approach 500 if you examine a sufficiently large of number of samples. Consequently, the probability that a nucleus has disintegrated is 500 ÷ 1,000 = 0.5. Notice that this is exactly the result you would achieve using our simplest definition: on average, one-half of the nuclei of each sample has disintegrated. On the other hand, suppose you have a large number of samples, each containing only a single atom. Examine each sample at the end of one half-life; if the atom has disintegrated, assign a value of 1; if the atom has **not** disintegrated, assign a value of 0. Adding the assigned values and then dividing by the total number of samples examined will give a value that approaches 0.5 as the total number of samples becomes infinitely large. Hence, the average probability that a nucleus has disintegrated is 0.5, matching our definition of half-life in terms of *probability*.

Example 2.8: Suppose that a pure sample initially contains 325.00 grams of a radioactive element whose half-life is 1.500 hours. After 270.0 minutes have elapsed, how many grams of the original element remain in the sample?

Answer:

**Organize:** Pure sample of a radioactive element
Total starting mass is 325.00 grams
The half-life is 1.500 hours

**Unknown:** How many **grams** of the original element remain at the end of 270.0 minutes?

**Translate:** To provide an answer to the unknown, you must determine how many half-lives have passed in 270.0 minutes.

Because one-half of the mass of the radioactive element disintegrates during **each** half-life, you must divide by 2 **once for each half-life**.

**Solve:** Convert half-life time from hours to minutes:
Half-life = (1.500 hours) × (60 minutes/hour)
= 90.00 minutes
(Note: No loss of significant figures)

$$\text{Number of half-lives} = (270.0 \text{ minutes}) \div (90.00 \text{ minutes}) = 3.000$$

$$\text{Amount remaining} = (325.00 \text{ g}) \div (2 \times 2 \times 2) = 40.625 \text{ g}$$

(Note: Five significant figures)

It has been estimated that there are more than three thousand distinct nuclei that are radioactive, and they exhibit a tremendous range of half-lives, ranging from approximately $10^{-23}$ s to more than $10^{24}$ y. Table 2.4 lists a subset of the radioactive nuclei and their associated half-lives. There is a direct correlation between the magnitude of the half-life of a radioactive nucleus and its stability: longer half-lives correspond to more stable nuclei.

## Mass-Energy Calculations and Nuclear Energy

With the discovery of radioactivity by Becquerel, the identification the various types of radiation associated with radioactivity, and the achievement of transmuting one element into another by Rutherford (correctly interpreted by Blackett, see equations 2.1 and 2.2), it had become increasingly clear that the atomic nucleus is a source of energy (radioactivity) and that the nucleus can be altered (achieving the alchemist's dream) by adding sufficient energy to the nucleus (transmutation). Further, we have already noted on several occasions that in 1905 Albert Einstein articulated a fundamental relationship between matter and energy: $E = mc^2$. This relationship implies that a very small amount of mass is equivalent to a gigantic quantity of energy. The $c$ in this equation is the speed of light, where $c = 2.99792458 \times 10^8$ $ms^{-1}$, making the quantity $c^2$ an enormous number.

Table 2.4: Radioactive Nuclei and Their Associated Half-Lives

| Nucleus | Half-Life |
|---|---|
| Lithium-5 | $3.24 \times 10^{-22}$ s |
| Beryllium-6 | $5 \times 10^{-21}$ s |
| Copernicum-277 | $2.40 \times 10^{-4}$ s |
| Bohrium-266 | $1.02 \times 10^{-1}$ s |
| Fluorine-18 | $6.586 \times 10^{3}$ s (1.8295 h) |
| Chromium-51 | $2.39350 \times 10^{6}$ s (27.7025 d) |
| Cobalt-60 | $1.6635 \times 10^{8}$ s (5.2714 y) |
| Carbon-14 | $1.804 \times 10^{11}$ s (5 715 y) |
| Plutonium-239 | $7.61 \times 10^{11}$ s (24 110 y) |
| Uranium-235 | $2.221 \times 10^{15}$ s ($7.038 \times 10^{8}$ y) |
| Uranium-238 | $1.41 \times 10^{17}$ s ($4.468 \times 10^{9}$ y) |
| Bismuth-209 | $6.00 \pm .63 \times 10^{26}$ s ($1.9 \pm .2 \times 10^{19}$ y) |

Example 2.9: Use Einstein's fundamental relation between mass and energy to calculate the amount of energy (in units of joules) produced if 3.750 g of mass are converted **completely** to energy.

Answer:

**Organize:** Mass = 3.750 g
Mass completely converted to energy
Speed of light = $2.99792458 \times 10^8$ m $s^{-1}$

**Unknown:** How much energy, in joules, is produced?

**Translate:** Key equation: $E = mc^2$
Convert mass to kilograms (review the definition of the term "joule").

**Solve:** $E = (3.750 \times 10^{-3}\ kg) \times (2.99792458 \times 10^8\ m\ s^{-1})^2$
$E = 3.370 \times 10^{14}$ J
(Note: Four significant figures)

By the early 1930s it had become clear that, in addition to altering the nucleus by bombarding it with radiation (this is the transmutation process achieved by Rutherford, in which a much more massive nucleus is bombarded by a relatively less massive particle, such as an alpha or beta particle), heavier nuclei could be created by combining several low-mass nuclei (usually nearly equivalent nuclei) into a single, more massive and more stable nucleus while releasing energy along with **possibly** one or more neutrons. This new process is called **nuclear fusion.** While fusion appears superficially to be simply another type of transmutation, it differs significantly in that nearly equivalent nuclei are combined in the fusion process, and the amount of energy released is substantially greater. It was proposed at this time (by Hans Bethe) that the primary mechanism powering the stars is a fusion process that combines hydrogen nuclei to form a helium nucleus. During this same period, experimental investigations in both Italy and Germany (although Enrico Fermi in Italy did not recognize the significance of his work) were exploring a completely different phenomenon. In 1938 it was realized (by the Austrian physicist Lise Meitner and her nephew, Otto Frisch) that the uranium nucleus had been fragmented (colloquially "split") into two less massive nuclei accompanied by the release of energy and several neutrons. Because the nucleus was fragmented, this process is called **nuclear fission.**

In the fusion process a new, more massive nucleus is produced by combining less massive, lighter nuclei. However, if the rest masses of each individual nucleon (see table 2.2 for the rest masses of the proton and neutron) in the new nucleus are added together, the resulting sum **differs** from the measured mass of the new nucleus. In fact, the mass of the new nucleus is **less than** the sum of the masses of the individual nucleons. This difference is called the **mass deficit**, or, as a result of Einstein's equation $E = mc^2$ equating mass and energy, the corresponding energy is called the **nuclear binding energy**. Because the resulting new nucleus is more stable than the individual starting nuclei, the nuclear binding energy is the energy **released** as a result of the fusion process. In effect, fusion creates new nuclei that are more stable and converts a small amount of mass **completely** into energy that is then released to the surrounding environment. From the reverse point of view, the nuclear binding energy is simply a measure of the amount of energy required to completely fragment a nucleus into its constituent protons and neutrons, converting the energy into the mass required by the free nucleons.

In the case of nuclear fission, a more massive, unstable (i.e., radioactive) nucleus is fragmented, or split, into two less massive nuclei. The less massive nuclei are **more** stable than the original massive nucleus, and, again, there is a mass deficit that is the difference between the mass of the original nucleus and the sum of the masses of the two lighter nuclei along with the masses of the neutrons released during the fission process. Applying the Einstein equation, $E = mc^2$, the corresponding energy is again the nuclear binding energy. Consequently, just as in the case of fusion, the fission process converts a small amount of mass **completely** into energy that is released to the surrounding environment.

Example 2.10: Suppose two deuterons (deuterium is a form of hydrogen with one proton and one neutron in its nucleus; the nucleus of a deuterium atom is called a **deuteron)** fuse to produce a helium atom (2 protons and 2 neutrons in the nucleus). Calculate the **mass deficit** (in kilograms) and the **nuclear binding energy** (in joules). The mass of a deuterium is 2.014102 u. The mass of helium is 4.002602 u.

Answer:

**Organize:**
Deuterium mass = 2.014102 u
Helium mass = 4.002602 u
Require the mass of one unified atomic unit in kilograms
Require the Einstein equation: $E = mc^2$
Speed of light = $2.99792458 \times 10^8$ m s$^{-1}$;
The number of electrons does not change; this may be ignored.

**Unknown:**
Mass deficit (in kilograms)
Nuclear binding energy (in joules)

**Translate:**
Calculate mass deficit in unified atomic mass units
Convert mass deficit to kilograms
Compute the nuclear binding energy

**Solve:**
Mass deficit = (4.002602 u) – (2 × 2.014102 u)
Mass deficit = – 0.025602 u
(The negative sign simply indicates that the mass of helium is **smaller** than the sum of the masses of the 2 deuterons. Report the positive value.)
Mass deficit = 0.025602 u
Now convert to kilograms:
$1\text{u} = (1/12) \times (0.012 \text{ kg}/6.022 \times 10^{23})$
Mass deficit = $(0.025602 \text{ u}) \times (1/12) \times (0.012 \text{ kg}/6.022 \times 10^{23})$
Mass deficit = $4.2514115 \times 10^{-29}$ kg
(No rounding!)
$E = m \times c^2$
$E = (4.2514115 \times 10^{-29} \text{ kg}) \times (2.99792458 \times 10^8 \text{ m s}^{-1})^2$
Nuclear binding energy = $3.821 \times 10^{-12}$ J
(Note: Four significant figures)

The nuclear fission process is most efficiently initiated by bombarding a massive, radioactive nucleus with neutrons. Because neutrons are **uncharged**, they are not deflected by the negatively charged electrons or the positively charged protons, making the neutron an efficient initiator of the fission process. However, because the fission process often produces two or three additional free neutrons, these free neutrons can initiate another fission cycle. Such a repetitive process is called a **chain reaction**. In order for a chain reaction to continue, a minimum amount of the radioactive material is required so that the neutrons produced by each fission cycle will encounter another radioactive nucleus **before** escaping from the radioactive material. If the amount of radioactive material is too small, the probability that a neutron will fragment another radioactive nucleus is too small to sustain the repetitive fission cycle. Consequently, the cycle of fission processes will cease. This minimum amount of radioactive material varies, depending on the individual radioisotope, and is called the **critical mass**. Table 2.5 lists the approximate critical mass amounts (assuming a spherical geometry) for some representative radioactive materials.

Both nuclear fission and nuclear fusion are at the center of several technologies that have dramatically affected the course of human history over a period of nearly seventy-five years. Fission has been applied very successfully to generate between 10% and 15% of the world's electrical power; more than four hundred nuclear reactors were operational worldwide in 2013. Several regions rely on nuclear power reactors to meet well over 50% of their demand for electricity. While nuclear reactors do not contribute to climate change caused by atmospheric emissions (greenhouse gases), there are significant challenges created by the continued use of this technology. The nuclear disasters at Chernobyl, USSR, in 1986 and at Fukushima Daiichi, Japan, in 2011, along with a less serious partial melting of a reactor core at Three Mile Island in the United States in 1979, were all significant failures of safety protocols. Even though the Three Mile Island event released radioactive contaminants into the environment, the

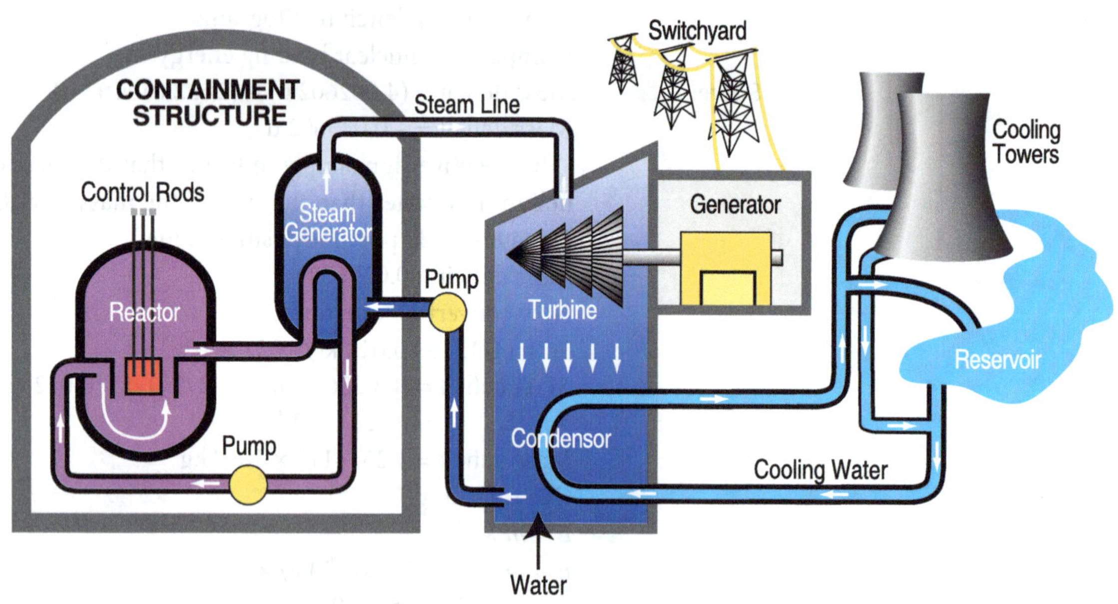


Figure 2.7: Schematic of a Fission Power Reactor

Table 2.5: Critical Masses (Spherical Shapes)

| Nucleus | Critical Mass (kg) |
|---|---|
| Uranium-233 | 15.8 |
| Uranium-235 | 46.7 |
| Neptunium-237 | 63.6 |
| Plutonium-238 | 9.5 |
| Plutonium-239 | 10.0 |
| Americium-241 | 57.6 |
| Curium-245 | 9.1 |
| Californium-251 | 5.5 |

disasters at Chernobyl and Fukushima Daiichi were far worse in terms of both the numbers of people affected and the total land area contaminated.

But beyond these notable reactor failures, the continued used of fission reactors to generate electrical power immediately raises several questions, whose answers have proven to be discouragingly elusive: What should be done with the highly radioactive waste generated by the fission process? If the waste is sequestered, can we guarantee its security and isolation for the thousands of years required for radioisotopes with very long half-lives to decay to acceptable radiation levels? In addition to the radioisotopes generated by the fission process in the nuclear fuel (uranium, plutonium, or thorium), the reactor itself, the coolant pumps, and the control mechanisms will also be highly radioactive, further complicating the decommissioning of the plant. Finally, if the radioactive waste is reprocessed to extract residual materials that can be used to fuel other reactors, how do we ensure the security of this reprocessed nuclear material?

This last question surrounding the use of nuclear fission to generate electricity directly touches the second significant technology that employs the fission process: nuclear weapons. The use of a fission chain reaction to build devastating atomic weapons began during World War II and culminated in the use of two of those weapons at the end of that war on the cities of Hiroshima and Nagasaki. Beginning with nuclear material that is suitable for a power reactor, the material can highly refined (the technical term is *highly enriched*) to become the key to awesome atomic weapons. Consequently, the question about the security of reprocessed nuclear material is highly relevant. But even more insidious would be the use of the rich cocktail of radioisotopes found in the nuclear waste shipped to a reprocessing location to prepare a so-called "dirty bomb" whose goal is not to destroy by explosive force, but simply to contaminate a large region with radioactive materials.

What about nuclear fusion? As noted earlier, the first experimental studies of the fusion process occurred in the 1930s. To date (2014), the one successful technological application of the fusion process has been the construction of thermonuclear weapons. The fusion reaction involving the element hydrogen generates thousands of times more energy per reaction than the energy produced per fission reaction. Consequently, thermonuclear weapons (colloquially called "hydrogen bombs" or "H-bombs") are much more destructive than the original atomic weapons. In fact, these weapons are called *thermonuclear* because they use an atomic explosion (with its accompanying extreme temperatures) as the trigger mechanism for the fusion reaction among the hydrogen atoms. These weapons literally use the power of stars.

While the somber pall of humankind's achievement as a weapons builder has cast an ominous shadow over more than half a century of human events, a parallel drama has been unfolding in the effort to harness the power of the stars in the service of humanity's needs. With the beginning of concerted fusion research after the end of World War II and the realization of fusion's tremendous energy potential, there has been a persistent dedication to the dream of controlled fusion. A variety of machines and approaches have been investigated, and the most promising devices are the **tokamaks**, magnetic plasma confinement machines (a *plasma* is an exotic state of matter, distinctly different from solids, liquids, and gases) that are descendants of a design invented by Soviet physicists in the 1950s. Today (2014), the most ambitious project is the International Thermonuclear Experimental Reactor (ITER). Construction began in 2007 at Cadarache in southern France, and it is expected to produce its first plasma in 2020. The device is designed to produce 500 megawatts of output power using just 50 megawatts of input power, a ten-to-one ratio. This would successfully demonstrate the ability to produce more power from the fusion process than was used to initiate the process; this goal has never been achieved by any previous fusion project. Figure 2.8 shows the diagram of an experimental tokamak, the Alcator C-Mod; the torroidal plasma is shown in pink, the vacuum chamber is shown in yellow, and the magnetic coils are represented by the green and purple regions.

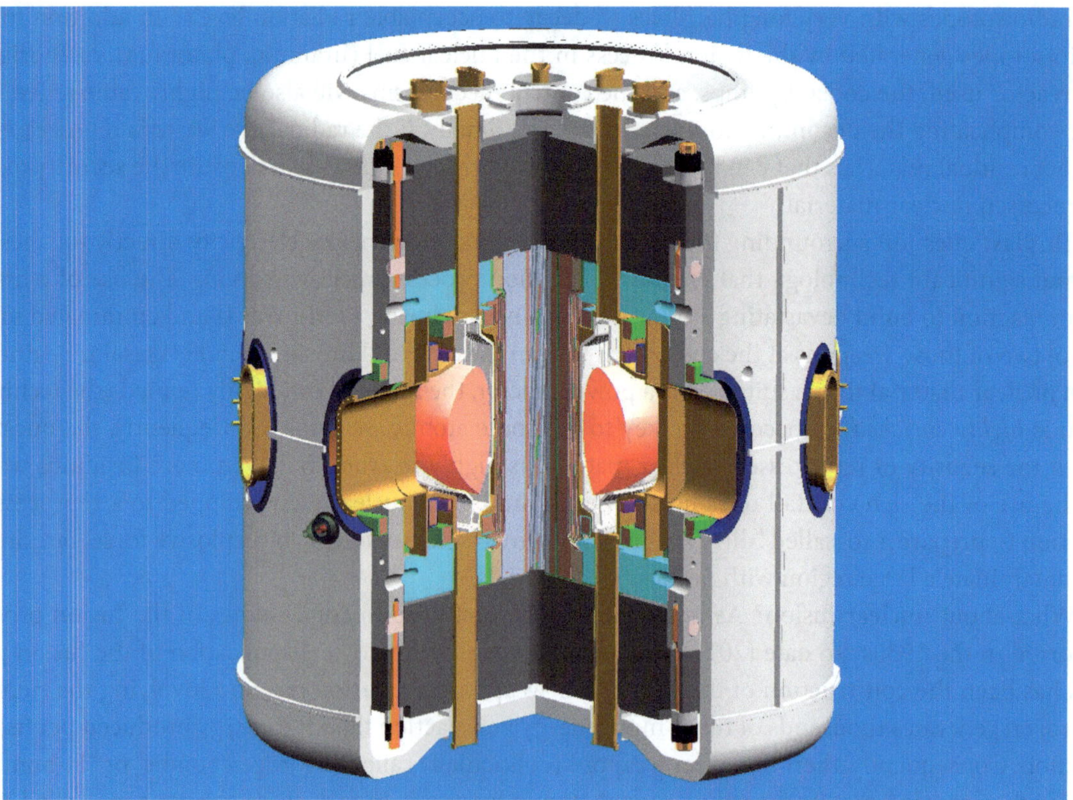

Figure 2.8: Diagram of the Alcator C-Mod Tokamak

## Transmutation and the Particle Accelerator

We have already seen that the first successful transmutation of an element was achieved by Rutherford in the period between 1917 and 1919 by using naturally occurring α radiation to bombard the nitrogen nucleus. (See equations 2.1 and 2.2; the correct interpretation had to wait for Blackett's work.) In 1935 Frederic and Irene Joliet-Curie used naturally occurring α radiation to bombard an aluminum target, producing for the first time a radioactive isotope, phosphorus-30. However, at this same time, physicists began to develop the first generation of machines designed to accelerate subatomic particles artificially. Beginning with the inventions of Robert Van de Graaff in 1929 (the Van de Graaff electrostatic generator) and of John Cockcroft and Ernest Walton in 1932 (the CW generator, which is sometimes called a voltage multiplier), experimenters were no longer limited to the energy levels offered by naturally occurring radioactive materials. Rather, subatomic particles could be artificially accelerated, thereby gaining kinetic energies (recall that kinetic energy depends on the mass **and** the velocity, $E_k = ½\ mv^2$; see equation 1.16) that greatly exceeded the energies possessed by the α and β particles produced by naturally occurring radioactive materials. In 1932 Cockcroft and Walton irradiated a lithium-7 (an isotope of lithium) target with protons, yielding two alpha particles. This was the first example of a nuclear transmutations produced by an artificially accelerated subatomic particle.

The late 1920s and early 1930s saw the emergence of both linear accelerators and a circular design proposed by E. O. Lawrence in 1930. Lawrence's cyclotron accelerated charged particles in a spiral path using two D-shaped vacuum chambers carrying alternating positive and negative charges and positioned between the poles of a large magnet. Each time the charged particle approached the gap between the two D-shaped regions, the charge (either positive or negative) on the D-shaped region would change to be **identical** to that of the particle itself, thereby **repelling** the particle. At the same instant, the charge on the other D would change to **attract** the charged particle. This rapid alternating

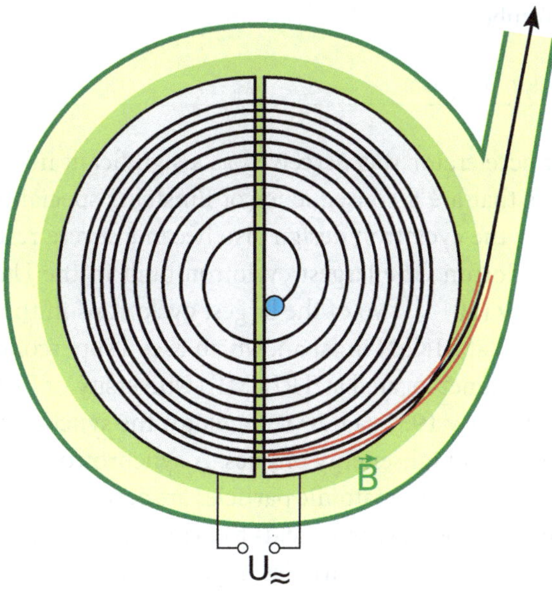


Figure 2.9: Cyclotron Diagram

"pushing" and "pulling" of the charged particle **increases** its velocity and causes the charged particle to follow a spiral path within the constant field of the magnet. Finally, the high-velocity charged particle collides with a stationary target. Figure 2.9 is a schematic diagram of a cyclotron; the magnetic field is symbolized by B and the electrical potential by U.

The same type of alternating "pushing" and "pulling" of the charged particles is the underlying principle of linear particle accelerators, which were invented at nearly the same time as the cyclotron. However, the need for extremely long linear accelerators was viewed as a major disadvantage, making the cyclotron the dominant technology for accelerating subatomic particles for nearly twenty years.

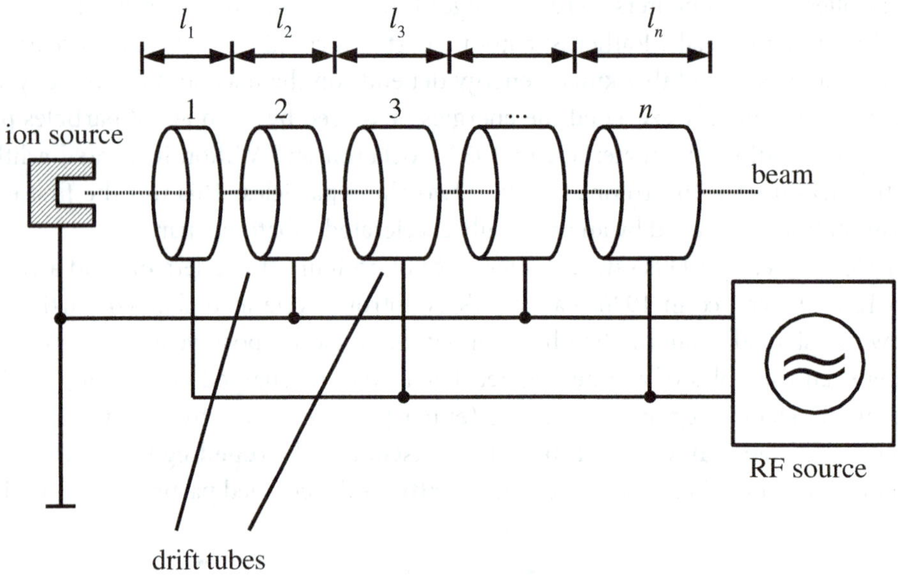


Figure 2.10: Linear Accelerator Diagram

Just as the length of a linear accelerator was perceived as a significant impediment to the widespread use of this technology, effects that are a consequence of Einstein's special theory of relativity and the size of the magnet required by the cyclotron design also became severe restrictions on the maximum energies achievable with a cyclotron. The largest cyclotron built in the United States used a magnet with a diameter of 4.7 m. Today (2014), two of the largest cyclotrons in the world are a machine with a 19-m diameter magnet located at RIKEN in Japan, which uses a superconducting magnet, and a machine with an 18-m diameter magnet located at TRIUMF (University of British Columbia) in Canada. However, by the late 1940s and early 1950s a new technology, the synchrotron, replaced the cyclotron as the principal tool for research in high-energy physics. A synchrotron is an evacuated hollow donut (technically called a *torus*) in which the subatomic particles are accelerated along a path with a constant radius (the donut) using time-varying magnetic fields to maintain the particles' path and time-varying electromagnetic fields to add energy to the particles. Figure 2.11 provides a schematic diagram for a synchrotron with the torus labeled (1) and the circulating particles labeled (2). The vertical path leads to a stationary target.

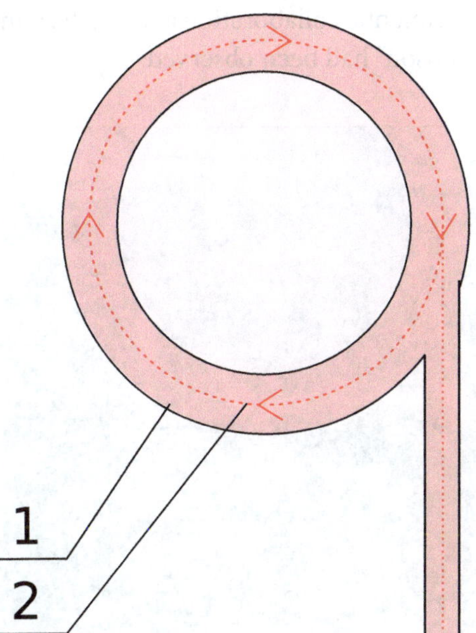


Figure 2.11: Synchrotron Diagram

While the synchrotron design successfully circumvented the physical restrictions of the early cyclotrons and enabled experimental physicists to accelerate subatomic particles to much higher energies (the unit of energy is the *electronvolt* (eV), where 1 eV approximately equals $1.602176565 \times 10^{-19}$ J; the unit GeV refers to billions of electronvolts, and the unit TeV refers to trillions of electronvolts), the first synchrotrons directed their beams of accelerated particles at stationary targets. Consequently, they are referred to as "fixed-target" machines. However, there is another route to achieving much higher collision energies: allow two beams of accelerated particles to collide head-on. This is the basic idea underlying the **collider**, the most powerful contemporary particle accelerator.

A collider produces two beams of accelerated particles, one rotating in a clockwise direction and one rotating in a counterclockwise direction. At specific regions, the two paths cross, allowing the accelerated particles to collide head-on. Alternatively, the beams of accelerated particles can be stored in a magnetic storage ring (a large torus like that of a synchrotron), one traveling in a clockwise direction, the other in a counterclockwise direction. Again, the two stored beams intersect at specified regions to create head-on collisions of the particles. The Tevatron at Femilab (outside of Chicago) is a collider whose circumference is 6.86 km (4.26 mi) and was capable of producing two counter-rotating beams of particles (one consisting of protons, the other of antiprotons), each with an energy of approximately 1 TeV. The Tevatron ceased operation in 2011. Today (2014), the Tevatron's scientific successor is the Large Hadron Collider (LHC) located on the border of France and Switzerland and operated by the European Council for Nuclear Research (CERN, derived from the French phrase *Conseil Européen pour la Recherche Nucléaire*). The LHC is a collider with a circumference of 27 km (17 mi) and was designed to produce two counter-rotating beams of protons, each accelerated to 7 TeV. Consequently, the total energy available in the head-on collisions is 14 TeV. The LHC began operating in 2008, and in

July 2012, as we noted earlier, the experimental collaboration at the LHC announced that a new particle "consistent with the theorized Higgs boson" had been observed.

Figure 2.12: Aerial View of the Tevatron at Fermilab

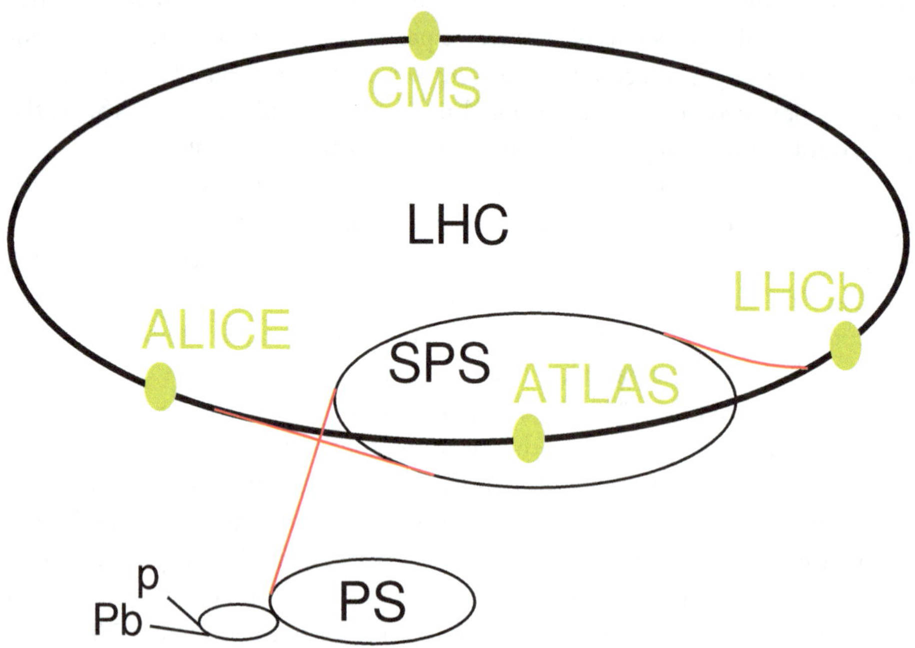


Figure 2.13: LHC Diagram

## Quantum Theory

Our exploration of the developments of the last two hundred years, from the work of Dalton in the early nineteenth century to the remarkable discovery of the Higgs particle at the LHC in the early twenty-first century by a collaboration of more than six thousand physicists, has revealed that the universe that we share is a remarkably diverse and surprisingly multifaceted place. During these two centuries an amazing number of human beings have made astounding observations, have followed the path first articulated by Galileo, and then have attempted to use the principles invented by Isaac Newton to understand this universe. To their great amazement, not only is the universe filled with an incredible range of diverse phenomena, but humankind's attempt to comprehend it has required a new and totally different mode of thought. The conceptual framework that Newton had introduced at the end of the seventeenth century was unable both to describe the newly observed phenomena and to make quantitatively (or even qualitatively!) consistent predictions that agree with further experimental observations. The very heart of science as a model-building process—the ability to describe observations and the ability to predict new phenomena—was challenged by these significant failures of the Newtonian model. As a result, humankind began a journey of creative invention that continues even today (2014) with the development of an entirely new way of thinking about the universe: quantum theory.

The first critical insight took place in the waning months of 1900 as Max Planck, a professor of physics at the University of Berlin, attempted to answer an unresolved question about the radiation emitted by a "black-body," a hypothetical object that is both a perfect emitter and a perfect absorber of radiation. Planck created in his mind an ideal model of the black-body by imagining that the energy was radiated and absorbed by a collection of small masses connected to vibrating springs (the technical term is *mechanical oscillators*). Figure 2.14 shows a simple mechanical oscillator system consisting of two masses connected to three springs. It is permitted to oscillate in only one dimension (horizontally). Planck's model consisted of a three-dimensional cube of connected masses and springs that he permitted to oscillate in three dimensions.

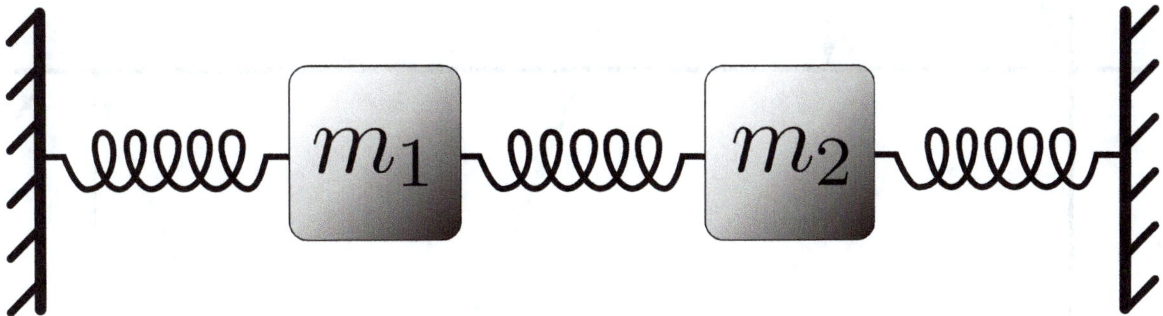


Figure 2.14: Mechanical Oscillator

Although he had begun examining this question in 1894 and had been applying the well-understood principles of thermodynamics (a branch of physics dating from approximately the mid-seventeenth century) to resolve a discrepancy between experimental observations and theoretical predictions, it was not until December 1900 that he made a crucial assumption. Rather than allowing the energy of each oscillator of the black-body to change **continuously** (that is, to emit or absorb **any** amount of energy, ranging from a very small quantity to a very large quantity), Planck postulated that the energy can be transferred only in **discrete** quantities that are **integer** multiples of a fundamental quantity, $E = h\nu$, where $\nu$ is the *frequency* of the transferred radiation (see figure 2.15 for a graphic definition of the terms frequency, wavelength, and amplitude) and $h$ is a fundamental constant of proportionality (later to be called **Planck's constant**).

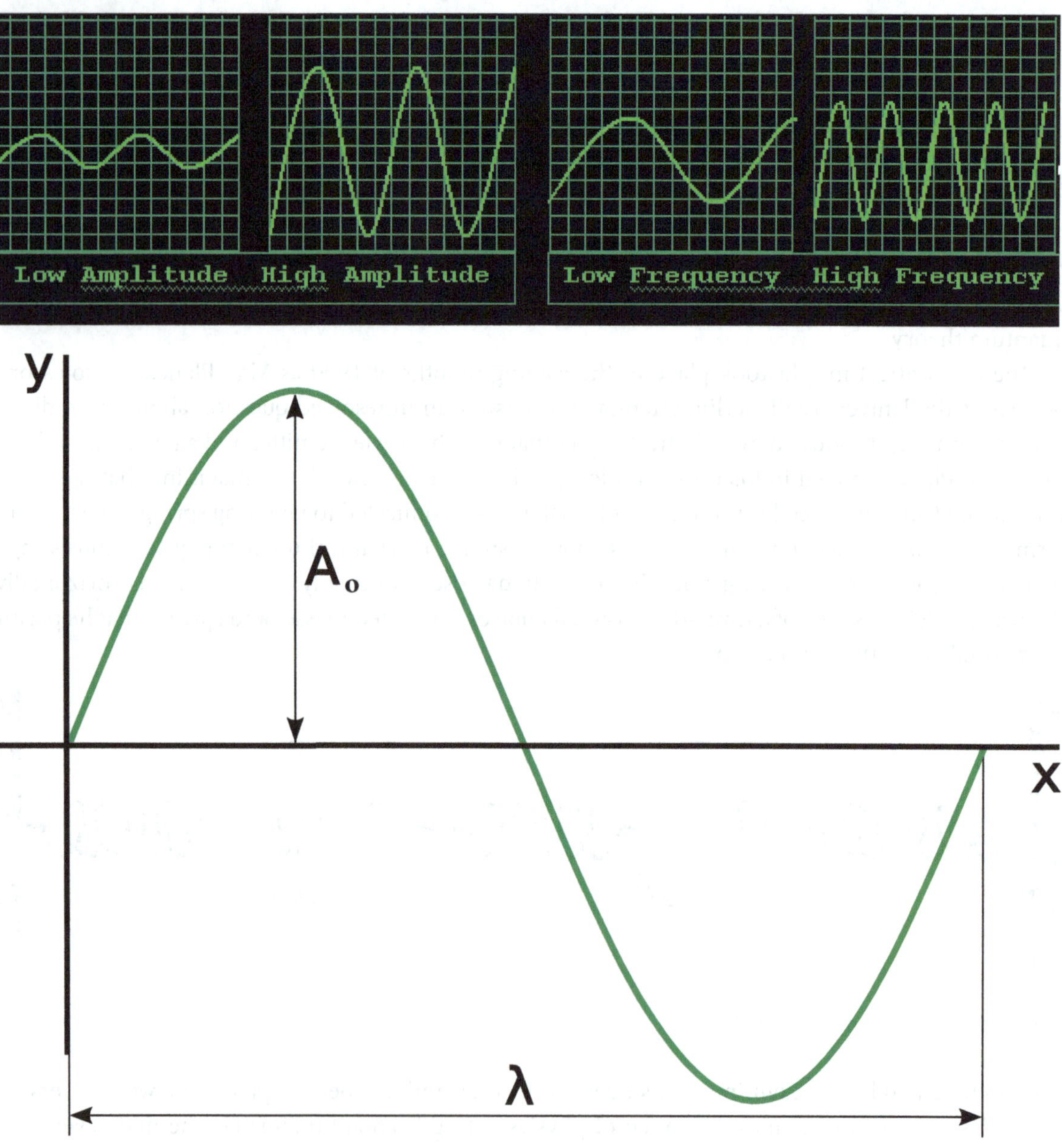


Figure 2.15: Amplitude ($A_0$), Frequency ($\nu$), and Wavelength ($\lambda$)

This postulate means that energy can be exchanged among the oscillators (or transferred between the black-body and the surrounding environment) **only in very *specific amounts*** or **definite units** called **quanta**. We say that the energy is **quantized**. The suggestion that energy can **only** be transferred in specified, discrete quantities contradicted the understanding of energy that had been developed over the more than two hundred years since the time of Isaac Newton.

The ideas of Planck lay dormant for almost five years. In fact, Planck himself viewed his quantization of energy to be simply a **mathematical** technique that allowed him to make numerical predictions that matched experimental observations. His thinking was still firmly rooted in the view that energy is a **physical** characteristic of our universe that can assume **any** continuous value. This perspective was consistent with Maxwell's model of light (the culmination of nearly two centuries of investigations supporting the *wave* model of light), which suggested that a light wave is a physical component of the universe produced by **continuously** oscillating electric and magnetic fields. Because the energy of such a light wave depended on its intensity and the intensity could be varied **continuously**, the energy itself must also be a **continuous** phenomenon. This model was effectively supported by the experimental observation of electromagnetic waves by Heinrich Hertz in 1888.

However, in 1905 Albert Einstein argued persuasively that Planck's quantum postulate (the existence of discrete energy units) applied not simply to the energy of the mechanical oscillators that Planck had idealized in his model of a black-body, but rather to the **entire** spectrum of electromagnetic radiation (see figure 2.16). Using an argument that was grounded in thermodynamic concepts (here, there is a similarity to the work of Planck), Einstein concluded that light also behaved as if it were composed of discrete energy units. (Like Planck, Einstein used the term "quanta" for these units of energy. In 1926 the American physical chemist G. N. Lewis coined the term "photons" to describe the quanta of electromagnetic radiation.) Further, the energy associated with each quantum of electromagnetic radiation depended on the radiation's **frequency**, not its intensity.

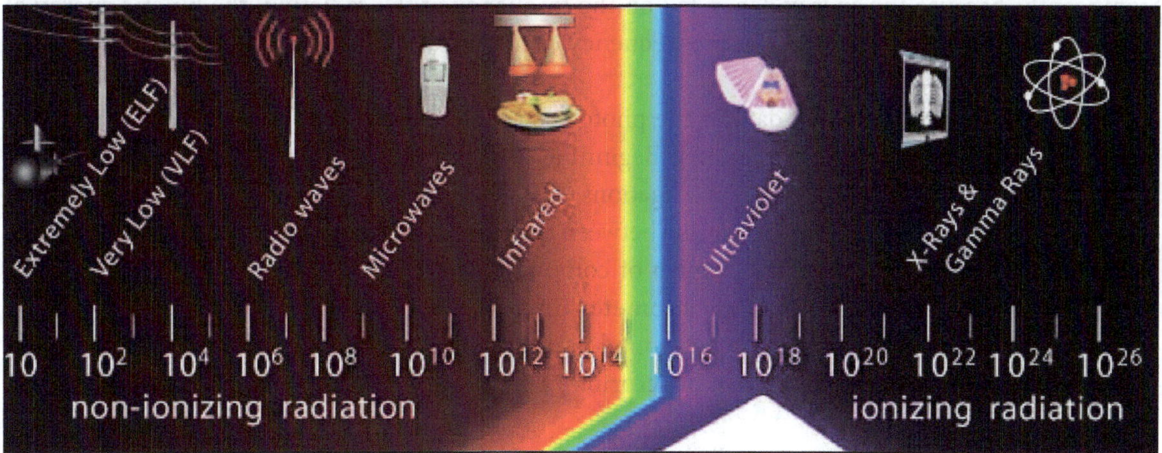


Figure 2.16: Electromagnetic Spectrum (Increasing Frequency)

The conclusions reached by Einstein were significant because during the last half of the nineteenth century, experimental studies had observed that certain metals, when irradiated with visible or

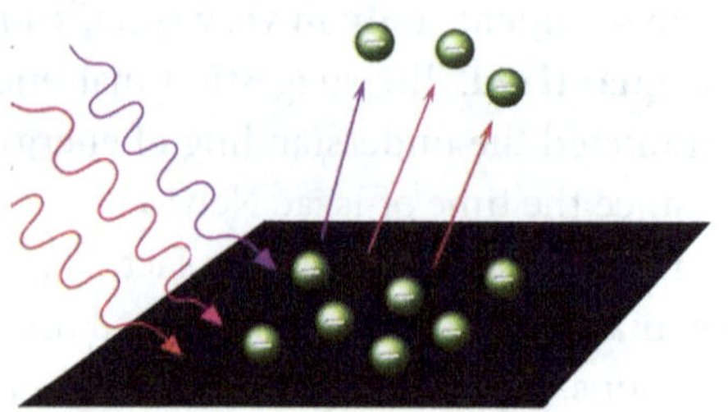

Figure 2.17: Photoelectric Effect

ultraviolet radiation, would produce an electrical current (see figure 2.17). This behavior was called the **photoelectric effect**, and the electric current was interpreted as a stream of cathode rays (in more modern language, a stream of *photoelectrons*).

However, the experimental results could **not** be explained by Maxwell's model of light. Most critically, while the experiments showed that the kinetic energy of the photoelectrons was **independent** of the radiation's *intensity*, this kinetic energy **was dependent** on the *frequency* of the incident radiation. Einstein first applied Planck's postulate, $E = h\nu$, calculating the energy of the quanta to be an integer multiple of $h\nu$ ($nh\nu$), and then used the idea that electromagnetic radiation is quantized to explain the puzzling photoelectric effect. Hence, the kinetic energy of the photoelectrons was a direct result of transferring the energy of the quanta to the photoelectrons by means of a simple collision. (The energy of the quanta is equal to $nh\nu$, where the integer $n$ is called a **quantum number**.) Significantly, because the energy depended on the frequency of the radiation, the kinetic energy of the photoelectrons also depended on the frequency of the radiation, consistent with the experimental observations!

The perplexing experimental data of the photoelectric effect were not the only unexplained observations made in the nineteenth century. Beginning with the early work of Joseph von Fraunhofer and continuing with the subsequent investigations of Robert Bunsen, Gustav Kirchhoff, and Anders Ångström, chemists identified an association between an element or compound and a pattern of colored lines produced when samples of the element or compound were heated and the glowing sample was viewed using an instrument called a **spectrometer** (invented by Fraunhofer). The pattern of colored lines associated with each element or compound is **unique** to that particular element or compound (analogous to the uniqueness of human fingerprints) and is called the element's or compound's **spectrum** (see figure 2.18).

While hundreds of spectra were collected throughout the nineteenth century, the **discrete sharpness** of the spectral lines remained an unexplained puzzle. Using Maxwell's wave model of light (as noted earlier, this was the most widely accepted understanding of light in the nineteenth century), it was impossible to explain both the discrete sharpness of the spectral lines and the uniqueness of the patterns.

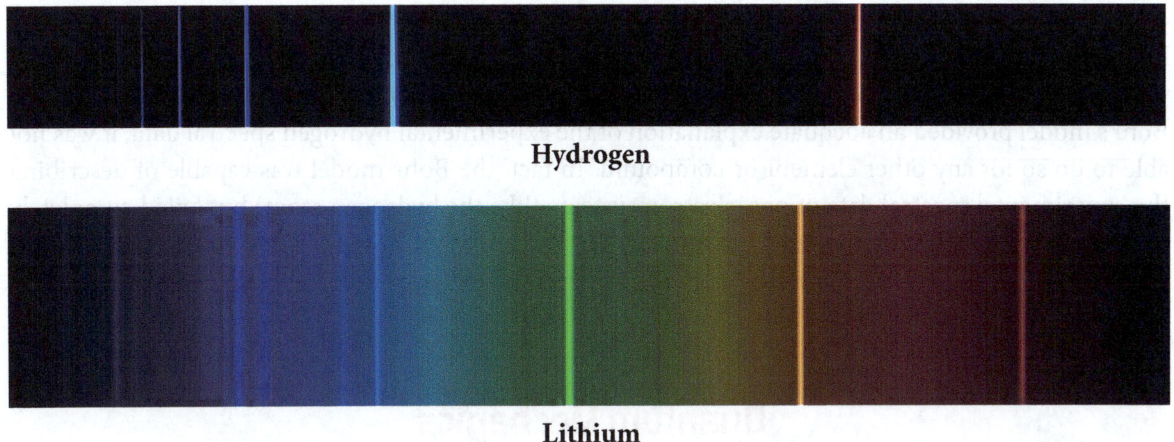


Figure 2.18: Visible Spectra

In our earlier discussion of the three-particle model of the atom, we noted that Niels Bohr published in 1913 a series of the three manuscripts proposing a radically new atomic model. (Bohr had recently returned to Denmark after working in England with both J. J. Thomson and Ernest Rutherford.) Using a combination of established Newtonian physics and the new quantum ideas applied by Planck and Einstein, Bohr suggested that the negatively charged electrons were confined to circular atomic orbits possessing **specific** energy values and that the electrons surrounded the dense positive core (the nucleus) of the atom in a manner analogous to the planets surrounding the sun in the solar system. Between any two successive orbits, **no additional** orbits with associated energy values existed for an electron: energy **did not vary continuously** between two successive orbits. Bohr further proposed that an electron confined to a specific orbit **did not emit** radiation; Maxwell's theory required that a charge moving in a circular (or elliptical) orbit must radiate and thereby **lose** energy. Such an energy loss would make an atom unstable.

With this proposal, Bohr had both **quantized** the possible orbits of the electrons in an atom and had provided a model of a stable nuclear atom. Further, as an electron changed from an initial orbit with energy $E_i$ to a final orbit with energy $E_f$, the difference, $E_f - E_i$, is a **specific** energy (because both $E_f$ and $E_i$ are specific energy values, with $E_f > E_i$), and, for the case of the hydrogen atom, Bohr was able to show that $E_f - E_i = h\nu$, where $\nu$ is one of the characteristic frequencies of the hydrogen spectrum. Consequently, by quantizing the electron orbits and requiring that each orbit possess a single, specific energy value, Bohr was able to calculate the frequencies of the sharp spectral lines in the hydrogen spectrum. Because the energy of each electron orbit is a specific value, the calculated frequency is also specific, thereby accounting for the discrete sharpness of the spectral lines.

While we have examined only one aspect of the Bohr atom (sometimes called the Rutherford-Bohr atom) and have intentionally avoided the many additional details of its characteristics, it is crucial to note that Bohr's work represents the first application of quantum ideas to matter. Most importantly, the utilization of quantum ideas provided an explanation of the experimental hydrogen spectral data as well as a model of a dynamically stable nuclear atom. However, these were not unqualified successes. Bohr could provide no fundamental rationale for the existence of stable orbits and their associated energies; at this point in the development of the quantum theory, this requirement was simply an ad hoc theoretical postulate. In fact, to simplify the model even more, Bohr had assumed that the orbits were circular and that the revolving electrons obeyed Newtonian dynamics. More significantly, while Bohr's model provided an adequate explanation of the experimental hydrogen spectral data, it was not able to do so for any other element or compound. In fact, the Bohr model was capable of describing the experimental spectral data of one-electron systems (like the hydrogen atom) but failed completely when it was applied to the spectral data of multi-electron systems. Finally, experimental observations had shown that the presence of a magnetic field altered the recorded spectra; Bohr's model was silent about the role of magnetism.

## Quantum Mechanics

For more than a decade following Bohr's three remarkable papers, physicists struggled either to extend or to modify Bohr's model in an attempt to provide a consistent theoretical basis from which to understand the microscopic universe. During this same period a virtual flood of experimental studies continued to support the quantum theoretic interpretation of the way the universe appears to behave. In the fall of 1925 and the spring of 1926 two seminal developments took place that dramatically changed the quantum landscape. Utilizing the large collection of experimental spectroscopic data that had been collected over more than a century, Werner Heisenberg, a young German physicist working with Bohr in Copenhagen, proposed a new quantum theory called **matrix mechanics**. It was an approach rooted in a particle interpretation of microscopic behavior that used a very abstract branch of mathematics (matrix theory) to interpret and explain the experimental data. The vast majority of physicists were unfamiliar with Heisenberg's mathematics, thus restricting the accessibility of the new theory and limiting its adoption. Less than six months later, Erwin Schrödinger, an Austrian physicist who was unusually old (Heisenberg's senior by nearly fourteen years) to propose a revolutionary and groundbreaking theory, formulated **wave mechanics**. This approach was based on a wave interpretation of microscopic behavior and used the very familiar mathematics of differential equations. Consequently, the Schrödinger theory was more readily embraced and utilized by the physics community of the time.

While the two quantum theories initially appeared to be dramatically different, within a span of two years, they were shown to be entirely equivalent and represented simply different mathematical formulations of a single theory. Additional work by mathematician Max Born and physicist Pascual Jordan, both working at the University of Göttingen, further elaborated and clarified the mathematical characteristics of the two formulations. The fact that Heisenberg's proposal focused on a particle interpretation of microscopic behavior while Schrödinger's model interpreted the same behavior using a wave theoretic approach signaled that the new **quantum mechanics** could **not** interpret the experimental microscopic data using **solely** a particle interpretation or **solely** a wave interpretation. The first attempt to resolve the tension (and apparent contradiction) between a particle approach and a wave approach was the articulation of a new physical principle: **wave-particle duality**. This idea suggested that all microscopic transformations of the universe exhibit both particle-like and wave-like characteristics. Early experimental procedures seemed to confirm that any single experiment can exhibit either particle-like characteristics or wave-like characteristics, but not **both** simultaneously. In light of this early experimental evidence, Bohr formulated his **complementarity principle** (1928), postulating that even though particle-like and wave-like characteristics are mutually exclusive categories of thought that predate quantum mechanics, they are both necessary in order to provide a complete description of our universe. That is, characteristics of the universe always occur in complementary pairs. (Bohr used the term *conjugate pairs.*) This principle forms a key component of the **Copenhagen interpretation** of quantum mechanics. The most recent development of **quantum field theory** (the modern successor to the quantum theories of 1925–1926) has integrated the nonlocality of a field (wave-like characteristic) with quantization (particle-like characteristic), thereby apparently eliminating the paradox of wave-particle duality. Perhaps the most insightful approach we can take is to admit that ascribing either wave-like or particle-like characteristics to microscopic phenomena is not appropriate. These phenomena are so different from our daily experience of the universe that categories of thought derived from that everyday experience are completely inappropriate tools with which to describe and interpret microscopic phenomena.

Without providing detailed explanations, it is important to note that the Copenhagen interpretation of quantum mechanics is but one of many different interpretations that have been proposed over nearly a century in an effort to understand the implications of quantum theory. While there is no dispute about the consistency between the experimental data and the calculated predictions of quantum mechanics (the agreement is astonishing!), there is no agreement among quantum theorists on the best interpretation of quantum mechanics. We live in a time (2014) that is too close to the origin of the theory; no final synthesis of the "meaning" of quantum theory has yet emerged.

## Schrödinger Equation

In an attempt to understand several of the chemically significant dimensions of quantum theory, we will utilize the mathematically familiar Schrödinger approach. Although the theory was familiar to the physicists of the early twentieth century, it was by no means mathematically simple. Its fundamental postulate is a complicated differential equation that involves both **time** and the **three spatial dimensions**:

$$i\hbar\frac{\partial}{\partial t}\Psi(\mathbf{r},t) = \left[\frac{-\hbar^2}{2m}\nabla^2 + V(\mathbf{r},t)\right]\Psi(\mathbf{r},t),$$

where the Laplacian operator applied to an appropriate function, $f$, is defined as

$$\nabla \cdot \nabla f = \nabla^2 f = \frac{\partial^2 f}{\partial x^2} + \frac{\partial^2 f}{\partial y^2} + \frac{\partial^2 f}{\partial z^2}.$$

We will **not** attempt to solve this equation; mathematically, it requires techniques that lie well beyond the scope of this text. Suffice it to say that the solutions to this equation are called **wave functions**, $\psi(x, y, z, t)$. Each solution (a wave function for a single electron) is called an **orbital** (not an orbit!). The

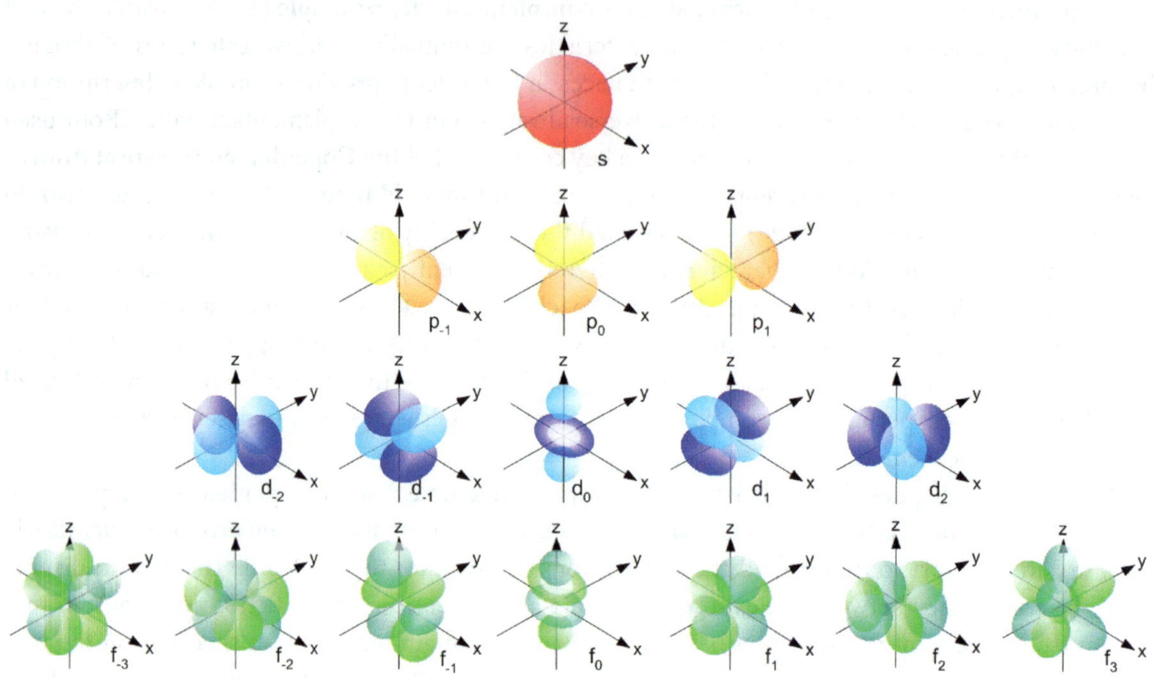


Figure 2.19: Atomic Orbital Probability Densities

reader should note carefully that the wave function is a solution of the Schrödinger equation and, as such, is itself a mathematical object and not a physically measurable quantity. Further, the difference in terminology is significant because the new quantum mechanics of Heisenberg and Schrödinger views an electron as a quantum object whose spatial location can **only** be predicted probabilistically. In late 1926, Max Born proposed an approach (often called Born's rule) that utilized the electron's wave function to predict its spatial probability distribution. Consequently, the new quantum mechanics does not specify a definite spatial location of an electron (like a planetary orbit around the sun) but only the **probability** of locating an electron at specific spatial coordinates. The results of this probabilistic prediction are the characteristic orbital **shapes** seen in figure 2.19. These shapes represent the spatial properties of the probability distribution associated with the wave function solutions of the Schrödinger equation.

Because quantum theory's probability prediction of an electron's spatial location is nonzero **anywhere** in space, it is not possible to depict the **entire** probability distribution. Consequently, the images in figure 2.19 are approximate representations of contour surfaces; typically, the contour surfaces are drawn to identify the spatial region in which there is a 90% probability of finding an electron.

## Quantum Numbers

Prior to the work of Heisenberg and Schrödinger, we have seen that only a single quantum number was necessary to explain the energy of the quantum systems explored by Planck, Einstein, and Bohr. In contrast, the solution of the Schrödinger equation (along with its extension by Paul Dirac to include the effects of Einstein's special theory of relativity) generates **four** distinct quantum numbers. Let's examine some of the characteristics of each of these quantum numbers.

The first of the four quantum numbers is called the **principal quantum number** and is symbolized by a lowercase ***n***. The principal quantum number can have **only integer values** (1, 2, 3, …), and the principal quantum number has **no largest value.** In the case of the hydrogen atom, the **energy** of the element's single electron (i.e., the energy associated with an orbital) depends **solely** on the value of the principal quantum number; this is **not** the case for many-electron atoms. Finally, the principal quantum number is a measure of an electron's **average distance from the nucleus.**

The second quantum number is called the **angular momentum** (sometimes **azimuthal**) **quantum number** and is symbolized by $\ell$ (a lowercase L). The name of this quantum number refers to an important physical property of the universe: angular momentum. Like energy, angular momentum is a **conserved** quantity; it is an invariant of a physical system that depends on the **distribution of mass.** The simplest example of the conservation of angular momentum is demonstrated by an ice skater spinning rapidly in one place; as she pulls in or extends her arms (changing the distribution of her mass in space), her rotational velocity changes in order to keep her **angular momentum constant.** As in the case of the first quantum number, the angular momentum quantum number takes on integer values (0, 1, 2, 3, … $n - 1$); but note that the values begin with zero and that there **is a maximum value,** $n - 1$. This means that the angular momentum quantum number **depends on** the principal quantum number; it is **not independent** of the principal quantum number. The angular momentum quantum number is associated with the **shape** of an orbital; more precisely, it identifies the **spatial probability distribution** of each particular orbital.

Historically, several $\ell$ values have come to be associated with specific names of the corresponding atomic orbitals. The orbitals with values of $\ell = 0$ are called **s orbitals.** Those with values of $\ell = 1$ are called **p orbitals.** Orbitals having values of $\ell = 2$ are called **d orbitals.** Finally, those with values of $\ell = 3$ are called **f orbitals.** Because the current (2014) periodic table includes only 118 elements, the maximum value of the angular momentum quantum number required to describe all the elements in the periodic table is $\ell = 3$. However, there are mathematical approaches used to model the properties of the elements and that require values of $\ell > 3$. In this case, the orbitals used in the mathematical techniques are assigned names in alphabetical order, beginning with **g** and skipping the letter **j.** Figure 2.20 (which redisplays figure 2.19 in order to identify the name of each orbital) depicts the **shapes** of the probability distributions associated with the **s, p, d, and f orbitals.**

The third quantum number required by the solutions of the Schrödinger equation is known as the **magnetic quantum number**, and the symbol $m_\ell$ is used to designate it. The name of this quantum number was selected because its properties directly correlate with the experimental behavior of atoms in the presence of a magnetic field. This third quantum number again can have **only** integer values, but in this case half of them are **negative integers** ($-\ell$ $-\ell + 1$, … $-1, 0, 1, 2,$ … $\ell - 1, \ell$). (Note that the subscripts in figure 2.20 are the corresponding values of the $m_\ell$ quantum number.) Another way of stating the possible values of the magnetic quantum number is to observe that they range from $-\ell$ to $+\ell$ in integer steps. Because this range includes zero, there are $2\ell + 1$ possible values of the magnetic quantum number ($\ell$ negative values, $\ell$ positive values, and zero). It is also important to note again (unlike the principal quantum number) that the number of values of the magnetic quantum number

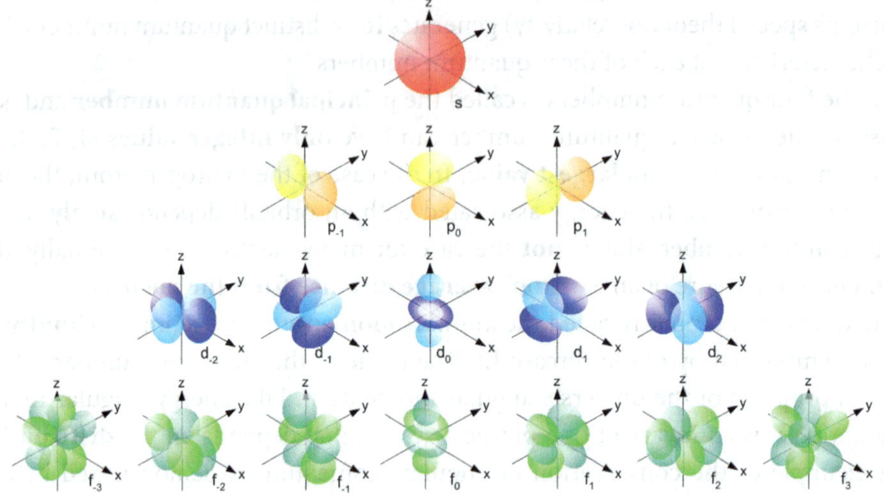


Figure 2.20: s, p, d, and f Orbital Images to Demonstrate l Values

is limited and that limit is determined by $\ell$, the value of the angular momentum quantum number. Hence, the magnetic quantum number **depends on** the angular momentum quantum number. Finally, this quantum number describes the **orientation in space** of the probability distribution associated with each individual orbital.

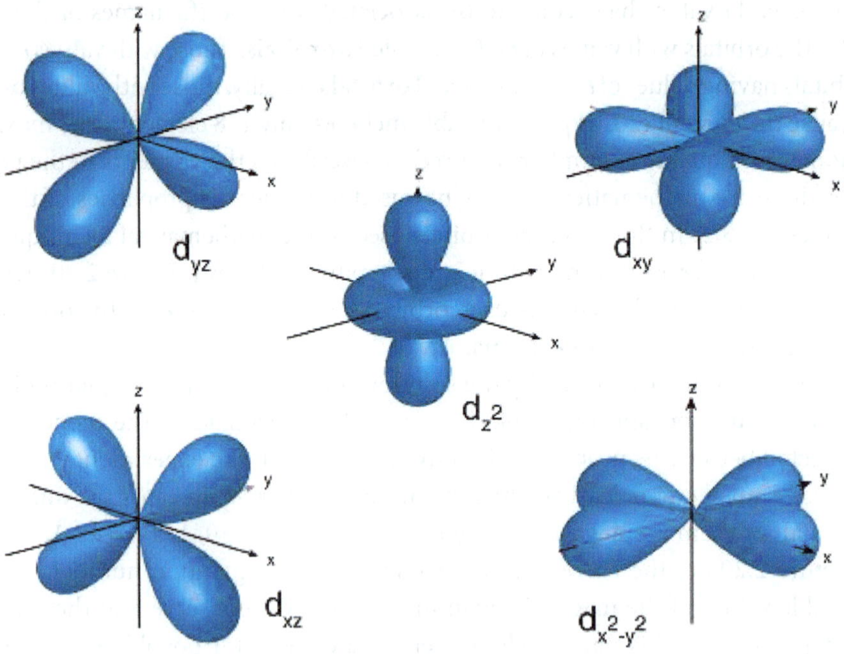


Figure 2.21: d Orbital Images (Five Possible $m_\ell$ Values)

The fourth and last quantum number is called the **spin quantum number** and is assigned the symbol $m_s$. This quantum number, unlike the previous three, is not a consequence of the original equation proposed by Schrödinger in early 1926. The first hint that nature requires another quantum number actually **predates** the work by Schrödinger and was the result of an experiment performed in 1922 by Otto Stern and Walther Gerlach. In 1925, two Dutch physicists, George Uhlenbeck and Samuel Goudsmit, and, independently, a German physicist, Ralph Kronig, suggested that an electron seemed to behave as if it were rotating on its axis in space. This physical image was the origin of the concept of **electron spin**. While the physicist Wolfgang Pauli, in an effort to understand a variety of experimental data, also proposed in 1925 a new, two-valued quantum number that would eventually be associated with the concept of spin, the first mathematically rigorous derivation of this new quantum number did not occur until Paul Dirac's application of Einstein's special theory of relativity to the Schrödinger equation in 1928.

The spin quantum number is unusual in several ways. When applied to the case of the electron, it can have only **two** values, and the values are **half-integers** (+½ and −½), rather than the integer values of the three previous quantum numbers. Curiously, the concept of **spin** (it applies not only to the electron but to other quantum objects as well, although with different values) behaves like another type of angular momentum. This second type of angular momentum is called **intrinsic angular momentum**. However, it is important to note that the picture of an electron spinning on an axis **is not supported either by the experimental data or by the mathematics of quantum theory**. The image of the electron as a miniature ball rotating on an axis is a purely macroscopic picture that has no place in the quantum description of the electron's behavior.

## Quantum Mechanical Consequences

We have already encountered a number of the dramatic changes in perspective that were required by the introduction of quantum ideas, beginning with Planck in 1900 and continuing through the development of the new quantum mechanics by Heisenberg and Schrödinger in the third decade of the twentieth century. The revolution in human thinking that was initiated by these pioneers continues as the twenty-first century unfolds with new experimental and theoretical explorations at the very frontiers of human understanding. It is important to understand that the fundamental research questions that are being asked today (2014) not only utilize the most sophisticated incarnations of the quantum theory when probing the universe in surprisingly novel ways, but also challenge the very foundations of the theory itself in an attempt to add understanding to the theory's powerful computational capabilities.

While recognizing these ongoing research efforts and unresolved questions, several significant consequences of the quantum theory that was developed in the first third of the twentieth century have been repeatedly verified and play important roles in understanding modern chemistry. In particular, as the science of chemistry entered the twentieth century, the central focus was on atomic and molecular structure and the prediction of chemical reactivity based on a detailed knowledge of those structures. The effort to link chemical behavior to an understanding of specific chemical structural relationships had become a central theme of chemistry in the nineteenth century, and the advent of the new quantum mechanics propelled models of chemical structure to new levels of sophistication. However, the new quantum mechanics would fundamentally change the manner in which both chemists and physicists understood microscopic structures.

One of the earliest consequences of the new quantum theory was the recognition of the **Heisenberg uncertainty relation**, or, more accurately stated, the **Heisenberg indeterminacy relation.** We have already seen that Bohr formulated his complementarity principle to emphasize the apparent fact that characteristics of the universe exist in complementary or conjugate pairs, and that both members of each pair must be treated **equally** in order to achieve a complete description of our universe. In 1928 Heisenberg recognized that as a result of the complementarity principle, the macroscopic variables of momentum and position are quantum mechanically conjugate variables and that the attempt to determine microscopic values for them is subject to a specific limit. He enunciated this limit in the statement of the **indeterminacy relation**: it is impossible to know **simultaneously** both the momentum and the position of a particle with **unlimited** accuracy. In fact, if we allow the symbol "$\Delta x$" to represent the uncertainty of a position measurement and the symbol "$\Delta p_x$" to represent the uncertainty of a momentum measurement (here we limit our discussion to a single dimension), the following inequality defines the specific limit Heisenberg recognized: $(\Delta x) \times (\Delta p_x) \geq h \div (4\pi)$, where $h$ is Planck's constant. (Again, we have written the expression for one dimension; it applies equally well to a particle moving in three dimensions.) The limit imposed by the Heisenberg indeterminacy relation means that microscopic structures must be understood as being fundamentally different from the structures of the macroscopic world.

As we noted above, Wolfgang Pauli had proposed a new quantum number in 1925, significantly before Paul Dirac completed the work in which he applied the theory of special relativity to Schrödinger's equation. Pauli's work rested on an interpretation of a collection of experimental data and resulted in the introduction of a two-valued, fourth quantum number, the spin quantum number. Based on these considerations Pauli formulated a concise statement about the behavior of electrons in any atom that has come to be called the **Pauli exclusion principle**: no two electrons in an atom can simultaneously have the **same** four values for each of the quantum numbers $\mathbf{n}$, $\ell$, $\mathbf{m}_\ell$, and $\mathbf{m}_s$. In effect, the four quantum numbers act like "a four number house address" of each electron in an atom; every electron has a **unique** "house address." Because the spin quantum number can have only two values, +½ and –½, the Pauli exclusion principle imposes a limit of **two** on the number of electrons associated with an orbital specified by the three quantum numbers $\mathbf{n}$, $\ell$, and $\mathbf{m}_\ell$. This is the origin of the terminology of "exclusion"; **all** electrons after the first two are **excluded** from being associated with a given orbital, and the two electrons that are associated with an orbital must have **distinct** values of the spin quantum number.

Although the concept of electron spin came to be understood in increasingly abstract terms as the quantum theory underwent further development in the twentieth and twenty-first centuries, the initial articulation of the Pauli exclusion principle immediately affected the efforts of chemists to understand chemical structures from the quantum point of view. Because the exclusion principle enforced a limit of associating two electrons per atomic orbital, it was **not** possible for all the electrons in a multi-electron atom to be associated with a **single** orbital. Consequently, atoms with differing numbers of electrons **must** require differing volumes of space. That is, different atoms must be different sizes, consistent with the experimental data! Extending the exclusion principle to the electronic arrangement of molecules, it became clear that the size of a molecule was a property determined not only by the total **number** of atoms in the molecule but also by the **different kinds** of atoms constituting the molecule.

Finally, a third consequence of the new quantum mechanics of Heisenberg and Schrödinger was identified by the German physicist Friedrich Hund, following an analysis of spectroscopic data. Although Hund proposed three distinct rules and considered five separate cases while analyzing the behavior of angular momentum in molecules, we will focus on only the first of Hund's rules: **Hund's rule of maximum multiplicity**. Because of its importance to understanding the arrangement of electrons in an atom or molecule, it is usually referred to as simply **Hund's rule**. What does this rule state?

Recall that for any given value of the angular momentum quantum number, $\ell$, there are $2\ell + 1$ possible values of the magnetic quantum number, $\mathbf{m}_\ell$. This means that there are $\mathbf{m}_\ell$ orbitals, each with a specific **orientation** in space. For example, when $\ell = 1$, the possible values of $\mathbf{m}_\ell$ are −1, 0, and +1; consequently, there are three p-type orbitals (because $\ell = 1$), usually designated as $p_x$, $p_y$, and $p_z$. The three p orbitals can be associated with a maximum of six electrons, two electrons per orbital as required by the Pauli exclusion principle. Suppose that the total number of electrons associated with the p orbitals is **less than six**. How, then, should the electrons be distributed? Figure 2.22 depicts the possible associations of two electrons with the three p orbitals. (The +/- symbols represent the +½ and -½ values of the spin quantum number.)

| 1s | 2s | $2p_x$ | $2p_y$ | $2p_z$ |
|---|---|---|---|---|
| + − | + − | + − | | |
| + − | + − | + | − | |
| + − | + − | + | + | |

Figure 2.22: Possible Arrangements of p Electrons in the Configuration $1s^2s^2p^2$

Hund observed that the lowest energy (most stable arrangement) is achieved if the electrons are associated with the orbitals **singly** rather than in pairs. Further, all the singly associated electrons **must** have the **same value** of the spin quantum number. The algebraic sum of the spin quantum number values is called the **multiplicity** of the arrangement. Note that if the maximum number of electrons is associated with a set of orbitals, each orbital has exactly two electrons associated with it, one with a spin quantum number value of +½ and other with a spin quantum number value of −½. In this case the multiplicity is zero. Using this terminology, Hund's rule can be restated as follows: the electron configuration with the **maximum** multiplicity is the **most** stable (**lowest**-energy) configuration. In figure 2.22, the last example possesses the lowest energy.

## The Aufbau Principle and Electron Configurations

The exclusion principle also offers an insight into the organization of the periodic table of elements. While Dmitri Mendeleev and Julius Lothar Meyer published their periodic tables in 1869 and 1870, respectively, basing their proposals on the available experimental data, they did not possess the theoretical tools to suggest a more fundamental organizing principle for the table of elements. They used the empirical chemical properties of the elements and their ability to form various compounds to construct these initial tables. The iconic shape of the modern periodic table of elements can be understood as a consequence of using these experimental observations. However, slightly more than a half-century later, the Pauli exclusion principle provided a quantum mechanical basis both for the arrangement of the electrons in the atoms of each element and for the distinctive shape of the periodic table.

In order to understand the explanatory role of the Pauli exclusion principle, we will view each atom from the structural perspective that dominated chemical thinking at the end of the nineteenth and at the beginning of the twentieth centuries. The introduction of the atomic concepts of Rutherford and Bohr and the experimental work of Moseley made it clear that the elements of the periodic table should be ordered by **increasing** atomic number, hence, by **increasing** positive charge in the nucleus. The electrical neutrality of matter immediately requires an equal number of electrons in each atom. But in view of the new quantum mechanics, how are the electrons distributed in an atom? We will imagine a hypothetical process (that is, this process is an **idealization** that we formulate in our mind), in which, as each positive charge is added to the nucleus (increasing the atomic number by one unit), a negative charge (represented by an electron) is added to the atom with each added electron being associated with an atomic orbital. The process proceeds such that, at each step, the most stable (lowest-energy) arrangement (an association of electrons and atomic orbitals) of the electrons is achieved.

Let's begin with the elements of the two leftmost columns of the periodic table. In this case, the last electrons added in our hypothetical process of adding protons and electrons and associating the electrons with orbitals means that we are associating the electrons with s-type orbitals. We are focused on **two** columns, because for each principal quantum number (which label the **rows** of the periodic table), there is only **one** s-type orbital, and, according to the Pauli exclusion principle, only **two** electrons are associated with it. Now consider the last six columns (13–18) of the periodic table. For these elements our hypothetical process of adding protons and electrons associates the last electrons added with p-type orbitals. Note that there are three p-type orbitals; each one can be associated with two electrons, and $2 \times 3 = 6$ is consistent with six columns. For columns 3–12, the last electrons added by the process are associated with d-type orbitals. There are five d-type orbitals; the Pauli exclusion principle means that only two electrons can associated with each orbital, and $2 \times 5 = 10$ is again consistent with ten columns. The elements in columns 3–12 are known collectively as the **transition metals**. Finally, the two rows depicted under the main table (we will see shortly that these elements are called the **lanthanides** and **actinides**) have fourteen columns. For both of these rows, our hypothetical processes of adding protons and electrons associates the electrons with f-type orbitals. There are seven f-type orbitals; the Pauli exclusion principle means that only two electrons can associated with each orbital, and $2 \times 7 = 14$ is consistent with fourteen columns. Figure 2.23 shows the traditional periodic table of elements and its correlation with the quantum mechanical orbitals.

The hypothetical process we have described is called the **Aufbau principle** (the word *aufbau* is German for "building up"). The meaning is clear: we have described a hypothetical process that allows us to "build up" the atoms of the elements in the periodic table in a stepwise process. Because there are four quantum numbers, specifying values for each of them (that is, associating the electron with a specific orbital type and providing a value of the spin quantum number) completely specifies the

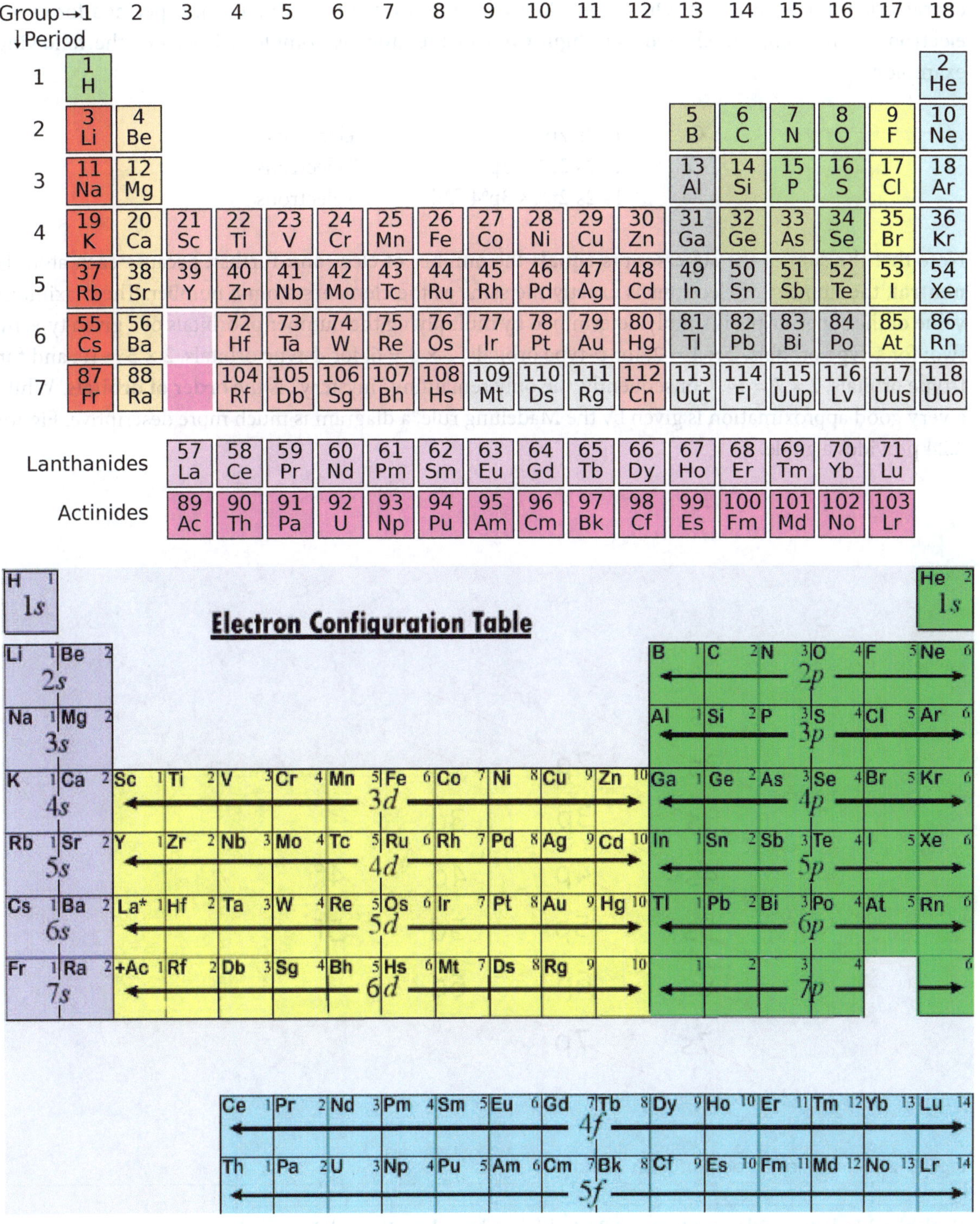


Figure 2.23: Block Diagram of the Traditional Periodic Table

characteristics of an electron. Once the values of the four quantum numbers are specified for **every** electron in an atom, the **electron configuration** of the atom is complete. Consider the following examples:

| | | |
|---|---|---|
| O | $1s^2 2s^2 2p^4$ | 8 electrons |
| Cl | $1s^2 2s^2 2p^6 3s^2 3p^5$ | 17 electrons |
| Fe | $1s^2 2s^2 2p^6 3s^2 3p^6 4s^2 3d^6$ | 26 electrons |

Note that the sum of the superscripts **equals** the number of electrons; further, because each atom is **neutral**, the number of electrons is exactly the same as the element's atomic number. The maximum value of the superscripts is simply determined by multiplying the number of orbitals of a given type by two: for s-type orbitals, $2 \times 1 = 2$; for p-type orbitals, $2 \times 3 = 6$; for d-type orbitals, $2 \times 5 = 10$; and for f-type orbitals, $2 \times 7 = 14$. Consequently, the only remaining "mystery" is the **order of orbitals**. While a very good approximation is given by the Madelung rule, a diagram is much more descriptive. Figure 2.24 provides a guide.

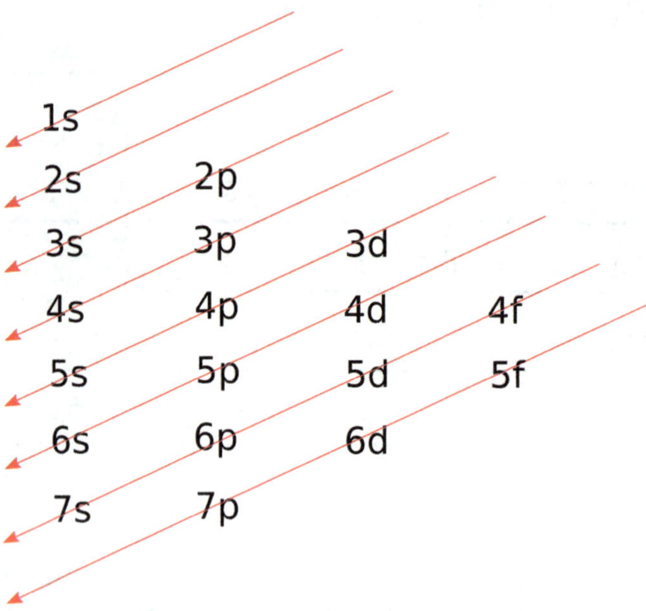


Figure 2.24: Electron Configuration Chart

Both the Madelung rule and figure 2.24 (which is based on the rule) are only **approximately** correct. To make highly accurate energetic distinctions, it is necessary to solve the multi-electron Schrödinger equation, a task that we have chosen to avoid in this text!

# Chapter 2 Exercises

1. What is the name of the cosmological model that incorporates the observation of an expanding universe?
2. Because only hydrogen, helium, lithium, and beryllium existed at the origin of our universe, how were elements such as carbon, oxygen, phosphorous, and iron produced?
3. As the twenty-first century begins, it appears that the universe is composed of three distinct components. What are they?
4. Who was John Dalton?
5. List Dalton's hypotheses.
6. What is the difference between the law of definite proportions and the law of multiple proportions?
7. The observations of Röntgen, Becquerel, and the Curies suggested a very different understanding of an atom when compared to Dalton's hypotheses, which suggested that atoms are both eternal and indivisible. As a result of the new observations, how were atoms viewed?
8. What is J. J. Thomson's most important insight?
9. The experiment of Rutherford's students, Geiger and Marsden, suggested that the apparently solid gold foil was not, in fact, solid. What was the major component of the gold foil?
10. As a result of his experiments, Rutherford concluded that most of the mass of atom was concentrated in a dense central core. What is the name given to this dense central core?
11. In the three-particle model of the atom, what is the charge of a proton?
12. In the three-particle model of the atom, what is the charge of a neutron?
13. In the three-particle model of the atom, what is the charge of an electron?
14. Of the three particles, proton, electron, and neutron, which has the smallest mass?
15. What is the atomic number? What symbol is used to represent it?
16. What is the atomic mass number? What symbol is used to represent it?
17. Determine the number of protons, neutrons, and electrons in the following atoms:

    (a) ${}^{12}_{6}C$ (b) ${}^{56}_{26}Fe$ (c) ${}^{32}_{16}S$ (d) ${}^{39}_{19}K$

18. What is an isotope?
19. Define the unified atomic mass unit. How is this definition related to an amu and a dalton?
20. What is the average atomic mass of an element?
21. What is a mole?
22. For the following elements, calculate the mass of 7.5 moles:
    (a) C (b) W (c) P (d) N (e) Fe
23. Does the numerical magnitude of Avogadro's number change?
24. What is the molar mass of an element?
25. In the standard model of particle physics, what are the two major classes of subatomic particles?
26. To which class of particles does the electron belong?
27. Do all the quarks carry a unit electromagnetic charge?
28. Because protons and neutrons are now understood to be composite particles, what are their constituents? Identify the number and type of hadrons that make up protons and neutrons.

29. Which particle is responsible for the fact that most of the particles of the standard model exhibit a nonzero mass?
30. What are the names of the three types of radiation that had been identified by 1900? Two additional types of radiation were identified. What are they?
31. How is radioactivity different from the cathode rays studied by Thomson?
32. What is a curie? How is a curie related to a becquerel?
33. What is an **absorbed dose** of radiation? What units are used to measure the **absorbed dose** of radiation?
34. What is an **equivalent dose** of radiation, and what units are used to measure it?
35. What is the half-life of a radioactive substance?
36. If the amount of radioactive cesium-131 in a sample decreases from 0.800 g to 0.100 g in 697.608 hours, what is the half-life (in days) of cesium-131?
37. Suppose that a student begins with 625 g of radioactive phosphorus-33, whose half-life is 25.34 days. How much radioactive phosphorus remains after 1824.48 hours?
38. What is nuclear fission?
39. What is nuclear fusion?
40. Define the following terms:
    (a) Binding energy (b) Critical mass (c) Chain reaction (d) Mass deficit
41. Give the names of three types of machines used to artificially accelerate subatomic particles.
42. What is a black-body?
43. What are quanta?
44. What is a photon?
45. True or false? According to the quantum theory of Planck (also used by Einstein), the energy associated with each quantum of electromagnetic radiation depends on the radiation's intensity.
46. What is the photoelectric effect?
47. By using quantum ideas, Bohr suggested a model of the hydrogen atom. What were two important accomplishments of this model?
48. In the new quantum mechanics, what is wave-particle duality?
49. Write the statement of the Heisenberg indeterminacy relation.
50. Write the statement of the Pauli exclusion principle.
51. State Hund's rule.
52. Write the electron configuration for the following:
    (a) Co (b) Cl (c) N (d) S

# CHAPTER THREE
## The Classical Chemical Paradigm
### Part I (Pre-Twentieth Century)

The previous chapter examined a variety of concepts that apply to vastly different time periods in the history of the universe as well as to dramatically different dimensional scales, ranging from galaxies to the subatomic structure of matter. Perhaps the most significant portion of this wide-ranging story is the dramatic reorientation of human thought caused by the introduction of the quantum theory at the start of the twentieth century. Despite this magnificent panorama of ideas, we should not lose sight of the fact that the origin of modern chemistry stems from the pioneering work of Cavendish, Priestly, Lavoisier, and Dalton at the end of the eighteenth and beginning of the nineteenth centuries. Unlike so much of the empirical work at the end of the nineteenth and throughout the twentieth centuries, this pioneering work depended on macroscopic observations for which microscopic explanations (chemical theories) were constructed. However, these chemical theories simply translated macroscopically observed physical properties to the microscopic explanations, assuming that the macroscopic properties behaved **in an identical manner** when used to construct a microscopic model. There was no indication or anticipation of the impending revolution in thought that appeared with the introduction of quantum theory.

We now turn our attention to those specific chemical terms and concepts whose origin lies in the macroscopic observations of the pre-twentieth century world, a world not dominated by the quantum theory. While the terms we will encounter originated in an intellectual climate very different from that of the twenty-first century, this does not mean that these terms should be ignored or discarded or that the chemistry accomplished using these terms was, in some sense, invalid. Quite to the contrary, the terms derived from the pre-twentieth century intellectual milieu formed a highly effective framework for chemical practice. They simply represent a model that is **different** from the contemporary model.

We embrace the contemporary model to the extent that its explanatory and predictive capabilities surpass those of earlier models. Finally, it is crucial to distinguish between the **observations** (whether made in the eighteenth century or the twenty-first century) and the **interpretations** provided by a particular chemical model.

## Early Terminology

Let's begin by looking at a series of terms first articulated prior to the twentieth century. The first two terms we encounter are very old; in fact, they predate the rise of modern science in the seventeenth century.

### Substance and Mixture

A **pure substance** is a form of matter that has a **definite** or **constant composition** and **distinct properties**. Simple examples include water, silver, gold, or iron. Humankind's observations have identified more than fifty million different substances. In contrast, a **mixture** is a combination of **two or more substances** in which the substances **retain their distinct properties**. Common examples are air, soda, and milk.

### Physical and Chemical Processes

After making these very general distinctions between components of our world, the next two terms make general observations about the types of *transformations or changes* that occur in our universe. A **physical process** is any transformation of matter that does not change the **chemical identity** of a substance. Simple examples of physical processes include melting and freezing (more generally, fusion), vaporization and condensation, and sublimation (the change from a solid directly to a vapor). It is interesting to note that this idea rests on an implicit model (one that may have not been consciously developed or stated) in which substances possess certain properties that depend on their microscopic organization. These properties (dependent on this microscopic organization) constitute the **chemical identity** of a substance. Any mixture can be separated into its **pure components** by physical processes. On the other hand, a **chemical process** is any transformation of matter that **does change** the chemical identity of a substance. Again, there are many simple examples, with combustion (rapid oxidation) and digestion being only two.

### Element

An **element** is a substance that **cannot** be separated into **simpler** substances by **chemical processes**. Common examples include gold, silver, copper, and iron. Currently (2014), 118 distinct elements have been identified, and each element in the periodic table of elements has been given a **unique** symbol. (We'll examine these symbols in more detail very shortly.) It is important to notice that this definition of an element does **not** depend on the details of the quantum theory that we met in the previous chapter. It is a definition that rests entirely on pre-twentieth century macroscopic observations yet is certainly **consistent** with the extensive developments of the twentieth century.

## Atom

An **atom** is the most basic or fundamental unit of an **element** that can participate in **chemical combinations** or **chemical transformations.** Again, it is important to note that this understanding of the atom is tied to the **macroscopic** understanding of an element and, hence, is a concept rooted in **macroscopic** observations. As defined here, the concept of an **atom** does not require a fully developed quantum description of the universe. The most that can be said is that elements are composed of basic units (objects beyond human observation) that possess the characteristic properties of each element. Like the idea of a substance, the concept of rests on an **implicit** model in which atoms possess certain properties that depend on their microscopic structure. However, this microscopic structure is never articulated or expanded upon in any detail; it is left as an **undefined** state that **implicitly accompanies macroscopic behavior.**

## Compound

A **compound** is a substance composed of **two or more** elements **chemically united in fixed** (that is, specific and well-defined) **proportions.** It is important to notice the requirement that **two or more** different elements must be present in a compound. This means that commonly occurring gases, such as $O_2$, $N_2$, and $H_2$, **are not compounds.** These diatomic species represent the **normal** form (pressure of 1 atm and room temperature) of the elements oxygen, nitrogen, and hydrogen.

## Molecule

A **molecule** is a distinct unit of a substance (that is, it possesses a **definite** or **constant** composition and distinct properties) that is composed of two or more atoms linked together chemically. Examples of molecules include oxygen ($O_2$), nitrogen ($N_2$), methane ($CH_4$), and methanol ($CH_3OH$). Again, the reader should note that while oxygen ($O_2$) is a **molecule,** it is **not a compound**!

In the previous two paragraphs we have begun using the **chemical symbols** of the elements from the periodic table in order to represent simple molecules. The complete chemical symbol for a substance that gives both the **number** and **identity** of each atom in a discrete unit of the substance is called a **molecular formula.** The symbols identify the elements, while the subscripts in each molecular formula indicate the **number of a particular type of atom.** For example, the molecular formula $CH_4$ uses the "C" of carbon and the "H" of hydrogen. The subscript "4" indicates that each molecule contains four hydrogen atoms. If only a single atom of an element is present, by convention (that is, by general agreement), we choose **not** to write a subscript of "1." Hence, the "C" without a subscript means that here is a single atom of carbon in the $CH_4$ molecule.

## Molecular Mass

The **molecular mass** of a substance is the sum of the atomic masses of all the atoms in the molecular formula of the substance. It is common to use the term "molecular weight" to designate the sum of the atomic masses of all the atoms in a molecule. However, the reader will recall that the terms "mass" and "weight" were clearly differentiated in chapter 1, noting that the term "weight" refers to a *force.* As a force, the "weight" of an object depends on its acceleration, in particular, its gravitational acceleration. Consequently, in this text we will always use the terms **atomic mass** and **molecular mass** to be clear that the quantity of matter being discussed is **independent** of any local gravitational acceleration. As we saw in chapter 2, the **molar mass** of any element in the periodic table is a sample **whose mass measured in grams** is **numerically** equal to the element's average atomic mass. A sample whose mass is the element's molar mass constitutes one mole of the element. In the same manner, the **molar mass of a molecule** is a sample **whose mass measured in grams** is **numerically** equal to the molecule's molecular mass. A sample whose mass is a molecule's molar mass constitutes one mole of the molecule.

Example 3.1: Calculate the molecular mass of water, $H_2O$.

Answer:

**Organize:** The given molecular formula is $H_2O$
Two atoms of hydrogen
One atom of oxygen

**Unknown:** Molecular mass

**Translate:** Mass of hydrogen is 1.008 u
Mass of oxygen is 16.00 u

**Solve:** Molecular mass = 2 × (Atomic mass H) + (Atomic mass O)
= 2 × (1.008 u) + (16.00 u)
= 18.02 u

(Note: Four significant figures)

## Molecular and Empirical Formulas

The concepts of a **molecule** and a **molecular mass** have their roots in the structural focus of nineteenth-century chemistry. Throughout this time period chemists attempted to link chemical reactivity and the elemental composition of the substances involved in chemical reactions. Early experimental studies of substances were able to establish the **mole ratios** of the elements present in a substance, but the concept of a molecular formula, which identifies **both** the number and type of each element present in a molecule, was not introduced until the latter half of the nineteenth century. Another term was also introduced in the latter half of the nineteenth century that distinguished the new structural idea of a molecular formula from the earlier experimental observations of mole ratios: **empirical formula**. The empirical formula provides the **simplest** ratio of the elements present in a compound. For example, he **molecular formula** of glucose is $C_6H_{12}O_6$, while the **empirical formula** is $CH_2O$. However, in some instances, the molecular formula and the empirical formula are identical: $H_2O$. Associated with the empirical formula is the concept of a **formula mass**: the **formula mass** of a substance is the sum of the atomic masses of all the atoms in the empirical formula of the substance.

## States of Matter

Many of the concepts and terms that are part of the modern chemistry are very old and originated from some of humankind's earliest observations of the world around us. This is particularly true of the term **states of matter**. In chapter 1, we took great care to understand the concept of matter (distinguishing the terms **matter** and **mass**) because the most elementary observation about the universe is that it is filled with matter. Further, a significantly important observation about the matter in our universe is that it appears in a variety of **different** forms, the three most elementary being given the names *solid*, *liquid*, and *gas*. Solids possess a definite volume and defined shape; in terms of atomic concepts, this means that solids possess a very stable arrangement of microscopic components whose positions in space do not easily change. In contrast, while liquids possess a definite volume, the do **not** possess a definite shape. The same volume of liquid easily "flows" into a variety of different shapes. While the density of a liquid can be very large (consider the liquid metal mercury) or very small (evidenced by less dense liquids "floating" on more dense liquids), in terms of atomic ideas, a liquid's microscopic components are not constrained to specific locations as in a solid. In colloquial language, they are "free" to move. However, these microscopic components are close enough to one another to be identified as a single object: a puddle, a glass of water, a lake or ocean. Finally, we briefly met gases in chapter 1 as we discussed some of the earliest quantitative chemical experiments and the resulting mathematical relationships that are known collectively as "the gas laws." The atmosphere of our planet is a gaseous mixture, consisting primarily of nitrogen and oxygen. Gases readily fill a container of any shape or volume, subject, as we have already seen, only to the constraints of temperature and pressure.

However, solids, liquids, and gases are **not** the only possible guises for matter in our universe. There are several more exotic states of matter that have been identified as a result of fascinating experimental and theoretical work in the twentieth and twenty-first centuries. However, for the purposes of this introduction to the chemical sciences, the three states—solids, liquids, and gases—provide a sufficiently rich range of diversity to understand vast stretches of the chemical landscape. There is, however, one important point to note. The earliest descriptions of these three states of matter were primarily **qualitative**. It was the effort to understand gases that led to a **quantitative** description and the realization that more abstract concepts such as temperature and pressure are critical to understanding a state of matter. We shall pursue this direction in a little more detail when we discuss the science of thermodynamics and its contribution to our understanding of chemistry.

## Chemical Reactions

We have been referring repeatedly the idea of a chemical transformation. We now formalize the definition of a **chemical reaction** as any transformation that changes the chemical identity of one or more substances participating in the transformation. In order to express clearly and succinctly the characteristics of a chemical reaction, a chemist uses a convenient symbolic representation (much like an algebraic equation) called a **chemical equation** to show what happens during a chemical reaction. A chemical equation employs the symbols of the elements from the periodic table to represent the atoms/molecules that participate in the chemical reaction. By **convention**, the reacting species (known as the **reactants** or **reagents**) are written on the left-hand side of the expression, while the results of the chemical reaction (known as the **products**) are written on the right-hand side of the expression. For example, the metabolically significant reaction between glucose and oxygen producing carbon dioxide and water is represented by the following chemical equation:

$$C_6H_{12}O_6 + 6O_2 \rightarrow 6CO_2 + 6H_2O$$

In this example, the reactants ($C_6H_{12}O_6$ and $O_2$) are written on the left-hand side of the arrow (which plays the role of an "equal sign" in the equation and can be read as "yields"), while the products ($CO_2$ and $H_2O$) are written on the right-hand side of the arrow. The numbers written to the left of each molecular formula are called the **stoichiometric coefficients** (from the Greek words *stoicheion*, meaning "element," and *metron*, meaning "measure"); they represent the relative quantities (or numbers) of reactant and product atoms/molecules that participate in the specific chemical reaction. As in the case of the subscripts of a molecular formula (or empirical formula), if a "1" occurs as a stoichiometric coefficient, it is **not** written explicitly. Consequently, in the above example the number 1 is omitted before $C_6H_{12}O_6$.

The stoichiometric coefficients can be readily interpreted two ways. First, from an atomic/molecular point of view, they specify the **exact numbers** of reactant and product atoms/molecules involved in a chemical reaction. In the above example, one **molecule** of $C_6H_{12}O_6$ reacts with six **molecules** of $O_2$ to yield six **molecules** of $CO_2$ and six **molecules** of $H_2O$. This is a **microscopic** interpretation. Second, as a result of the definition of the mole given in chapter 2, the equation can also be interpreted as one **mole** of $C_6H_{12}O_6$ reacts with six **moles** of $O_2$ to yield six **moles** of $CO_2$ and six **moles** of $H_2O$. Because the mole represents a macroscopic quantity, the second point of view is a **macroscopic** interpretation. Further, because the concept of the mole is directly linked to the SI unit of mass, the macroscopic interpretation implies a relationship among the **reacting masses.** In fact, long before the emergence of chemistry's focus on structures and their subsequent reactivity, the earliest chemical experiments were entirely concerned with the possible reactions among various **masses** of reactants that produced specific **masses** of each product.

Finally, the reader should pay particular attention to the fact that a chemical equation is indeed an equation. That is, the **total number** and **kind** of atoms on the left side of the equation **are exactly equal** to the **number** and **kind** of atoms on the right side of the equation. We often call such an equation a **balanced** equation to emphasize this equality of number and kind of atoms. Consistent with Dalton's hypothesis, a **chemical reaction** neither **creates** nor **destroys** atoms (see hypothesis 3 in chapter 2). This is simply a statement of the principle of the **conservation of mass** (which we have already extended in chapter 2 to the principle of the **conservation of mass-energy** using Einstein's 1905 equation). To the limit that chemists can routinely measure mass in the laboratory, the principle of the conservation of mass has never failed. Further, because chemical equations are first and foremost equations, any **multiple** of a chemical equation represents the **same** reaction. That is, multiplying the stoichiometric coefficients on **both** sides of a balanced equation by the **same** factor **does not change the equation.** The following two examples express the **same** reaction:

$$C_6H_{12}O_6 + 6O_2 \rightarrow 6CO_2 + 6H_2O$$

$$3C_6H_{12}O_6 + 18O_2 \rightarrow 18CO_2 + 18H_2O$$

By convention, the form of the chemical equation in which the stoichiometric coefficients are the **smallest integers** needed to express the quantities (numbers) of the reactant atoms/molecules and of the product atoms/molecules is the preferred chemical equation. It is very important to distinguish between the **stoichiometric coefficients** in a chemical equation and the **subscripts** of the molecular species in the equation. Notice very carefully that multiplying both sides of the equation by the same factor **does not change the subscripts in** any manner. The subscripts determine the reacting substances and the product substances produced by the reaction. **If** the subscripts **were changed in any manner**, an entirely different reaction would be represented by the resulting chemical equation.

## Solutions

We have already made the distinction between **pure substances** and **mixtures**. While a pure substance is either a single element or a single compound, mixtures come in two varieties. A **homogeneous** mixture is a mixture that is, by some measure, **uniform**; that is, when the mixture is observed, it appears to be *the same everywhere*. A convenient example is the mixing of sucrose in water. The mixture of the two substances (water and sucrose) looks completely uniform. In contrast, a **heterogeneous** mixture is a mixture whose composition is **not uniform**. For example, a mixture of salt and pepper is heterogeneous. The reader should note that the terms "homogeneous" and "heterogeneous" are old terms that entered the chemical vocabulary very early in the history of chemistry, probably sometime after the middle of the seventeenth century. The terms are **scale dependent**. What does this mean? It means that a physical mixture can change its appearance from being "homogeneous" to being "heterogeneous" if a different instrument is used to view the mixture. A mixture that may appear very uniform to the unaided eye may appear very non-uniform when viewed through a powerful microscope, that is, at a very different length scale.

We will now focus our attention on a very special type mixture called a **solution**. A **solution** is a homogeneous mixture of two or more substances. While the simple English word "solution" connotes the idea that, in some sense, the solution involves a *liquid* (such as water at room temperature), this is not required. Our atmosphere is a solution of *gases*, while, at room temperature, bronze is a *solid* solution of copper and tin and brass is a *solid* solution of copper and zinc. At this point we are still using the terms *solid*, *liquid*, and *gas* in the very qualitative fashion used earlier; it simply suffices to note that the word "solution" covers a very wide range of phenomena in our universe. Because a solution is a mixture, it is composed of two or more distinct substances; the substance present in the largest amount is called the **solvent**, while the substances present in the smaller amounts are called the **solutes**. In the case of a **binary solution** there are only **two** components; the one present in the larger amount is called the **solvent,** and the one present in the smaller amount is called the **solute**. In the case of a binary solution, if the two substances are present in **equal amounts**, either may be identified as the solvent, with the second one being the solute. The terms *solvent* and *solute* are clearly only **qualitative** descriptions of a solution and do not attempt to provide a **quantitative** understanding of the solution.

There are two additional expressions that are often used to describe physical systems that appear to be similar to a solution but are, in fact, distinctly different: **colloids** and **suspensions**. These systems are like solutions because both colloids and suspensions are mixtures, but they exhibit distinctly different characteristics. A **colloid** is a mixture in which **large** particles ("large" when compared to molecular dimensions, ranging in size from approximately 1 nanometer to 1,000 nanometers in diameter) are distributed throughout the solvent. The particles are so small that they are invisible to the common optical microscope. However, a colloid exhibits the qualitative light-scattering effect known as the Tyndall effect. Solutions **do not** exhibit the Tyndall effect. On the other hand, while a **suspension** is also a mixture, it is a heterogeneous mixture; both solutions and colloids are homogeneous mixtures. Because a suspension is a heterogeneous mixture of relatively large particles that are visible with a common optical microscope (the **minimum** size is approximately 1 micrometer in diameter, which is equal to the **largest** particle present in a colloid), the solvent and the particles of a suspension will separate into two distinct regions over time. Such a separation does **not** occur in either a solution or a colloid.

The development of a more quantitative understanding of solutions begins by adopting the elementary microscopic viewpoint with a focus on structural explanations that were emerging as the nineteenth century drew to a close. **Solvation** can be understood as the process in which a solute particle is surrounded by solvent molecules arranged in a specific manner. While solvation is seen as a dynamic event, the central point of this initial understanding of the formation of a solution is the **specific arrangement** of the solvent molecules. It would take sustained effort throughout the twentieth and into the twenty-first centuries for chemists to formulate a dynamical understanding of the solvation process. However, imagining the specific structural arrangement that produces a solution proved to be a much simpler task than understanding the complex chemical dynamics that take place once a solution has formed! If the solvent is water, the solvation process is called **hydration**, and the resulting solution is called an **aqueous solution**.

While a quantitative understanding of solution chemical dynamics lay in the future, early chemists did have access to certain macroscopic properties with which to begin a quantitative understanding of solutions. **Solubility** is defined as the maximum number of **grams** of a **solute** present in a given amount of a solvent at a specified temperature. Measured in units of grams per cubic centimeter or grams per milliliter (in general, mass per volume), the solubility is clearly a quantitative description of one characteristic of a solute that could be macroscopically measured.

Example 3.2: If a maximum of 1.252 kilograms of glucose dissolve in 1.375 liter of water while the temperature is held constant at 327.15 K, calculate the solubility of glucose in units of grams per milliliter.

Answer:

**Organize:**
1.252 kg of glucose
1.375 L of water
Temperature kept constant

**Unknown:**
Solubility in units of grams per milliliter

**Translate:**
Convert 1.375 L to mL
1.375 L = 1375 mL
Convert 1.252 kg to grams
1.252 kg = 1252 g

**Solve:**
Solubility = (Mass of glucose) ÷ (Volume of water)
= (1252 g) ÷ (1375 mL)
= $9.105 \times 10^{-1}$ g $mL^{-1}$
(Note: Four significant figures)

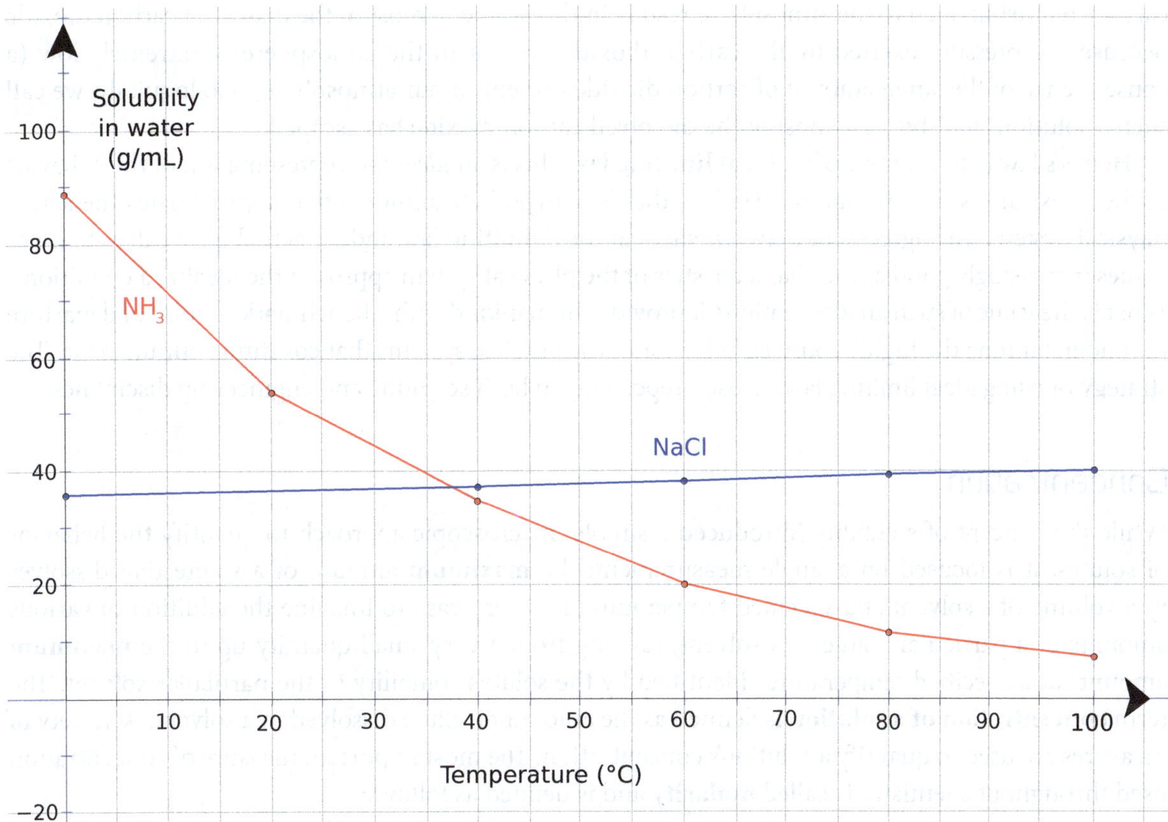


Figure 3.1: Solubility of a Solid (NaCl) and a Gas ($NH_3$) as a Function of Temperature

It is important to pay particular attention to the role of temperature. It is an everyday experience to discover that, as a cup of hot chocolate cools, a layer of chocolate forms on the bottom of the cup. That is, the solubility of the chocolate changes as the temperature of the hot chocolate changes. In general, the solubility of most **solids** *increases* with **increasing** temperature. In contrast, the solubility of **gases dissolved in water** *decreases* with **increasing** temperature. While this statement reflects the general *trends*, there are exceptions among solids.

But temperature is not the only macroscopic property that affects the solubility of a solute in a given solvent. An observation (called **Henry's law**) by an English chemist, William Henry, early in the nineteenth century describes the effect of pressure on the solubility of gases in liquids. The most basic statement of Henry's law (which ignores the many complexities of a liquid solution) notes that the solubility of a gas in a liquid (forming a solution in which the gas is the solute) is **directly proportional** to the **pressure exerted by that gas** on the liquid. Several simple examples serve to demonstrate the practical consequences of this observation. Open a completely sealed container of a carbonated drink (a solution that includes carbon dioxide, $CO_2$, as one of the solutes). The audible hiss and bubbling of the carbonated drink accompanying the removal of the cap indicates the escape of carbon dioxide from the solution as the pressure over the solution decreases with the removal of the cap. Similarly, leaving a carbonated drink open to the atmosphere results in the escape of most of the dissolved carbon dioxide because the pressure exerted by the carbon dioxide present in the atmosphere is extremely low (a consequence of the small amount of carbon dioxide present in our atmosphere). Colloquially, we call such a solution "flat" because most of the dissolved carbon dioxide has escaped.

Henry's law is an example of an **ideal limiting law**; this is an *idealized* representation of the behavior of the physical system. It is an idealization, that is, a hypothetical model that approximates the actual physical system. The agreement between such an ideal limiting law and an actual physical system becomes increasingly good as the characteristics of the physical system approach the idealized conditions. What is the value of such an idealization? It provides humankind with a benchmark, a point of departure for understanding the highly complex behavior of the physical systems that constitute our universe. The strategy of using ideal limiting laws is used repeatedly in both scientific and engineering disciplines.

## Concentration

While the concept of solubility introduced a simple, macroscopic approach to quantify the behavior of solutes, it is focused on a single measurement: the **maximum** amount of a solute that dissolves in a volume of a solvent at a specified temperature. It is very easy to imagine the addition of various amounts of a particular solute to a solvent, ranging from a very small quantity up to the **maximum** amount (at a specified temperature) identified by the solute's solubility in the particular solvent. The term **concentration of a solution** is defined as the amount of solute dissolved in a solvent. A variety of measures are used to quantify a solution's concentration. The most important measure of concentration used throughout chemistry is called **molarity** and is defined as follows:

$$\text{Molarity} = (\text{Moles of solute}) \div (\text{Liters of solution}). \tag{3.1}$$

**Molarity** is symbolized either by ***M*** or by placing square brackets ([ ]) around a chemical formula. Each part of the definition is crucial: note that the definition does **not** specify simply "moles"; it requires ***moles of solute***. Similarly, the volume specified in the definition is ***liters of solution***. Consequently, the statement that molarity means that "moles per liter" is **not** correct! Occasionally, the **millimole** is used, particularly in medical applications. Recall that a millimole is simply one-thousandth of a mole.

Example 3.3: Suppose that 1.775 moles of $C_{12}H_{22}O_{11}$ (sucrose) is dissolved in a sufficient quantity of water to produce exactly 750.0 mL of a solution. What is the molarity of this solution?

Answer:

**Organize:** 1.775 moles of sucrose
Solution volume is 750.0 mL

**Unknown:** Because sucrose is the solute, the question asks you to calculate the concentration of the sucrose in the solution. The measure of the sucrose concentration is the molarity of the sucrose.

**Translate:** Moles of solute = 1.775 moles
Convert 750.0 mL to liters
750.0 ml = 0.7500 L

**Solve:** Molarity = (Moles of solute) ÷ (Liters of solution)
= (1.775 moles) ÷ (0.7500 L)
= 2.367 mol $L^{-1}$
(Note: Four significant figures)

Example 3.4: Suppose that 325 grams of $C_2H_5OH$ (ethanol) is dissolved in a sufficient quantity of water to produce exactly 2250.0 mL of a solution. What is the molarity of this solution?

Answer: **Organize:** 325 grams of ethanol
Solution volume is 2250.0 mL

**Unknown:** Because ethanol is the solute, the question asks you to calculate the concentration of the ethanol in the solution. The measure of the ethanol concentration is the molarity of the ethanol.

**Translate:** Calculate the molar mass of ethanol.
Molar mass = 2 × (12.01 g $mol^{-1}$) +
6 × (1.008 g $mol^{-1}$) +
1 × (16.00 g $mol^{-1}$)
= 46.068 g $mol^{-1}$
Calculate moles of ethanol
Mol (ethanol) = 325 g ÷ 46.068 g $mol^{-1}$
= 7.055 mol
Convert 2250.0 mL to liters
2250.0 ml = 2.2500 L

**Solve:** Molarity = (Moles of solute) ÷ (Liters of solution)
= (7.055 moles) ÷ (2.2500 L)
= 3.14 mol $L^{-1}$
(Note: Three significant figures)

While molarity is the most commonly used measure of concentration in chemistry, it is **not** the only way to measure the concentration of solutions. Concentrations are often measured by identifying the ratios of the mass or volume of the solute to the total mass of the solution, and reporting the ratio as a percent. **Percent by mass** is defined as follows:

$$\text{Solute \% (m/m)} = [(\text{Mass of solute}) \div (\text{Mass of solution})] \times 100\%. \qquad (3.2a)$$

But, because (Mass of solution) = (Mass of solute) + (Mass of solvent), we can write

$$\text{Solute \%} = [(\text{Mass of solute}) \div \{(\text{Mass of solute}) + (\text{Mass of solvent})\}] \times 100\%. \qquad (3.2b)$$

In the same way, **percent by volume** is defined as follows:

$$\text{Solute \% (vol/vol)} = [\,(\text{Volume of solute}) \div (\text{Volume of solution})\,] \times 100\%. \qquad (3.3)$$

Both the percent by mass and the percent by volume measures of concentration have the advantage that the (Mass ÷ Mass) ratio and the (Volume ÷ Volume) ratio are **dimensionless** (that is, the units of mass or the units of volume should exactly cancel).

Example 3.5: Household vinegar is a solution composed of water and acetic acid ($CH_3COOH$). Suppose that 1.750 L of vinegar contain 70.00 mL of acetic acid. What is the percent by volume of the acetic acid in the vinegar?

Answer:

**Organize:** Total volume of vinegar is 1.750 L
Volume of acetic acid is 70.00 mL

**Unknown:** Percent by volume of acetic acid

**Translate:** The percent by volume requires knowing the volume of the solute and the total volume of the solution.
Convert volume of solution to millilters.
$1.750\text{ L} = 1.750 \times 10^3\text{ mL}$

**Solve:** Acetic acid % = [ (Volume of solute) ÷ (Volume of solution)] × 100%
$= [(70.00\text{ mL}) \div (1.750 \times 10^3\text{ mL})] \times 100\%$
$= 4.000\%$
(Note: Four significant figures)

Example 3.6: Suppose that the concentration of glucose in a glucose and water solution is 4.75% by mass (percent by mass). If the mass of the glucose in the solution is 435 g, what is the total mass of the water in the solution?

Answer:

**Organize:** Concentration of glucose is 4.75% by mass
The mass of the glucose in the solution is 435g

**Unknown:** Find the mass of the water. The water is the solvent.

**Translate:** Recall that
(Mass of solution) =
[ (Mass of solute) + (Mass of solvent) ]
Use equation (3.2b).

**Solve:** Glucose % = [(Mass of solute) ÷ {(Mass of solute) + (Mass of solvent)}] × 100%
4.75 % = [(435 g) ÷ {(435 g) + (Mass of solvent)}] × 100%
Solve for the mass of the solvent.
Mass of solvent = $8.72 \times 10^3$ g
(Note: Three significant figures)

## Dilution

The concepts of solubility and concentration provide convenient categories (depending on physical properties—masses, volumes, and temperature—that can be measured macroscopically) with which to understand significant characteristics of solutions. While solubility identifies a single, temperature-dependent maximum, the concept of concentration provides a route by which we can *manipulate* the character of a solution. Given access to quantities of solute and solvent, we can *produce* solutions with a wide variety of concentrations. But how is this *manipulation* accomplished?

**Dilution** is a procedure for preparing a **less concentrated** solution from a **more concentrated** solution. This means that, given a concentrated starting solution, we can produce any variety of solutions of **lesser concentrations**. We are **not limited** to beginning only with **pure** solutes and **pure** solvents; we can begin with simply a *concentrated solution*! The crucial activity in producing a less concentrated solution form a more concentrated solution is the addition of **only pure solvent** to the starting solution; **no additional solute** is added. This means that the number of moles of solute (the quantity of solute) **does not change**. This is exactly like making lemonade from concentrate. The directions are to add **only** water (the solvent) to the concentrated lemonade mix. The realization that the dilution procedure requires only solvent to be added to the starting solution provides a simple mathematical relationship that allows us to *calculate* the volume of additional solvent needed to achieve a specific **less concentrated** solution. This relationship is as follows:

$$M_1 \times vol_1(\text{L}) = M_2 \times vol_2(\text{L}). \tag{3.4}$$

In equation 3.4, $M_1$ is the molarity of the starting solution, $M_2$ is the molarity of the **less concentrated** solution, and $vol_1$ and $vol_2$ are the volumes of the respective solutions. The units for the volumes specified in equation 3.4 are "liters"; this is instructive because **when** the volumes are reported in liters, both the left-hand and the right-hand sides of the equation have units of "moles," emphasizing that the number of moles **before** dilution must equal the number of moles **after** dilution. In fact, the units used for the volumes in equation 3.4 may be **any unit of volume** as long as $\text{vol}_1$ and $\text{vol}_2$ are reported in the **same units**. However, pay close attention to the fact that **if** the volumes are not reported in liters, the quantities ($M_1 \times vol_1$) and ($M_2 \times vol_2$) **do not** represent the number of moles of solute in each solution! Consider the following examples.

Example 3.7: Suppose that a student has 1000.0 mL of a starting hydrochloric acid (HCl) solution (where the solvent is water) with a concentration of 0.750 M. What volume of this starting solution must be used to prepare 500.0 mL of a solution with an acid concentration of 0.357 M?

Answer: **Organize:**

Starting solution has a concentration of 0.750 M.

A total volume of 1000.0 mL of the starting solution is available to the student.

The student is asked to make 500.0 mL of a new (less concentrated) solution.

The concentration of the new, less concentrated solution is 0.357 M.

**Unknown:** Find the amount (volume) of the starting solution that must be used by the student.
This is $vol_1$. Note that $vol_1$ is **not** the 1000.0 mL!

**Translate:** Use equation 3.4
$M_1 \times vol_1(L) = M_2 \times vol_2(L)$
In this example, convert all the required volumes to liters.
500.0 ml = 0.5000 L
$M_1 = 0.750$ M
$M_2 = 0.357$ M
$vol_2 = 0.5000$ L

**Solve:** $M_1 \times vol_1 = M_2 \times vol_2$
Hence,
$vol_1 = (M_2 \times vol_2) \div M_1$
$vol_1 = (0.357\ M \times 0.5000\ \text{L}) \div 0.750$ M
$vol_1 = 0.238$ L = 238 mL
This means that 238 mL of the starting solution must be combined with 262 ml of pure water (because 238 + 262 = 500) to make 500.0 mL of a solution with a concentration of 0.357 M.
(Note: Three significant figures)

Example 3.8: Suppose that 250.0 mL of a starting nitric acid ($HNO_3$) solution (where the solvent is water) with a concentration of 0.150 M is used by a student to make 630.0 mL of a less concentrated solution. What is the concentration (molarity) of the less concentrated solution?

Answer:

**Organize:** Starting solution has a concentration of 0.150 M.
The student used 250.0 mL of the starting solution.
A total volume of 630.0 mL of the less concentrated solution was made by the student.

**Unknown:** Find the concentration of the less concentrated solution.

**Translate:** Use equation 3.4.
$M_1 \times vol_1(L) = M_2 \times vol_2(L)$
In this example, volume units remain milliliters.
Note that the units of $vol_1$ and $vol_2$ are the **same**.
$M_1 = 0.150$ M
$vol_1 = 250.0$ mL
$vol_2 = 630.0$ mL

**Solve:** $M_1 \times vol_1 = M_2 \times vol_2$
Hence,
$M_2 = (M_1 \times vol_1) \div vol_2$

$M_2$ = (0.150 M × 250.0 mL) ÷ 630.0 mL
$M_2$ = 5.95 × $10^{-2}$ M
(Note: Three significant figures)

Example 3.9: Suppose that 250.0 mL of a starting nitric acid ($HNO_3$) solution (where the solvent is water) with a concentration of 0.150 M is used by a student to make a less concentrated solution with a concentration of 5.95 × $10^{-2}$ M. How much water (in mL) did the student add to the starting volume of 250.0 mL?

Answer: **Organize:** Starting solution has a concentration of 0.150 M.
The student used 250.0 mL of the starting solution.
The concentration of the less concentrated solution is 5.95 × $10^{-2}$ M.

**Unknown:** Find the **additional amount** of solvent added by the student.

**Translate:** Use equation 3.4.
$M_1 \times vol_1$(L) = $M_2 \times vol_2$(L)
In this example, volume units remain milliliters. Note that the units of $vol_1$ and $vol_2$ are the **same**.
$M_1$ = 0.150 M
$M_2$ = 5.95 × $10^{-2}$ M
$vol_1$ = 250.0 mL

**Solve:** $M_1 \times vol_1 = M_2 \times vol_2$
Hence,
$vol_2 = (M_1 \times vol_1) \div M_2$
$vol_2$ = (0.150 M × 250.0 mL) ÷ 5.95 × $10^{-2}$ M
$vol_2$ = 630 mL
(Note: Three significant figures)
Water added = 630 mL – 250 mL
Water added = 380 mL

## Solution Categories

The introduction of the concepts of solubility and solution concentration has allowed us to investigate quantitatively the characteristic behavior of solutions. In fact, as a result of these concepts, not only can we *describe* solutions; we also are able to *manipulate* directly the makeup of solutions using the procedure of dilution. However, while the thrust of every science is the movement from a qualitative understanding of phenomena to a quantitative understanding, we should not overlook important insights that do occur at the qualitative level. This is particularly true in the case of solutions. Every solution can be **qualitatively** assigned to one of three categories: **unsaturated**, **saturated**, and **supersaturated**. This tripartite categorization does **not** require any numerical quantitation; it depends on only a very qualitative assignment simply using the observed properties of solutions.

But this apparently simple observational division of solutions into three groups fundamentally rests on a much more sophisticated principle: **equilibrium**. While we will save a detailed investigation of this powerful microscopic idea until we discuss the behavior of acids and bases, it suffices to note that at equilibrium, the concentrations of reactants and products **do not change** with time. This definition gives the false impression that the state of equilibrium is a **static** state, a chemical condition in which nothing is changing. This misunderstanding ignores the discovery made by Claude Berthollet in the early years of the nineteenth century that some chemical reactions are **reversible**; they proceed in the forward direction, from reactants to products, **and** in the reverse direction, from products back to reactants. As we will shortly see, by the middle of the nineteenth century two chemists, Cato Guldberg and Peter Waage, proposed a quantitative description of the reversibility originally observed by Berthollet. Consequently, equilibrium is a **dynamic** state in which there is a constant interchange between reactants and products, even though the concentrations of reactants and products, after the chemical system reaches equilibrium, ***do not change***.

What do the three categories of solution mean? An **unsaturated** solution is a solution containing **less than the maximum** amount of solute in a specified amount of solvent at a defined temperature. We can easily determine this maximum by simply observing the solution's behavior; we are not even required to make a quantitative determination! A **saturated** solution is a solution containing the maximum amount of solute in a specified amount of solvent at a defined temperature. This is a state of *stable equilibrium*, which means that the dynamic equilibrium state can be **disturbed by small changes** and the system will **return** to the stable state. Finally, a **supersaturated solution** is an unstable solution that contains **more solute than is present in a saturated solution** made from a specified amount of solvent at a defined temperature. This a state of *unstable equilibrium*, which means that if the dynamic equilibrium state is **disturbed by small changes**, the system will **never return** to the state that existed before the disturbance. An effective mental image of an *unstable equilibrium* is to imagine a pencil standing precariously balanced on its sharpened point. The slightest touch will cause the pencil to fall over, never to return to its previously balanced state.

# Chapter 3 Exercises

1. What is a substance?
2. What is a mixture?
3. Distinguish a chemical process from a physical process.
4. Identify each of the following as either a chemical process or a physical process:
   a. Melting water
   b. Burning a wooden log
   c. Mixing salt and pepper
   d. Digesting food
   e. Boiling water
   f. Rusting iron
5. How many grams of each of the following compounds constitute one mole of the compound?
   a. $CO_2$ (carbon dioxide)
   b. $CH_4$ (methane)
   c. $NH_3$ (ammonia)
   d. $CH_3OH$ (methanol)
6. Give the empirical formula for the following compounds:
   a. $H_2O_2$ (hydrogen peroxide)
   b. $N_2O_4$ (nitrogen tetroxide)
   c. HCl (hydrogen chloride)
   d. $NH_3$ (ammonia)
   e. $C_6H_{12}O_6$ (glucose)
7. List the three elementary states of matter.
8. Which characteristic of a gas distinguishes it from a liquid?
9. Define the term "chemical reaction."
10. What is a chemical equation?
11. What are the stoichiometric coefficients?
12. Give two interpretations of the stoichiometric coefficients.
13. What is a solution?
14. Define the two components of a solution.
15. Give an example of each of the following:
    a. A solid solution
    b. A gaseous solution
    c. A solution composed of two liquids
16. What are the differences that distinguish a solution, a colloid, and a suspension?
17. What is solvation?
18. What is solubility?
19. Suppose that the solubility of sucrose in water at 298.15 K is $2.000 \times 10^3$ g $L^{-1}$. What is the maximum amount of sucrose (measured in grams) that will dissolve in 335.0 mL of water?

20. Suppose that the maximum amount of NaCl (common table salt) that will dissolve in a volume of water at 298.15 K is $1.3875 \times 10^2$ g. What is the volume of water (measured in mL) if the solubility of NaCl in water at that temperature is 395 g $L^{-1}$?
21. What is meant by "the concentration of a solution"?
22. What is molarity?
23. Calculate the molarity of 526 g KCl in a water solution whose total volume is 1.500 L. The density of water is 1.000 g $mL^{-1}$. The molar mass of K is 39.10 g $mol^{-1}$, and the molar mass of Cl is 35.45 g $mol^{-1}$.
24. What is dilution?
25. Suppose that 250.0 mL of an acid solution with a concentration of 6.250 M is diluted to make an acid solution whose concentration is 3.450 M. What is the volume of the new solution?
26. Suppose that a solution with a density of 1.175 g $mL^{-1}$ at 25.00°C and a total volume of 3750 mL was made by mixing a sample of $NH_4Cl$ with 3690 mL of water. What is the molarity of the solution? The molar mass of H is 1.008 g $mol^{-1}$, the molar mass of N is 14.01 g $mol^{-1}$, and the molar mass Cl is 35.45 g $mol^{-1}$.
27. Suppose that the concentration of sucrose in a sucrose and water solution is 8.75% by mass (percent by mass). If the mass of the sucrose in the solution is 335 g, what is the total mass of the water in the solution?
28. White vinegar is a solution composed of water and acetic acid ($CH_3COOH$). Suppose that a student has 2.250 L of vinegar whose concentration is 4% by volume. What volume (measured in mL) of acetic acid is present in the solution?
29. Define the following terms:
    a. Saturated solution
    b. Unsaturated solution
    c. Supersaturated solution
30. What is a reversible reaction?
31. Define the term "equilibrium."

# CHAPTER FOUR

## The Classical Chemical Paradigm

### Part II (Pre-Twentieth Century)

## Periodic Table

We continue now with our focus on those specific chemical concepts that are based on macroscopic observations made before the principles of quantum theory came to dominate chemical thinking. Throughout the nineteenth century, as the number of distinct elements with well-characterized properties continued to grow, there were a number of separate attempts to organize the elements into definite patterns that reflected either the experimentally observed properties or underlying chemical principles responsible for their reactivity. Early in the nineteenth century it was observed that the elements seemed to occur in groups of three, with common chemical properties identifying each group. Interestingly, the atomic masses of the middle member of each group of three appeared to be approximately equal to the mean value of the atomic masses of the other two members of the group. This identification of groups of elements based on the averaging of atomic masses became a central focus of attention, so much so that by midcentury this behavior came to be known as the **law of triads**. However, several groups curiously seemed to contain four elements, and at least one group had five. Because the most physically significant property that differentiates each element is its atomic mass, arranging the elements in a monotonic sequence of ever-increasing atomic mass became another obvious starting point around which to organize the elements. Several such arrangements appeared by the early 1860s, one proposed by the French geologist Alexandre-Émile Béguyer de Chancourtois, and a second by the English chemist William Odling. While neither of these proposals made a significant impact on the science of chemistry, it is interesting to note that both men were aware that the chemical properties of the elements recurred cyclically; when arranged by increasing atomic mass, the chemical properties seemed to exhibit a periodicity. Odling commented on this behavior, but did not capitalize on its significance; the unclear explanation given by the geologist, de Chancourtois, was ignored by the chemists of the time.

In addition to the specific chemical properties of the elements (the most important being the atomic mass of each element), chemists had noted, beginning as early as the late eighteenth century, that each element seemed to exhibit a characteristic combining ability (by 1884 this was called the "combining power of an element") in its reactions. This recognition was embodied in the theory of **chemical valence** that arose to explain the reactivity of the elements. In recognition of this organizing principle, attempts (including a proposal by Odling in 1870) were made to arrange the elements using their assigned valences, again hoping to recognize a unifying order to the growing list of elements.

During the middle 1860s, an English chemist, John Newlands, noted in a series of papers that the physical and chemical properties of the elements seemed to exhibit a **periodicity** if the elements were arranged by increasing atomic mass. However, unlike the earlier law of triads, Newland identified a cyclic recurrence of physical and chemical properties in groups of seven, with the eighth element exhibiting properties similar to those of the first element of the previous group of seven. In an analogy to the periodicity of Western music, Newlands proposed a **law of octaves**. (At this point in time, none of the noble gases had been identified. With their discovery, the cyclic recurrence of the physical and chemical properties of the elements occurred in groups of *eight*.) Unfortunately, he was severely ridiculed by his contemporaries, and because the Chemical Society (founded as the Chemical Society of London in 1841) had decided to avoid publishing purely theoretical proposals, it declined to publish his idea. Newlands did produce a draft of his table of elements and used it to predict the element **germanium**, an element not discovered experimentally until 1886.

It was the nearly simultaneous and independent proposals of Dmitri Mendeleev (1869) and Julius Lothar Meyer (1870) that became the basis of the modern periodic table of elements. Both Lothar Meyer and Mendeleev, already established professors of chemistry in Germany and Russia, respectively, ordered the elements by increasing atomic mass, the key physical parameter utilized in the earlier attempts to organize the elements. Because Lothar Meyer and Mendeleev had attended the first international chemistry conference, the Karlsruhe Congress of 1860, they benefited greatly from the pivotal efforts of the Italian chemist Stanislao Cannizzaro to reform the definition of atomic and molecular masses. While the Congress adopted no formal resolution addressing the perplexing issue of the masses assigned to chemical species, the paper distributed by Cannizzaro immediately revolutionized and systematized the assignment of mass values in chemistry. This in turn significantly aided Lothar Meyer and Mendeleev. Additionally, both men possessed a deep knowledge of the physical and chemical properties of the elements, which enabled them to recognize the periodicity of these properties. While Lothar Meyer focused more heavily on *physical* properties (for example, atomic volumes, densities, melting points), Mendeleev demonstrated a virtuoso command of the *chemical* properties of the elements and used this knowledge to make crucial choices in his arrangement. Both Lothar Meyer and Mendeleev left gaps in their tables, recognizing the possibility of undiscovered elements; Mendeleev, however, made bolder predictions of the characteristics of these missing elements.

This brief overview of the development of the periodic table clearly indicates that there were many co-discovers of the periodicity principle in the short span of only a decade. Mendeleev's system had the greatest impact on the chemical sciences; it was the most complete model, and he worked both longer and harder for its general acceptance, publicly demonstrating its ability to predict undiscovered elements. However, despite the rapid acceptance of Mendeleev's table as the nineteenth century drew to a close, it was not until the 1920s, following the work of Moseley and his successors, that chemists recognized the atomic number of an element as the central organizing factor of chemical periodicity. The modern periodic table is arranged by atomic number, beginning with hydrogen (atomic number = 1) and monotonically increasing in steps of one unit to ununoctium (atomic number = 118).

The horizontal rows of the periodic table are called **periods**; the vertical columns are identified as **groups** or **families**. Elements belonging to the **same** group exhibit very similar chemical properties. From a broader perspective, the elements can be segregated into three general categories: (1) the **metals**, which are good conductors of both heat and electricity; (2) the **nonmetals**, which are poor conductors of both heat and electricity; and (3) the **metalloids**, which exhibit properties that are intermediate between the metals and the nonmetals. With the exception of copernicium (element 112), the chemical properties of the elements beyond hassium (element 108) have yet to be determined. Figure 4.1 displays the medium-long form of the periodic table, in which two rows of fourteen elements each are displayed below the main body of the table, while figure 4.2 displays the long form of the periodic table (elemental symbols are not included).

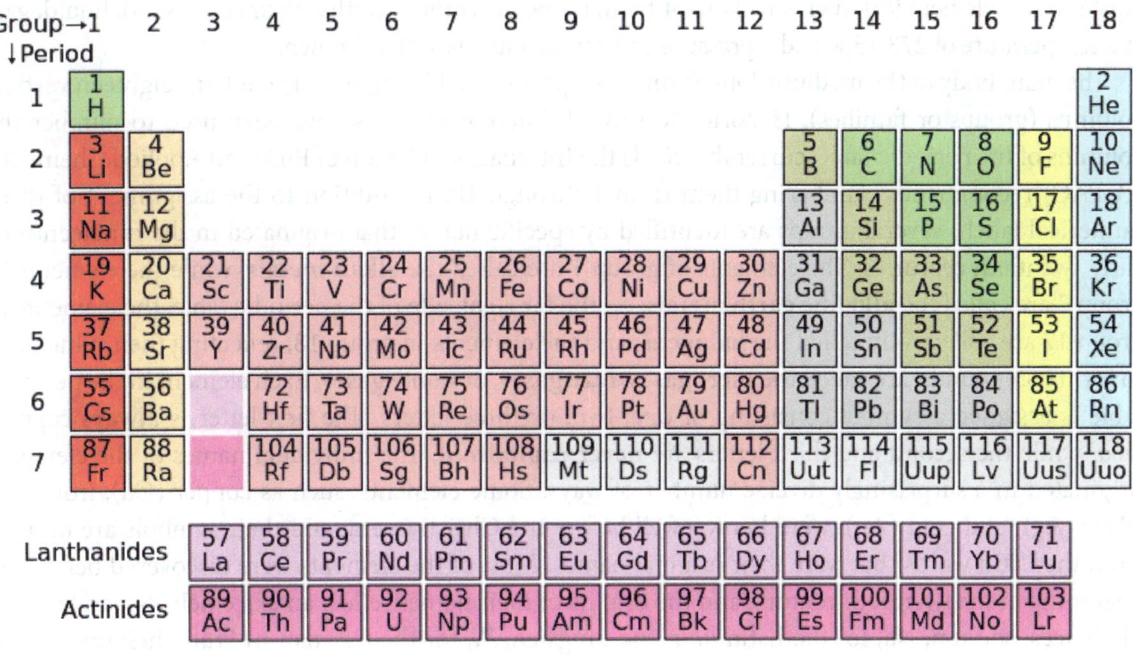


Figure 4.1: Medium-Long Form of the Periodic Table

Beginning in the leftmost column of any period, there is a gradual transition from the metals, through the metalloids, and finally, to the nonmetals. As the figures indicate, the metals constitute the majority (slightly less than 78%) of the elements in the table, while the nonmetals make up only approximately 16% of the elements and the metalloids represent the remaining approximately 6% of the elements. At a temperature of 273.15 K and a pressure of 1 atmosphere, the metals vary from soft and rather dull solids

Figure 4.2: Long Form of the Periodic Table

(sodium and potassium, for example) to hard and lustrous substances, such as chromium, gold, and silver. Under the same conditions, the nonmetals are, in general, either solids or gases. At these conditions, the liquid metal mercury (element 80) and the liquid nonmetal bromine (element 35) are clear exceptions; in the cases of astatine (element 85), francium (element 87), and all the elements following einsteinium (element 99), such small quantities have been synthesized that their state (solid, liquid, gas) at a temperature of 273.15 K and a pressure of 1 atmosphere is still unknown.

The main body of the medium-long form of the periodic table (figure 4.1) contains eighteen vertical columns (groups or families). Historically, several different schemes have been used to number the columns of the periodic table; currently (2014) the International Union of Pure and Applied Chemistry (IUPAC) recommends numbering them from 1 through 18. In addition to the assignment of these numerical labels, several groups are identified by specific names that originated in the nineteenth or early twentieth centuries. The elements of group 1 are called the **alkali metals**, while the elements in group 2 are called the **alkaline earth metals**; on the far right side of the periodic table, the elements in group 17 are collectively called the **halogens**; and the elements in group 18, reflecting their reluctance to form compounds, are called the **inert gases**, **rare gases**, or **noble gases**. Each element in the periodic table is assigned a symbol consisting of one, two, or three letters. The first letter is **always** capitalized, while the second and third letters are **never capitalized**. The individual names of the elements originated in a surprisingly diverse number of ways. Some elements, such as copper (Cu), iron (Fe), silver (Ag), and gold (Au), were known well before the Christian era, and their symbols are derived from the Latin words that were their original names. Most of the elements were discovered before the beginning of the twentieth century, and the origins of their names reflect an extremely disparate range of choices: sodium, Na, for the Latin *natrium*; tungsten, W, for the German *wolfram*; helium, He, for the Greek *helios*; and erbium, Er, named for the Swedish town of Ytterby, to list only a few. With the dawn of nuclear transmutation, newly identified elements were assigned a name and symbol that are based on the Latin name for the digits of the new element's atomic number. Since the early 1950s, all such candidates have atomic numbers greater than 100 and, hence, have been assigned a three-letter symbol. Element 113 has been assigned the provisional name "ununtrium" and the symbol Uut; element 118 has been assigned the provisional name "ununoctium," along with the symbol Uuo. These names and symbols are provisional because the discovery is awaiting independent confirmation. Once the discovery of a new element has been independently confirmed, the original discoverers are given the privilege of proposing a permanent name and two-letter symbol for the new element. After acceptance by IUPAC, the proposed name and symbol become the new element's permanent designation. The three most recently assigned permanent names and symbols are copernicium (element 112), Cn (2010); flerovium (element 114), Fl (2012); and livermorium (element 116), Lv (2012).

Example 4.1: A student has samples of four elements: sodium (Na), magnesium (Mg), calcium (Ca), and beryllium (Be). Which element does not belong?

Answer:

| | |
|---|---|
| **Organize:** | Given Na, Mg, Ca, and Be |
| **Unknown:** | Identify the relationships among the elements. |
| **Translate:** | Review periodic properties of the elements. |
| **Solve:** | Be, Mg, and Ca are alkaline earth metals;<br>Na is an alkali metal.<br>Therefore, Na does not belong. |

Example 4.2: A student has samples of five elements: sodium (Na), iron (Fe), mercury (Hg), bromine (Br), and tin (Sn). Which element does not belong?

Answer:

| | |
|---|---|
| **Organize:** | Given Na, Fe, Hg, Br, and Sn |
| **Unknown:** | Identify the relationships among the elements. |
| **Translate:** | Review periodic properties of the elements. |
| **Solve:** | Na, Fe, Hg, and Sn are metals;<br>Br is a nonmetal.<br>Therefore, Br does not belong. |

Example 4.3: A student has samples of five elements: nitrogen (N), oxygen (O), iron (Fe), neon (Ne), and chlorine (Cl) at a temperature of 273.15 K and a pressure of 1 atmosphere. Which element does not belong?

Answer:

| | |
|---|---|
| **Organize:** | Given N, O, Fe, Ne, and Cl<br>Temperature of 0° C<br>Pressure of 1 atmosphere |
| **Unknown:** | Identify the relationships among the elements. |
| **Translate:** | Review periodic properties of the elements. |
| **Solve:** | N, O, Ne, and Cl are all nonmetals;<br>only Fe is a metal.<br>Therefore, Fe does not belong. |

The images presented in figures 4.1 and 4.2 are the traditional chemist's configurations of the table of elements. However, these do not exhaust the possibilities for displaying the numerous relationships among the elements. In fact, more than one thousand different configurations of the periodic table have been created, many of which reflect a particular specialized application. In the late 1920s Charles Janet proposed an organization based on the order of orbital filling by electrons. Janet's table exactly reflects the Madelung rule that we encountered in chapter 2, in which the sum of the first two quantum numbers, $n + \ell$, determines each row of the table of elements. This particular form of the periodic table is sometimes called the left-step table, as is shown in figure 4.3.

| | f-block | d-block | | | | | | | | | | p-block | | | | | | s-block | |
|---|---|---|---|---|---|---|---|---|---|---|---|---|---|---|---|---|---|---|---|
| 1s | | | | | | | | | | | | | | | | | | 1 H | 2 He |
| 2s | | | | | | | | | | | | | | | | | | 3 Li | 4 Be |
| 2p 3s | | | | | | | | | | | | 5 B | 6 C | 7 N | 8 O | 9 F | 10 Ne | 11 Na | 12 Mg |
| 3p 4s | | | | | | | | | | | | 13 Al | 14 Si | 15 P | 16 S | 17 Cl | 18 Ar | 19 K | 20 Ca |
| 3d 4p 5s | | 21 Sc | 22 Ti | 23 V | 24 Cr | 25 Mn | 26 Fe | 27 Co | 28 Ni | 29 Cu | 30 Zn | 31 Ga | 32 Ge | 33 As | 34 Se | 35 Br | 36 Kr | 37 Rb | 38 Sr |
| 4d 5p 6s | | 39 Y | 40 Zr | 41 Nb | 42 Mo | 43 Tc | 44 Ru | 45 Rh | 46 Pd | 47 Ag | 48 Cd | 49 In | 50 Sn | 51 Sb | 52 Te | 53 I | 54 Xe | 55 Cs | 56 Ba |
| 4f 5d 6p 7s | 57–70 | 71 Lu | 72 Hf | 73 Ta | 74 W | 75 Re | 76 Os | 77 Ir | 78 Pt | 79 Au | 80 Hg | 81 Tl | 82 Pb | 83 Bi | 84 Po | 85 At | 86 Rn | 87 Fr | 88 Ra |
| 5f 6d 7p 8s | 89–102 | 103 Lr | 104 Rf | 105 Db | 106 Sg | 107 Bh | 108 Hs | 109 Mt | 110 Ds | 111 Rg | 112 Cn | 113 Uut | 114 Fl | 115 Uup | 116 Lv | 117 Uus | 118 Uuo | 119 Uue | 120 Ubn |
| | | | | 57 La | 58 Ce | 59 Pr | 60 Nd | 61 Pm | 62 Sm | 63 Eu | 64 Gd | 65 Tb | 66 Dy | 67 Ho | 68 Er | 69 Tm | 70 Yb | | |
| | | | | 89 Ac | 90 Th | 91 Pa | 92 U | 93 Np | 94 Pu | 95 Am | 96 Cm | 97 Bk | 98 Cf | 99 Es | 100 Fm | 101 Md | 102 No | | |

Figure 4.3: Janet's Left-Step Periodic Table (compact form)

## Solutions and Electricity

Paralleling the recognition of the profound periodicity of the elements summarized by Mendeleev's table, a second physical phenomenon emerged from the shadows in the nineteenth century and soon came to dominate the attention of both investigators and innovators around the world. Building on the quantitative investigation of **electrostatics** that was completed as the eighteenth century ended, **electricity** and a plethora of electrical phenomena now became the focus of numerous experimental observations. Investigations by Luigi Galvani and Alessandro Volta through the 1790s culminated with Volta's construction of the first primitive battery in 1800, making electricity in the form of a continuous flow (direct current, or **DC** current) a readily available tool for experimentation. Using Volta's invention as a source of electricity, Nicolson and Carlisle demonstrated that electricity could break down (**decompose**) water into hydrogen and oxygen. Additionally, during the first third of the nineteenth century, Michael Faraday, first as an assistant to Humphry Davy, and then as an independent investigator, made several important contributions to the field of **electrochemistry**, which studies the role of electrical phenomena in chemical reactions.

While the chemistry of pure substances can certainly be affected by electrical phenomena (for example, the decomposition of water into hydrogen gas and oxygen gas), we will focus our attention on solutions. Solutes can be divided into two general categories: **electrolytes** and **nonelectrolytes**. An **electrolyte** is any substance that, when dissolved in a solvent, yields a solution that **conducts electricity**. In contrast, a **nonelectrolyte** is any substance that, when dissolved in a solvent, yields a solution that does **not conduct electricity**. Note that while these definitions apply to any solvent, we will remain focused on **aqueous** solutions, that is, solutions in which the solvent is water. But what do these definitions mean from an observational point of view? What do we actually see, experimentally?

Well before 500 BCE, the Greeks observed a physical property called **charge** that appears to be present in two distinct types, one called *positive* and the other called *negative*. In chapter 2 we saw that this very same concept of charge played a fundamental role in the development of our understanding of the microscopic structure of atoms. As we noted earlier, Thomson's critical insight in 1897 was the linking of matter and electricity (soon to be identified as mobile charge) in his work with the corpuscles that were later associated with the name "electrons." Following this work, the three-particle model of the atom, which included the electron, the proton, and the neutron, incorporated the concept of charge as a fundamental characteristic of the microscopic structure of matter. Further, using the primitive concepts of acceleration and force introduced in chapter 1, positive charges and negative charges exert forces on one another. The observational evidence is that an object possessing the property of charge (we call it a **charged object**) **accelerates** when it is near to another charged object. If the two charges are the same (either **two** positive or **two** negative), the objects **repel** one another; if the two charges are opposite (**one** positive and **one** negative), the objects **attract** one another. (Note the similarity to the behavior of the north and south poles of a magnet.) By the end of the eighteenth century the French physicist Charles-Augustin de Coulomb had developed a mathematical statement (called Coulomb's law) that described the behavior of large concentrations of positive and negative charge (often called **static** charges before they initiated a sudden acceleration or the dramatically discontinuous phenomenon we call lightning). With the development of Volta's primitive battery, it was now possible to study an **electrical current**, that is, a continuous movement of charge through space and over time.

It is at this point, the first third of the nineteenth century, that we meet Michael Faraday, who observed the effects of adding an electrolyte to pure water to produce an aqueous solution. Prior to adding the electrolyte, the pure water solvent acted as a barrier to the movement of charge; it is an **insulator**, not a **conductor** of charge, like the metals copper, aluminum, or mercury. However, once the electrolyte was added, the resulting solution became a conductor of charge. In effect, if two wires (more generally, two conductors of charge) were connected to one of Volta's batteries (one connection designated *positive*, the other connection designated *negative*; like its modern counterpart, Volta's battery had two ends, one positive, the other negative), Faraday observed that the metallic component of a dissolved electrolyte would move through the solution and deposit on one of the conductors. This indicated two things to Faraday: (1) because the substance was attracted to one of the conductors, it must carry an **opposite** charge; and (2) because measurable amounts of metal were deposited on the conductor, the species moving through the solution possessed the property of **mass**. While Faraday did not know the structure or the exact composition of the charge carriers, he did conclude that they must be **matter with the property of charge**. He called the charge carriers **ions** (from the Greek word meaning "to go"). The ions that collected at the positive conductor (anode) were called **anions** (they have a negative charge); those that collected at the negative conductor (cathode) were called **cations** (they have a positive charge).

Electrolytes themselves are classified as either **strong** or **weak**. A **strong electrolyte** is a substance that, when dissolved in water, produces an aqueous solution that is a **very good conductor** of an electrical current. A solution that is a good conductor of an electric current must contain a large number of mobile charge units (ions). Consequently, from another viewpoint, a strong electrolyte must produce a large number of ions in the solution. That is, in a solution of a strong electrolyte, the concentration of ions must be very large. On the other hand, a **weak electrolyte** is a substance that, when dissolved in water, produces an aqueous solution that is a **poor conductor of electricity**. A weak electrolyte **cannot** produce a solution with a large concentration of ions.

The discovery of electrolytes uncovered an unexpectedly rich variety of charged matter. It was possible to verify experimentally that the **quantity** of charge associated with the ions was not limited to a single value; in fact, a range of integer values is possible. If the charge is positive and the symbol 1+ is assigned to the basic unit of positive charge, typical values of positive charge range from 1+ to 8+. Similarly, if the basic unit of negative charge is assigned the symbol 1–, the typical values of negative charge range from 1– to 4–. Examining the periodic table, the alkali metal atoms are found to produce 1+ ions, the alkaline earth metal atoms yield 2+ ions, and many of the transition metal atoms as well as several heavy metals (for example, Sn, Pb, and Au) are capable of producing positively charged ions that have several different values. The nonmetals yield ions with negative charges: the halogens produce 1– ions, oxygen and sulfur 2– ions, and nitrogen and phosphorous 3– ions. A partial list of observed **monatomic ions** is contained in table 4.1. The naming convention for the monatomic ions is very simple—for the cations formed by the metals, the name is the elemental name followed by the word *ion*: $Na^+$ is named "sodium ion"; for the anions formed by the nonmetals, the suffix *-ide* is added to the elemental root: $O^{2-}$ is named "oxide" and $Cl^-$ is named "chloride." In the case of metals capable of forming ions that have several different values, the ions are distinguished by including the charge, written as a Roman numeral, as part of the name: $Fe^{2+}$ is named "iron(II) ion" and $Fe^{3+}$ is named "iron(III) ion." The systematic naming procedure that uses Roman numerals is called the **Stock system**.

Even more amazing, it is possible for a molecule composed of two or more atoms of distinct elements to produce ions in solution. In this case, the electrolytes are more complex substances, and the ions are identified as **polyatomic ions**. Table 4.2 contains a list of some of the polyatomic ions that have been experimentally observed, along with their names.

Table 4.2: Selected Monatomic Ions

| Atom | Ion Symbol | Atom | Ion Symbol |
|---|---|---|---|
| H | $H^+$ | Cd | $Cd^{2+}$ |
| Li | $Li^+$ | Ag | $Ag^+$ |
| Na | $Na^+$ | Au | $Au^+$, $Au^{3+}$ |
| K | $K^+$ | Cu | $Cu^+$, $Cu^{2+}$ |
| Rb | $Rb^+$ | Cr | $Cr^{2+}$, $Cr^{3+}$ |
| Cs | $Cs^+$ | Fe | $Fe^{2+}$, $Fe^{3+}$ |
| Mg | $Mg^{2+}$ | N | $N^{3-}$ |
| Ca | $Ca^{2+}$ | P | $P^{3-}$ |
| Sr | $Sr^{2+}$ | O | $O^{2-}$ |
| Ba | $Ba^{2+}$ | S | $S^{2-}$ |
| Al | $Al^{3+}$ | F | $F^-$ |

| Sn | $Sn^{2+}$, $Sn^{4+}$ | Cl | $Cl^-$ |
|---|---|---|---|
| Pb | $Pb^{2+}$, $Pb^{4+}$ | Br | $Br^-$ |
| Zn | $Zn^{2+}$ | I | $I^-$ |

Note that, by convention, the symbols for ions with either a 1+ charge or a 1− charge omit the numeral 1.

## Table 4.2: Selected Polyatomic Ions

| Characteristic Atom | Ion Symbol | Name |
|---|---|---|
| Hydrogen | $OH^-$ | Hydroxide |
| Nitrogen | $NH_4^+$ | Ammonium |
| Nitrogen | $NO_3^-$ | Nitrate |
| Nitrogen | $NO_2^-$ | Nitrite |
| Chlorine | $ClO_3^-$ | Chlorate |
| Chlorine | $ClO_2^-$ | Chlorite |
| Carbon | $CO_3^{2-}$ | Carbonate |
| Carbon | $CN^-$ | Cyanide |
| Carbon | $C_2H_3O_2^-$ | Acetate |
| Carbon | $C_2O_4^{2-}$ | Oxalate |
| Sulfur | $SO_4^{2-}$ | Sulfate |
| Sulfur | $SO_3^{2-}$ | Sulfite |
| Phosphorous | $PO_4^{3-}$ | Phosphate |
| Phosphorous | $PO_3^{3-}$ | Phosphite |
| Chromium | $CrO_4^{2-}$ | Chromate |
| Chromium | $Cr_2O_7^{2-}$ | Dichromate |

Note that, by convention, the symbols for ions with either a 1+ charge or a 1− charge omit the numeral 1. Also notice that many polyatomic ions end with *-ate*; an ion having one fewer oxygen atom ends in *-ite*: compare, for example, chlorate and chlorite.

# Why Do Substances Exist?

The identification of positively charged and negatively charged ions in solutions and the recognition of the role that electricity plays in chemical processes stimulated the first attempts to propose a comprehensive model describing the diverse ways in which the elements combined to form the substances of our world. We have already seen that Dalton's reintroduction of the idea of the atom provided a powerful microscopic interpretation of chemical compounds and chemical reactivity. Dalton's contemporary, the Swedish chemist Jöns Jacob Berzelius (who is often identified along with Dalton and Robert Boyle as one of the founders of modern chemistry), proposed that chemical compounds were composed of two parts, one positively charged and the other negatively charged. A chemical compound remained stable (consequently, the substances of our world exist) as a result of the mutual attraction of the opposite charges. This theory, called **chemical dualism**, while serving as an organizing principle during the first third of the nineteenth century, was largely discredited after 1830. Rapidly accumulating new empirical data could not be adequately described and interpreted by this **Berzelian dualism**.

As we have already seen, the experimental evidence collected during the second half of the nineteenth century led to the emergence of the periodic table and, as the century ended, to the dramatic demonstration of subatomic matter and the fact that atoms can decay. Atoms are not eternal, as the Greeks had suggested. As noted in chapter 2, physicists were concerned with the dynamical interactions between the negatively charged electrons and the positively charged regions of the atom. Thomson's "plum pudding" model and then the Rutherford-Bohr model of the nuclear atom both suggested that the electrons move rapidly either in a region of positive charge or in orbits surrounding a central nucleus. Further, the quantum theory proposed by Bohr (often called the *old quantum theory* to distinguish it from the later work of Heisenberg and Schrödinger) was stunningly successful in its ability to reproduce the experimental spectra of hydrogen by assuming that electrons radiate energy only when they *"jump" between stationary energy states in the atom.* All of these interpretations of the electron's behavior viewed it as a *dynamic* microscopic component.

In contrast, the chemists of this period increasingly focused their attention on the idea of chemical structures, the arrangement of atoms that produced the macroscopic substances of our world and determined the great variety of chemical reactions among these distinct substances. Because chemistry is concerned with the properties of stable substances and transformations among stable substances that produce new stable substances, chemists have repeatedly struggled to understand the fundamental *static* structures that sustained the observed stability of substances. Recognizing that the Berzelian dualism was unable to provide a comprehensive and quantitative understanding of chemical structures, a number of successor models were proposed throughout the latter half of the nineteenth century. Very significantly, these models increasingly attempted to link the *static* characteristics that effectively describe the stability of chemical substances with the reactivity that connected stable reactants to stable products. However, for chemists, the structure of molecules as formulated in terms of the *static* characteristics that ensured molecular stability remained the central theme of their science.

Early in the twentieth century a new proposal was made by Gilbert Newton Lewis (an American chemist) that offered a model of the microscopic and united the most recent discoveries of the physicists with the chemistry's focus on structures. Being acutely aware of the periodic chemical properties so elegantly displayed in the periodic table, Lewis noted that these properties of the elements in the second and third periods, when arranged by increasing atomic number, began to repeat after each group of *eight* elements. That is, elements 3 and 11 exhibited similar chemical properties as did elements 4 and 12; this pattern persisted through elements 10 and 18. The newly proposed models that were emerging from the work of Thomson, Rutherford, Bohr, and Moseley suggested to chemists that it was the *outermost* electrons that were central to the chemistry of the elements. This view was consistent with the theory of chemical valence that had emerged in the second half of the nineteenth century. Consequently, Lewis argued that an atom is composed of an *inner kernel* surrounded by a cubical *outer atom*. The eight corners of the outer cube coincided with the observation of the cyclic recurrence of chemical properties for groups of eight elements. Utilizing our idealized aufbau principle (see chapter 2), as each proton and electron are added to build electrically neutral atoms, each added electron occupies one of the eight cubical corners. After all eight cubical corners were filled with electrons (corresponding to the configuration of the noble gases in the periodic table), a new outer cube would form with the next added electron. This new outer cube, which contains a single electron, appears identical to the outer cube of the element whose atomic number is *eight units less*. Because the outermost electrons (often called **valence electrons**) are responsible for chemical reactivity and chemical properties, this model accounts for the periodic recurrence of chemical properties after each group of *eight* elements. Further, as noted earlier, because the filling of all eight cubical corners by electrons corresponds to the configuration of the noble gases, the unusual stability of these gases is attributed to the stability arising from the completely filled cubical arrangement.

## Understanding Electrolyte Compounds

But this conceptualization was not limited to the elements alone. Lewis was also keenly aware of the chemistry of compounds and sensitive to the distinctly different solution properties of the substances that we have called electrolytes and nonelectrolytes. As we have already noted, electrolytes produce ions in an aqueous solution, which enable the solutions to conduct electricity. While Faraday recognized ions as **matter with the property of charge**, he offered no description or detailed explanation of the structure of ions. Lewis used his model of the outer cube of the atom to suggest that electrons can be either **added** to or **lost** from an atom to leave the atom with a **completely filled outer cube containing exactly eight electrons**. This is the highly stable configuration of the noble gases and came to be known as the **octet rule**. As a result, alkali metal elements tended to form ions with a single unit of positive charge (1+ ions) by **losing an electron**; elements from the halogen group tended to form ions with a single unit of negative charge (1– ions) by **adding an electron**. In general, in an attempt to satisfy the octet rule, elements from the left side of the periodic table (metals) tended to **lose** electrons, while elements from the right side of the periodic table (nonmetals) tended to **add** electrons. From this perspective, the binary electrolyte compounds (composed of one metal element and one nonmetal element) are observed as macroscopic substances as the result of the electrostatic attraction (Coulomb's law) between the positively charged and negatively charged matter. However, in order to understand the behavior of electrolytes that involve polyatomic ions (table 4.2), Lewis was required to conceptualize his model of the cubical outer atom in a more general setting.

## What about Nonelectrolyte Compounds?

Throughout the second half of the nineteenth century, chemistry had struggled to understand the structure of the vast array of nonelectrolyte compounds that were being synthesized and characterized. The rapid growth of the chemical subdiscipline known as *organic chemistry* (primarily focused on the chemistry of the carbon atom) and its industrial application after 1860 in the making of dyes made it essential to understand the structures and the related reactivity of these compounds. As noted earlier, the failure of the Berzelian dualistic theory led to the development of a number of explanatory models for these compounds. Lewis now offered another viewpoint, recognizing that the behavior of the binary electrolytes (composed of one metal and one nonmetal) represented the extreme end of a range of behavior. Instead of either adding or losing and electron at one of the corners of the outer atom, Lewis suggested that the atoms of two nonmetal elements, for example, could have a *pair* of electrons *simultaneously* occupying two cubical corners of the outer atom of each element. In effect, the two atoms would *share a pair of electrons*. The fundamental physical reason (rooted in the chemical properties of these substances) that any sharing would occur is the same reason that Lewis proposed for the electrolytes: achieving the stability present in the outer atom configuration of the noble gases, the octet rule. The sharing of electron pairs became the foundation for understanding the chemical linkages that produced macroscopic amounts of nonelectrolyte substances. The sharing of electron pairs meant that these compounds in solution do **not** exhibit the property of mobile charges and, consequently, the aqueous solutions of these compounds do **not** conduct electricity.

## Polyatomic Ions

This sharing of pairs of electrons also allowed Lewis to understand the behavior of the polyatomic ions. With the exceptions of the last two entries of table 4.2, all the remaining entries of the table include only nonmetal atoms (in fact, nonmetals from **only** periods 2 and 3). Consequently, the atoms are bound as a unit by the sharing of electron pairs. However, the entire unit attains maximum stability by also satisfying the octet rule; this is only achievable by either adding or losing electrons. Consequently, while the chemical linkages between the constituent atoms of the polyatomic moiety requires a *sharing of electron pairs*, the total number of electrons required to satisfy the octet rule leaves the polyatomic unit with a charge. Hence, it is an ion. Finally, the reader may wonder about the last entries of table 4.2; the polyatomic unit includes a metal and a nonmetal; how does the Lewis proposal apply? We have already noted that the proposal Lewis made was guided by the observed behavior of the elements in the second and third periods. Because chromium is a transition metal and is a member of the fourth period, we should not be surprised that the chromium polyatomic ions appear to lie beyond the explanatory power of Lewis's work. In fact, the Lewis model possesses a number of limitations; we will briefly look at some of these shortcomings, remembering that Lewis formulated his model very early in the twentieth century, significantly before the evolution of the modern quantum theory described in chapter 2.

## Multiple Bonds

Before examining the limitations of Lewis's model, we pause to take note of one last significant success achieved by Lewis. During the last half of the nineteenth century, experimental evidence collected by chemists studying the behavior of the carbon atom identified several remarkable characteristics of the compounds of carbon (**organic compounds**). The carbon atom appeared to be capable of forming four linkages—today called **chemical bonds**—with other atoms. In the terminology of the valence theory, carbon is **tetravalent**. Additionally, carbon routinely formed molecules composed of long chains of atoms linked together by chemical bonds; these structures are one of the central characteristics that distinguish the amazingly diverse chemistry of carbon. But, even more remarkably, the bonds linking carbon atoms and carbon to other atoms were **not** experimentally identical; in fact, it was possible to identify empirically three distinct types of carbon-carbon bonds as well as three distinct types of bonds linking carbon to other atoms. These bond types came to called **single**, **double**, and **triple** bonds, with the double and triples bonds known collectively as **multiple bonds**.

The experimentally identified single bonds, in the framework of Lewis's model, were interpreted to be two atoms sharing a *single pair of electrons* (one *side* of the outer cubical atom). An experimental double bond required two atoms to share *two pairs of electrons* (one *face* of Lewis's outer cubical atom). Finally, the existence of a triple bond required two atoms to share *three pairs of electrons*, at feat achievable only if Lewis now abandoned his model of an outer cubical atom, replacing it with a **tetrahedral** arrangement of the eight outer electrons. To accomplish this rearrangement, Lewis proposed that *pairs of electrons* interacted magnetically to align themselves at the corners of a tetrahedron (a pyramid made up of four identical triangular faces). Then a triple bond could be interpreted as two atoms *sharing a tetrahedral face*.

## Lewis Structures

In the 1916 paper that introduced his model of chemical bonding, Lewis employed a simple two-dimensional representation of his ideas that came to be known as a **Lewis structure** (also known as a **Lewis dot diagram, electron dot diagram**, or **Lewis diagram**). Lewis structures use either a *pair of dots* or a simple *straight line* to symbolize the electron pair constituting the chemical bond between two atoms in a molecule. Figure 4.4 contains several examples of Lewis structures.

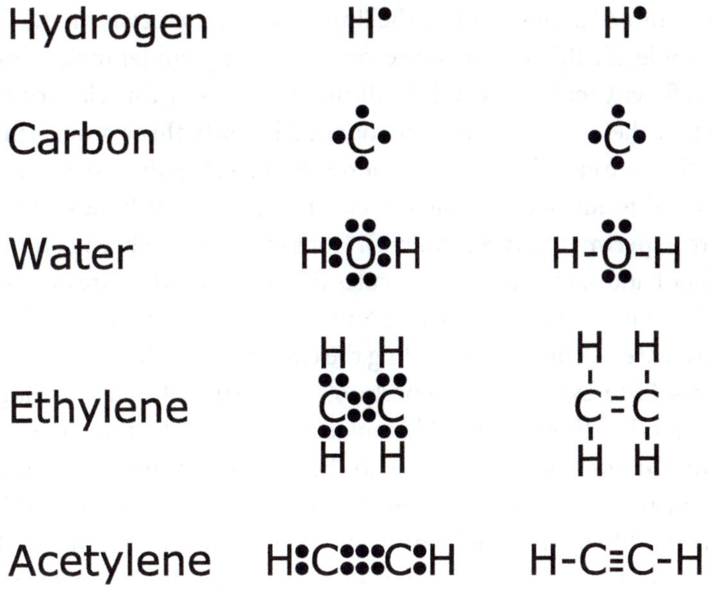

| | | |
|---|---|---|
| Hydrogen | H• | H• |
| Carbon | •C• (with dots above and below) | •C• (with dots above and below) |
| Water | H:Ö:H | H-Ö-H |
| Ethylene | H H / C::C / H H | H H / C=C / H H |
| Acetylene | H:C:::C:H | H-C≡C-H |

Figure 4.4: Representative Lewis Structures

Even though Lewis clearly understood that chemical molecules are **three-dimensional** (in his 1916 paper, he refers to the *cubical atom* and a *tetrahedral* arrangement of electron pairs), it is a curious quirk of history that his **two-dimensional** symbols (the Lewis structures) became ubiquitously adopted by chemists to represent the connectivity between atoms in a molecule. While Lewis structures are still widely used today (we will meet them again in our discussion of organic and biochemistry), it is not our goal to become proficient in writing Lewis structures to represent molecules but only to recognize them as steps along the road to a more encompassing understanding of molecular structure.

It is critical to note that Lewis structures do **not** convey **geometric** information about the structure of molecules. For example, the methane molecule ($CH_4$) is represented as a *flat, two-dimensional* object in which the angle between two adjacent C-H bonds is equal to 90°. The physical methane molecule is a three-dimensional object with the H-atoms arrayed *symmetrically* around the central carbon atom, and the angle between any two C-H bonds is approximately equal to 109.47°. In fact, the four hydrogen atoms occupy the corners of a *tetrahedron*.

## Limitations of the Lewis Model

The crucial concepts that lie at the heart of Lewis's model are the **octet rule** and the role of **shared electron pairs** in establishing chemical bonds between individual atoms in a molecule. While the idea of the octet rule (reflected in the symmetry of Lewis's cubical atom) provided an initial chemical impetus to understanding the bonds among the atoms in a molecule, the concept of the shared electron pair has proven to be the more subtle, profound, and persistent of the ideas proposed by Lewis. As we shall see later in a more detailed examination of the chemical bond, it is the concept of the shared electron pair that has played a significant role in the evolution of modern quantum theory. The importance of this concept was, ironically, hinted at in several of the limitations of the Lewis model.

Consider now the molecule $BF_3$. It is but one example of many similar molecules that are collectively known as **electron-deficient** molecules. While all the members of this class of molecules are stable, the reader will note that the central atom is surrounded by **only three** atoms (in this instance, three fluorine atoms), and they are each linked to the boron by a *single pair of shared electrons*. If any of the B-F bonds were assumed to involve two pairs of electrons, the three bonds would **not** be equivalent (even though they are **experimentally** equivalent). Consequently, the boron atom does **not** satisfy the octet rule! On the other hand, consider the molecules $PCl_5$ and $SF_6$. They are but two examples of stable molecules in which the central atom (either P or S in our examples) is surrounded five or six electron pairs (that is, ten or twelve electrons) far exceeding the eight required by the octet rule. The phosphorus and sulfur atoms are said to exhibit the property of an **expanded octet**, adding a new category of bonding to the Lewis model. Third, in a stable molecule such as NO, the nitrogen and oxygen atoms contribute an **odd number** of valence electrons to the chemical bonds of the molecule (five valence electrons from nitrogen and six valence electrons from oxygen). Consequently, NO and other so-called *radicals* are often called **odd-electron molecules**, and they cannot satisfy the octet rule. Because such radicals are highly reactive species, this limitation was not viewed with surprise, because Lewis's model was focused on the structure of stable substances; nevertheless, the Lewis model fails to provide a compelling rationale for the existence of these radical species. Finally, molecules such as $SO_2$ and $C_6H_6$ pose a special challenge to the Lewis model's effort to describe effectively the relationships among the atoms comprising stable substances. For $SO_2$, each atom contributes six valence electrons (that is, a total of eighteen electrons) to the molecular structure. But in this case, there is no longer a *unique* way to arrange the electrons to satisfy Lewis's criteria. Figure 4.5 depicts three possible structures for the $SO_2$ molecule, each of which meets the requirements of the Lewis model. In all three images, the rods connecting the atoms represent *electron pairs*; electron pairs that do not participate in bonding (**nonbonding electrons** or **lone pairs**) are **not** shown in the figure. The second and third structures both satisfy the octet rule (remember that each atom, in addition to the pairs of electrons participating in bonding, also has associated with it nonbonding pairs sufficient in number to satisfy the octet rule). However, this means there are **two** equally valid diagrams representing the molecule, not just **one**. But the two S-O bonds are experimentally *identical*; the two equivalent diagrams suggest that one bond is a *single* bond and the other a *double* bond. (The **formal charge** designations, *plus* and *minus*, in the first and third structures attempt to evaluate the charge distribution in the molecule, *assuming* that the pair of electrons in each bond is shared equally by S and O. Because there is one plus charge and one minus charge, the sum is zero for $SO_2$.) But this is not the end of the story! The first structure is also a possibility if S is assumed to be capable of possessing an **expanded octet**. In this instance, the two S-O bonds **are** *equivalent* (consistent with the experiment), but the octet rule is violated.

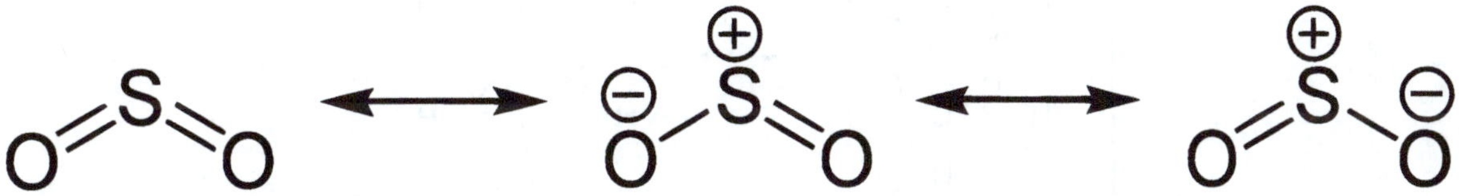

Figure 4.5: Resonance Structures for $SO_2$

Fundamentally, *none* of the three structures in figure 4.5 accurately represent the $SO_2$ molecule, nor does $SO_2$ *oscillate* among the three structures, spending a moment of time in each of the three forms. The essential failure of the three representations of $SO_2$ in figure 4.5 lies in the assumed *static character* of the representation used by the Lewis model. In fact, the electrons critical to the bonding in $SO_2$ are *delocalized* and belong to the entire molecule. They are *not localized*; they must be treated as *dynamic* objects that cannot be adequately described without using the ideas of the new quantum mechanics we met in chapter 2 (remember Bohr's complementarity principle). But it would be another decade before these ideas were invented.

In an attempt to account for molecules such as $SO_2$, the chemists of the early twentieth century invented the concept of **resonance**. Each structure in figure 4.5 is called a **resonance structure** or a **canonical form**, and the $SO_2$ molecule that is part of the physical universe is considered to be an approximate intermediate form with *simultaneous* contributions from *all* the resonance structures. The approximate intermediate is more stable (lower in energy) than all the contributing canonical forms and is called a **resonance hybrid**. As conceived early in the twentieth century, resonance was an attempt to elaborate and expand the static approach to chemical structure inherent in the Lewis model. We will see resonance in another guise in a later chapter, when the new quantum mechanics is utilized to describe the structure of molecules.

The benzene molecule, $C_6H_6$, is another example for which a single Lewis structure provides an inadequate representation. Figure 4.6 shows two resonance structures (often called Kekulé structures, after the German chemist Friedrich August Kekulé (later known as Friedrich August Kekulé von Stradonitz), which contribute to the benzene resonance hybrid. Each of the rods connecting the atoms represents a pair of electrons. Notice that the octet rule is satisfied for each carbon atom, while each hydrogen atom is associated with two electrons. The two electrons represent a special case of the octet rule applied to hydrogen because the hydrogen can form only *one* bond (two shared electrons) with any other atom. The two resonance structures depict alternating single and double bonds between the carbon atoms in the ring. First note that, because there are two possible ways to arrange the single and double bonds linking the carbon atoms, there are **two** equally valid Lewis diagrams. Further, because both representations include single and double bonds between carbon atoms in the ring, the diagrams suggest that there are **differences** (single bond versus double bond) in the linkages between the carbon atoms. Experimentally, **all** of the C-C bonds are **identical**, suggesting that *delocalized* electrons are central to the bonding in benzene. Consequently, from the perspective of the Lewis model, benzene is a resonance hybrid.

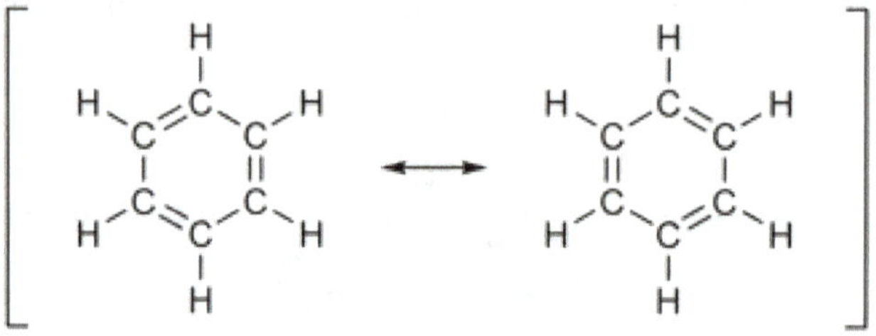

Figure 4.6: Resonance Structures for Benzene

Historically, the Lewis model of bonding in the case of electrolytes came to be called **ionic bonding**, while the bonding characterized by a shared pair of electrons (nonelectrolytes and the *internal* structure of the polyatomic ions) came to be called **covalent bonding**. Chemists associated with these bonding modalities the concepts of an **ionic bond** and a **covalent bond**, as if these two types of bonds are, in fact, constituents of the physical universe. However, they are better understood to be convenient fictions, or, from another perspective, limiting idealizations that provide only an efficient *model* summarizing a collection of experimental observations. Our brief introduction to the quantum mechanics that was constructed after 1925 (chapter 2) makes it clear that an electron is unlike any macroscopic object. Bohr's complementarity principle and Heisenberg's indeterminacy principle both make the *static* and *localized* model of the electron imagined by Lewis untenable. The strong focus on chemical structure through the last half of the nineteenth century and at the beginning of the twentieth century completely overshadowed the *dynamic* characteristics of the electron that dominated the developments in physics during this period. It will only be in our later study of more sophisticated descriptions of chemical bonding based on the evolution of modern quantum theory that a more complete and sophisticated image of the electron and its role in bonding will emerge.

Despite the limitations of the Lewis model, it has proven to be a highly effective tool in the interpretation of experimental data. Its ability to make powerful predictions, using the convenient fictions of *ionic bond* and *covalent bond*, is a testament to this effectiveness. Consequently, even though the model utilizes bonding idealizations and was unable to account for the *geometric shapes* associated with chemical structures, it has proven to be a useful route to achieving macroscopic consequences. As such, the Lewis model should not be dismissed as "unreal" but rather recognized for what it is: a facile model that is capable of describing experimental results, making accurate macroscopic predictions, but limited by the tools and concepts used in its construction.

## Putting a Name to the Face

We have now met a variety of chemical species, beginning with the elements and continuing through the substances classified as electrolytes and nonelectrolytes. While the names of the elements have originated in a variety of ways, they are conveniently tabulated in the periodic table, along with their unique symbols. Additionally, we have already seen that the monatomic ions of both metals and nonmetals follow a simple naming procedure, while the polyatomic ions listed in table 4.2 are given names related to their central atom, often (but not always) a nonmetal.

How do we now give names to the binary electrolytes? For a binary electrolyte we simply combine the names of the two ions, always listing the cation first, but dropping the word "ion" from the name. Consequently, NaCl is named *sodium chloride*, and BaO is named *barium oxide*. In the case of a cation that can occur with various charges (often transition metals) whose name includes a Roman numeral, the binary compound also includes the Roman numeral: FeO is named *iron(II) oxide*, CuO is named *copper(II) oxide*, and CuCl is named *copper(I) chloride*. In case of a polyatomic ion, simply use the ions' names: $NaNO_3$ is named *sodium nitrate*, $MgSO_4$ is named *magnesium sulfate*, and $NH_4Cl$ is named *ammonium chloride*.

Finally, what are the names of the nonelectrolytes? The class of nonelectrolytes contains an immense number of substances, the vast majority of which are compounds built around the carbon atom and known as **organic** compounds. We will see later that there is a very systematic (and intricate, but easy to use) approach to naming organic compounds; here we will limit our discussion to the naming of binary inorganic (compounds that, with only a few historical exceptions, do **not** contain carbon). There is a simple, three-step process to name a simple binary inorganic compound:

1. Name the first element in the formula
2. **Name the second element in the formula, changing the ending to -ide**
3. Designate the **number** of atoms of each element present in the formula by using the **Greek** prefixes found in table 4.3

Consequently, the molecule $N_2O_4$ is named *dinitrogen tetroxide*, the molecule $CO_2$ is named *carbon dioxide*, and the molecule $Cl_2O$ is named *dichlorine monoxide*. There are some common exceptions: $H_2O$ is commonly named *water* rather than *dihydrogen oxide*, and $NH_3$ is commonly named *ammonia* rather than *nitrogen trihydride*.

Table 4.3: Greek Prefixes

| Prefix | Meaning |
|---|---|
| Mono- | 1 |
| Di- | 2 |
| Tri- | 3 |
| Tetra- | 4 |
| Penta- | 5 |
| Hexa- | 6 |
| Hepta- | 7 |
| Octa- | 8 |
| Nona- | 9 |
| Deca- | 10 |

## Electronegativity

The *static*, structure-dominated view of the chemical bond that conditioned the model proposed by Lewis continued to influence chemical thinking during the first several decades of the twentieth century. In 1932, the American chemist Linus Pauling proposed a new property called **electronegativity** that describes the behavior of atoms *linked in a chemical bond*. To define this new property, Pauling combined the convenient fictions of the *ionic bond* and the *covalent bond* and compared the bond energies of homonuclear diatomic molecules with those of a large group of heteronuclear diatomic molecules (for example, $H_2$, $Cl_2$, $F_2$, and $Br_2$ compared with HCl, HF, and HBr). Pauling noted that the bonds formed between elements from opposite sides of the periodic table were more stable (that is, required more energy to rupture them) than the bonds formed either between identical elements or between elements located in close proximity to one another. He suggested that the increased stability was the result of an *ionic* contribution to the bond. He argued that if two dissimilar elements (element A and element B) were linked in a molecule AB, an estimate of the *covalent* contribution to the bond could be made using the bond strengths of the homonuclear diatomic molecules $A_2$ and $B_2$. On the other hand, there was an *ionic* contribution to the A-B bond because of some *transfer of charge between A and B*. This *ionic* contribution to the bond should arise from a *difference in the electronegativity of element A and element B in the A-B bond*. Consequently, Pauling defined an element's electronegativity as the tendency for an atom involved in a chemical bond to draw charge (interpreted to mean "electrons") to itself.

Electronegativity is **not** a directly measurable physical property. Using the constructs of an *ionic bond* and a *covalent bond*, Pauling created an arbitrary scale (unitless) to quantify the property of electronegativity. It **is significant** that the values of electronegativity scale exhibit general trends consistent with the conventional layout of the periodic table. Figure 4.7 displays the electronegativity of the elements using Pauling's scale. The central point is the **trends in electronegativity** exhibited by the elements: electronegativity *tends* to **increase** in a period from left to right and to **decrease** in a group from top to bottom.

| Group (vertical) / Period (horizontal) | 1 | 2 | 3 | 4 | 5 | 6 | 7 | 8 | 9 | 10 | 11 | 12 | 13 | 14 | 15 | 16 | 17 | 18 |
|---|---|---|---|---|---|---|---|---|---|---|---|---|---|---|---|---|---|---|
| 1 | H 2.20 | | | | | | | | | | | | | | | | | He |
| 2 | Li 0.98 | Be 1.57 | | | | | | | | | | | B 2.04 | C 2.55 | N 3.04 | O 3.44 | F 3.98 | Ne |
| 3 | Na 0.93 | Mg 1.31 | | | | | | | | | | | Al 1.61 | Si 1.90 | P 2.19 | S 2.58 | Cl 3.16 | Ar |
| 4 | K 0.82 | Ca 1.00 | Sc 1.36 | Ti 1.54 | V 1.63 | Cr 1.66 | Mn 1.55 | Fe 1.83 | Co 1.88 | Ni 1.91 | Cu 1.90 | Zn 1.65 | Ga 1.81 | Ge 2.01 | As 2.18 | Se 2.55 | Br 2.96 | Kr 3.00 |
| 5 | Rb 0.82 | Sr 0.95 | Y 1.22 | Zr 1.33 | Nb 1.6 | Mo 2.16 | Tc 1.9 | Ru 2.2 | Rh 2.28 | Pd 2.20 | Ag 1.93 | Cd 1.69 | In 1.78 | Sn 1.96 | Sb 2.05 | Te 2.1 | I 2.66 | Xe 2.60 |
| 6 | Cs 0.79 | Ba 0.89 | * | Hf 1.3 | Ta 1.5 | W 2.36 | Re 1.9 | Os 2.2 | Ir 2.20 | Pt 2.28 | Au 2.54 | Hg 2.00 | Tl 1.62 | Pb 2.33 | Bi 2.02 | Po 2.0 | At 2.2 | Rn 2.2 |
| 7 | Fr 0.7 | Ra 0.9 | ** | Rf | Db | Sg | Bh | Hs | Mt | Ds | Rg | Uub | Uut | Uuq | Uup | Uuh | Uus | Uuo |
| Lanthanides | * | La 1.1 | Ce 1.12 | Pr 1.13 | Nd 1.14 | Pm 1.13 | Sm 1.17 | Eu 1.2 | Gd 1.2 | Tb 1.1 | Dy 1.22 | Ho 1.23 | Er 1.24 | Tm 1.25 | Yb 1.1 | Lu 1.27 | | |
| Actinides | ** | Ac 1.1 | Th 1.3 | Pa 1.5 | U 1.38 | Np 1.36 | Pu 1.28 | Am 1.13 | Cm 1.28 | Bk 1.3 | Cf 1.3 | Es 1.3 | Fm 1.3 | Md 1.3 | No 1.3 | Lr 1.291 | | |

Periodic table of electronegativity using the Pauling scale

Figure 4.7: Periodic Table Using Pauling's Electronegativity Scale

## Polarity

Earlier we introduced two very general categories of solution: the electrolytes and the nonelectrolytes. The simple macroscopic observation of the ability of a solution to conduct electricity distinguishes an electrolyte solution from a nonelectrolyte solution. As we have seen, it is the presence of mobile carriers of charge that is the key to this distinction. Another very old macroscopic observation has led to the discovery of another important role for the property of charge. It has been long noted that water and oil do **not** form a homogeneous solution; they do not mix. More precisely, if a small amount of oil (the solute) is added to a larger volume of water (the solvent), the water does not hydrate the oil. There are literally thousands of distinct substances that exhibit the same pairwise behavior; they do not mix.

Proceeding from the *static*, structure-dominated mind-set that was so characteristic of chemical thinking, it was a simple matter to apply the Lewis model to molecules and conclude that the arrangement of electrons often produces a **separation of positive and negative charge**. From the perspective of the Lewis model this meant that one region of a molecule could be relatively positively charged, while another region could be relatively negatively charged. A molecule that exhibits such a **charge separation** is called **polar**. If a molecule does **not** exhibit a **charge separation**, it is called **nonpolar**.

As we have already noted, the *static* perspective that characterized such a large segment of chemical thinking as the nineteenth century ended and the twentieth century began simply did not include the *dynamic* perspective of physics. Further, the model that Lewis proposed and that was readily applied as an organizing principle in chemistry simply predated the evolution of modern quantum mechanics. It is important to remember that the empirical observations are accurate; polar and nonpolar substances are part of our universe. Further, the concept of **charge separation** plays a critical role in a multitude of chemical processes. What is required is a more sophisticated perspective, one that incorporates our best microscopic models, in order to describe the significance of charge separation in our world. Finally, as we have stated, the Lewis model and the models for which it is a foundation should not be discounted; they have provided powerful descriptions with predictive capabilities of the order of our world. They are simply limited.

# Chapter 4 Exercises

1. What are the horizontal rows of the periodic table called?
2. What are the two terms that are used to designate the columns of the periodic table?
3. The elements in the periodic tables of Mendeleev and Lothar Meyer were arranged using which physical property?
4. In the modern periodic table, which physical property of the elements is used to arrange the elements?
5. What are the names of the three general categories of elements in the periodic table?
6. A student has samples of four elements: sodium (Na), potassium (K), francium (Fr), and beryllium (Be). Which element does not belong?
7. A student has samples of four elements: fluorine (F), chlorine (Cl), iodine (I), and oxygen (O). Which element does not belong?
8. What is electrochemistry?
9. How are electrolytes and nonelectrolytes different?
10. The physical property called *charge* appears to be present in our universe as two distinct types. What are these two types of charge?
11. What are the charges (*positive* or *negative*) on an anion and a cation?
12. What is a strong electrolyte?
13. What is a weak electrolyte?
14. Name the following monatomic ions:
    (a) $Cl^-$ (b) Na+ (c) $Ca^2$+ (d) $S^{2-}$ (e) $N^{3-}$
15. Name the following polyatomic ions:
    (a) $NO_3^-$ (b) $CO_3^{2-}$ (c) $NH_4$+ (d) $C_2H_3O_2^-$ (e) $PO_4^{3-}$
16. What are valence electrons?
17. Using Lewis's approach, explain the octet rule.
18. How many chemical bonds can the carbon atom form?
19. Are the bonds formed by the carbon atom either with itself or with atoms of other elements identical?
20. What are the names given to the multiple bonds that the carbon atom can form?
21. What is the name given to the two-dimensional representations used by Lewis to explain his theory of chemical bonding?
22. Identify the terms *ionic bond* and *covalent bond.*
23. What is a resonance hybrid?
24. Name the following compounds:
    (a) HBr (b) KCl (c) $MgSO_4$ (d) $CaCl_2$ (e) MgO
25. Name the following compounds:
    (a) $CaCO_3$ (b) $NH_4Cl$ (c) $K_3PO_4$ (d) NaCN (e) KOH
26. Name the following compounds:
    (a) $PbO_2$ (b) $Fe_2O_3$ (c) $CuCl_2$ (d) CrO (e) $AuCl_3$
27. What is electronegativity?

28. What is the *trend* of electronegativity values in the conventional layout of the periodic table?
29. In the following molecules, which element is more electronegative?
    a. HCl
    b. $BaCl_2$
    c. $PbO_2$
    d. $PCl_3$
30. What is a *polar* molecule?
31. Which of the following molecules are polar?
    a. HBr
    b. $CH_4$
    c. $CCl_4$

# CHAPTER FIVE

# Transition from the Classical Chemical Paradigm to the Quantum Paradigm (Early Twentieth Century)—The Dance of Charges

## Redox, Acid-Base Chemistry, and Buffers

As we have already seen, the property of charge appears to be a fundamental characteristic of our universe, appearing in two distinct types labeled *positive* and *negative*. Following Thomson's crucial insight linking matter and charge in 1897 and the identification of charged, subatomic pieces of matter, the concept of charge played a critical role in our understanding of both electrolytes and nonelectrolytes and polar and nonpolar substances. Further, the concept of charge was central to Lewis's model of chemical structure and to the subsequent elaboration by Pauling of his notion of electronegativity. The goal of this chapter is to examine in greater detail the important *charge transfer processes* in chemistry; in particular we shall encounter for the first time two new players in the chemistry arena: **acids** and **bases**.

### Oxidation and Reduction

The extensive empirical data accumulated particularly during the second half of the nineteenth century when combined with the evolution of the new quantum mechanics after 1925 have strongly demonstrated that the unique chemical properties of an element depend on two key parameters: (1) the number of protons in the atomic nucleus of an element and (2) the configuration of the electrons surrounding the nucleus. Modern quantum mechanics clearly demonstrates that these two parameters are not independent of one another. However, in terms of the three-particle model of the atom, it is the atomic number that distinguishes each element; remember, since the work of Moseley and his successors, the elements in the periodic table have been arranged in terms of *increasing* atomic number. Because the nuclei of the elements do not participate in chemical reactions, it is the rearrangement of the electrons (the negatively charged species surrounding the nucleus) that ultimately determines the chemical bonds in a molecule. Consequently, while acknowledging the role of the atomic nucleus in defining the identity of each atom, a particularly instructive view of all chemical reactions is to treat them as *electron rearrangement* or *charge transfer* processes.

Remembering the broad distinction we made between electrolytes and nonelectrolytes, it is reasonable to suppose that the chemical reaction responsible for the formation of an electrolyte involves some type of electron rearrangement or charge transfer process, because the dissolution of an electrolyte by the solvent produces a solution with mobile charges (ions). The solution conducts electricity. Because the world is, in general, **neutral** (that is, possessing neither a net positive charge nor a net negative charge), as the elements form an electrolyte, they must undergo some rearrangement of their electrons to make it possible to produce ions in solution. While the model articulated by Lewis is intuitively attractive, the later development of quantum mechanics reminds us that the Lewis model is much too limited to account for the rich diversity of our world.

The case of the nonelectrolytes is much more problematic. The formation of these substances from the elements certainly involves a rearrangement of electrons; the experimental data and the framework of modern quantum mechanics require such an electron rearrangement. But, the dissolution of a nonelectrolyte does **not** produce a solution with mobile charges! How are we to understand this?

In order to describe the rearrangement of electrons that takes place **whenever** a molecule is formed (either an electrolyte or a nonelectrolyte), we can assign a *fictitious number*, called an **oxidation number**, to each atom in a molecule. The values of these fictitious numbers range from –4 to +8, usually in *integer* steps. (There are examples of fractional oxidation numbers. These cases will not be considered here.) The value of the fictitious number assigned to each atom in a molecule identifies the **oxidation state** of the atom; an atom assigned an oxidation number of +3 is then said to be in a +3 oxidation states. In the course of a chemical reaction, an atom starting as a reactant (or part of a reactant molecule) with an assigned oxidation number appears as a product (or part of a product molecule). Participation in a reaction can either **increase** or **decrease** the value of an atom's oxidation number by an *integer* amount, signifying a changes in the atom's oxidation states as a result of the reaction. If the oxidation number becomes *more positive*, the process can be interpreted as a *loss of negative values*—for example, an oxidation number changing from a –2 to a +3 can be viewed as a *loss of* –5, because $-2 - (-5) = +3$. Similarly, if an oxidation number becomes *more negative*, the process can be interpreted as a *gain of negative values*—again, for example, an oxidation number changing from a +4 to a –1 can be viewed as a *gain of* –5, because $+4 + (-5) = -1$. By interpreting the change in the value of the oxidation number as either the *loss or gain of negative values*, we can make a very facile association between the rearrangement of electrons (matter associated with a *negative* charge) in a chemical reaction and the change in the value of an atom's oxidation number.

We make this association by making several fundamental definitions. **Oxidation** is defined as a **chemical process in which the value of an atom's oxidation number becomes *more positive*.** Because this is viewed as the *loss of negative values*, Oxidation is thought of as the *loss of electrons*. Correspondingly, **reduction** is defined as a **chemical process in which the value of an atom's oxidation number becomes *more negative*.** Because this is viewed as the *gain of negative values*, reduction is thought of as the *gain of electrons*. It is critical to observe that neither the oxidation process nor the reduction process occur in isolation; they *always* occur together, much like two sides of a single coin.

These definitions give rise to a simple mnemonic phrase that describes the joint oxidation-reduction process: **LEO** goes **GER**, **L**ose **E**lectrons **O**xidation, **G**ain **E**lectrons **R**eduction. Because the word "Leo" is the Latin word for "lion," the image suggested by the mnemonic is that of a growling lion.

The reader should be careful to note that we have defined the oxidation-reduction process in terms of the *value changes of our* ***fictitious*** *number, the oxidation number*, and then *interpreted* these changes as either a loss or gain of electrons. This approach does not require a specific detailed microscopic model of the electron or a description of the actual rearrangement or electron transfer process that takes place during a chemical reaction. We only require the observation of the macroscopic reactants and products whose constituent atoms are represented by chemical symbols. We then assign to each atom in the reactants and products the values of the fictitious oxidation numbers. (There are detailed rules describing the assignment procedure. We will not be concerned with the details of this procedure in this course.) Finally, we observe the changes in the values of the assigned numbers. This approach to the rearrangement or electron transfer process describe by oxidation-reduction (often concisely called **redox**) avoids the temptation to treat the electron as if it is a macroscopic object (like a ping-pong ball) and to forget that modern quantum mechanics imposes important constraints on the behavior of an electron (recall the complementarity principle and the Heisenberg indeterminacy principle).

The above definition of redox has produced a set of terms that are routinely used:

1. **Reducing agent:** This is the species that causes the value of the oxidation number of another atom to become *more negative*. In terms of the above interpretations, the reducing agent *transfers electrons to the atom whose oxidation number value becomes more negative.*
2. **Oxidizing agent:** This is the species that causes the value of the oxidation number of another atom to become *more positive*. In terms of the above interpretations, the oxidizing agent *removes electrons from the atom whose oxidation number value becomes more positive.*

Note that the reducing agent is **always oxidized**, and the oxidizing agent is **always reduced**.

There are literally hundreds of thousands of redox reactions in our world. The commonly observed green patina on the roofs and gutters of many public building is the consequence of a redox reaction between copper metal and the carbon dioxide in the atmosphere to produce copper(II) carbonate. This is the origin of the green appearance of the Statue of Liberty. More importantly, as we shall see later, redox is critical to the very processes that sustain life itself. The most important step in the production of high-energy biological molecules depends on redox.

## Acids and Bases

Substances identified by the terms *acid* (from the Latin words *acidus* and *acere*, meaning "sour") and *base* (from the Latin word *basis*, meaning "base" or "pedestal") were known long before chemistry began to emerge as a modern science at the end of the eighteenth and beginning of the nineteenth centuries. In writings dating from the first century of the Christian era, in the works of Islamic alchemists dating from the eighth to the thirteenth centuries, and in the writings of medieval alchemists, there appear to be references to the substance we now know as sulfuric acid. Lavoisier, based on the study of a very small number of acid compounds, proposed (in the mid-1770s) that acids are substances that contain oxygen (in fact, the term *oxygen* comes from two Greek words meaning "acid-former"). In the mid-nineteenth century Justus von Liebig, building on the investigations of Humphry Davy, which characterized the hydrohalic acids (HF, HCl, HBr, and HI) along with $H_2S$ and $H_2Te$, suggested that an acid is a substance containing hydrogen in which the hydrogen can be replaced by a metallic element. Liebig's description of an acid grew out of his extensive study of organic acids and remained the accepted description of acids for nearly half a century. Importantly, it marked a shift from a focus on oxygen (Lavoisier's explanation) to a focus on hydrogen.

In 1884, the Swedish chemist Svante Arrhenius offered what is recognized as the first modern definitions of the terms *acid* and *base*. The proposal Arrhenius made clearly reflected the nineteenth century's focus on electricity and the property of charge recognized by Faraday in his description of anions and cations. Arrhenius defined an **acid** as a substance that, when dissolved in water, increases the concentration of **$H^+$ ions**. A **base** is a substance that, when dissolved in water, increases the concentration of **$OH^-$ ions**. (Note that Arrhenius used the term "ions" and did not call the $H^+$ ions *protons*.) There are two key ideas to note about Arrhenius's definitions. First, his definitions emphasize that the formation of ions (mobile carriers of charge) does **not** require the application of an electric current. The simple process of dissolution (of an acid or base in water) is sufficient to produce ions in the resulting solution, and this is consistent with the observations of ions made by Faraday. Faraday's conclusions about ions was the result of applying an electric current to an aqueous electrolyte solution, but Faraday's experiments did **not** exclude the possibility that the ions were formed as a consequence of the electric current itself. Arrhenius, on the other hand, made the claim that the existence of ions in an aqueous solution is **independent** of an electric current. Second, Arrhenius's definitions apply specifically to **aqueous** solutions. The solvent **must** be water.

Approximately forty years after the proposal by Arrhenius, two chemists, Johannes Brønsted (Danish) and Thomas Lowry (British), independently provided new definitions of the terms *acid* and *base* that continued the focus on hydrogen but removed Arrhenius's restriction of water as the solvent. A **Brønsted-Lowry acid** (or, more simply, a **Brønsted acid)** is defined as a substance that is able to **lose** (that is, "**donate**" or "give up") a proton, while a **Brønsted-Lowry base (Brønsted base)** is defined as a substance that is able to **gain** (that is, "accept") a proton. Hence, a Brønsted-Lowry acid is often called a **proton donor**, and a Brønsted-Lowry base is called a **proton acceptor**. (Note that Brønsted and Lowry use the term *proton*, a twentieth-century term, to describe the behavior of acids and bases.) The Brønsted-Lowry definitions do **not** require water to be the solvent; in addition, the effect of a base is no longer defined in terms of a changing $OH^-$ ion concentration. From a broader perspective, these definitions reformulate the concepts of acids and bases in terms of a single unifying theme: a **proton transfer processes.** While the Brønsted-Lowry definitions are more general, encompassing a wide range of both solvents and species that satisfy the new definitions, the reader should not forget that **every** Arrhenius acid is also a Brønsted-Lowry acid, and **every** Arrhenius base is also a Brønsted-Lowry base.

The view of acids and bases as proton transfer processes parallels our earlier discussion of oxidation and reduction as electron transfer processes. The two types of processes exhibit a symmetrical relationship. Both are understood in terms of the transfer of charges—negative charges in the case of oxidation and reduction, positive charges in the case of acids and bases. Just as redox processes occur in pairs, we can think of acids and bases as occurring in pairs. In the Brønsted-Lowry model, these pairs are called **conjugate pairs**. The **conjugate base of a Brønsted acid** is the species that remains after a proton is donated by the Brønsted acid. The **conjugate acid of a Brønsted base** is the species that is formed by the addition of a proton to a Brønsted base. These definitions mean that acids and bases in the Brønsted-Lowry model always occur as **conjugate acid-base pairs**; figure 5.1 illustrates the symmetry of this relationship.

$$CH_3COOH(aq) \quad + \quad H_2O(l) \quad \Leftrightarrow \quad CH_3COO^-(aq) \quad + \quad H_3O^+(aq)$$

```
    *             *             *             *
    *             *             *             *
    *             *             *             *
    ***************************************   *
  Acid      Conjugatepair      Base           *
                  *                           *
                  *                           *
                  *****************************
                Base       Conjugate pair    Acid
```

Note: This figure contains a new symbol, the double-headed arrow, ⇔, which will be described below when we explore the concept of chemical equilibrium.

Figure 5.1: Brønsted-Lowry Conjugate Acid-Base Pairs

The models of acids and bases proposed by Arrhenius, Brønsted, and Lowry represent only two of the numerous proposals made throughout the twentieth century to describe the rich and varied chemical reactions that have been recognized over a period of several millennia as acid-base reactions. Edward Franklin explored nonaqueous solvents as early as 1905; G. N. Lewis in 1923 (contemporaneously with Brønsted and Lowry and fully elaborated in 1938) suggested an acid-base model that relied on the transfer of electron pairs; in 1938 Mikhail Usanovich proposed a model of acids that is not restricted to hydrogen-containing compounds; at the same time, the German chemist Lux revived an oxygen-based model of acids and bases that was further supported by the work of the Norwegian chemist Flood in 1947; in the early 1960s Pearson advanced the principle of hard and soft acids and bases (the HSAB model), which was made quantitative in collaborative work with Parr in 1984. All of this activity is a testament to the enduring fascination that chemistry has had with the concepts of acids and bases. Note that each of the acid-base models utilized concepts based on the terminology and discoveries of the twentieth century.

The Brønsted-Lowry model provides sufficient power and generality for our study of acids and bases, and it is the model we shall use throughout the remainder of this text. Consequently, we shall treat acids as **proton donors** and bases as **proton acceptors**, thereby emphasizing that the chemistry of acids and bases can be interpreted as a **proton transfer process**. Further, we shall focus our attention, as we did with electrolytes and nonelectrolytes, on the acid and base chemistry of **aqueous** solutions. In doing so it is important to note that naked protons ($H^+$ ions) do not exist independently in such solutions. Rather, the protons are hydrated; that is, they are surrounded by water molecules forming complexes of the form: $H_3O^+$, $H_5O_2^+$, and $H_9O_4^+$. Hydrated protons in water are complex species and are not simply explained. In this text we shall symbolize the hydrated protons by the single formula $H_3O^+$, which is called the **hydronium ion**. Because this symbol will represent **all** the hydrated protons in an aqueous solution, we will eventually use the concentration of $H_3O^+$ ions as a fundamental descriptor of an acid in an aqueous solution.

## Strong and Weak Acids and Bases

Earlier, we met the nineteenth-century concepts of a **strong electrolyte** and a **weak electrolyte**; the distinction between the two ideas was defined in terms of the ability of the resulting solution to conduct electricity (chapter 4). We further noted that an aqueous solution that is a very good conductor of electricity must have a large concentration of mobile charges, while an aqueous solution that is a poor conductor of electricity **cannot** have a large concentration of mobile charges. The developments of the twentieth century have given us a microscopic language with which to describe acid and base molecules as proton donors and proton acceptors. Using the Brønsted-Lowry model, an acid produces mobile charges, $H_3O^+$ ions, in solution by "donating" protons; hence, Brønsted-Lowry acids are electrolytes. Similarly, because a Brønsted-Lowry base is a proton acceptor, in an aqueous environment it gains protons from the ubiquitous water ($H_2O$) molecules, leaving an excess of $OH^-$ ions in solution. Consequently, a Brønsted-Lowry base is an electrolyte. But, like all other electrolytes, Brønsted-Lowry acids and bases can be either strong electrolytes or weak electrolytes. How should we think about Brønsted-Lowry acids and bases as strong and weak electrolytes?

Because a **strong** electrolyte produces a solution with a large concentration of ions, this means that, in an aqueous solution, virtually **every** acid molecule has **donated** a proton to form an $H_3O^+$ ion. Such an acid is called a **strong** acid. In the case of a base, because virtually **every** base molecule has **accepted** a proton, resulting in a large excess of $OH^-$ ions in solution, we call the base a **strong** base. On the other hand, if only a small number of acid molecules **donate** a proton, or only a small number of base molecules **accept** a proton, the acid or base is called **weak**. In discussing acids and bases, the donation or acceptance of a proton is often called **ionization** (because ions are produced). Using this language, virtually **all** the molecules of a **strong** acid or a **strong** base are ionized, while **very few** (in fact, a vanishingly small number!) of the molecules of a **weak** acid or a **weak** base are ionized. The distinction between the concepts of *strong* and *weak* is consistent with a wide range of empirical data, but, as we shall now see, this is **not** the complete story.

Figure 5.2 depicts the qualitative difference in the behavior of strong and weak acids. (The applicability of the figure is limited to solutions whose initial acid concentration is less than or equal to approximately 1 M. At higher concentration, strong acids do not exhibit 100% dissociation.) The figure also shows the curious fact that the extent of ionization of a weak acid **depends** on its initial concentration, while the ionization of a strong acid is **independent** of its initial concentration. This seems to suggest that weak acids are in some way "sensitive to" or "aware of" the concentration of the ions in a solution. If initially there are very few weak acid molecules present (that is, the initial concentration of the acid is **low**) in the solution, almost all of them will ionize. As an increasing number of weak acid molecules are present, fewer and fewer of these molecules ionize. In contrast, the maximum number of strong acid molecules **always** ionizes; the percent ionization is **independent** of the initial acid concentration of the solution (limited to solutions whose initial acid concentration is less than or equal to approximately 1 M). This dichotomy in the behavior of *strong* and *weak* species indicates that a more fundamental physical principle is at work here. This principle is called **chemical equilibrium.**

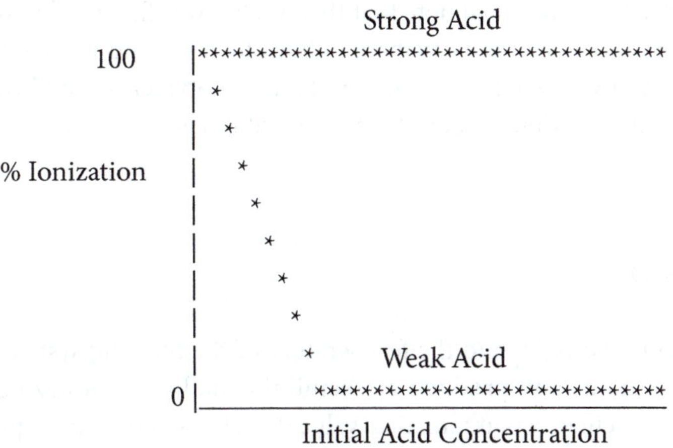


Figure 5.2: Qualitative Ionization of Strong and Weak Acids

## Chemical Equilibrium

In chapter 2, when we first met chemical reactions and the symbolic equations that we use to represent them, the simple suggestion was made that reactions proceed in *only one* direction, from starting materials, the reactants, to chemical products. Further, the implication of a **balanced equation** was that **all** the reactant molecules were converted to product molecules. By the early nineteenth century, however, it had become clear that chemistry is much more interesting and complex than this simple picture implies. In actuality, virtually **no** chemical reaction proceeds either to *completion* or in *only one* direction. Chemical reactions are not simply one-way tunnels through which all the starting materials pass and are rearranged to form products. Chemical reactions are complex systems in which, at any instant in time after a reaction has begun, there is a **mixture** of reactants, products, reactants becoming products, and products returning to reactants. We define the **state of chemical equilibrium** to be the state of a chemical system involved in a reaction when the **concentrations of reactants and products *no longer change with time*.** However, it is important to understand that at equilibrium, even though the concentrations of reactants and products no longer change with time, reactants are still **forming** products, and products are still **reforming** reactants. Chemical equilibrium is a **dynamic** state!

The reader should note that there is a difference between chemical equilibrium and a much simpler **physical equilibrium**. A physical equilibrium involves no chemistry; that is, the chemical characteristics of a substance do not change. For example, an ice cube at 273.15 K is in physical equilibrium with a pool of water at 273.15 K. The solid water changes *phase* to become liquid water, and the liquid water changes *phase* to become solid water (ice), but water and ice are chemically identical.

Consider now the following chemical equilibrium:

$$N_2O_4(g) \Leftrightarrow 2NO_2(g). \tag{5.1}$$

First, notice that equation 5.1 makes use of the new double-headed arrow symbol $\Leftrightarrow$,which is used to indicate that the reactants (conventionally written on the left side of a chemical equation) form products (forward direction), but the products can reform reactants (reverse reaction). That is, the reaction can proceed in both directions. The reader will note that this symbol was first used in figure 5.1 and indicates that the conjugate pairs in the figure participate in both forward and reverse chemical reactions. The molecule in figure 5.1, $CH_3COOH$, is *acetic acid* and is a prototypical example of a *weak acid.* The reacting system symbolized in equation 5.1 admits three possibilities:

1. Initially pure $NO_2$
2. Initially pure $N_2O_4$
3. Initially a mixture of $NO_2$ and $N_2O_4$

Both the initial concentrations of $NO_2$ and $N_2O_4$ and the temperature of the reacting system will determine the concentrations of the reactants and products at chemical equilibrium. Amazingly, for a specified temperature, it is possible to calculate a mathematical value that characterizes the equilibrium state of the system. This mathematical quantity is called the **equilibrium constant**, and it is a true constant at the specified temperature. The mathematical definition of the equilibrium constant is given in equation 5.2:

$$K_{eq} = \frac{[NO_2]_{eq}{}^2}{[N_2O_4]_{eq}}. \tag{5.2}$$

Note in equation 5.2 that the equilibrium concentration (measured in molarity) of the products appears in the numerator, while the equilibrium concentration (again measured in molarity) of reactants appears in the denominator. The subscript "eq" is used in equation 5.2 to emphasize that the equilibrium constant is defined in terms of *equilibrium concentrations.* Further, and very importantly, each concentration is raised to the power of the stoichiometric coefficients from the balanced equation. The product stoichiometric coefficient is 2, while the reactant stoichiometric coefficient is 1. The numerical value for $K_{eq}$ is reported, by convention, **without** units. Hence, it is considered to be a dimensionless quantity whose meaning is specified by the accompanying chemical equation. As a result, an equilibrium constant value **must always** be accompanied by the chemical equation used to calculate it.

Consider the following *general* chemical equation (which we assume to be balanced):

$$a\,A + b\,B \Leftrightarrow c\,C + d\,D. \tag{5.3}$$

In equation 5.3, *A* and *B* are reactants, *C* and *D* are products, and the stoichiometric coefficients are *a*, *b*, *c*, and *d*. We can now write the equilibrium constant expression for this general equation:

$$K_{eq} = \frac{[C]_{eq}{}^c [D]_{eq}{}^d}{[A]_{eq}{}^a [B]_{eq}{}^b}. \tag{5.4}$$

The reader should take special note that the equilibrium concentration of each reactant and each product is raised to the power of its corresponding stoichiometric coefficient.

On the one hand, the numerical value of the equilibrium constant calculated in equation 5.2 provides a *quantitative* characterization of a chemical reaction's equilibrium state at a specific temperature. However, the magnitude of $K_{eq}$, because it is a ratio of product concentrations (numerator) and reactant concentrations (denominator), also provides an important *approximate qualitative* understanding of the chemical system. At a specific temperature, if $K_{eq} > 1$, when the reaction reaches an equilibrium state, the concentration of the **products is greater than the concentration of the reactants** (the value of the numerator is **greater** than the value of the denominator); at a specific temperature, if $K_{eq} < 1$, when the reaction reaches an equilibrium state, the concentration of the **products is less than the concentration of the reactants** (the value of the numerator is **less** than the value of the denominator). Finally, again at a specific temperature, if $K_{eq} \approx 1$, when the reaction reaches an equilibrium state, the concentrations of the reactants and the products are approximately equal; the equilibrium state of the reaction favors **neither** the formation of products from the reactants **nor** the reformation of reactants from the products.

Example 5.1: Assume that the following reaction is in its equilibrium state at $T = 525$ K.

$N_2(g) + 3H_2(g) \Leftrightarrow 2NH_3(g)$

The equilibrium concentrations are as follows:

$[N_2] = 0.125$ mol $L^{-1}$

$[H_2] = 0.115$ mol $L^{-1}$

$[NH_3] = 0.459$ mol $L^{-1}$

Calculate the equilibrium constant for this reaction at $T = 525$ K.

Answer:

**Organize:** Balanced chemical equation:

$N_2(g) + 3H_2(g) \Leftrightarrow 2NH_3(g)$

Equilibrium concentrations of all three species at $T = 525$ K.

**Unknown:** Equilibrium constant, $K_{eq}$

**Translate:** Use equation 5.4.

Pay close attention to the superscripts in the equation.

**Solve:**

$$K_{eq} = \frac{[NH_3]_{eq}^{\ 2}}{[N_2]_{eq}[H_2]_{eq}^{\ 3}}$$

$$K_{eq} = \frac{(0.459 \text{ mol L}^{-1})^2}{(0.125 \text{ mol L}^{-1}) \times (0.115 \text{ mol L}^{-1})^3}$$

$K_{eq} = 1.11 \times 10^3$

(Note: Three significant figures; dimensionless)

## Forward or Reverse: Which Way Does the Reaction Go?

Because the process of attaining chemical equilibrium is fundamentally a process of achieving the most probable arrangement of reactants and products, as well as the most probable distribution of the energy associated with the reaction, we may anticipate that this process is highly complex. Indeed, it is not surprising that the study of the intricate dynamics of chemical systems as they achieve equilibrium has occupied humankind's attention for well over a century. We seek to ask a simpler question: If a system that has achieved chemical equilibrium is disturbed (for example, by changing the pressure, the volume, the number of moles of reactants and/or products, or the temperature of the reacting system), can we predict, using the balanced equation, how the system will respond to the disturbance? In the mid-1880s, a French chemist, Henry Luis Le Châtelier, and a German chemist, Karl Ferdinand Braun, independently enunciated a principle based on their empirical observations. (While Braun did advance a proof of this principle, the proof contained flaws that were later identified by Paul Ehrenfest in a paper published in 1911.) Known as Le Châtelier's principle (or, more accurately, as the Le Châtelier-Braun principle), it is often used to analyze the response of a chemical system to one of the disturbances noted above. However, because the usual formulation of Le Châtelier's principle lacks sufficient precision (this is the formulation originally used by both Le Châtelier and Braun), it is often possible to use it to make **incorrect** predictions. There is a better approach.

Consider, again, the general chemical equation in equation 5.3. We now **define** a new quantity called the **reaction quotient, *Q***:

$$Q = \frac{[C]^c[D]^d}{[A]^a[B]^b}. \tag{5.5}$$

Note that the *reaction quotient* defined in equation 5.5 looks exactly like the *equilibrium constant* defined in equation 5.4, **except** that the *reaction quotient* does **not** use the equilibrium concentrations of the reactants and products. This means that the reaction quotient can be calculated at **any time** during the course of a chemical reaction; there is **no requirement** that the chemical reaction must attain equilibrium before the reaction quotient is calculated. Consequently, unlike the *equilibrium constant*, which has a **unique** value determined by the *temperature* and *equilibrium concentrations* of the reactants and the products, the *reaction quotient* has a **different** value at each instant of time during the course of a chemical reaction. In particular, the reaction quotient can be calculated at the instant in time when a disturbance is introduced in a system currently at equilibrium. (Because the system is at equilibrium, the value of $K_{eq}$ is known.)

At the instant of time when the disturbance is introduced, $Q$ can be calculated. If $Q < K_{eq}$, the reaction will proceed in the direction of products (that is, producing more products from the reactants) until a **new** equilibrium is attained by the system. (We say that the reaction proceeds in the "forward" direction.) If the temperature at the new equilibrium is the **same** as the temperature of the initial equilibrium, the concentration of reactants and products at the new equilibrium will satisfy equation 5.4, yielding the same $K_{eq}$ as the original equilibrium. On the other hand, if $Q > K_{eq}$ after the disturbance, the reaction will proceed in the direction of reactants (that is, producing more reactants from the products) until a **new** equilibrium is attained by the system. (We say that the reaction proceeds in the "reverse" direction.) Again, if the temperature at the new equilibrium is the **same** as the temperature of the initial equilibrium, the concentration of reactants and products at the new equilibrium will satisfy equation 5.4, yielding the same $K_{eq}$ as the original equilibrium. Consequently, by examining the relationship between $Q$ and $K_{eq}$ we can *predict* the behavior of the reacting system. Clearly, if $Q = K_{eq}$ (and assuming that there has been **no** temperature change), the system is still at equilibrium and the disturbance will cause **no** change in the reacting system.

The reader will note that in the above description of the relationship between $Q$ and $K_{eq}$, we have been very careful to require that the temperature **does not change**; the above analysis depends on the constancy of the temperature, because a change in the temperature **will change the value of $K_{eq}$**. The science of thermodynamics provides a powerful tool, the van't Hoff Equation, which quantitatively determines the effect of a temperature change on the value of $K_{eq}$ if the reaction takes place under conditions of constant pressure. We shall not investigate this additional level of complexity further; an understanding the relationship between $Q$ and $K_{eq}$ when the temperature remains constant provides a powerful insight into the equilibrium behavior of a sufficiently large class of chemical reactions for the purposes of this text.

Example 5.2: Assume that the following reaction has reached its equilibrium state at $T = 373.15$ K.

$$SO_2Cl_2(g) \Leftrightarrow SO_2(g) + Cl_2(g)$$

The equilibrium constant, $K_{eq,}$ at 373.15 K is 0.078.

Suppose that the following concentrations of reactants and products are combined in a flask and heated to 373.15 K:

$[SO_2] = 0.195$ mol L$^{-1}$
$[Cl_2] = 0.140$ mol L$^{-1}$
$[SO_2Cl_2] = 0.335$ mol L$^{-1}$

What is the value of $Q$ **before** the reaction begins?

Will the reaction produce more products or more reactants to reestablish the equilibrium state?

Answer:

**Organize:** Balanced chemical equation:

$$SO_2Cl_2(g) \Leftrightarrow SO_2(g) + Cl_2(g)$$

New concentrations of all three species combined and heated to $T = 373.15$ K.

**Unknown:** What is the value of Q, the reaction quotient?
Does the reaction produce more products or more reactants to reestablish the equilibrium state?

**Translate:** Use equation 5.5.
Pay close attention to the superscripts in the equation.

**Solve:**

$$Q = \frac{[SO_2][Cl_2]}{[SO_2Cl_2]}$$

$$Q = \frac{(0.195 \text{ mol L}^{-1}) \times (0.140 \text{ mol L}^{-1})}{(0.335 \text{ mol L}^{-1})}$$

$Q = 8.1 \times 10^{-2}$
(Note: Two significant figures; dimensionless)
$K_{eq} = 7.8 \times 10^{-2}$
Because $Q > K_{eq}$, the reaction will produce more **reactants** in order to reestablish the equilibrium state. That is, the reaction will **decrease** concentrations of $SO_2$ and $Cl_2$ and **increase** the concentration of $SO_2Cl_2$ in order to reestablish the equilibrium state at $T = 373.15$ K.

## Weak Acid Ionization Constants

Now that we have briefly explored some of the amazing characteristics of chemical equilibrium and the calculation of the equilibrium constant, why are these ideas important to the study of acids and bases? We have already noted that virtually **all** the molecules of a strong acid or a strong base in a solution are ionized, while **very few** (remember, a vanishingly small number!) of the molecules of a weak acid or a weak base in a solution are ionized. We can now understand this dichotomous behavior by using the framework of chemical equilibrium. Let's begin by considering a monoprotic weak acid, which we symbolize by HA (the "A" symbolizes the conjugate base of the acid) and examining its behavior in an aqueous solution:

$$HA(aq) + H_2O(\ell) \Leftrightarrow H_3O^{+}(aq) + A^{-}(aq). \tag{5.6}$$

Because the behavior of HA in an aqueous solution is an **equilibrium process**, we can write the equilibrium constant associated with equation 5.6 as follows:

$$K_{eq} = \frac{[H_3O^+]_{eq}[A^-]_{eq}}{[HA]_{eq}[H_2O]_{eq}}. \tag{5.7}$$

However, because HA is a **weak acid** and only a vanishingly small number of the HA molecules dissociate in the solution, the $[H_2O]_{eq}$ is essentially the same as the original $[H_2O]$. That is, because $[H_2O]_{eq} = [H_2O]$, the $[H_2O]$ is effectively a **constant**. Because the $[H_2O]$ is effectively a constant, it is conventional to make the following **definition**:

$$K_a = K_{eq} \times [H_2O]_{eq}. \tag{5.8}$$

The newly defined quantity, $K_a$, is called the **acid ionization constant**, and it possesses all the characteristics of any other equilibrium constant. In particular, it is **temperature-dependent**. Using this new definition, equation 5.7 can be rewritten as follows:

$$K_a = \frac{[H_3O^+]_{eq}[A^-]_{eq}}{[HA]_{eq}}. \tag{5.9}$$

The quantity $K_a$ is indeed a **constant**, and at a specific temperature, it is a measure of the strength of the weak acid because it is a direct measure of the weak acid's degree of ionization. The reader should carefully note that, except for *extremely* dilute solutions (see figure 5.2), at equilibrium, the predominant species in an aqueous solution is the un-ionized acid (note that the typical values of $K_a$ in table 5.1 are significantly less than 1.0). Further, the behavior of weak acids in an *extremely* dilute solution (again, see figure 5.2) can be understood within the framework of chemical equilibrium. The ionization constant for a weak acid limits the number of ionized acid molecules to a vanishingly small number; if there are only a vanishingly small number of acid molecules to begin with (the definition of an *extremely* dilute solution), nearly 100% of them are ionized. But even as the number of acid molecules increases (increasing concentration), the ionization constant still limits the number of ionized acid molecules in the solution. Consequently, the percent ionization for a weak acid drops to nearly zero as the concentration of the acid in an aqueous solution increases, consistent with figure 5.2.

Example 5.3: Use information from table 5.1 to calculate the $[H_3O^+]$ at 298.15 K after the equilibrium state has been reached for the following reaction:

$$C_6H_5COOH(aq) + H_2O(\ell) \Leftrightarrow C_6H_5COO^-(aq) + H_3O^+(aq)$$

Assume that the equilibrium concentration of $C_6H_5COOH$ is 0.175 M.

Answer:

**Organize:** Balanced chemical equation:

$$C_6H_5COOH(aq) + H_2O(\ell) \Leftrightarrow C_6H_5COO^-(aq) + H_3O^+(aq).$$

$[C_6H_5COOH]_{eq} = 0.175$ M; $T = 298.15$ K

**Unknown:** $[H_3O^+]_{eq}$

**Translate:** Use equation 5.9.

**Solve:**

$$K_a = \frac{[H_3O^+]_{eq}[A^-]_{eq}}{[HA]_{eq}}$$

$$\left[H_3O^+\right]_{eq} = \frac{K_a \times [C_6H_5COOH]_{eq}}{[C_6H_5COO^-]_{eq}}$$

But each time a $H_3O+$ ion is formed, a $C_6H_5COO^-$ ion is also formed.

Hence, $[H_3O^+]_{eq} = [C_6H_5COO^-]_{eq}$.

We can use this to rewrite the last equation:

$[H_3O^+]_{eq}^{\ 2} = K_a \times [C_6H_5COOH]_{eq}$

$[H_3O^+]_{eq}^{\ 2} = (6.46 \times 10^{-5})(0.175)$

$[H_3O^+]_{eq} = [\ (6.46 \times 10^{-5})(0.175)\ ]^{½}$

$[H_3O^+]_{eq} = 3.36 \times 10^{-3}$ M

(Note: Three significant figures)

Table 5.1: Weak Acid Ionization Constants at 298.15 K

| Acid | Formula | Conjugate Base | Ka |
|---|---|---|---|
| Hydrofluoric | HF | $F^-$ | $6.6 \times 10^{-4}$ |
| Formic | HCOOH | $HCOO^-$ | $1.77 \times 10^{-4}$ |
| Benzoic | $C_6H_5COOH$ | $C_6H_5COO^-$ | $6.46 \times 10^{-5}$ |
| Acetic | $CH_3COOH$ | $CH_3COO^-$ | $1.76 \times 10^{-5}$ |
| Carbonic[1] | $H_2CO_3$ | $HCO_3^-$ | $4.3 \times 10^{-7}$ |
| Hydrocyanic | HCN | $CN^-$ | $6.17 \times 10^{-10}$ |
| Ammonium Ion | $NH_4^+$ | $NH_3$ | $5.6 \times 10^{-10}$ |
| Carbonic[2] | $HCO_3^-$ | $CO_3^{2-}$ | $4.8 \times 10^{-11}$ |
| Water | $H_2O$ | $OH^-$ | $1.8 \times 10^{-16}$ |

[1] Carbonic acid is the only polyprotic acid in this list; it is capable of losing two $H^+$ ions. This ionization constant is the value for the loss of the first $H^+$ ion.

[2] This ionization constant is the value for the loss of the second $H^+$ ion.

## Weak Base Ionization Constants

Now consider the behavior of a weak base. We will symbolize our prototypical weak base by B and examine the following reaction:

$$B(aq) + H_2O(\ell) \Leftrightarrow HB^+(aq) + OH^-(aq). \tag{5.10}$$

Because the behavior of the base, B, in an aqueous solution is an **equilibrium process**, we can write the equilibrium constant associated with equation 5.10 as follows:

$$K_{eq} = \frac{[HB^+]_{eq}[OH^-]_{eq}}{[B]_{eq}[H_2O]_{eq}}. \tag{5.11}$$

Because B is a **weak base**, only a vanishingly small number of the B molecules accept a $H^+$ ion from a $H_2O$ molecule, making the $[H_2O]_{eq}$ essentially the same as the original $[H_2O]$. Just as in the case of a weak acid, because $[H_2O]_{eq} = [H_2O]$, the $[H_2O]$ is effectively a **constant**. Because the $[H_2O]$ is effectively a constant, it is conventional to make the following **definition**:

$$K_b = K_{eq} \times [H_2O]_{eq}. \tag{5.12}$$

The newly defined quantity, $K_b$, is called the **base ionization constant**, and it possesses all the characteristics of any other equilibrium constant. In particular, it is **temperature-dependent**. Using this new definition, equation 5.11 can be rewritten as follows:

$$K_b = \frac{[HB^+]_{eq}[OH^-]_{eq}}{[B]_{eq}}. \tag{5.13}$$

The quantity $K_b$ is indeed a **constant**, and at a specific temperature, it is a measure of the strength of the weak base because it is a direct measure of the weak base's ability to accept a proton.

## The Two Faces of Water

Up to this point in our discussion, we have focused our attention on aqueous solutions, solutions in which the solvent is water. This is not to say that other solvent systems are unimportant, for that certainly is not the case! Many important industrial processes depend critically on nonaqueous solvent systems. Further, in focusing on aqueous solutions, virtually nothing has been said about the characteristics of water itself. In fact, water is a remarkable substance that both dominates our environment (more than 70% of the Earth's surface is covered by water) and is essential to the life processes of all living organisms on the planet. As we shall soon see, the water molecule is central to **all** the biochemical processes that define life as we know it.

In the context of our current discussion examining the properties of acids and bases, water is distinguished by simultaneously being **both** a weak acid and a weak base. We say that water is **amphoteric.** Consider the following reaction:

$$H_2O(\ell) + H_2O(\ell) \Leftrightarrow H_3O^+(aq) + OH^-(aq). \tag{5.14}$$

In equation 5.14 note that one of the water molecules acts as a **proton donor** (an acid), while the other water molecules acts as a **proton acceptor** (a base). The process described by equation 5.14 is often called the **autoionization** of water because two charged species, one positively charged and one negatively charged (represented by the symbols $H_3O^+$ and $OH^-$), are produced. Significantly, water is a **weak electrolyte** and, consequently, behaves as both a **weak acid** and a **weak base.** We can write an expression for the water equilibrium constant that accompanies equation 5.14:

$$K_{eq} = \frac{[H_3O^+]_{eq}[OH^-]_{eq}}{[H_2O]_{eq}^2}. \tag{5.15}$$

However, because water is a **weak acid (base)** and, as a result, only a vanishingly small number of the water molecules either accept or donate a proton, the $[H_2O]_{eq}$ is essentially the same as the original $[H_2O]$. That is, because $[H_2O]_{eq} = [H_2O]$, the $[H_2O]$ is effectively a **constant.** Because the $[H_2O]$ is effectively a constant, it is conventional to make the following **definition**:

$$K_w = K_{eq} \times [H_2O]_{eq}^{\ 2}. \tag{5.16}$$

The newly defined quantity, $K_w$, is called the **ion product constant of water**, and it possesses all the characteristics of any other equilibrium constant. In particular, it is **temperature-dependent**. Using this new definition, equation 5.15 can be rewritten as follows:

$$K_w = [H_3O^+]_{eq}[OH^-]_{eq}. \tag{5.17}$$

The form of equation 5.17 explains the name given to $K_w$; in pure water at 298.15 K, the ion-product constant of water, $K_w$, is equal to $1.00 \times 10^{-14}$.

## How Are $K_a$, $K_b$, and $K_w$ Related?

We have now defined the three different quantities, $K_a$, $K_b$, and $K_w$, which characterize the behavior of weak acids and weak bases in aqueous solutions as well as the behavior of the solvent (water) itself. On the one hand, the fact that we need **three** different mathematical quantities to characterize aqueous solutions may seem intimidating and even overwhelming. You might be tempted to throw up your hands and shout, "This is too complicated!" The first thing to keep in mind is that all three, $K_a$, $K_b$, and $K_w$, are nothing more than equilibrium constants that inform us about a molecule's ability to either donate or accept a proton. Remember, because all three are equilibrium constants, **all three** depend on temperature.

However, there is an amazing relationship connecting the three different equilibrium constants that characterize the behavior of weak acids, weak bases, and water in aqueous solutions. To explore this relationship we will use a prototypical conjugate acid-base pair, $NH_4$+/$NH_3$. We begin by writing the weak acid equilibrium chemical equation:

$$NH_4^+(aq) + H_2O(\ell) \Leftrightarrow NH_3(aq) + H_3O^+(aq). \tag{5.18}$$

(Note that in equation 5.18 the water solvent is acting as a base.) The expression for the weak acid ionization constant, $K_a$, is as follows:

$$K_a = \frac{[NH_3]_{eq}[H_3O^+]_{eq}}{[NH_4^+]_{eq}}. \tag{5.19}$$

How does the conjugate base, $NH_3$, behave in an aqueous solution? The chemistry is described by the following equation:

$$NH_3(aq) + H_2O(\ell) \Leftrightarrow NH_4^+(aq) + OH^-(aq). \tag{5.20}$$

(This time, the water molecule is acting as an acid.) The expression for the weak base ionization constant, $K_b$, is as follows:

$$K_b = \frac{[NH_4^+]_{eq}[OH^-]_{eq}}{[NH_3]_{eq}} \tag{5.21}$$

Now, using equation 5.19 and equation 5.21, we can calculate the quantity $K_a \times K_b$:

$$K_a \times K_b = \frac{[NH_3]_{eq}[H_3O^+]_{eq}}{[NH_4^+]_{eq}} \times \frac{[NH_4^+]_{eq}[OH^-]_{eq}}{[NH_3]_{eq}}. \quad (5.22)$$

Simplifying equation 5.22 yields a remarkably simple relationship:

$$K_w = K_a \times K_b = [H_3O^+]_{eq}[OH^-]_{eq}. \quad (5.23)$$

While we have demonstrated the relationship among $K_a$, $K_b$, and $K_w$ by using a prototypical conjugate pair, one weak acid and one weak base, we must emphasize that this relationship is a completely **general**; it is accurate for **all** conjugate pairs of weak acids and weak bases in an aqueous solution. In particular, equation 5.23 applies to pure water at 298.15 K.

Example 5.4: Use the information from table 5.1 to calculate the $[OH^-]$ at 298.15 K after the equilibrium state has been reached for the following reaction:

$$NH_3(aq) + H_2O(\ell) \Leftrightarrow NH_4^+(aq) + OH^-(aq)$$

Assume that the equilibrium concentration of $NH_3$ is 0.315 M.

Answer: **Organize:** Balanced chemical equation:

$NH_3(aq) + H_2O(\ell) \Leftrightarrow NH_4^+(aq) + OH^-(aq)$

$T = 298.15$ K

$[NH_3]_{eq} = 0.315$ M

**Unknown:** $[OH^-]_{eq}$

**Translate:** Use equation 5.21 along with equation 5.23.

**Solve:**

$$K_b = \frac{[NH_4^+]_{eq}[OH^-]_{eq}}{[NH_3]_{eq}}$$

From equation 5.23 we have $K_w = K_a \times K_b$

Hence, $K_b = K_w \div K_a$

Substituting for $K_b$ gives

$$K_w \div K_a = \frac{[NH_4^+]_{eq}[OH^-]_{eq}}{[NH_3]_{eq}}$$

Rearrange this last equation to get

$$[OH^-]_{eq} = \frac{K_w \times [NH_3]_{eq}}{K_a \times [NH_4^+]_{eq}}$$

But each time an $OH^-$ ion is formed, an $NH_4^+$ ion is also formed.

Hence, $[OH^-]_{eq} = [NH_4^+]_{eq}$

We can use this to rewrite the last equation:

$$[OH^-]_{eq}^{\ 2} = \frac{K_w \times [NH_3]_{eq}}{K_a}$$

$[OH^-]_{eq} = [(1.00 \times 10^{-14})(0.315) \div (5.6 \times 10^{-10})]^{½}$
$[OH^-]_{eq} = 2.37 \times 10^{-3}$ M
(Note: Three significant figures)

## Acid and Base Concentrations

In our discussion up to this point, we have categorized acids and bases using the terms *strong* and *weak*; the distinction between the two terms rests on the ability of an aqueous solution to conduct electricity. An aqueous solution of a strong acid or a strong base is a **very good** conductor of electricity, while an aqueous solution of a weak acid or a weak base is a **poor** conductor of electricity. However, the Brønsted-Lowry model of acids and bases (and the Arrhenius model as well), along with the concept of chemical equilibrium, suggests another approach to the classification of acids and bases. This alternate classification approach is based on the **concentration of $H_3O^+$ ions** in an aqueous solution of an acid or base. This approach offers several attractive characteristics.

First, by focusing on the **$H_3O^+$ ion concentration** in an aqueous solution, we have the opportunity to use **pure water at a specified temperature** as the reference state of the new classification system. Consequently, we are using a uniquely ubiquitous substance as the foundation of this method of classification. From equation 5.14 we have the following chemistry:

$$H_2O(\ell) + H_2O(\ell) \Leftrightarrow H_3O^+(aq) + OH^-(aq). \tag{5.24}$$

This means that, in a sample of **absolutely pure** water at 298.15 K, the numbers of $H_3O^+$ ions and $OH^-$ ions are equal; that is, $[H_3O^+]_{eq} = [OH^-]_{eq}$. This is a unique state that is then used to make the following **definitions**:

1. If $[H_3O^+] = [OH^-]$, the solution is **NEUTRAL.**
2. If $[H_3O^+] > [OH^-]$, the solution is **ACIDIC.** (5.25)
3. If $[H_3O^+] < [OH^-]$, the solution is **BASIC.**

From equation 5.23 we also have the relationship $K_w = [H_3O^+]_{eq}[OH^-]_{eq}$, where, in pure water at 298.15 K, $K_w$ is equal to $1.00 \times 10^{-14}$ and $[H_3O^+]_{eq} = [OH^-]_{eq}$. Consequently, under these specified conditions for water, we can conclude that

$$[H_3O^+] = 1.00 \times 10^{-7}\ M$$

and

$$[OH^-] = 1.00 \times 10^{-7}\ M\,. \tag{5.26}$$

The ability to assign quantitative values to the $H_3O^+$ and $OH^-$ ion concentrations of the unique state of pure water at 298.15 K means that we now have the opportunity to define a quantitative scale by which to measure the acidity of any aqueous solution. This is exactly the same procedure that was used to define temperature scales in chapter 1.

## The pH Scale

At the beginning of the twentieth century a quantitative scale to measure the acidity of aqueous solutions was proposed by the Danish chemist Søren Sørensen; it was further refined in 1924 to incorporate measurements based on electrochemical cells (chemical systems in which either electrical currents were produced by the chemistry or electrical currents were necessary for a chemical reaction to take place). Using the base-10 logarithm (see chapter 1), the **pH** of an aqueous solution is defined as follows:

$$\mathrm{pH} = -\log[\mathrm{H_3O^+}]. \tag{5.27}$$

In equation 5.27, pH is a **dimensionless** quantity, and the molarity units of $[H_3O^+]$ are ignored. (More precisely, the $[H_3O^+]$ is divided by a reference state of 1 M, so the units in the logarithm cancel but the value of $[H_3O^+]$ remains unchanged. Even this description is a simplification, but one that is sufficient for the purposes of this text.) This means that at 298.15 K

For a **neutral** aqueous solution, $[H_3O^+] = 1.00 \times 10^{-7}$ M → pH = 7.00.
For an **acidic** aqueous solution, $[H_3O^+] > 1.00 \times 10^{-7}$ M → pH < 7.00.
For a **basic** aqueous solution, $[H_3O^+] < 1.00 \times 10^{-7}$ M → pH > 7.00. (5.28)

In a similar fashion, we can define the **pOH** of an aqueous solution as follows:

$$\mathrm{pOH} = -\log[\mathrm{OH^-}]. \tag{5.29}$$

Just like the pH, the pOH is a **dimensionless** quantity. Further note that equation 5.30 describes the relationship between the pH and the pOH of an aqueous solution at 298.15 K:

$$\mathrm{pH} + \mathrm{pOH} = 14.00. \tag{5.30}$$

In an aqueous solution, the definition of pH establishes a scale whose values range from 0 to 14. (Recall that in our discussion of figure 5.2, we specified that the maximum initial strong acid concentration of a solution must be less than or equal to approximately 1 M. Because strong acids do **not** exhibit 100% dissociation at concentrations greater than approximately 1M, it is inappropriate to apply the pH scale to such highly concentrated solutions of strong acids. In fact, the Brønsted-Lowry model of acids and bases fails to describe accurately the chemistry of such solutions. While it is possible to apply the mathematical definition of pH to more concentrated strong acid solutions and calculate a **negative** pH value, only **positive** pH values carry significant meaning. Consequently, ***never* report negative**

**pH values.**) Figure 5.3 displays the pH scale, as well as the approximate pH of a number of commonly occurring substances.

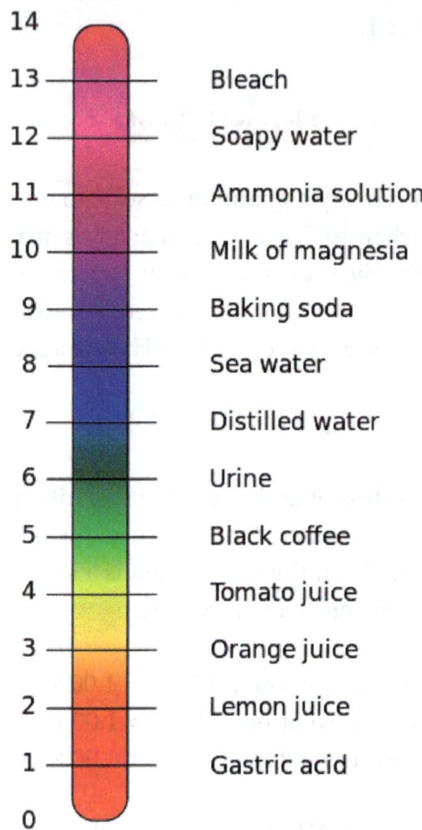


Figure 5.3: pH Scale

Example 5.5: For each of the following equilibrium concentrations in an aqueous solution at 298.15 K, calculate the pH of the solution:

a. $[H_3O^+]_{eq} = 0.257$ M
b. $[H_3O^+]_{eq} = 1.58 \times 10^{-10}$ M
c. $[OH^-]_{eq} = 0.557$ M

Answer:

**Organize:** Three concentrations given.

**Unknown:** The corresponding pH of the solution.

**Translate:** Use equation 5.27.

**Solve:** $pH = -\log[H_3O^+]$

a. $[H_3O+]_{eq} = 0.257$ M
$pH = -\log(0.257)$
$pH = 0.590$

(Note: Three digits to the right of the decimal)

b. $[H_3O^+]_{eq} = 1.58 \times 10^{-10}$ M

$pH = -\log(1.58 \times 10^{-10})$
$pH = 9.801$
(Note: Three digits to the right of the decimal)

c. $[OH^-]_{eq} = 0.557$ M

We must use equation 5.23 because we need $[H_3O^+]$ to calculate pH.

$K_w = [H_3O^+]_{eq}[OH^-]_{eq}$
$[H_3O^+]_{eq} = K_w \div \mathbf{[OH^-]}_{eq}$
$[H_3O^+]_{eq} = 1.00 \times 10^{-14} \div 0.557$ M
$[H_3O^+]_{eq} = 1.80 \times 10^{-14}$
$pH = 13.746$
(Note: Three digits to the right of the decimal)

Example 5.6: For each of the following pH values (measured for the equilibrium state in an aqueous solution at 298.15 K), calculate the $[H_3O^+]_{eq}$ of the solution:

a. pH = 8.15
b. pH = 3.75
c. pH = 4.651

Answer:

**Organize:** Three pH values given
**Unknown:** The corresponding $[H_3O^+]_{eq}$ of the solution
**Translate:** Use the inverse of equation 5.27; see chapter 1 for a discussion of logarithms.
**Solve:** $[H_3O^+] = 10^{-pH}$

a. pH = 8.15
$[H_3O^+]_{eq} = 10^{-8.15}$
$[H_3O^+]_{eq} = 7.1 \times 10^{-9}$ M
(Note: Two significant figures)

b. pH = 3.75
$[H_3O^+]_{eq} = 10^{-3.75}$
$[H_3O^+]_{eq} = 1.8 \times 10^{-4}$ M
(Note: Two significant figures)

c. pH = 4.651
$[H_3O^+]_{eq} = 10^{-4.651}$
$[H_3O^+]_{eq} = 2.23 \times 10^{-5}$ M
(Note: Three significant figures)

Example 5.7: For each of the following pH values (measured for the equilibrium state in an aqueous solution at 298.15 K), calculate the pOH of the solution:

a. pH = 11.45
b. pH = 1.89
c. pH = 6.90

Answer:

**Organize:** Three pH values given
**Unknown:** The corresponding pOH of the solution
**Translate:** Use equation 5.30
**Solve:** pH + pOH = 14.00

a. pH = 11.45
pOH = 14.00 – 11.45
pOH = 2.55
(Note: Three significant figures)

b. pH = 1.89
pOH = 14.00 – 1.89
pOH = 12.11
(Note: Four significant figures)

c. pH = 6.90
pOH = 14.00 – 6.90
pOH = 7.10
(Note: Three significant figures)

In the same spirit as the logarithmic definition of pH, using the acid ionization constant, $K_a$, we can define a quantity called the ***p*$K_a$** in the following way:

$$pK_a = -\log[K_a]. \tag{5.31}$$

The $pK_a$ is dimensionless and provides a simple and direct way to compare the strengths of acids; the **smaller** the $pK_a$ value, the **stronger** the acid. Unlike pH, the values of $pK_a$ **can be negative.**

## Buffers

The most elementary chemical reaction that occurs when an acid and a base are mixed is called a **neutralization** reaction. That is,

$$\text{Acid} + \text{Base} \rightarrow \text{Salt} + \text{Water}.$$

A **salt** is defined as a compound composed of a **cation** (other than $H^+$) and an **anion** (other than $OH^-$ or $O^{2-}$). The prototypical neutralization reaction is

$$HCl(aq) + NaOH(aq) \rightarrow NaCl(aq) + H_2O(\ell). \quad (5.32)$$

A neutralization reaction in an aqueous solution involves the generation of **both** hydronium ions ($H_3O^+$) from the acid and hydroxide ions ($OH^-$) from the base, which then react to form $H_2O(\ell)$ (see equation 5.24). The final pH of the solution (after neutralization) depends on the behavior of the salt; acidic (pH < 7), basic (pH > 7), and neutral (pH = 7) solutions are all possible results. (While the three possibilities are easily explained by either weak acid or weak base equilibrium reactions, in the interest of brevity, we omit the details.)

However, suppose that **only** $H_3O^+$ ions or **only** $OH^-$ ions are added to an aqueous solution. What can happen to the solution's pH? The obvious answer (based on the definitions of pH and pOH, equation 5.27 and equation 5.29) is that the pH of the solution will either **dramatically** decrease (that is, become more acidic because of an increase in the $[H_3O^+]$) or **dramatically** increase (that is, become more basic because of an increase in $[OH^-]$). But there is a third possibility! The solution to which **only** $H_3O^+$ ions or **only** $OH^-$ ions are added may be able to keep its pH value nearly unchanged. Such a solution is called a **buffer.**

Formally, a **buffer** is defined to be a solution (here we continue to restrict our focus to aqueous solutions) that **resists** changes to its pH when **small** amounts of an acid or a base are added to the solution. A **buffer** (**buffer solution**) is a solution composed of *two* parts:

1. A weak conjugate acid-base pair
2. A salt derived from the weak acid or the weak base

Both components **must** be present in the solution. For example, a buffer could be an aqueous solution containing $CH_3COOH/CH_3COO^-$ (acetic acid/acetate ion, a weak conjugate acid-base pair) and $NaCH_3COO$ (sodium acetate), the salt formed from the conjugate base of the weak acid. Another possible buffer could be formed from $NH_3/NH_4+$ (ammonia/ammonium ion, a weak conjugate acid-base pair) and $(NH_4)_2SO_4$ (ammonium sulfate), the salt formed from the conjugate acid of the weak base. It is important to note that buffers can be overwhelmed; the definition states that a buffer will **resist** changes to its pH following the addition of only ***small*** amounts of either an acid or a base. If large quantities of either an acid or a base are added to a buffer solution, it may be impossible for the buffer to prevent a change in its pH. Second, note that a buffer ***resists*** changes to its pH following the addition of only **small** amounts of either an acid or a base. That is, the pH *will* change, but the change in pH **will be small** (remember that the quantity of added acid or added base is small). Buffers are most effective when the desired pH nearly equals the $pK_a$ of the weak acid of the weak conjugate acid-base pair used to construct a buffer. In fact, a buffer continues to be effective if the pH = $pK_a \pm 1$. (A buffer constructed from a weak base and its conjugate acid possesses a pOH nearly equal to the $pK_b$ of the weak base.)

Because thousands of the chemical reactions occurring in aqueous solutions are pH-dependent, the role of buffers cannot be overstated; the choice of an appropriate buffer is often crucial to completing successfully either a single or a complex series of chemical reactions. Many life forms on this planet are highly susceptible either to the pH of their external environments or to their internal physiological pH. The acidification of lakes in Eastern North America in the twentieth century as a result of acid rain (rainwater with very low pH values) and the consequential eradication of life forms in those lakes are well-documented. Human physiology critically depends on several chemical buffer systems as well as the response of several organ systems to maintain the pH of both the extracellular fluid (ECF) and the intracellular fluid (ICF). The most important chemical buffer that maintains the blood's pH at a narrow range of 7.35 to 7.45 (with a nominal value of approximately 7.40) is the carbonic acid/bicarbonate ion buffer that participates in the multiple equilibria depicted in equation 5.33:

$$2H_2O(\ell) + CO_2(aq) \Leftrightarrow H_2CO_3(aq) + H_2O(\ell) \Leftrightarrow H_3O^+(aq) + HCO_3^-(aq). \tag{5.33}$$

If the blood's pH falls below approximately 7.35, the condition is called **acidosis**, while a pH value above approximately 7.45 is consistent with the condition known as **alkalosis**. The pH range consistent with stable human biochemical processes is extremely narrow; if the blood's pH drops below approximately 6.8 to 7.0 or rises above approximately 7.8, the possibility of death is very high.

While it is not immediately obvious why this particular buffer came to play a central role in human physiology, we shall eventually see that the production of $CO_2$ is a central component of human catabolism (the breakdown of complex molecules). Consequently, a buffer that includes $CO_2$ is completely consistent with the life processes in carbon-based life forms. The pK associated with the multiple equilibria of equation 5.33 (not simply the $pK_a$ of carbonic acid) is 6.1 at normal body temperature (37°C or 310.15 K). The reader may wonder how such a buffer system is effective in maintaining a pH in the range from 7.35 to 7.45. The answer lies in the cooperative effect of several processes: the chemical buffering described by equation 5.33; the participation of the respiratory system to remove $CO_2$ from the blood; and the activity of the renal system to excrete $H_3O^+$ ions. Working in concert, these three processes are capable of maintaining the very narrow pH range required by the human biochemistry.

# Chapter 5 Exercises

1. What are the two parameters that determine the unique chemical properties of each element?
2. What does an element's oxidation number identify?
3. What is oxidation?
4. What is reduction?
5. Define the reducing agent.
6. Define the oxidizing agent.
7. State the definitions of an acid and a base proposed by Arrhenius.
8. What is a Brønsted acid?
9. What is a Brønsted base?
10. What is conjugate base of a Brønsted acid?
11. What is conjugate acid of a Brønsted base?
12. What is a strong acid (or base), and how is the concept of a strong acid (or base) related to the concept of a strong electrolyte?
13. What is a weak acid (or base), and how is the concept of a weak acid (or base) related to the concept of a weak electrolyte?
14. What is chemical equilibrium?
15. How does chemical equilibrium differ from physical equilibrium?
16. Assume that the following reaction has reached its equilibrium state at $T = 373.15$ K:

    $$3ClO^-(aq) \Leftrightarrow 2Cl^-(aq) + ClO_3^-(aq)$$

    Write the expression for the equilibrium constant of this decomposition reaction.

17. Assume that the equilibrium concentrations of the species in the reaction

    $$3ClO^-(aq) \Leftrightarrow 2Cl^-(aq) + ClO_3^-(aq)$$

    at a temperature of 373.15 K are

    $$[ClO^-]_{eq} = 0.395 \text{ mol L}^{-1}$$
    $$[Cl^-]_{eq} = 0.240 \text{ mol L}^{-1}$$
    $$[ClO_3^-]_{eq} = 0.635 \text{ mol L}^{-1}$$

    Calculate the equilibrium constant for this decomposition at 373.15 K.

18. What is the reaction quotient of a chemical reaction? How does it differ from the equilibrium constant for the same reaction?
19. Define "acid ionization constant" and "base ionization constant."
20. What is the definition of "ion product constant of water"?
21. Give the mathematical definitions of "pH" and "pOH."
22. For each of the following equilibrium concentrations in an aqueous solution at 298.15 K, calculate the pH of the solution:

    a. $[H_3O^+]_{eq} = 0.457$ M
    b. $[H_3O^+]_{eq} = 4.58 \times 10^{-10}$ M
    c. $[OH^-]_{eq} = 0.757$ M

23. For each of the following pH values (measured for the equilibrium state in an aqueous solution at 298.15 K), calculate the $[H_3O^+]_{eq}$ of the solution:

    a. pH = 11.15
    b. pH = 2.35
    c. pH = 3.851

24. For each of the following pH values (measured for the equilibrium state in an aqueous solution at 298.15 K, calculate the pOH of the solution:

    a. pH = 9.45
    b. pH = 3.29
    c. pH = 7.10

25. What is a neutralization reaction?
26. What is a salt?
27. What is the definition of a buffer?
28. What are the two components found in a buffer solution?

# CHAPTER SIX

## Modern Chemical Bonding Models

### "Why Is There Something and Not Nothing?"

Beginning in antiquity and, more recently, from the time of its emergence as a modern science at the end of the eighteenth and the beginning of the nineteenth century, chemistry has been fascinated by substances. The initial questions studied by chemists centered on the interaction of one distinct substance with another, what we now call a *chemical reaction*. But increasingly throughout the nineteenth century, the science struggled with the concept of *structure*, attempting to understand essentially *static* relationships that seemed to characterize the multitude of substances carefully cataloged by chemists. The importance of structural relationships at the macroscopic level was emphasized by the recognition that not every structure is either stable or aesthetically pleasing. Further, humankind recognized recurring patterns throughout the natural world that suggested an underlying order; this observation was reinforced with the invention of the microscope and the recognition of order and patterns at the microscopic level.

The earliest attempts to translate these observations into a model based on Dalton's revival of Greek atomism achieved only limited success (recall the model of chemical dualism proposed by Berzelius that we met in chapter 4). While no completely satisfactory model of chemical structure emerged during the nineteenth century, a strong link was forged between classes of substances defined by an assumed *structural similarity* and the *chemical reactivity* of these various classes. The fact that all the members of a particular class of molecules reacted in a similar fashion *suggested* that there is a common structural component shared by all members of a given class. The development of a cohesive branch of chemistry, called *organic chemistry*, and the parallel growth of a chemical industry based on the production of dyes stand as a testament to the relationship between structure and reactivity.

The reasoning used by chemists, and in fact, by all physical scientists throughout the nineteenth century, made the fundamental assumption (usually an *implicit assumption*) that the patterns and interactions observed at the macroscopic level were directly applicable to a microscopic world that was being revealed by new instruments and novel experimental techniques. It was *assumed* that the paradigm created by Isaac Newton in the seventeenth century provided a complete framework for understanding the natural universe at *all* levels, ranging from the stars and planets of astronomy to the microscopic structure of cells and even to the *submicroscopic* level of the atoms and molecules of chemistry. However, as we have already seen (in chapter 2), the emergence of the quantum theory at the beginning of the twentieth century and the subsequent development of quantum mechanics and quantum field theory have led to an entirely different model with which to understand the most elementary components and interactions of our world. The new quantum perspective required humankind to utilize an entirely different set of concepts when interpreting and comprehending the body of stunning experimental data that began accumulating as the twentieth century dawned.

Of course, as is often the case with novel ideas, the adoption of the new quantum mechanics occurred over a period of time and was characterized by contentious discussions and the all-too-human tendency to continue using familiar and well-understood ideas rather than to embrace revolutionary and conceptually challenging points of view. We have already seen in chapter 4 that chemistry's concern with the static character of molecular structure led Lewis to propose his cubical atom model with electrons localized at the corners of a cube or tetrahedron. This static conceptualization of a localized electron pair as the basis for chemical bonding was articulated in the second decade of the twentieth century, well after the quantum revolution had begun and after the physicist's model of the atom required the existence of a *dynamic* electron. Interestingly, Lewis's suggestion that an electron pair is central to chemical bonding survived as a key component of a quantum mechanical description of molecular bonds.

While it is often the case that earlier ideas are borrowed and successfully incorporated in subsequent models, a more distinguishing characteristic of the model-building conceptualization of science is the simultaneous persistence of *multiple* models. Sometimes this occurs because the development of a new model takes place over a lengthy span of time, as has been the case with quantum mechanics and quantum field theory. The complexity of a new model may simply require intense effort by a large number of investigators over an extended time period. Frequently, however, *two or more* models may coexist for a time (a period of time that may well stretch into decades), not only because an earlier model is familiar or a new model is dauntingly complex or there is a reluctance to adopt a revolutionary perspective, but also because an earlier model possesses a pragmatic utility: with only moderate effort it is capable of making powerful predictions that are consistent with either experimental observations or macroscopic manipulations. We are all familiar with space probes sent to various parts of our solar system along orbital trajectories computed with Newton's gravitational dynamics, even though Einstein's general theory of relativity (now nearly a century old!) is humankind's best description of astronomical gravitational dynamics.

We now begin an examination of three models of chemical bonding, two of which rely on the new quantum mechanics, while the third finds its roots in the static structural relationships that motivated so much chemical thought throughout the nineteenth century and at the start of the twentieth century. Curiously, while this third model, known as the **valence shell electron pair repulsion (VSEPR)** model, grew out of chemistry's focus on static structures and was initially proposed as a link between Lewis's valence electrons and molecular geometry in 1940 (by Sidgwick), a mature form of the model only appeared in the work of Gillespie and Nyholm during the mid-1950s, after quantum mechanics had undergone more than a half-century of development. As a result, the VSEPR model does not employ the concepts of modern quantum mechanics but instead relies on the idea of a localized electron pair enunciated by Lewis and the simple electrostatic repulsion of identical charges. (Recall from chapter 4 that two charges that are the **same**, either two positive charges or two negative charges, **repel** one another. This is an observation that dates from Coulomb's work in the eighteenth century.) Despite its relatively unsophisticated and nonmathematical approach (in the sense that it does *not* employ modern quantum theory) and its central concern with static chemical structure, VSEPR is quite successful in making powerful predictions about the spatial arrangement of nuclei. Let's begin.

## Valence Shell Electron Pair Repulsion (VSEPR)

The VSEPR model begins by utilizing the framework proposed by Lewis and identifying regions of concentrated negative charge located around the central atom in a molecule. From the perspective of Lewis's model, this means cataloging the bonding electron pairs as well as the nonbonding pairs arranged around the central atom to satisfy Lewis's octet rule. Each bonding pair and each nonbonding pair is counted as one region of concentrated negative charge; multiple bonds (double bonds or triple bonds) are counted as only **one** region of concentrated negative charge. Because the Lewis model has been extended to include both deficient and expanded octets, the number of regions of negative charge surrounding a central atom may be **less** than four or **greater** than four. Once the regions of concentrated negative charge have been identified, the geometric arrangement of these regions around the central atom of the molecule is determined by **minimizing the electrostatic repulsion** among the regions of concentrated negative charge. Note that this is the origin of the VSEPR name: the regions of negative charge arise from the pairs of **valence electrons**, and the geometric arrangement of these regions of negative charge depends on **minimizing the repulsion** among them, hence the name **valence shell electron pair repulsion**. This simple idea gives rise to only **six** geometric arrangements of the negatively charged regions around the central atom: (1) linear, (2) trigonal planar, (3) tetrahedral, (4) trigonal bipyramidal, (5) octahedral, and (6) pentagonal bipyramidal. Figure 6.1 shows these six geometries.

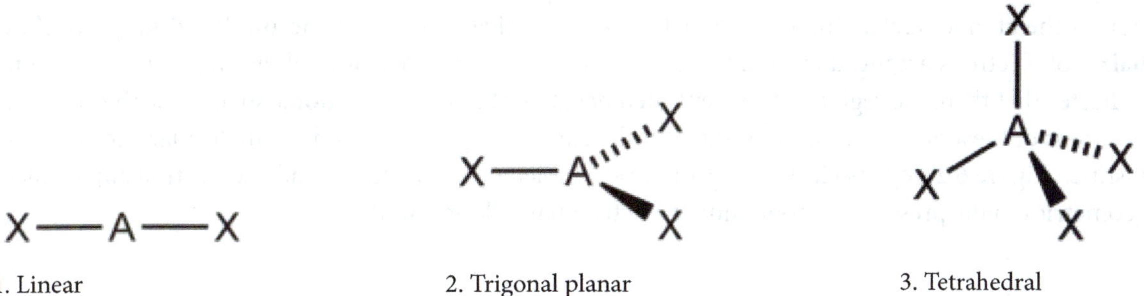


1. Linear  2. Trigonal planar  3. Tetrahedral

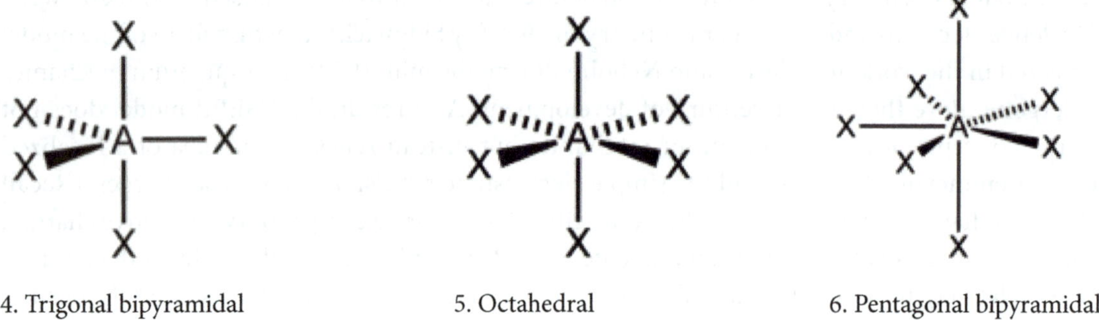


4. Trigonal bipyramidal 5. Octahedral 6. Pentagonal bipyramidal

Figure 6.1: The Six Geometries of VSEPR

The fact that the VSEPR theory can successfully predict the *geometry* of molecules is a very important attribute of this theory of molecular structure. The reader will recall that the very early effort by Lewis (1916) provided **no** *geometric* information about a molecule's structure. The symbols (Lewis structures) that became so widely used by chemists left the unfortunate *impression* that molecules are only two-dimensional objects existing in our three-dimensional universe. While the analysis completed above refers to "the central atom" of a molecule, suggesting that we are limited to molecules with a single "central" atom. This is not the case; the above analysis can be applied repeatedly to each nonhydrogen atom in a molecule to identify the geometric arrangement of the regions of negative charge at each atom. (We exclude hydrogen because the hydrogen atom forms only one bond with any other element in a molecule. As a result, it cannot be the "central" atom of a molecule.)

## Molecular Geometry and Electronic Geometry in VSEPR

There is, however, an important subtle point in the analysis we have just completed. While the previous VSEPR analysis can successfully identify the geometric arrangement of the regions of concentrated negative charge, this geometric configuration does **not necessarily** correspond to the geometry of the molecule. The molecule's geometry is determined by the positions of the atomic **nuclei** in space, **not** the geometry of the regions of negative charge. Consequently, the VSEPR model distinguishes two distinct geometries: the **electronic geometry**, which is determined by the spatial arrangement of the negative charge concentrations, and the **molecular geometry**, which is determined by the spatial configuration of the atomic nuclei. The two geometries are **identical** if there are **no nonbonding pairs (lone pairs)** of electrons arranged around the central atom. The presence of nonbonding pairs (lone pairs) indicates that there are regions of concentrated negative charge that are **not** associated with the spatial positions of the atomic nuclei. Consequently, the electronic geometry and the molecular geometry are distinct. Figure 6.2 depicts electronic geometries, molecular geometries, and the relationship of these geometries to the presence of nonbonding electron pairs (lone pairs).

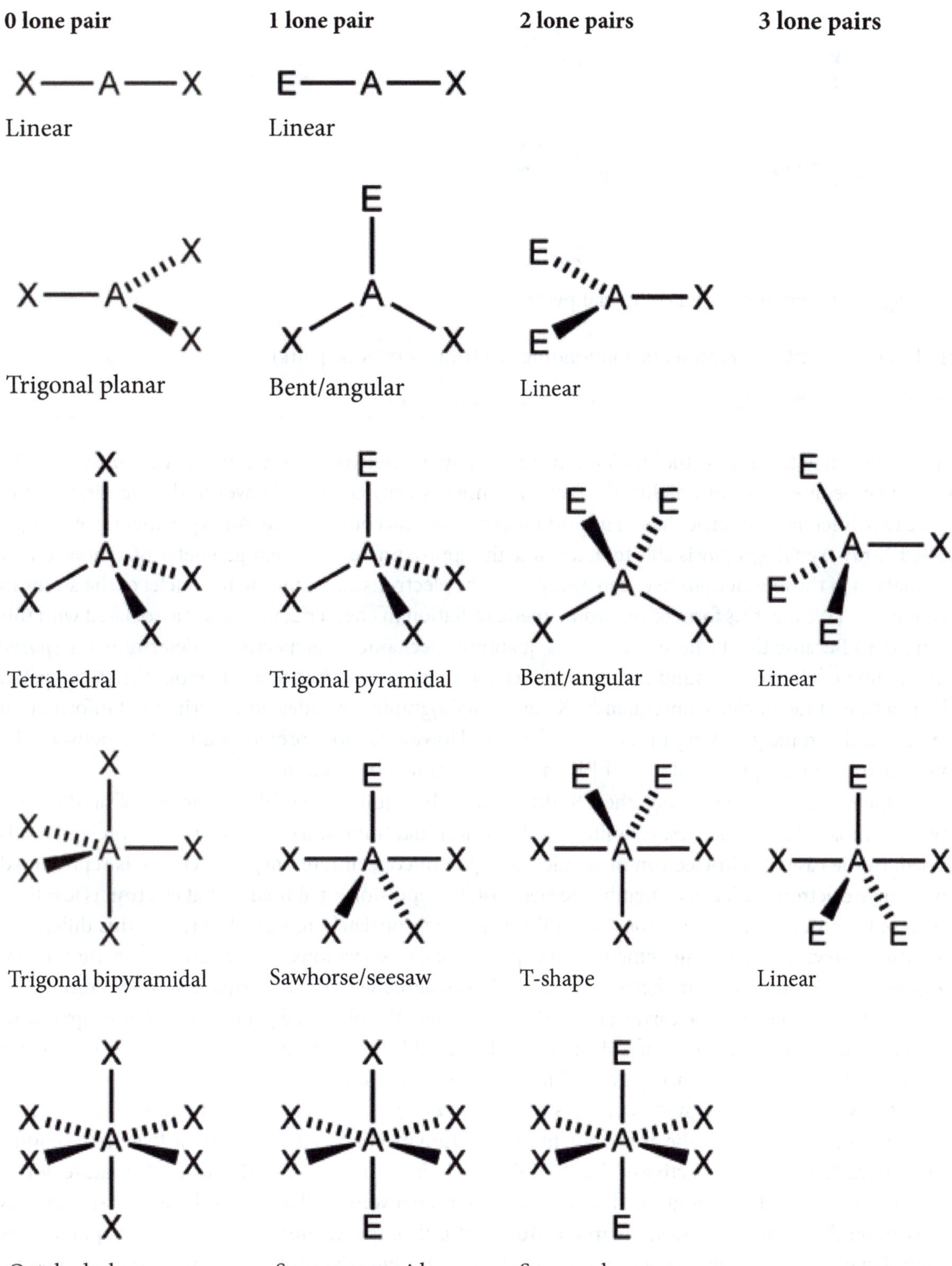
0 lone pair
1 lone pair
2 lone pairs
3 lone pairs
X—A—X
Linear
E—A—X
Linear
Trigonal planar
Bent/angular
Linear
Tetrahedral
Trigonal pyramidal
Bent/angular
Linear
Trigonal bipyramidal
Sawhorse/seesaw
T-shape
Linear
Octahedral
Square pyramid
Square planar

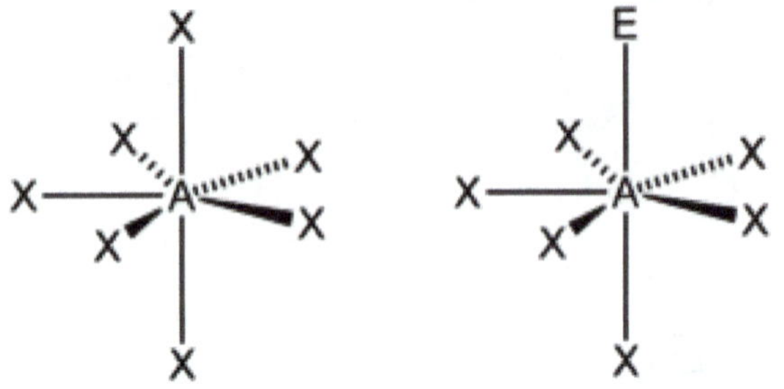

Pentagonal bipyramidal Pentagonal pyramidal

In the above chart, "E" represents nonbonding electron pairs (lone pairs).

Figure 6.2: Electronic Geometries and Molecular Geometries

The reader may think that the distinction we have made between the electronic geometry and the molecular geometry is both highly abstract and unnecessarily esoteric. However, the identification of these two different geometries has an important experimental consequence. An experimental technique called *X-ray crystallography* is able to determine the approximate molecular geometry of a molecule by an analysis of the interactions between X-rays and the electrons surrounding the nuclei of the atoms in a molecule. (We met this form of electromagnetic radiation in chapter 2; no mass is associated with this radiation.) Because this technique utilizes a quantum mechanical perspective to determine the *spatial distribution* of electrons around the atomic nuclei, it does provide information about the *static* spatial distribution of the nuclei. Consequently, X-ray crystallography provides no experimental information about the electronic geometry predicted by VSEPR. However it does provide a direct link between the molecular geometry predicted by VSEPR and experimental observations.

As the reader well may expect, the VSEPR model, while quite powerful and capable of predicting a range of molecular geometries, is limited by the conceptual framework and perspective on which it is based. In the case of odd-electron molecules, a region of concentrated negative charge is represented by a single electron, and, consequently, the electrostatic repulsion attributed to that electron is less than the repulsion due a pair of electrons. VSEPR must be appropriately modified to reflect this difference and can achieve qualitative agreement with experimental observations. In the case of transition metal complexes, the geometries predicted by VSEPR do not agree with the experimental observations. For heavier atoms of the alkaline earths group, the experimentally observed geometries do not agree with the linear geometry predictions of VSEPR; it may be possible that an interaction with the core electrons of these elements may influence the actual molecular geometries.

Just as we saw earlier with the concepts of *ionic bonding*, *covalent bonding*, and the *static, localized electron* of the Lewis model, the developments of quantum mechanics throughout the twentieth century directly challenge the perspective of the VSEPR model. Even though VSEPR reached its mature development in the mid-1950s, and the developers of the model were well aware of the nearly six decades of work on the quantum model, the model **does not** utilize the quantum perspective. In spite of this limitation of the VSEPR model, it has proven to be a highly effective tool for interpreting experimental data. Consequently, even though the VSEPR model utilizes concepts rooted in the limited Lewis model of bonding and the electrostatic observations of Coulomb that date from the eighteenth century, it has proven to be a useful tool for interpreting experimental observations. As such, just like our evaluation of the Lewis model, the VSEPR model should not be dismissed as "unreal" but rather accepted as a model whose pragmatic utility is capable of describing experimental results, making qualitatively correct predictions, but limited by the basic concepts used in its construction.

## Molecular Bonding: The Quantum Perspective

After more than a century of development, it is quite reasonable to ask about the contributions of quantum theory to our understanding of molecular bonding. The emergence of the *dynamic* electron in the physics of the first third of the twentieth century, accompanied by the probability interpretation of the wave function, Bohr's complementarity, and Heisenberg's indeterminacy principle (chapter 2), all demand an alternative to the *static structural* approach that dominated chemical thought from the middle of the nineteenth century into the beginning of the twentieth century. We shall now turn our attention to two descriptions of molecular bonding that overtly use quantum mechanical concepts in their description of the chemical bond: **valence bond (VB) theory** and **molecular orbital (MO) theory**. While a complete description of each theory, because of the complex mathematical structure of the theories and their use of subtle quantum concepts, individually merits a dedicated textbook of its own (and many have been written), our goal here is much more modest. We will **not** attempt a detailed exploration of the mathematical foundations of each theory; the required mathematical sophistication far exceeds the presuppositions of this text. We will only outline the conceptual quantum foundations of each theory, note the consequences for our understanding of atoms bonded together in molecules, and prepare a basis for understanding the reactions of organic, and later biochemical, molecules as well as the key structural characteristics of these molecules.

## The Questions

Up to this point in our discussion, we have not yet answered the question posed at the beginning of this chapter. While chapter 2 provided a panoramic view of the how and why of atoms, and chapters 3, 4, and 5 investigated a variety of the properties of chemical substances, the question still remains: Why does the vast array of chemical substances exist? Why is the universe not simply a collection (albeit a very *large* collection!) of individual atoms? Fundamentally, why do bonds exist at all? Why is there life as we know it? A cursory examination of two simple molecules highlights several particular disparities that are difficult, at first glance, to understand. Consider the two diatomic molecules $H_2$ and $F_2$. The first curious characteristic we note is a difference in the *bond energies*, that is, the energy liberated when the molecules form from separated atoms (equivalently, the energy required to decompose the diatomic molecules into individual atoms). For $H_2$ the experimentally measured bond energy is 436.4 kJ $mol^{-1}$. The measured bond energy for $F_2$ is 150.6 kJ $mol^{-1}$. Why should the bond energies for two simple, diatomic molecules differ by nearly a factor of 2.9? Now examine the *bond lengths*, that is, the mean distance between the nuclei in the diatomic molecules. For $H_2$ the bond length is 74 pm; for $F_2$ the bond length is 142 pm. Here, the difference is more than a factor of 1.9. Why should this disparity exist for two apparently simple, diatomic molecules?

## Valence Bond (VB) Theory

The **valence bond (VB) theory** was formulated at the end of the 1920s and represented a very early application of the rapidly developing field of quantum mechanics to important central questions in chemistry: Why do molecules exist? What is the *structure* of these molecules? The American chemist Linus Pauling, building on the work of two German physicists, Walter Heitler and Fritz London, and incorporating the suggestion made by Lewis almost fifteen years earlier that an electron pair is central to a molecular bond, formulated an **explicitly quantum mechanical model of chemical bonding**. In its initial incarnation, the VB theory made the fundamental assumption that the orbitals (the one-electron solutions of the Schrödinger equation that we met in chapter 2) associated with each electron are centered on **individual atoms**. Because the orbitals were interpreted by Born to be spatial probability distributions specifying the probability of locating at electron in some small but finite region of space, the assumption made by the VB theory means that these probability distributions are defined with respect to the nuclei of the **individual atoms** that make up a molecule. Consequently, we say that VB theory is **atom-centered**; it retains the point of view that **individual atoms** participate in chemical bonds.

So, how does VB theory conceptualize a bond? A chemical bond is defined as the **overlap of atomic orbitals**. But what does this mean? The phrase *overlap of atomic orbitals* means that orbitals associated with electrons from two distinct atoms (hence, they are atomic orbitals) delineate a probability distribution for a common region of space; this probability distribution specifies the probability of locating two electrons in some small but essentially finite region of space proximate to the two nuclei. Here we see the explicit use of Lewis's suggestion that a chemical bond depends on a pair of electrons. While this discussion is very qualitative and somewhat pictorial in its appeal (depending on the shapes of the probability distributions specified by each orbital type that we met in chapter 2), it can be made mathematically precise; the mathematical rigor is unnecessary for our purposes here.

The description of a chemical bond in VB theory provides a particularly facile understanding of the electron pair sharing that was identified with the term *covalent bond*. Because the common spatial probability distribution defined by the orbitals of individual atoms specifies the probability of locating two electrons in some small but essentially finite region of space proximate to the two nuclei, the electrons can be thought of as *being shared between the two atoms*. The VB theory has made Lewis's suggestion of a *shared electron pair* quantum mechanically precise. But how are we to understand the term *ionic bond*? When recounting Lewis's approach, the language that was used (chapter 4) employed the phrases *losing an electron* and *adding an electron* with the octet rule, providing the fundamental reason for the implied exchange of electrons. It was the attempt to satisfy the octet rule that caused elements from the left side of the periodic table (metals) to *lose* electrons while elements from the right side of the periodic table (nonmetals) tended to *add* electrons. How do we think of these processes in terms of spatial probability distribution (a key concept from quantum mechanics) and pairs of electrons (Lewis's persistent seminal idea)? The answer lies in understanding that the common spatial probability distribution defined by the orbitals of individual atoms for two electrons produces the experimentally observable phenomenon of *charge*. We observe an electrolyte solution conducting electricity! Consequently, from the VB perspective, an electron is **not** added to one atom or lost from another atom (the Heisenberg indeterminacy principle prohibits the use of such macroscopically inspired terms); rather, a pair of electron shares a spatial probability distribution whose experimental consequence is the observation of *charge carriers*. So, the key to understanding the chemical bonding again rests on the idea of a shared pair of electrons.

## The Answers

Earlier we noted that two very simple molecules, $H_2$ and $F_2$, exhibit curious differences in the experimentally measured values of their *bond energies* and *bond lengths*. The Lewis model of chemical structure does not use the term *energy* to discuss or explain chemical structures; consequently, it lacks the basic conceptual framework needed to address questions involving *bond energies*. Further, the qualitative description of the Lewis model made no quantitative predictions about the *lengths* of chemical bonds. Similarly, VSEPR, with its focus on minimizing repulsive *electrostatic forces*, does not provide quantitative estimates of either *bond energies* or *bond lengths*. In these respects VB theory is entirely different, because it utilizes the formalism of quantum mechanics, in which the concept of *energy* plays a central role. (Recall from chapter 2 that the solution of the Schrödinger equation yields a set of quantum numbers intimately related to a system's energy.) Consequently, VB theory predicts *bond energies* quantitatively, and further, demonstrates that stable molecules form from reacting atoms or molecules when the energy characterizing the system has **decreased to a minimum**. For the first time, a theory of molecular bonding tells us why molecules exist: **minimization of a system's characteristic energy**. Finally, VB theory is capable of explaining the quantitative difference in the *bond lengths* of the two molecules by noting that the common spatial probability distribution defined by the orbital overlap involves **different atomic orbitals** in $H_2$ compared to those used in $F_2$. (See chapter 2 and the probability distributions associated with **s**, **p**, **d**, and **f** orbitals.)

## Hybridization

In our brief exploration of modern quantum mechanics (chapter 2), we encountered the very puzzling experimental observation that quantum systems seem to exhibit both wave-like characteristics and particle-like characteristics. This gave rise to Bohr's suggestion of the **complementarity principle**, the integration of wave-like and particle-like characteristics in quantum field theory, and our ultimate conclusion that categories of thought derived from everyday experience (*wave* and *particle*) are completely inappropriate tools with which to describe and interpret microscopic phenomena. However, the highly successful quantum mechanical formalism that has evolved and is consistent with all of these puzzling experimental observations includes a mathematical procedure that links together (we say *couples*) multiple quantum states (the wave function solutions of Schrödinger's equation that describe a physical state) into a single descriptive mathematical object called a **quantum mechanical resonance** or simply a **resonance**. Resonances are both more descriptive and more predictive of our universe than individual quantum states, and therefore, are energetically more stable than the individual quantum states. (We have already met the term *resonance* when we discussed some of the limitations of and modifications to the Lewis model of chemical structure. However, as used here, *resonance* has a quantitative meaning that is distinctly different from the term's meaning in the context of Lewis's model. While the term is the same, the two applications rest on fundamentally different foundations and **the reader must be careful not to confuse them**.)

Wow! Why, you may ask, have we introduced this very **abstract** and apparently esoteric idea of a quantum mechanical resonance? The answer lies in the fact that Linus Pauling used the concept of a quantum mechanical resonance to make VB theory both highly descriptive of our observed universe and quantitatively predictive of many physical phenomena, the two key characteristics that are central to modern science. In particular, Pauling noted that the geometries associated with the pure (*hydrogen-like* is the technical term) atomic orbitals we have already discussed (**s**, **p**, **d**, and **f** orbitals from chapter 2) do not correspond to the observed geometries of molecules. By utilizing the resonance property of quantum mechanics, Pauling devised a procedure to *couple* together **nonequivalent atomic orbitals** (that is, orbitals characterized by **different** values of the *angular momentum quantum number*, see chapter 2) to form a new set of atomic orbitals. These new orbitals are called **hybrid atomic orbitals**, and the procedure used to construct them is called **hybridization**.

The hybrid atomic orbitals, just like the pure atomic orbitals we encountered in chapter 2, define probability distributions that specify the probability of locating an electron in some small but essentially finite region of space. As with the earlier assumption made by the VB theory, these probability distributions are defined with respect to the nuclei of the **individual atoms** that make up a molecule. Consequently, the hybridization procedure retains the point of view that it is the **individual atoms** that participate in chemical bonds. Most importantly, the hybridization procedure **conserves the *number* of orbitals**. That is, the ***number* of hybrid atomic orbitals** produced by the hybridization procedure ***exactly equals the number* of pure atomic orbitals** coupled together by the phenomenon of quantum mechanical resonance. Finally, the hybridization procedure does require energy. However, because the resulting set of hybrid atomic orbitals identifies quantum states that are energetically **more stable** than the quantum states associated with the pure atomic orbitals used by the hybridization procedure, the hybridization procedure is *advantageous*; the achievement of an energetically more stable state (identified by a hybrid atomic orbital) by a molecule compensates for the additional energy required by the hybridization procedure.

## sp Hybrid Orbitals

Let's now look at three particularly important sets of hybrid atomic orbitals. These orbital sets will play crucial roles in our future study of both organic and biochemical molecules by providing a powerful descriptive model for the chemical bonding in these molecules. Consider, first, the hybridization of a single pure **s** atomic orbital and a single pure **p** atomic orbital to form two (remember, conservation of orbital number!) new hybrid atomic orbitals that are called **sp hybrid orbitals**. The new hybrid orbitals are aligned along a straight line, 180° apart. Figure 6.3 depicts the hybridization process and the geometric relationship of the **sp** hybrid orbitals.

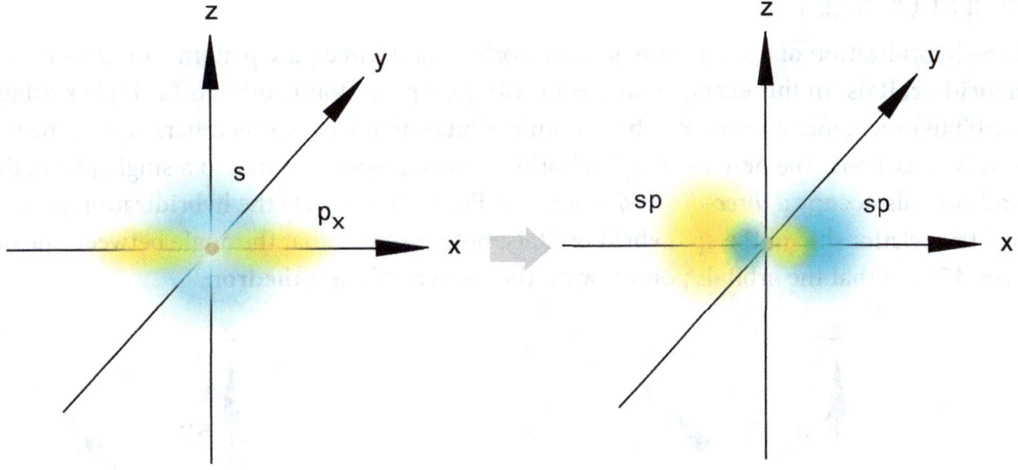


Figure 6.3: sp Hybrid Orbitals

## $sp^2$ Hybrid Orbitals

In our second example, a single pure **s** atomic orbital and two pure **p** atomic orbitals are hybridized to form a set of **$sp^2$ hybrid orbitals**. Note carefully that by beginning with *three* pure atomic orbitals, the set of hybrid atomic orbitals contains a total of *three* orbitals; the number of orbitals is *conserved*! The new hybrid orbitals are *coplanar* (that is, located in a *single* geometric plane) and are separated from one another by an angle of 120°. Figure 6.4 depicts the hybridization process and the geometric relationship of the **$sp^2$** hybrid orbitals.

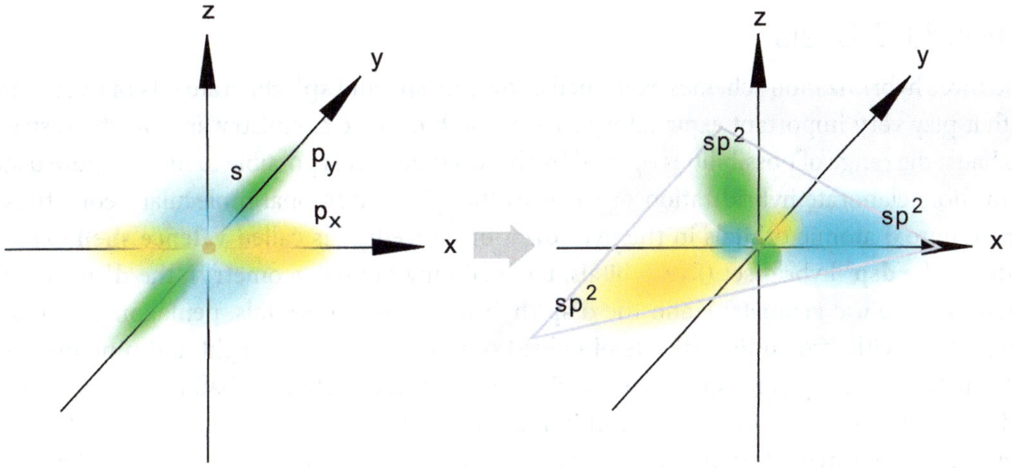


Figure 6.4: $sp^2$ Hybrid Orbitals

## $sp^3$ Hybrid Orbitals

Finally, the hybridization of a single pure **s** atomic orbital and three pure **p** atomic orbitals forms a set of **$sp^3$ hybrid orbitals**. In this example, we begin with *four* pure atomic orbitals (a single **s** orbital and three **p** orbitals) and generate a set of hybrid atomic orbitals that has *four* members; again, the number of orbitals is conserved. The new set of hybrid orbitals is *no longer* confined to a single plane; the *four* **$sp^3$** hybrid orbitals occupy a *three-dimensional space*. Figure 6.5 depicts the hybridization process and the geometric relationship of the **$sp^3$** hybrid orbitals; note carefully that the angle between the orbitals is now 109.47° and that the orbitals point toward the corners of a **tetrahedron**.

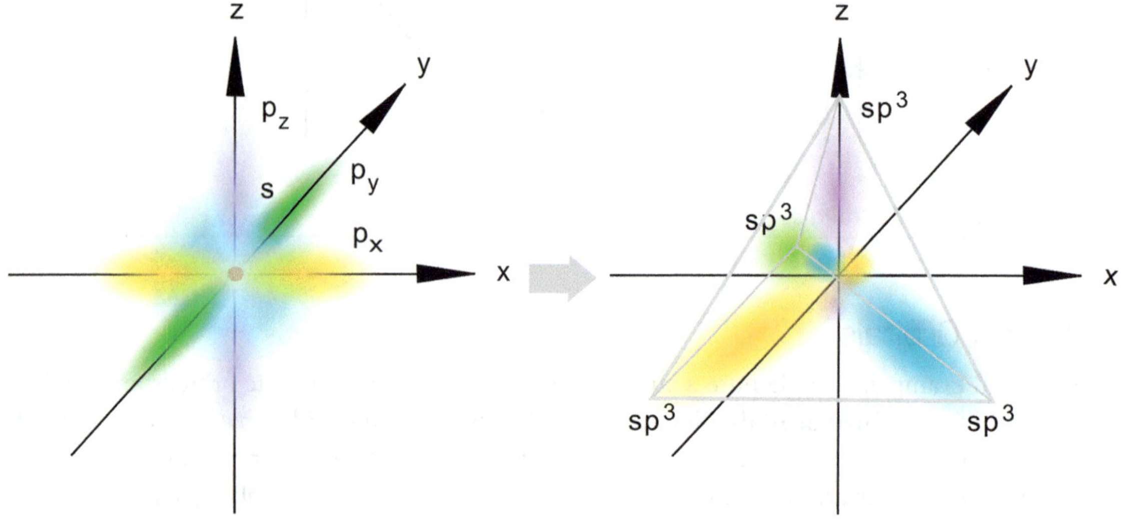


Figure 6.5: $sp^3$ Hybrid Orbitals

## More Hybrid Orbitals

While the three hybridization schemes examined above, **sp**, **$sp^2$**, and **$sp^3$**, generate sets of hybrid atomic orbitals that play very important explanatory roles in both organic chemistry and biochemistry, they do not exhaust the range of possibilities opened by the VB theory. It is possible to include pure **d** atomic orbitals in more elaborate hybridization procedures that yield additional molecular geometries. The inclusion of pure **d** atomic orbitals in the hybridization procedure is called **valence shell expansion** and produces the **$dsp^3$** hybrid set (five orbitals, trigonal bipyramidal geometry), the **$d^2sp^3$** hybrid set (six orbitals, octahedral geometry), and the **$d^3sp^3$** hybrid set (seven orbitals, pentagonal bipyramidal geometry), along with two additional sets of hybrid orbitals that contain eight and nine orbitals, respectively. Just as the **sp**, **$sp^2$**, and **$sp^3$** sets of hybrid atomic orbitals define probability distributions that specify the probability of locating an electron in regions of space with specific geometries (see figures 6.3, 6.4, and 6.5), the three sets of hybrid atomic orbitals that are the result of valence shell expansion, **$dsp^3$**, **$d^2sp^3$**, and **$d^3sp^3$**, also define probability distributions that specify the probability of locating an electron in regions of space with the specific geometries associated with each hybrid set.

Focusing our attention on the first six sets of hybrid orbitals, **sp**, **sp²**, **sp³**, **dsp³**, **d²sp³**, and **d³sp³**, and reviewing figure 6.1, an amazing correspondence emerges: the six geometries predicted by the VSEPR model **are exactly the same geometries** predicted by the VB theory. However, the VB theory is based on the new quantum mechanics introduced after 1925 that utilizes the *dynamic* electron accompanied by the probability interpretation of the wave function, Bohr's complementarity, and Heisenberg's indeterminacy principle. Rather than viewing chemical bonding in terms of minimizing the repulsive interaction between pairs of *static* electrons as did the VSEPR model, VB theory conceptualizes a chemical bond as the **overlap of either *two* hybrid atomic orbitals or *one* hybridized and *one* unhybridized atomic orbital, or even *two* unhybridized atomic orbitals**, meaning that orbitals associated with electrons from two distinct atoms delineate a probability distribution for a common region of space; this probability distribution specifies the probability of locating two electrons in a region of space possessing a characteristic geometry and proximate to the two nuclei. Figure 6.6 depicts the $C_2H_6$ molecule, which demonstrates the overlap of two hybrid atomic orbitals with one another (both carbon atoms are **sp³** hybridized) as well as the overlap of the six unhybridized pure **s** orbitals of the hydrogen atoms with the remaining six **sp³** hybridized atomic orbitals of the two carbon atoms.

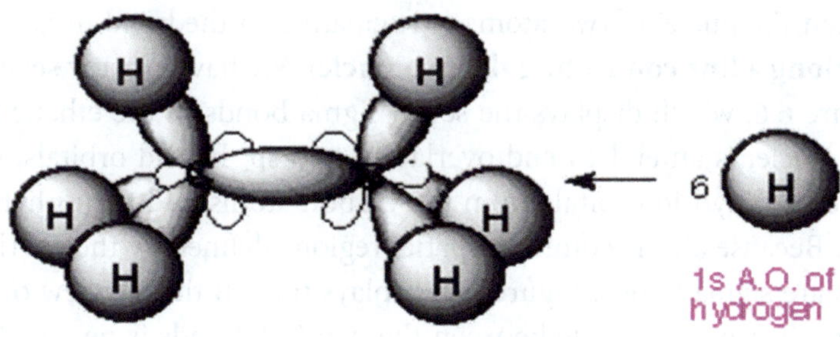


Figure 6.6: Ethane: sp³ Hybrid Orbitals on Carbon; Unhybridized Hydrogen Orbitals

## Molecular Geometry and Electronic Geometry in Valence Bond Theory

Just as a VSEPR analysis can successfully identify the geometric arrangement of the regions of concentrated negative charge, the VB theory identifies hybrid atomic orbitals *with a specific spatial geometric arrangement*. Like VSEPR, this geometric configuration does **not necessarily** correspond to the geometry of the molecule. The molecule's geometry is determined by the positions of the atomic **nuclei** in space, **not** the geometric arrangement of the hybrid orbitals. Consequently, the VB model (again, like VSEPR) distinguishes two distinct geometries: the electronic geometry, which is determined by the spatial arrangement of the hybrid atomic orbitals, and the molecular geometry, which is determined by the spatial configuration of the atomic nuclei. The two geometries are **identical** if the hybrid atomic orbitals of VB theory are **not** associated with any **nonbonding pairs** (**lone pairs**) of electrons around the central atom. Consequently, as in the VSEPR model, the electronic geometry and the molecular geometry are distinct. Just as figure 6.1 identifies molecular geometries common to both the VSEPR model (electron pairs) and the VB model (hybrid orbitals), figure 6.2 depicts electronic geometries, molecular geometries, and the relationship of these geometries to the presence of nonbonding electron pairs (lone pairs) in both the VSEPR model and the VB model.

## Sigma Bonds (σ)

VB theory identifies two distinct types of chemical bonds: **sigma bonds**, symbolized by the Greek character σ, and **pi bonds**, symbolized by the Greek character π. A **sigma bond** is a chemical bond that is a result of the end-to-end *overlap of atomic orbitals*. The overlap of orbitals is simply another example of the resonance phenomenon that characterizes modern quantum mechanics. The overlapping orbitals may be *two* **hybrid** atomic orbitals, *one* **hybrid** and *one* **unhybridized** atomic orbital, or even *two* unhybridized atomic orbitals. But what is the precise meaning of the *overlap* that characterizes a sigma bond? Like **all bonds** defined by VB theory, the overlapping orbitals participating in a sigma bond are associated with electrons from two distinct atoms and delineate a probability distribution for a common region of space. However, in the case of a sigma bond, this common region of space is located geometrically **between** the nuclei of two atoms participating in the bond. It is a region of space that lies **symmetrically along a line connecting the two nuclei**. We have already seen this very important characteristic in figure 6.6, which displays the seven sigma bonds in the ethane molecule. The bond between the carbon nuclei is an end-to-end overlap of two **$sp^3$** hybrid orbitals, while the remaining six bonds are between **$sp^3$** hybrid orbitals from the carbon atoms and the unhybridized **s** orbitals of the hydrogen atoms. Because all the common spatial regions defined by the overlaps lie **between** two nuclei, all the bonds are sigma bonds. Figure 6.7 displays the VB theory view of bonding in the $H_2S$ molecule. In this case, because the angle between the two H-S bonds is nearly 90° (92.1°), it is likely that no hybridization of the **s** and **p** orbitals on the sulfur atom occurs. Instead, the sigma bonds are the result of the overlap of unhybridized orbitals associated with electrons from the sulfur atom and unhybridized orbitals associated with the electrons of the hydrogen atoms.

## Pi Bonds (π)

In contrast to a sigma bond, a **pi bond** is a chemical bond that is formed by atomic orbitals *overlapping* ***above*** *and* ***below*** *the molecular plane* either of a planar molecule or of a planar segment of a larger molecule. A pi bond (or, more generally, a pi bond system, as in benzene) involves the overlap of *two or more* unhybridized **p** atomic orbitals. Again, like **all bonds** defined by VB theory, the overlapping orbitals participating in a pi bond are associated with electrons from two distinct atoms and delineate a probability distribution for a common region of space. However, in the case of a pi bond, this common region of space is located geometrically **above** and **below** the molecular plane either of a planar molecule or of a planar segment of a larger molecule. Figure 6.8 graphically display a generalized pi bond.

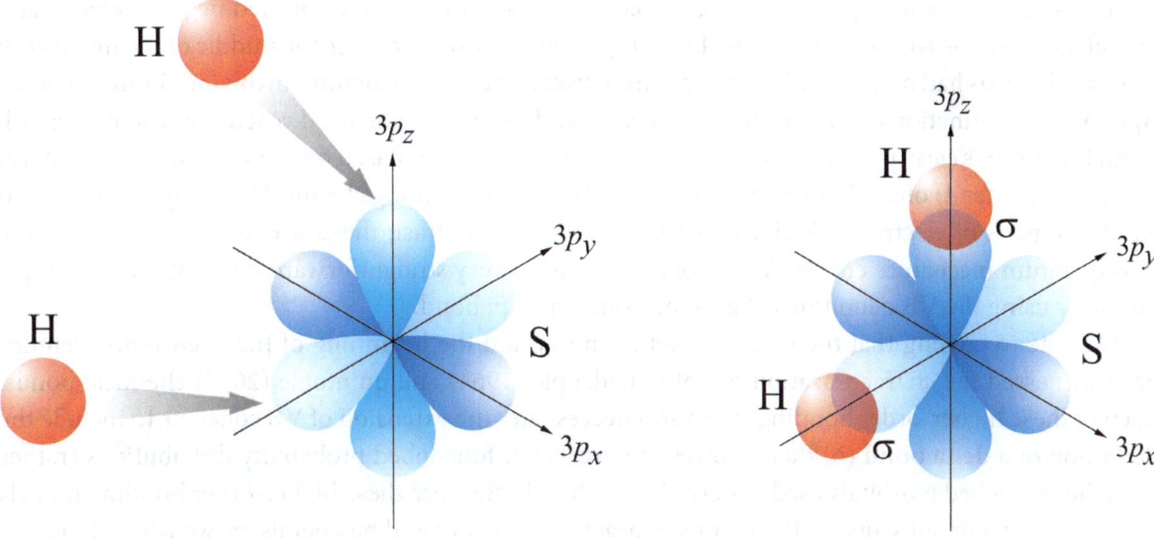


Figure 6.7: Sigma Bonds in the $H_2S$ Molecule

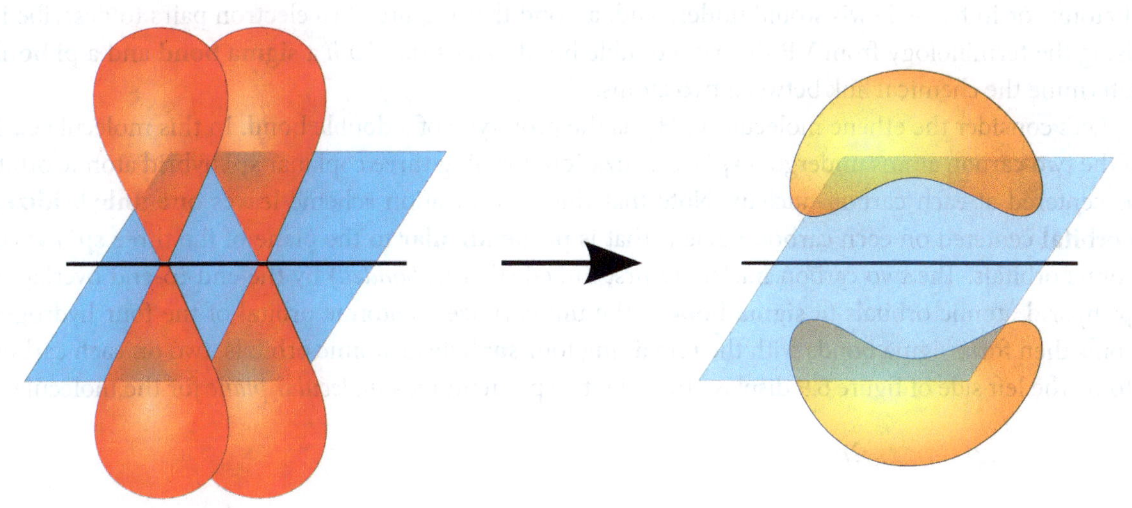

Figure 6.8: Representation of a Generalized Pi Bond

## Multiple Bonds

In our discussion of the VSEPR model, the concept of multiple bonds (**double bonds and triple bonds**, as well as higher order bonds: **quadruple bonds and quintuple bonds**) played a nearly nonexistent role; in order to determine a molecule's geometry, our discussion of the VSEPR model specified that both double bonds and triple bonds be counted as *single* regions of concentrated negative charge and was silent about the role of higher-order bonding schemes. However, since the middle of the nineteenth century chemists had made specific distinctions between chemical structures involving **single** and **multiple** bonds, distinctions that play critical roles in understanding chemical reactivity. The reader will recall that Lewis's early pre-quantum model of chemical structure (based on the suggestion of a cubical atom; see chapter 4) offered a preliminary approach to understanding chemical bonding involving two and three pairs of electrons (double and triple bonds). In contrast, the conceptual framework of the new quantum mechanics enables VB theory to provide a very straightforward description of multiple bonds by using the VB definitions of a sigma bond and a pi bond.

We note in passing that the mid-twentieth century and the beginning of the twenty-first century have witnessed the characterization of both quadruple (1964) and quintuple (2005) chemical bonds. Each of these higher-order bonding structures necessitates the extension of VB concepts to include the definition of a **delta bond** (δ) that involves the overlap of four-lobed probability distributions (rather than the two-lobed p orbitals used to describe a pi bond). Because these higher-order bonding models play no role in our subsequent discussions of organic chemistry and biochemistry, we will not present a detailed description of their characteristics in this text.

On the other hand, an understanding of both double and triple bonds from the perspective of VB theory will enrich and clarify the concepts we will encounter in our study of organic chemistry and the chemistry of life. As the name implies, a *double* bond represents *two* linkages between a pair of atoms, or in terms Lewis would understand, a bond that requires *two* electron pairs to describe it. Using the terminology from VB theory, a double bond means that *both* a **sigma bond** and a **pi bond** determine the chemical link between two atoms.

Let's consider the ethene molecule, $C_2H_4$, as the prototype of a double bond. In this molecule **each** of the two carbon atoms undergoes **sp²** hybridization, forming three coplanar **sp²** hybrid atomic orbitals centered at each carbon nucleus. Note that this hybridization scheme leaves **one unhybridized p orbital** centered on each carbon nucleus that is **perpendicular** to the plane of the three **sp²** hybrid atomic orbitals. The two carbon nuclei are first linked (that is, *bonded*) by the end-to-end overlap of **sp²** hybrid atomic orbitals (a **sigma bond**). The unhybridized **s** atomic orbital of the four hydrogen atoms then form sigma bonds with the remaining four **sp²** hybrid atomic orbitals, two on each carbon atom. The left side of figure 6.9 displays these first steps, defining a *molecular plane* for the molecule.

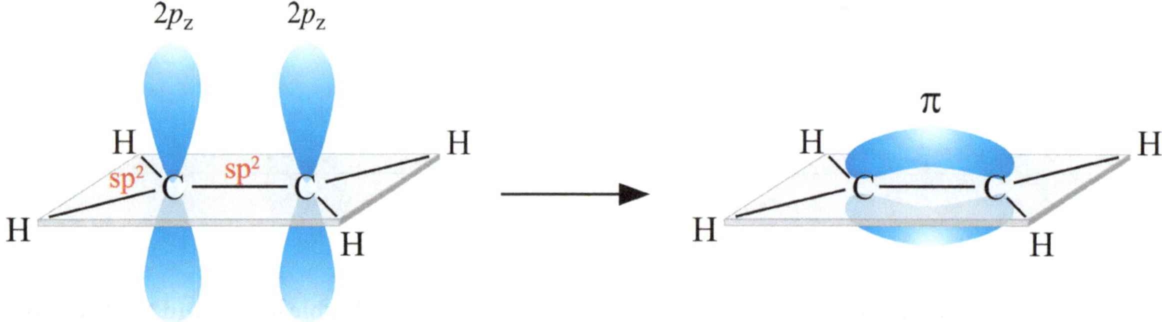


Figure 6.9: Ethene Described by Valence Bond Theory

The overlap **between** the two carbon nuclei is an end-to-end overlap of *two* **sp²** hybrid orbitals (black line between the carbon nuclei), while the remaining four black lines represent the end-to-end overlap of a **sp²** hybrid orbital with an unhybridized s atomic orbital from hydrogen. The unhybridized **p** orbitals (one on each carbon nucleus) are depicted as if they are **perpendicular** to the total planar sigma framework of the molecule. However, it is important to understand that if the unhybridized **p** orbitals are completely independent of one another (as shown in the left panel of figure 6.9), the molecule **is *not* constrained to remain planar**! However, the experimental data support the interpretation that $C_2H_4$ is indeed a planar molecule. This can be understood within the framework of VB theory by allowing the two perpendicular **p** orbitals to overlap and form a probability distribution that is located geometrically **above** and **below** the molecular plane. This is nothing more than the formation of a **pi bond**! Consequently, the presence of a *double* bond is understood as the formation of *one* **sigma bond** and *one* **pi bond** as depicted in the right panel of figure 6.9. Further, this model of the bonding in the $C_2H_4$ molecule is consistent with the experimental data, indicating a planar molecule whose carbon-carbon bond length is both **shorter** and **more stable** than the carbon-carbon bond in $C_2H_6$ (ethane).

As the reader may well expect, a **triple bond** requires *one* **sigma bond** and *two* **pi bonds** linking two atoms. We will use the $C_2H_2$ molecule (ethyne) as our prototype and will discuss the triple bond in the same qualitative and pictorial approach we used in the discussion of the double bond. (It goes without saying that both of these discussions can be made mathematically rigorous using quantum mechanics; our goal here is a conceptual understanding of the chemistry, not mathematical rigor.) In the $C_2H_2$ molecule, **each** of the two carbon atoms undergoes **sp** hybridization, forming two colinear **sp** hybrid atomic orbitals centered at each carbon nucleus. Note that this hybridization scheme leaves **two unhybridized p orbitals** centered on each carbon nucleus that are **perpendicular** to both the **sp** hybrid atomic orbitals and one another.

The two carbon nuclei are first linked (that is, *bonded*) by the end-to-end overlap of **sp** hybrid atomic orbitals (a **sigma bond**). The unhybridized **s** atomic orbital of the two hydrogen atoms then form sigma bonds with the remaining **sp** hybrid atomic orbitals on each carbon atom.

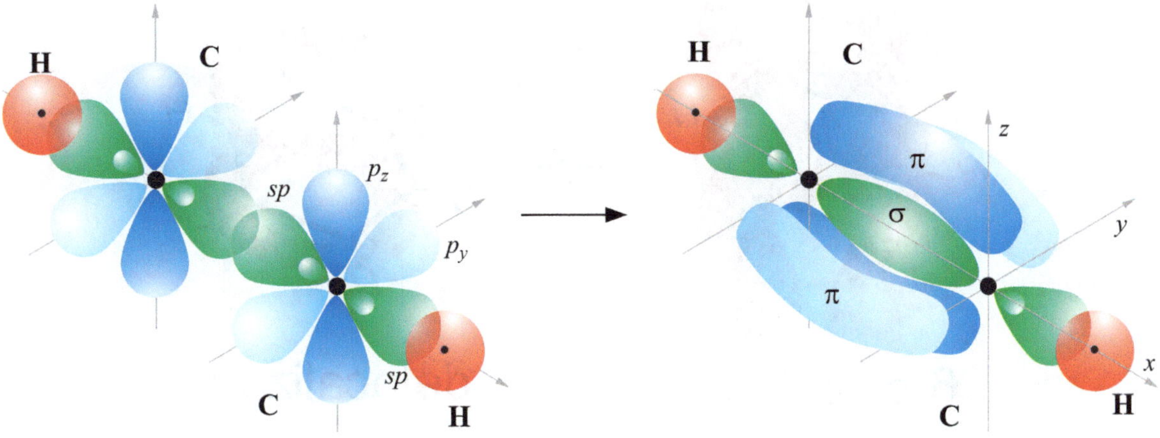


Figure 6.10: Ethyne Described by Valence Bond Theory

The left panel of figure 6.10 displays these first steps, defining a *linear* $C_2H_2$ molecule. The unhybridized **p** orbitals (two on each carbon nucleus) are **perpendicular** to both the **sp** hybrid atomic orbitals and one another. The experimental data support the interpretation that $C_2H_2$ is a linear molecule. This can be understood within the framework of VB theory by allowing the remaining two **p** orbitals centered at each carbon nucleus to overlap and form *four* probability distributions. These probability distributions specify the probability of locating two electrons in a region of space possessing a characteristic geometry and located symmetrically *around* the line joining the two nuclei (this is the spatial region defined by the sigma bond) **but which do not share the *same* space as the sigma bond.** This is nothing more than the formation of two **pi bonds**! Consequently, the presence of a *triple* bond is understood as the formation of *one* **sigma bond** and *two* **pi bonds**, as depicted in the right panel of figure 6.10. Further, this model of the bonding in the $C_2H_2$ molecule is consistent with the experimental data indicating a linear molecule whose carbon-carbon bond length is both **shorter** and **more stable** than the carbon-carbon bond in either $C_2H_6$ (ethane) or $C_2H_4$ (ethene).

The examples of pi bonds in both $C_2H_4$ and $C_2H_2$ molecules are but two examples of a more general phenomenon known as **electron delocalization** that is the result of multiple orbitals participating in a characteristic quantum mechanical resonance behavior. The benzene molecule, $C_6H_6$, provides a prototype of this general resonance behavior involving the overlap of multiple orbitals. In this molecule **each** of the six carbon atoms undergoes **sp²** hybridization, forming three coplanar **sp²** hybrid atomic orbitals centered at each carbon nucleus. Note that this hybridization scheme leaves **one unhybridized p orbital** centered on each carbon nucleus that is **perpendicular** to the plane of the three **sp²** hybrid atomic orbitals. The six carbon nuclei are first linked (that is, *bonded*) by the end-to-end overlap of **sp²** hybrid atomic orbitals (a **sigma bond**), with each carbon nucleus being linked to two other carbon nuclei. The unhybridized **s** atomic orbital of the six hydrogen atoms then form sigma bonds with the remaining six **sp²** hybrid atomic orbitals, one on each of the carbon atoms. The left panel of figure 6.11 displays these first steps, defining a *molecular plane* for the molecule.

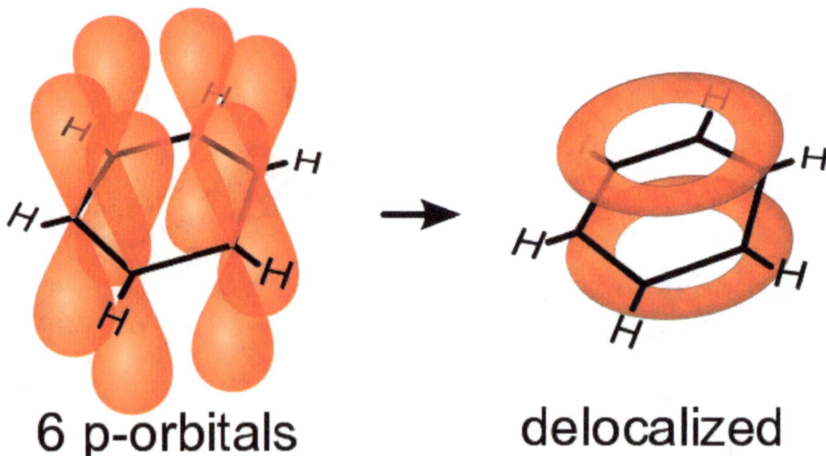


Figure 6.11: Benzene Described by Valence Bond Theory

The overlap **between** the carbon nuclei is an end-to-end overlap of *two* **sp**$^2$ hybrid orbitals (black lines between the carbon nuclei, which are symbolized by the six corners of the regular hexagon), while the remaining six black lines represent the end-to-end overlap of a **sp**$^2$ hybrid orbital with an unhybridized **s** atomic orbital from hydrogen. The unhybridized **p** orbitals (one on each carbon nucleus) are depicted as if they are **perpendicular** to the total planar sigma framework of the molecule. However, it is important to understand that if the unhybridized **p** orbitals are completely independent of one another (as shown in the left panel of figure 6.11), the molecule **is *not* necessarily constrained to be planar**! However, the experimental data support the interpretation that $C_6H_6$ is indeed a planar molecule and that the carbon nuclei occur at the corners of a regular hexagon. This can be understood within the framework of VB theory by allowing the six perpendicular **p** orbitals to overlap and form a probability distribution that is located geometrically **above** and **below** the molecular plane. Note that the **delocalization** of the probability distribution is emphasized by the two large ring regions depicted in the right panel of figure 6.11. This model of the bonding in the $C_6H_6$ molecule is consistent with the experimental data indicating a planar molecule in which each carbon-carbon bond is identical.

## Molecular Orbital (MO) Theory

Like VB theory, **molecular orbital (MO) theory** is an application of quantum mechanics to important central questions in chemistry: Again, why do molecules exist, and what is the *structure* of these molecules? Developed in the 1930s following the initial introduction of VB theory, MO theory is also an **explicitly quantum mechanical model of chemical bonding**. However, MO theory views the bonding in molecules from a distinctly different perspective than VB theory. While VB theory is **atom-centered**, retaining the point of view that it is the **individual atoms** that participate in chemical bonds and that the spatial probability distributions specifying the probability of locating electrons in some small but finite region of space are defined with respect to the nuclei of the **individual atoms**, MO theory is **molecule-centered**, making the fundamental assumption that the orbitals (the one-electron solutions of the Schrödinger equation that we met in chapter 2) associated with each electron are properties of the **entire molecule**. The molecular orbitals of MO theory are the result of the interactions of the atomic orbitals of the bonded atoms, but the molecular orbitals are **not** localized; they are associated with the **entire** molecule.

So, how does MO theory conceptualize a bond? A chemical bond is defined as the **overlap of molecular orbitals**. MO theory uses the phrase *overlap of molecular orbitals* to mean that orbitals **belonging to the entire molecule** and associated with electrons **belonging to the entire molecule** delineate probability distributions for common regions of space; these probability distributions specify the probability of locating pairs of electrons in small but essentially finite regions of space proximate to the nuclei of the molecule. While this discussion is very qualitative, (We have not even specified the **shapes** of these probability distributions.) it can be made mathematically precise; the mathematical rigor is unnecessary for our purposes here.

## The Question: Why is $O_2$ Paramagnetic?

The reader may wonder if the level of mathematical complexity implied by the above description is *absolutely* necessary in order to understand the chemistry of our world. However, mathematical elegance is not the sole reason to adopt the point of view represented by MO theory (although mathematical beauty has often played a significant role in the emergence of new scientific models). A simple experimental observation requires us to do more: oxygen in the liquid state is *magnetic*! Figure 6.12 shows liquid oxygen being poured between the poles of a magnet. Notice that the stream of liquid is attracted to the pole of the magnet, indicating that the liquid oxygen is indeed paramagnetic! (We say *paramagnetic* because the $O_2$ molecule responds to the presence of a magnetic field but does not **remain** magnetic once the magnetic field is removed. That is, $O_2$ cannot be magnetized; it does not exhibit the magnetic properties of iron, called *ferromagnetism.*)

Figure 6.12: Liquid Oxygen Attracted to a Pole of a Magnet

Important experimental discoveries and quantum mechanical formulations during the first third of the twentieth century demonstrated that the electron exhibits an intrinsic magnetism associated with its *spin*. (See chapter 2 for a discussion of the spin quantum number.) Because the spin quantum number has only two possible values, ±½, when a +½ spin is paired with a -½ spin, the resulting magnetic effects exactly cancel to zero. Applying the VB theory to describe the $O_2$ molecule requires all the electrons (an even number of them!) to be paired. Consequently, VB theory predicts that the $O_2$ molecule is **not** paramagnetic, directly contradicting the experimental observation! As we shall shortly see, MO theory provides a successful description of the nonzero magnetism of $O_2$.

## Bonding and Antibonding Molecular Orbitals

Within the conceptual framework of MO theory, the interaction of atomic orbitals (again, a quantum mechanical resonance phenomenon) produces two types of molecular orbitals, **bonding molecular orbitals** and **antibonding molecular orbitals**. Bonding molecular orbitals are **lower** in energy and exhibit **more stability** than the starting atomic orbitals, while antibonding molecular orbitals are **higher** in energy and exhibit **less stability** than the starting atomic orbitals. Just like VB theory, however, the quantum mechanical procedure for forming bonding and antibonding orbitals **conserves orbital number**. This means that the number of bonding and antibonding orbitals that form as a result of the quantum mechanical resonance is **exactly equal** to the number of interacting atomic orbitals. Each molecular orbital can accommodate a maximum of *two* electrons (the same as the case for atomic orbitals), which must have opposite spins (one with a +½ value for the spin quantum number and one with a -½ value for the spin quantum number), thus obeying the *Pauli exclusion principle*. When electrons are added to molecular orbitals of the same energy, the most stable arrangement is the one predicted by *Hund's rule*: Electrons fill molecular orbitals so as to maximize the number of parallel spins (electrons having the **same** value of the spin quantum number).

## Sigma and Pi Molecular Orbitals

In our discussion of VB theory a distinction was made between a *sigma bond* and a *pi bond* by noting that the probability distributions associated with these two types of bonds (which specify the probability of locating the pair of electrons responsible for each bond) identify **different geometric regions of space**. In the context of MO theory, the orbitals formed by the interaction of atomic orbitals may be either **sigma orbitals** or **pi orbitals**. A sigma molecular orbital defines a probability distribution that identifies a common region of space that is located geometrically **between** the nuclei of two atoms participating in the bond; it is a region of space that lies **symmetrically along a line connecting the two nuclei**. In contrast, a pi molecular orbital defines a probability distribution that identifies a common region of space that is located geometrically **above** and **below** the molecular plane either of a planar molecule or of a planar segment of a larger molecule. This spatial region is always located symmetrically *around* the line joining the two nuclei participating in a bond, **but it is *spatially disjoint* (sharing no common spatial coordinates) from any specified sigma orbitals**. Because the designations *sigma* and *pi* identify the geometric symmetry of the molecular orbitals, there are, in fact, **four** possibilities: **sigma bonding molecular orbitals** (symbolized by $\sigma$), **sigma antibonding molecular orbitals** (symbolized by $\sigma^*$), **pi bonding molecular orbitals** (symbolized by $\pi$), and **pi antibonding molecular orbitals** (symbolized by $\pi^*$).

## Bonding in MO Theory

Because MO theory is a molecule-centered approach to describing chemical bonding, the regions of space associated with the probability distributions specified by molecular orbitals are much more complex than the regions we encountered with the **s**, **p**, **d**, and **f** atomic orbitals. Consequently, in order to look at examples of chemical bonding from the MO point of view, we will simply draw energy diagrams that show the relationships among atomic and molecular orbitals, identifying the bonding and antibonding molecular orbitals as well as the association of electrons with the orbitals resulting from the application of the *Pauli exclusion principle* and *Hund's rule*. In a **stable molecule**, the number of electrons associated with bonding molecular orbitals is **always greater than** the number associated with antibonding molecular orbitals. Further, the *conservation of orbital number* means that the total number of electrons associated with **all** the molecular orbitals (*bonding* and *antibonding*) **must equal** the sum of all the electrons associated with all the atoms that compose the molecule. These requirements of MO theory permit us to calculate the **bond order** of any molecule (either a *stable* molecule or even a *hypothetical* molecule, as we shall see in the examples below). The bond order is defined as follows:

Bond order = (½) × [(Bonding valence electrons) – (Antibonding valence electrons )]. (6.1)

Equation 6.1 will be used to calculate the bond orders for several of the following examples. Let's begin with the most elementary example, the $H_2$ molecule. The molecular orbital diagram for $H_2$ is depicted in figure 6.13.

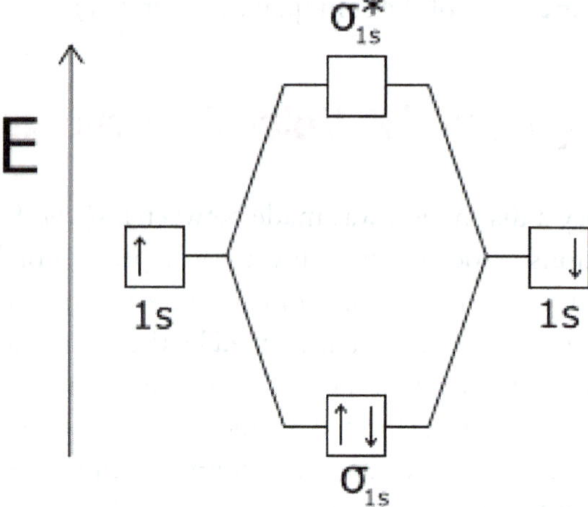


Figure 6.13: $H_2$ Molecular Orbital Diagram

The molecular orbital diagram, as we noted earlier, does **not** display the geometry of the molecular orbitals; only the qualitative energy relationships among the orbitals and the electrons associated with each orbital (atomic and molecular) are depicted. The left side and the right side of the image display the atomic orbitals of the individual (often called *separated*) hydrogen atoms. Notice that each hydrogen atom has a single electron associated with a **1s** atomic orbital (symbolized by the *up* and *down* arrows; they are energetically equivalent in the *separated* atoms). The center of the diagram represents the energetic relationship between the two *molecular* orbitals of the $H_2$ molecule, one a **sigma bonding orbital** (σ) (*lower in energy, more stable*) and the other a **sigma antibonding orbital** ($\sigma^*$) (*higher in energy, less stable*). Relative to the energy level of the two atomic orbitals, the sigma bonding orbital's energy is lower by the *same amount* as the antibonding orbital's energy is higher. There are two molecular orbitals because the **orbital number is conserved**: initially there were *two* **1s** atomic orbitals, which interact to form *two* sigma molecular orbitals. Finally, note that the two electrons are associated with the *lowest* energy molecular orbital, but the electrons obey the *Pauli exclusion principle*; the spins are *paired*, one *up* (+½, by convention) and one *down* (–½, by convention). Because electrons with two different spin quantum number values occupy the *same* orbital (molecular), the quantum mechanical process that formed the $H_2$ does require energy to pair the electrons in a single orbital. Applying equation 6.1 to calculate the bond order, we find that bond order = (½) × [2 – 0] = 1; this corresponds to a single bond, as we would expect for the $H_2$ molecule.

Let's now consider what happens when two helium atoms attempt to interact. Again, we display the simple molecular orbital diagram in figure 6.14.

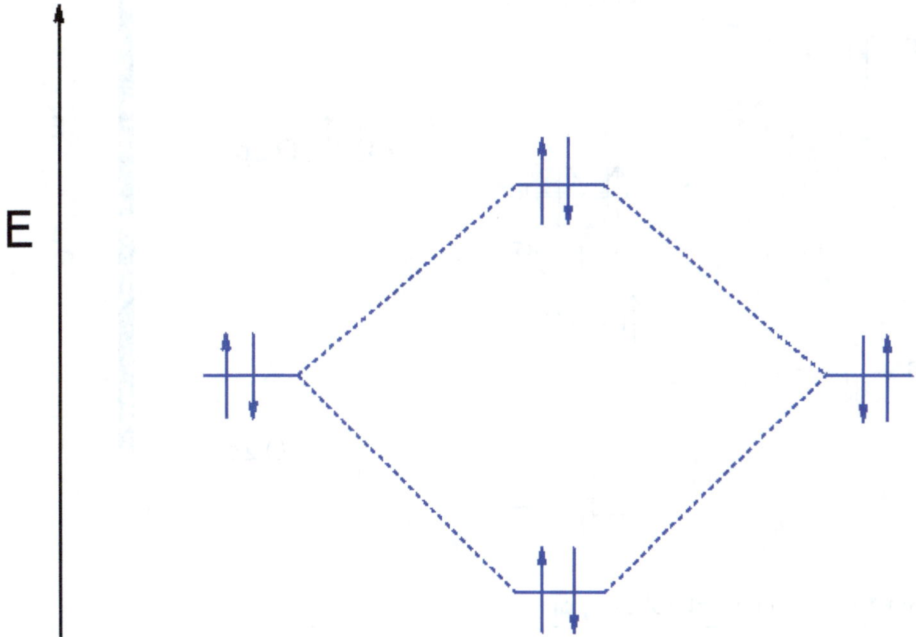


Figure 6.14: Molecular Orbital Diagram for Two Helium Atoms

Just as in figure 6.13, the left and right sides of the diagram represent schematically the *separated* helium atoms, while the center of the diagram represents the possible molecular orbitals. Again, the orbital number is conserved, the *Pauli exclusion principle* is respected, and relative to the energy level of the two atomic orbitals, the sigma bonding orbital's energy is lower by the *same amount* as the antibonding orbital's energy is higher. However, there is a significant difference. Let's calculate the predicted bond order using the possible molecular orbitals. Bond order = (½) × [ 0 - 0 ] = 0; that is, there is **no bond**! Consequently, the molecule $He_2$ **does not exist**.

The molecule CO is a simple example of a *heteronuclear diatomic* whose molecular orbital diagram is displayed in figure 6.15. While similar to the molecular orbital diagrams in figures 6.13 and 6.14, there are some specific differences to notice. Because both carbon and oxygen occupy positions in the second row of the periodic table, figure 6.15 includes **p** atomic orbitals; however, the diagram does **not** include the **1s** atomic orbitals of carbon and oxygen. The multiple lines on the right and left sides of the figure represent the *three* **2p** atomic orbitals of carbon and oxygen. Further, note that the interactions of the **p** atomic orbitals produce, for the first time, **pi molecular orbitals** (both *bonding* and *antibonding*).

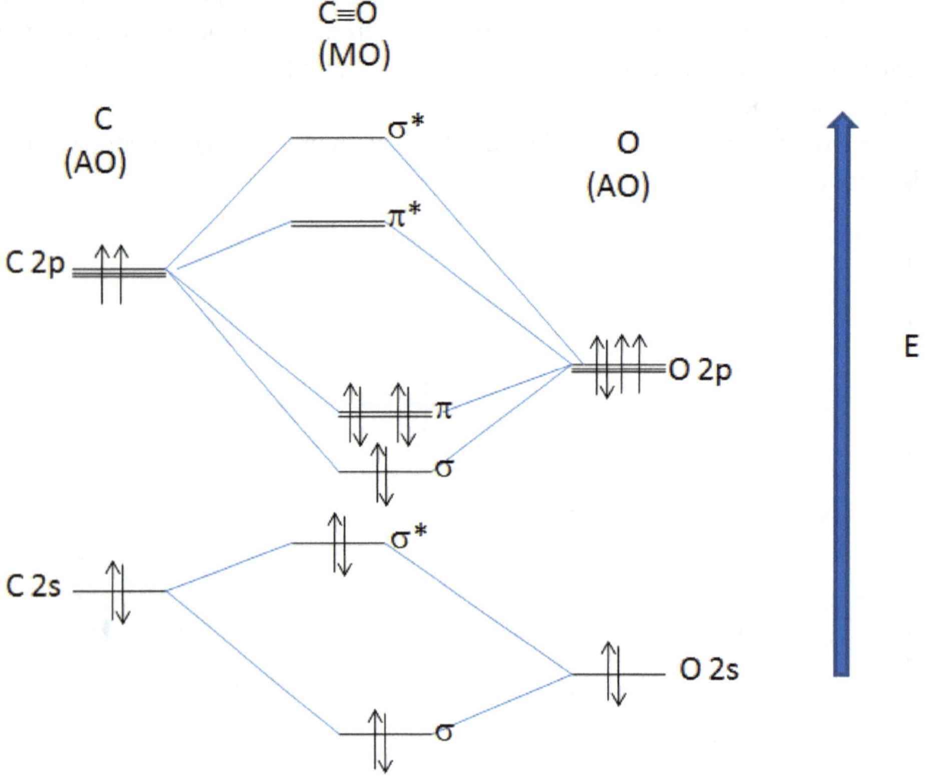


Figure 6.15: CO Molecular Orbital Diagram

It is particularly important to notice that the atomic orbitals of carbon and oxygen are energetically **different**, so that the relationships between the energies of the individual atomic orbitals and the molecular orbitals of the CO molecule orbitals are **no longer equal**. For example, the energy difference between the sigma bonding molecular orbital (formed from the **2s** atomic orbitals of carbon and oxygen) and the energy of the carbon **2s** atomic orbital is **significantly larger than** the energy difference between the same sigma bonding molecular orbital and the energy of the oxygen **2s** atomic orbital. Let's calculate the bond order for the CO molecule. Bond order = (½) × [ 8 - 2 ] = 6; that is, there is a *triple bond* linking the carbon and oxygen atoms in CO.

# The Answer: Why $O_2$ Is Paramagnetic

Figure 6.16 displays a molecular orbital diagram for the $O_2$ molecule, showing on the left and right sides **all** the electrons for each O atom. Beginning at the bottom of the diagram, *two* **1s** atomic orbitals interact to form *two* molecular orbitals, a *sigma bonding* orbital and a *sigma antibonding* orbital. At the next level in the diagram, *two* **2s** atomic orbitals interact again forming *two* molecular orbitals, a *sigma bonding* orbital and a *sigma antibonding* orbital. Finally, the **2p** atomic orbitals (three on each oxygen atom) interact to form *six* molecular orbitals, *two* of which are *sigma* orbitals (*bonding* and *antibonding*) and the remaining *four* of which are *pi* orbitals (*two bonding* and *two antibonding*). By its choice of colors, the diagram distinguishes the *sigma* orbitals (**blue**) from the *pi* orbitals (**red**). It is important to note that the interaction of the *six* **2p** atomic orbitals produces *two* molecular orbitals of *sigma symmetry* and *four* molecular orbitals of *pi symmetry*.

The *up* and *down* arrows represent the electrons associated with the orbitals in the diagram. There is a total of sixteen electrons contributed from the *separated* atoms, eight on the left side of the diagram and eight on the right side of thc diagram. In the center, the electrons are associated with the molecular orbitals, beginning with the lowest energy and satisfying the *Pauli exclusion principle* for each molecular. When the final two electrons are associated with the highest-energy orbitals (top of the diagram) *Hund's rule* requires them to be associated with **separate** orbitals and with **identical** values of the spin quantum number (*parallel spins*). Because the intrinsic magnetic properties of an electron are associated with its *spin*, all the magnetic properties add to **zero** except those contributed by the two electrons in the highest energy molecular orbitals. Consequently, MO theory predicts that the $O_2$ should exhibit paramagnetic properties, consistent with the experimental observations.

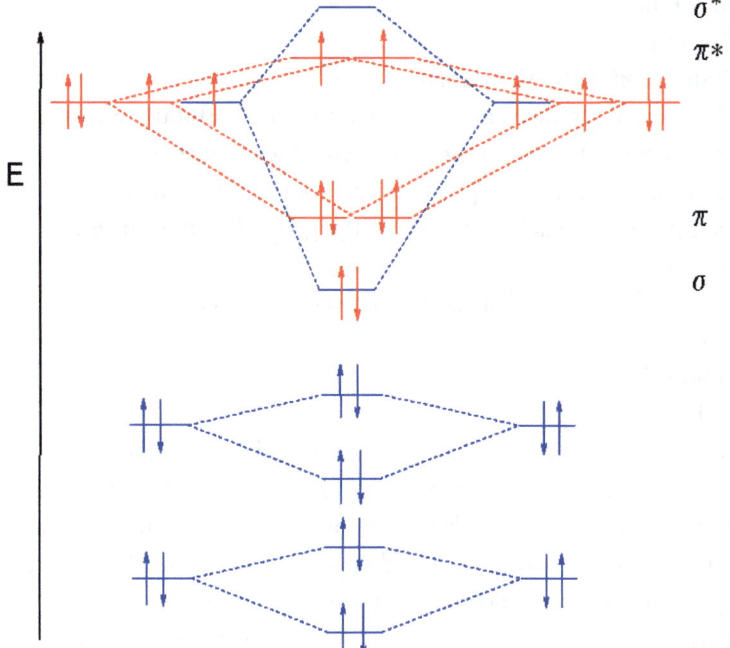


Figure 6.16: $O_2$ Molecular Orbital Diagram

# Chapter 6 Exercises

1. What does the acronym VSEPR stand for?
2. In the VSEPR model of chemical bonding, what is the central physical property that determines the arrangement of regions of negative charge around the central atom?
3. What are the six geometric arrangements of negative charged predicted by the VSEPR model?
4. What molecular attribute does the VSEPR model of chemical bonding predict that the Lewis model of 1916 could not correctly predict?
5. In the VSEPR model of bonding, distinguish between the electronic geometry and the molecular geometry of a molecule.
6. What is the role of nonbonding pairs (lone pairs) of electrons in the VSEPR model?
7. Identify the electronic geometry and the molecular geometry of the following molecules:
   (a) $HCl$ (b) $H_2O$ (c) $CH_4$ (d) $SO_2$
8. What are the factors that emerged in the twentieth century to challenge the static, structural view that dominated chemical thought in the nineteenth century?
9. True or false? The VB theory of bonding is an explicitly quantum mechanical model of chemical bonding.
10. True or false? VB theory is atom-centered; it retains the point of view that individual atoms participate in chemical bonds.
11. How does VB theory define a chemical bond?
12. From the point of view of VB theory, why do chemical bonds form?
13. What are hybrid orbitals?
14. What is hybridization?
15. Are hybrid orbitals *atomic* orbitals? Explain.
16. Does the hybridization procedure conserve the number of orbitals? Explain.
17. If the hybridization procedure requires energy, why does it occur?
18. Specify the pure atomic orbitals that produce each of the following hybrid orbital sets. How many hybrid orbitals are present in each set? What is the geometry of each set?
    (a) $sp$ (b) $sp^2$ (c) $sp^3$
19. Does the VB model of bonding distinguish between the electronic geometry and the molecular geometry? Explain.
20. What is a sigma bond?
21. What is a pi bond?
22. Explain a double bond? What roles do a sigma bond and a pi bond play in a double bond?
23. What is a triple bond? Explain the roles of sigma and pi bonds in a triple bond.
24. Is molecular orbital theory a chemical bonding theory based on quantum mechanics? Explain.
25. True or false? MO theory is an atom-centered approach to chemical bonding.
26. Define bonding and antibonding molecular orbitals by identifying their respective characteristics.
27. When bonding and antibonding orbitals are formed in the context of MO theory, is the number of orbitals conserved?

28. MO theory uses two types of molecular orbitals that are *geometrically* different. What are the two types of orbitals? Explain their distinctive geometries.
29. Calculate the bond order of $O_2$.
30. Liquid oxygen is paramagnetic. Explain how MO theory is able to account for oxygen's magnetic properties.

# CHAPTER SEVEN

## Structures in Organic Chemistry and Biochemistry

### The "Somethings" of Atoms Become the Molecules of Life

To this point in our exploration of the science of chemistry we have been primarily focused on the *inanimate* components of the world around us, on those dimensions of the universe that are classified as *nonliving*. With the exception of only a brief reference to the circulatory and respiratory systems (see the discussion of buffers, chapter 5), we have examined the concepts of time and space, matter and mass, and energy and charge, along with the experimental observations of solutions, acids and bases, and the phenomenon of electrical conductivity in those solutions. The brief excursions into quantum theory, the models of atomic structure, and the variety of theories that describe the linking of atoms into molecules made no mention of *life*, that ubiquitous phenomenon that seems to distinguish our planetary home from much of the rest of the universe. Is *life as we know it* unique? This question is perhaps the most profound unknown confronting humankind as the second decade of the twenty-first century continues to unfold. While the mathematical probabilities argue strongly for the existence of extraterrestrial life forms, the search continues, and the ultimate observational outcome is far from certain.

We turn now to address a significantly less profound question but one which is by no means uncomplicated: Using chemical tools developed from the mid-nineteenth century through the early twenty-first century, can we understand the phenomenon of *life* in terms of the atoms, simple molecules, and finally, the highly complex molecules that we can identify in our surrounding world? Put simply, beginning with the models developed in the earlier chapters of this text, can we understand how the "somethings" of our world (atoms and molecules) have found a route to become *living somethings*?

Interestingly, the answer to this question is indeed *yes*! During the last half of the nineteenth century, the science of chemistry came to be dominated by dramatic advances in a subdiscipline known today as organic chemistry. This rapid growth of organic chemistry as a discipline in its own right coupled with the broad application of organic chemistry in the German dye industry led to chemistry's focus on *structural questions* and the initial use of the *static* electron by Lewis at the dawn of the twentieth century to propose one of the first modern theories of chemical bonding. Utilizing this structural perspective and being better informed by the idea of the *dynamic* electron (used in our study of valence bond theory and molecular orbital theory), we are about to discover that organic chemistry's structural perspective has had two major consequences: (1) it has become a powerful tool with which to understand the *molecular structures* (that is, the links among the atoms constituting a molecule) of **both** organic molecules and, very significantly, biochemical molecules and (2) it has become a very fertile route by which to understand *chemical reactivity* in **both** organic molecules and biochemical molecules. In fact, we shall see the close linking of **structure** and **function**.

## First Steps to Understanding Organic Chemistry

The central focus of organic chemistry is the structure and reactivity of compounds in which carbon and hydrogen atoms are the most important components. Additionally, organic compounds may also contain atoms of oxygen, nitrogen, sulfur, phosphorous, and a few other minor constituents. However, it is the carbon atom that plays the most significant role, and it is carbon's characteristic bonding capability and versatile reactivity that largely determine the chemistry of organic compounds. Early in the study of carbon compounds a key observation was made: in stable compounds, the carbon atom appears to form **four chemical bonds**. That is, we say that carbon is **tetravalent**. Further, the carbon atom can form **multiple bonds** (both double bond and triple bonds). The tetravalency of the carbon atom along with its ability to form single, double, and triple bonds are key characteristics that are responsible for the diverse richness of both the structures and the reactions found in organic chemistry.

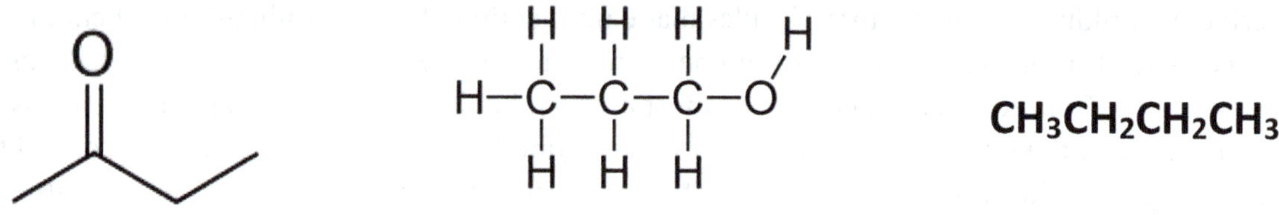


Butanone Skeletal Structure 1-Butanol Structural Formula Condensed Structure

Figure 7.1: Molecular Representations

An early practical issue faced by chemists during their exploration of the properties of carbon was the necessity to devise an efficient method of representing chemical substances that went beyond the initial efforts that Dalton had made at the beginning of the nineteenth century. In response to this practical necessity and driven by the focus on *structure*, several different presentation methods have developed. A simple **structural formula** shows how the atoms in a molecule are connected by representing each bond between atoms either as one line (single bond) or as two or three lines (two for a double bond, three for a triple bond). In a structural formula **all** the atoms are explicitly included. The center panel in figure 7.1 and the right panel of figure 7.2 show the structural formulas for 1-butanol and butanone, respectively. In contrast, a **skeletal formula** represents bonds as lines (one, two, or three), but the carbon atoms are **not** shown, and hydrogen atoms are drawn **only** when they are attached to atoms **other than carbon.** It is *assumed* that **each linear segment** of a skeletal formula (a single line, two single lines, or even three single lines) links **two carbon atoms, even though the carbon atoms are not explicitly shown.** Further, because the carbon atom is tetravalent, a skeletal formula implies that there is a sufficient number of hydrogen atoms at each carbon atom to ensure that the carbon atoms **always** have **four** bonds. The left panel in figure 7.1 shows the skeletal formula for butanone, while the right panel of figure 7.2 shows the skeletal formula for 1-butanol. Finally, in a **condensed structure**, while all the atoms in the molecule are still written explicitly (as in the case of a structural formula), **as few bonds as possible** (often none!) are displayed and atoms are grouped together. The right panel of figure 7.1 displays the condensed structure for n-butane.

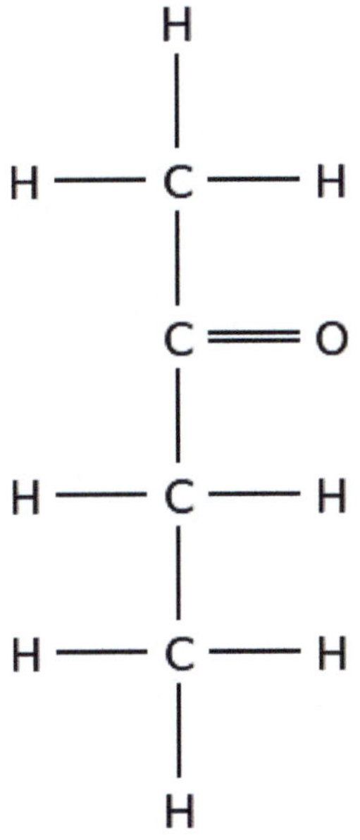

Butanone Structural Formula

1-Butanol Skeletal Formula

Figure 7.2: Molecular Representations

Example 7.1: Draw the skeletal formulas for the following molecules:

1. $CH_3CH_2CH_2CH_3$
2. Cyclobutane
3. $CH_3OH$
4. $CH_3CH_2CH_2CH_2CH_3$

Answer: **Organize:** The first molecule possesses four carbon Atoms.
The second molecule possesses four carbon atoms, but it is a **cyclic** molecule (that is, a molecular ring molecule; see text below).
The third molecule contains a noncarbon atom.
The first molecule possesses four carbon atoms.

**Unknown:** Four diagrams

**Translate:** Review the definition of a skeletal formula.

**Solve:**

1. $CH_3CH_2CH_2CH_3$

2. Cyclobutane

3. $CH_3OH$

4. $CH_3CH_2CH_2CH_2CH_3$

Example 7.2: Draw the structural formulas for the following molecules:

1. $CH_3CH_3$
2. Cyclobutane

Answer: **Organize:** The first molecule possesses two carbon atoms.
The second molecule possesses four carbon atoms, but it is a **cyclic** molecule (that is, a molecular ring molecule; see text below).

**Unknown:** Two diagrams

**Translate:** Review the definition of a skeletal formula.

**Solve:** 1. $CH_3CH_3$

```
   H  H
   |  |
H—C—C—H
   |  |
   H  H
```

2. Cyclobutane

```
    H   H
    |   |
H - C - C - H
    |   |
H - C - C - H
    |   |
    H   H
```

While structural formulas, skeletal formulas, and condensed structures are highly effective ways to represent the *structure* of molecules, it is critically important to remember that all three representations are only **two-dimensional**; they are well adapted for use on a sheet of paper, on a chalkboard or whiteboard, or on a video screen or projector screen. But the world we experience and visualize exhibits at least **three spatial dimensions** (in fact, there may be many additional spatial dimensions); the world is certainly much more than a two-dimensional surface. We use the term **conformation** to specify a particular three-dimensional arrangement of the atoms within a molecule. Figure 7.3 depicts a three-dimensional model (often called a *ball-and-stick model*) of the organic molecule $(CH_3)_2CO$ (propanone).

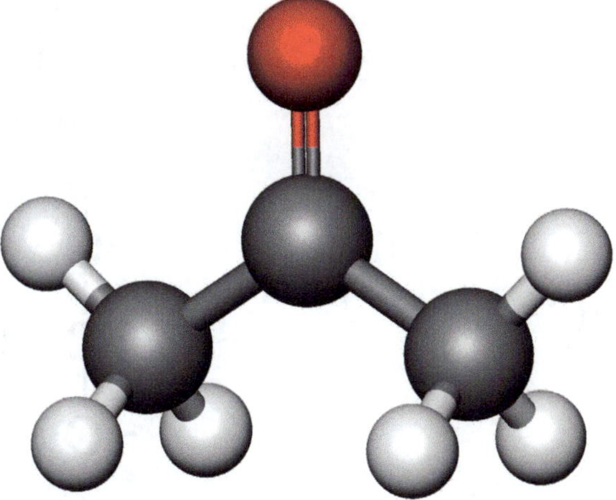

Figure 7.3: Ball-and-Stick Three-Dimensional Model

There is an alternate approach to representing the three-dimensionality of actual molecules. This approach is demonstrated in figure 7.4 for the methane molecule. The image is often called a *wedge-and-dashed model,* in which the *wedge* bond is used to indicate that the hydrogen atom is projecting *above* the plane of the textbook page, while the *dashed* bond indicates that the hydrogen atom projects *below* the plane of the textbook page. The remaining three atoms, one carbon atom and two hydrogen atoms, *lie in the plane* of the textbook page.

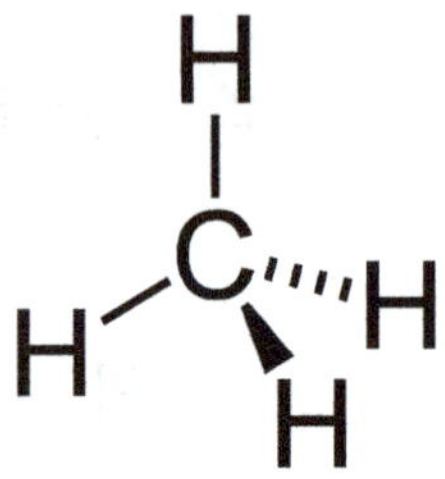

Figure 7.4: Wedge-and-Dashed Three-Dimensional Model

Because the carbon atom is tetravalent and capable of forming single, double, and triple bonds both with itself and with other elements, the surrounding environment of each carbon atom in a molecule is critical the molecule's behavior. Four terms are used to specify the environment of each carbon atom in a molecule. A **primary** carbon atom is a carbon atom with only one other carbon atom attached to it, a **secondary** carbon atom is a carbon atom with two other carbon atoms attached to it, a **tertiary** carbon atom is a carbon atom with three other carbon atoms attached to it, and a **quaternary** carbon atom is a carbon atom with four other carbon atoms attached to it. Figure 7.5 shows the four possibilities.

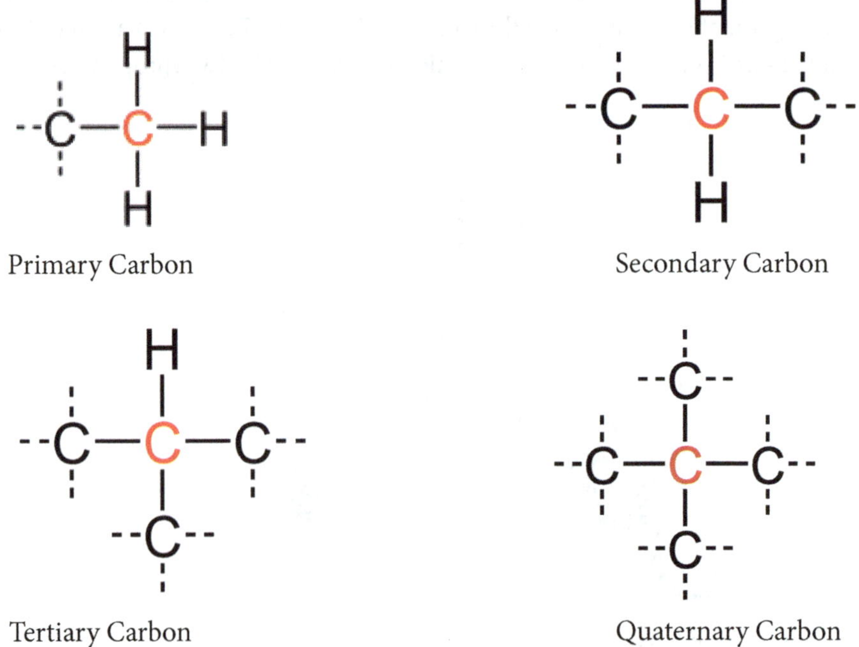


Figure 7.5: Primary, Secondary, Tertiary, and Quaternary Carbon Atoms

The ability of the carbon atom to form both single bonds and multiple (double and triple) bonds has such far-reaching consequences for both organic structures and reactivity, as well as significant implications for biochemical processes, that organic and biochemical molecules are easily divided into two general classes. An organic (and, of course, biochemical) compound is called **saturated** if it contains ***only*** single bonds; that is, all the atoms in the molecule are linked by single bonds. In contrast, a molecule in which ***one or more*** of the links between atoms in the molecule are ***multiple bonds*** is called **unsaturated**.

## The Bonds That Bind the Carbon Atom

The careful attention we paid to the quantum mechanical theories of chemical bonding in chapter 6 (valence bond theory and molecular orbital theory) now allows us to identify powerfully descriptive connections between these structural models and key experimental characteristics of the carbon atom. The fact that the carbon atom is *tetravalent*—that is, in stable compounds, the carbon atom invariably exhibits four distinct chemical bonds (either four single bonds or a combination of single and multiple bonds, totaling a *maximum* of four bonds)—can be directly correlated with the first three hybridization schemes encountered in chapter 6. The most elementary identification is between the set of *four* $\mathbf{sp^3}$ hybrid atomic orbitals and a tetravalent carbon atom exhibiting *four* single bonds. It is particularly important to note that such a carbon atom not only exhibits *four* bonds but also possesses an experimental geometry that is consistent with the *tetrahedral* geometry of $\mathbf{sp^3}$ hybridization.

The experimental settings in which the carbon atom exhibits *multiple* bonds directly correlate with our understanding of the $\mathbf{sp^2}$ and **sp** hybridization procedures. In the case of the $\mathbf{sp^2}$ hybridization of carbon, the *three* $\mathbf{sp^2}$ hybrid atomic orbitals that are centered on the carbon atom are *coplanar*, leaving a single unhybridized **p** atomic orbital centered on the same carbon atom and perpendicular to the plane defined by the three $\mathbf{sp^2}$ hybrid atomic orbitals. The end-to-end *resonance interaction* of *two* $\mathbf{sp^2}$ hybrid atomic orbitals (or of one $\mathbf{sp^2}$ hybrid atomic orbital with one unhybridized atomic orbital) to define a *single* bond as well as the *resonance interaction* of the unhybridized **p** atomic orbitals centered on **adjacent** atoms (carbon or noncarbon) directly correlate with the experimental observation of a *double* bond. Importantly, the experimental data are consistent with the planar geometry required by the *double* bond of VB theory and the necessary *four* bonds associated with each $\mathbf{sp^2}$ hybridized carbon atom.

The **sp** hybridization of the carbon atom produces a set of *two* **sp** hybrid atomic orbitals that exhibit a *linear* geometry while leaving *two* unhybridized **p** atomic orbitals, all centered on the same carbon atom. The unhybridized **p** atomic orbitals are *mutually perpendicular* as well as *perpendicular* to the imaginary line passing through the two **sp** hybrid atomic orbitals (because their geometry is *linear*). The end-to-end *resonance interaction* of *two* **sp** hybrid atomic orbitals (or of one **sp** hybrid atomic orbital with one unhybridized atomic orbital) to define a *single* bond as well as the *resonance interaction* of the two unhybridized **p** atomic orbitals centered on **adjacent** atoms (carbon or noncarbon) directly correlate with the experimental observation of a *triple* bond. Again, the linear geometry required by the *triple* bond of VB theory and the *four* bonds associated with each **sp** hybridized carbon atom are consistent with the experimental data.

How does the perspective offered by MO theory (again, see chapter 6) provide an avenue to understand the carbon atom? As the reader will recall, MO theory takes the point of view that the orbitals associated with the electrons of a molecule belong to the *entire* molecule rather than being characteristics of individual atoms. However, the orbitals described by MO theory still possess specific geometries and are identified as *sigma orbitals* and *pi orbitals*, leading immediately to a description of **both** single and multiple bonds. In fact, the calculation of the bond orders of in an organic molecule using MO theory is entirely consistent with the experimental observations. Further, the application of MO theory to organic molecules yields the same conclusion reached by VB theory: carbon is *tetravalent*. By focusing on orbitals that belong to the entire molecule, MO theory does, however, offer a significant conceptual change of perspective that will prove extremely useful in our study of a major **functional group** of molecules whose properties depend on *delocalized* electrons.

## Functional Groups

What is a **functional group**? The definition is actually quite simple: a **functional group** is a **part of a larger molecule** and is composed of an **atom**, a **group of atoms**, or a **bond** that has a **characteristic chemical behavior**. In a curious turn of events, the concept of a functional group, while focusing on unique patterns of atoms and bonds that are specifically *structural* concepts, spectacularly links these patterns to *chemical reactivity*. The enduring focus of chemistry on *structures* is about to become intimately linked to chemical behavior, to the *reactions* that characterize chemical substances. This is a very powerful combination, and it is one of the central organizing principles of organic chemistry. We are about to discover that *the characteristic chemistry of any organic molecule,* ***regardless*** *of its size or complexity, is fundamentally determined by the* ***functional groups*** *found in the molecule.*

Thirteen essential functional groups of organic chemistry can be arranged into three major families by noting critical common features that distinguish each of the three families. The first family is composed of *four* distinct functional groups with the common characteristic that all the molecules of this family are **hydrocarbons**. That is, every member of this first family of molecules is composed of only **two** elements: hydrogen and carbon. This is the origin of the name **hydrocarbon**. The four functional groups that belong to the hydrocarbon family are the **alkanes**, the **alkenes**, the **alkynes**, and the **aromatic hydrocarbons**. Often, the alkanes, alkenes, and alkynes are grouped together under the term **aliphatic** to distinguish them from the **aromatic hydrocarbons**. The hydrocarbon family of functional groups is displayed in figure 7.6. In the figure, the letter *R* stands for the phrase "the rest of the molecule," which, in the case of this first family of functional groups, can **only** contain hydrogen and carbon atoms. These *R* fragments simply represent *alkane chains.* (The term *alkane chain* means a group of carbon atoms linked by **single** bonds.) For the alkane functional group, even though the six *R*s are identical in the figure, they are **not** required to be identical; this is the meaning of the "primes" in the alkene and alkyne functional groups. In each of these cases, the $R^{i}$ components **may** be identical, but they are **not** required to be identical. Further, for the members of the alkane functional group, **all** the carbon-carbon bonds in each *R* must be single bonds. (It is possible for a single molecule to possess *multiple* functional groups so that the *R*s in figure 7.6 could contain additional functional groups. We call such a molecule **polyfunctional**. For the present, we shall assume that all of the organic molecules possess only a single, characteristic functional group. The case of **polyfunctional** molecules will be examined later.)

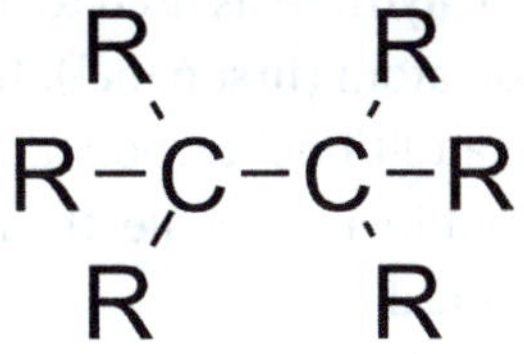

Alkane

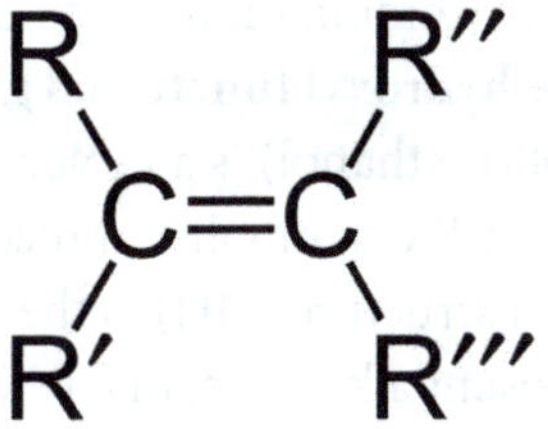

Alkene

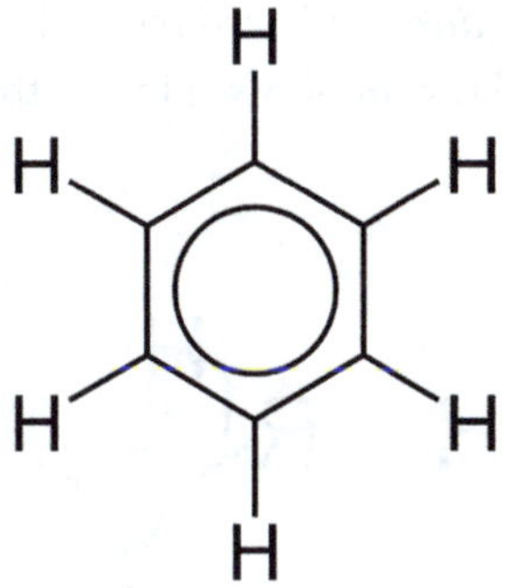

Alkyne

Aromatic

Figure 7.6: The Hydrocarbon Family of Functional Groups

Because all the members of the hydrocarbon family contain **only** the elements hydrogen and carbon, and **only** carbon is capable of making multiple bonds, the key identifying feature of each functional group is the nature of the carbon-carbon bond. In the **alkanes, all** the carbon-carbon bonds are **single bonds**; in the **alkenes, at least one** of the carbon-carbon bonds **must be a double bond**; in the **alkynes, at least one** of the carbon-carbon bonds **must be a triple bond.** The simplest alkane is an exception: methane consists of a single carbon atom bonded to four hydrogen atoms; there is **no** carbon-carbon bond. The molecules constituting the **aromatic hydrocarbon** functional group contain planar, benzene-like ring structures, and all the aromatic hydrocarbons exhibit the phenomenon of **delocalized electrons** that is characteristic of benzene. The single benzene structural formula represents the simplest aromatic hydrocarbon and is used in the figure to symbolize the entire aromatic hydrocarbon functional group. (Because we are focusing on molecules that **are not** polyfunctional, we will study alkenes, which contain **only** one double bond, and alkynes, which contain **only** one triple bond.)

The second family of functional groups consists of molecules in which a **single** bond links a carbon atom to an **electronegative** atom. (These are atoms that are assigned *large* values for their electronegativity; see chapter 4.) This second family is significant because it marks the appearance in organic molecules of several crucially important electronegative atoms. (These atoms are members of a class of atoms called **heteroatoms** by organic chemists. In general, a heteroatom is *any* atom that is ***not*** either hydrogen or carbon.) In particular, we see for the first time the inclusion of oxygen atoms, nitrogen atoms, and sulfur atoms in organic molecules. There are again four principal members of this family of functional groups: the **alcohols**, the **ethers**, the **sulfur-containing compounds**, and the **amines**. Representative structures of the alcohol functional group are displayed in figure 7.7. The five panels identify the *OH* atom combination is the characteristic molecular fragment of **all** alcohols. The *OH* group of atoms is the **hydroxyl functional group**, and often the adjective **hydroxy** is used to describe it. The simplest alcohol (methanol) is a molecule with only a **single** carbon atom (first panel). In the next three panels, $R^1$, $R^2$, and $R^3$ symbolize molecular fragments containing additional carbon atoms. If the characteristic alcohol structure (OH) is the *only* functional group present in a molecule, the fragments ($R^1$, $R^2$, and $R^3$) represent alkane chains, which **may** or **may not** be identical.

In the second panel, the *OH* group of atoms is bonded to a *primary* carbon atom, and the molecule is identified as a **primary alcohol.** In the third and fourth panels of the figure, the *OH* group of atoms is bonded to a *secondary* and *tertiary* carbon atom, respectively, and the molecules are identified as **secondary** and **tertiary alcohols.** Finally the fifth panel of the figure displays the simplest **aromatic alcohol** (phenol).

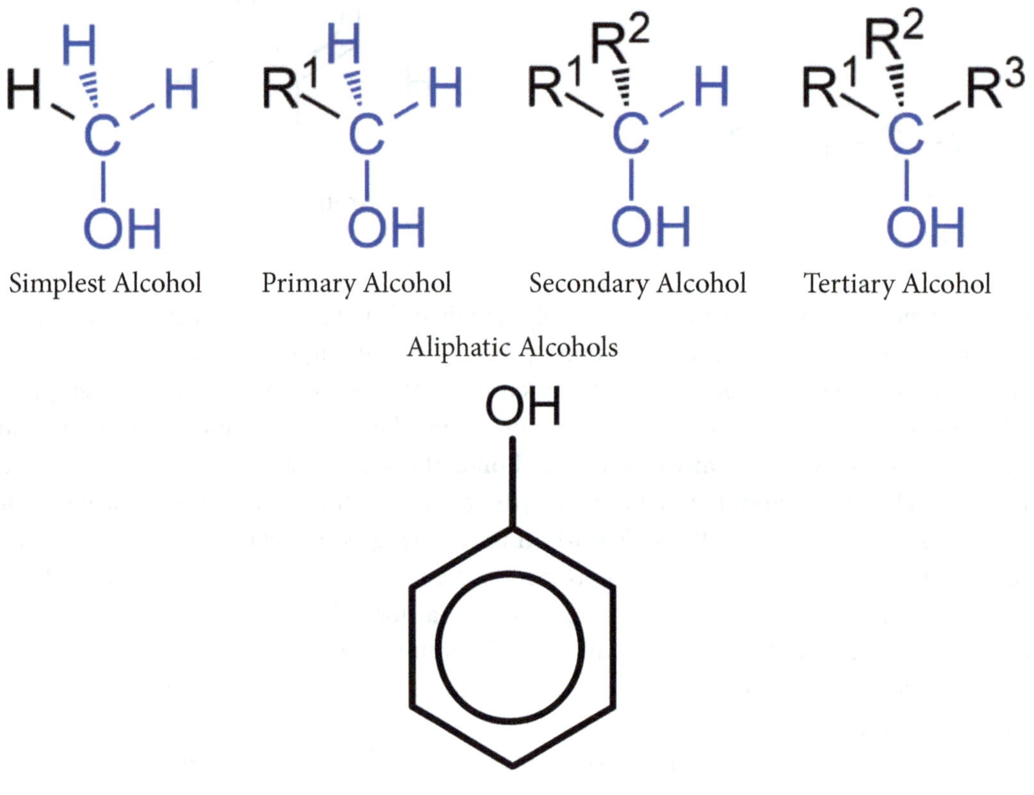


Figure 7.7: The Alcohol Functional Group

The **ethers** are a collection of molecules in which a single oxygen atom is linked to *two different* carbon atoms by **single** bonds. The representative structure of this class of molecules is displayed in figure 7.8. As in the case of the alcohols, if the ether molecule is *not polyfunctional*, the *R* and *R'* in figure 7.8 are simply alkane chains which, again, **may** or **may not** be identical.

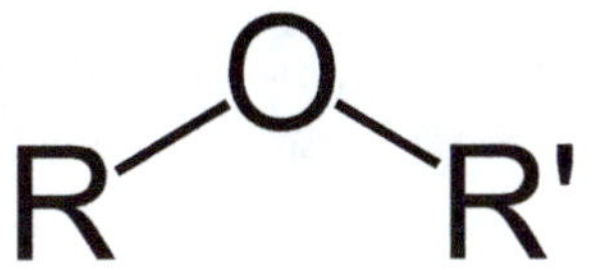

Figure 7.8: Ether Functional Group

The functional group of **sulfur-containing compounds** on which we will focus is made up of three broad categories of organic compounds that include at least one sulfur atom. The three panels of figure 7.9 display the representative structures of these three categories. The first two panels of the figure depict the **thiols** and the **thioethers** (the sulfur analogs of alcohols and ethers, in which the oxygen atom has been replaced by a sulfur atom). The prefix *thio* is from the Greek, meaning "sulfur," and explains the names in the left and center panels of figure 7.9. The name in the right panel of the figure reflects the presence of the two sulfur atoms. As we shall see later, the *disulfide* structure plays a critical role in determining the three-dimensional geometry of proteins. In each of the three panels of figure 7.9, if the sulfur compound is *not polyfunctional*, the $R^i$ components are simply alkane chains, which, again, **may** or **may not** be identical.

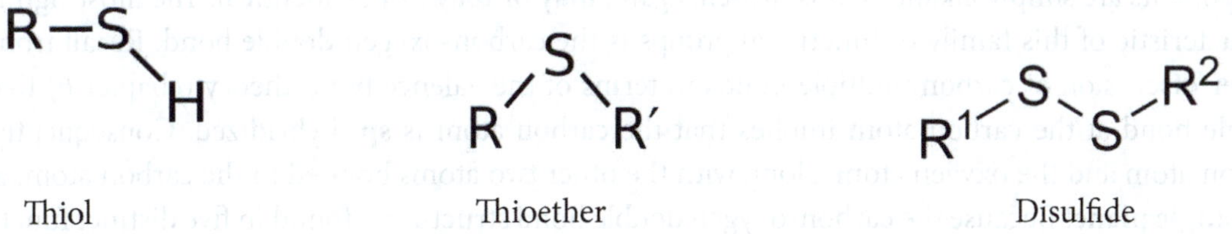

Figure 7.9: Sulfur-Containing Compounds

The final functional group that is a member of this second family consists of the **amines**. As figure 7.10 indicates, the amine functional group is composed of four general classes: the **primary amines**, the **secondary amines**, the **tertiary amines**, and the **quaternary amines**. The names *amine* is a reference to the ammonia molecule ($NH_3$), because **all** the amines can be viewed simply as substituted ammonia molecules. In each of the four panels of figure 7.10, if the amine is *not polyfunctional*, the $R^i$ components are simply alkane chains, which, again, **may** or **may not** be identical.

**Tertiary Amine** **Quaternary Amine**

Figure 7.10: Amine Functional Group

The third family of functional groups consists of five classes of molecules, **all** of which contain a carbon-oxygen **double** bond: the **aldehydes**, the **ketones**, the **carboxylic acids**, the **esters**, and the **amides**. The five panels of figure 7.11 display the representative structures of each of these five functional groups. In each of the panels of figure 7.11, if the indicated functional group is *not polyfunctional*, the $R^i$ components are simply alkane chains, which, again, **may** or **may not** be identical. The most significant characteristic of this family of functional groups is the carbon-oxygen double bond. Recall from our earlier discussion of carbon multiple bonds in terms of the valence bond theory (chapter 6) that the double bond at the carbon atom implies that the carbon atom is **sp²** hybridized. Consequently, the carbon atom and the oxygen atom, along with the other two atoms bonded to the carbon atom, **all lie in a single plane**! Because the carbon-oxygen double bond structure is found in five distinct functional groups and has important geometric consequences, it is given a special name. It is called a **carbonyl**, and the structure is highlighted in figure 7.12.

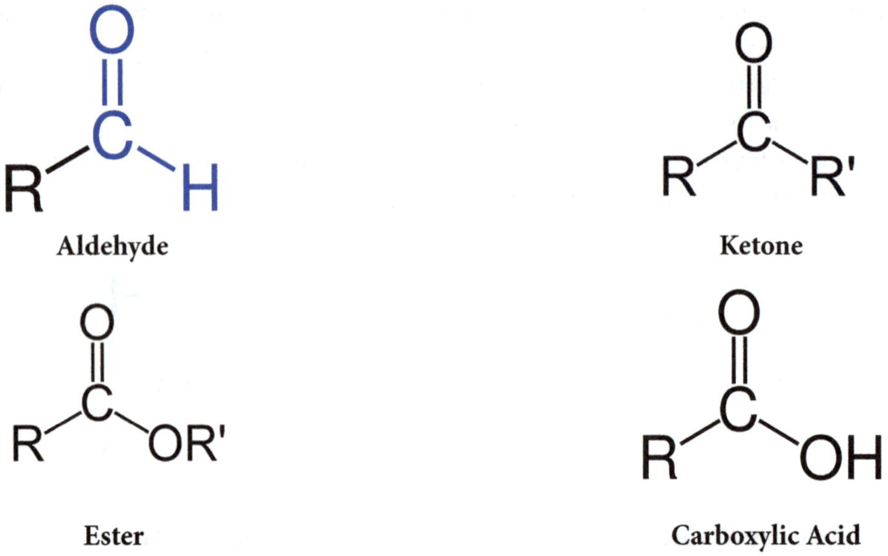

**Aldehyde** **Ketone**

**Ester** **Carboxylic Acid**

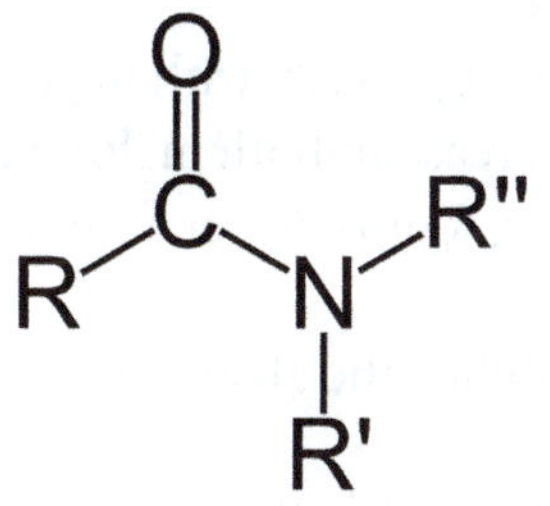

**Amide**

Figure 7.11: Functional Groups of the Third Family

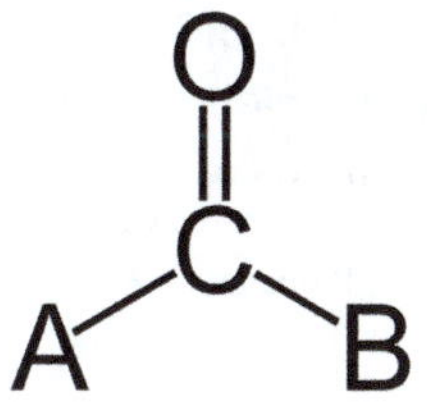

Figure 7.12: Carbonyl Group

## Hydrocarbons

As we have already noted, the hydrocarbons are organic molecules composed entirely of carbon and hydrogen atoms; they are often grouped into two large categories, the *aliphatics* and the *aromatics*. Interestingly, while members of the aromatic category often exhibit the basic ring structure of benzene, the aliphatics occur as **open-chain** molecules and as **cyclic molecules** (molecules with ring structures). In fact, members of the three functional groups, the *alkanes*, the *alkenes*, and the *alkynes*, can occur as **either** open-chain molecules **or** cyclic molecules.

### Alkanes and Cycloalkanes

The open-chain alkanes and the cycloalkanes are saturated hydrocarbons (alkanes are sometimes called **paraffins**, a historical name that has a more specific modern application); they are molecules containing **only** single carbon-carbon bonds (remember that **all** carbon-hydrogen bonds are **always single bonds**). Again, recall that methane is the exception. Figure 7.13 displays some representative examples of both open-chain alkanes and cycloalkanes. We summarize here some of the rather unremarkable properties that characterize the alkane functional group:

1. They are either odorless or possess a mild odor; they are colorless and tasteless.
2. The alkanes are nonpolar (see chapter 4) molecules; they are insoluble in water but soluble in nonpolar organic solvents.

3. The alkanes are less dense than water.
4. The alkanes **are** flammable, reacting vigorously with oxygen.
5. The alkanes exhibit a monotonic increase in both melting and boiling points as a function of their molecular weight; alkanes containing four or fewer carbon atoms are gases at 298.15 K and a pressure of 1 atmosphere.
6. With the exception of their flammability, the alkanes are not very reactive.

As the reader knows well, the *flammability* of the alkanes is one their most important economic properties. The combustion of hydrocarbon fuels (in particular, methane, propane, butane, and octane) is an energy source of crucial importance to every contemporary (early twenty-first-century) industrial and postindustrial economy.

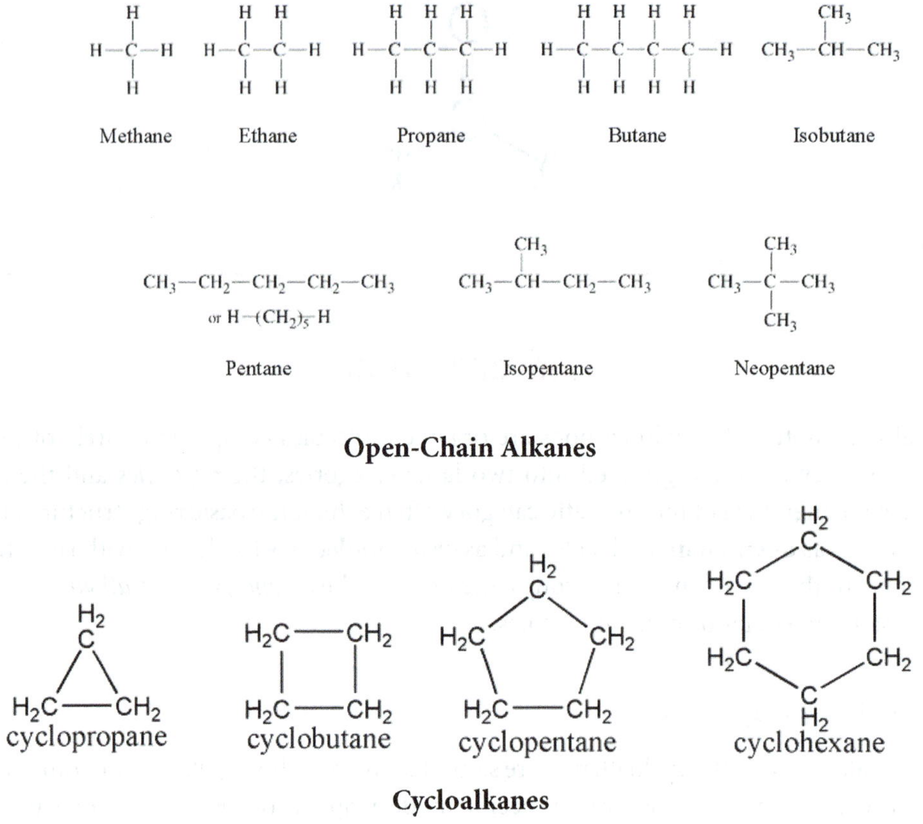

Methane Ethane Propane Butane Isobutane

$CH_3-CH_2-CH_2-CH_2-CH_3$ or $H\text{-}(CH_2)_5\text{-}H$

Pentane

$CH_3-CH(CH_3)-CH_2-CH_3$

Isopentane

$CH_3-C(CH_3)_2-CH_3$

Neopentane

**Open-Chain Alkanes**

cyclopropane cyclobutane cyclopentane cyclohexane

**Cycloalkanes**

Figure 7.13: Representative Alkanes

## Alkenes and Alkynes

As we noted earlier, the distinguishing feature of the *alkene* functional group is the presence of **one or more** double bonds in the molecule. Consequently, the alkenes are unsaturated hydrocarbons and, like the alkane family, occur as both open-chain molecules and cyclic molecules. The simplest alkene is

$CH_2CH_2$ and is named ethene. We shall see shortly that the geometric properties of the **$sp^2$** hybridization (from valence bond theory) associated with the carbon-carbon double bonds in alkenes have extremely important chemical consequences. Because we have chosen to focus primarily on organic molecules having only a single functional group (this will change when we examine polyfunctional biochemical molecules), the alkene molecules (both open-chain and cyclic) of interest here will have only **one** double bond. Figure 7.14 depicts the structural and skeletal formulas of typical examples of both open-chain and cyclic alkenes having only one double bond.

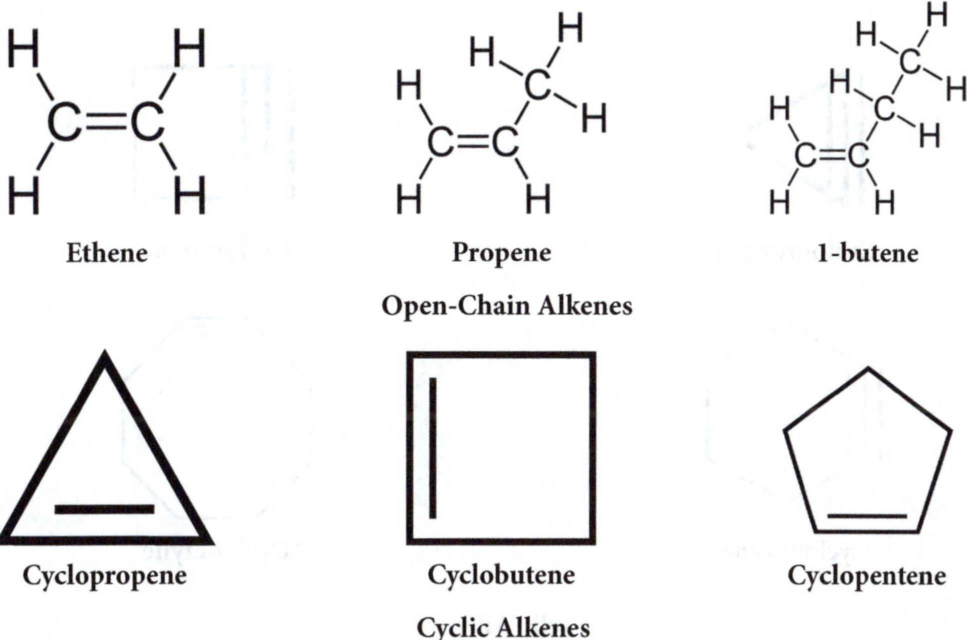


Figure 7.14: Alkene Molecules

Like the alkenes, the *alkynes* form a functional group whose distinguishing characteristic is, again, a specific type of carbon-carbon bond; in this case the bond is a **triple** bond. The simplest alkyne is CHCH ($C_2H_2$) and is called ethyne. Recall that a triple bond between two carbon atoms (again using the valence bond theory of chemical bonding) requires that each carbon atom must be **sp** hybridized, which imposes a **linear** geometry on the portion of the molecule participating in the triple bond. Figure 7.15 depicts several examples of both open-chain and cyclic alkynes having only **one** triple bond.

H—C≡C—H

**Ethyne**

H—C≡C—$CH_3$ (H—C≡C—C(H)(H)—H)

**Propyne**

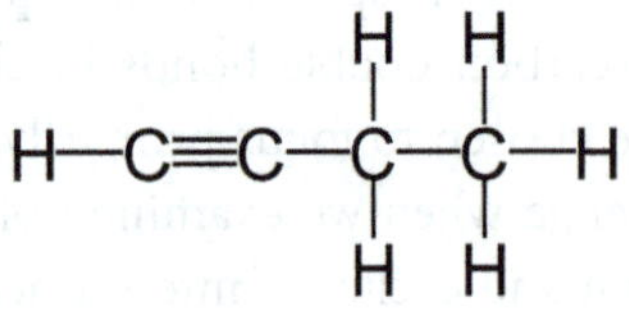

H H
H—C≡C—C—C—H
H H

1-butyne

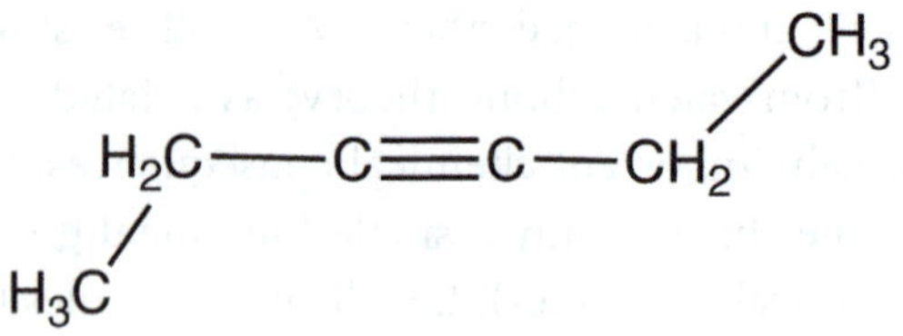

CH3
H2C—C≡C—CH2
H3C

3-hexyne

Open-Chain Alkynes

Cyclopropyne

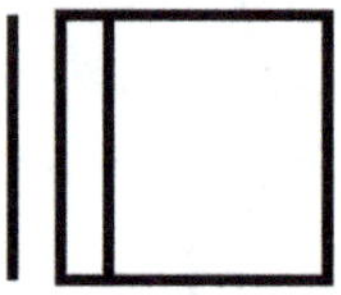

Cyclobutyne

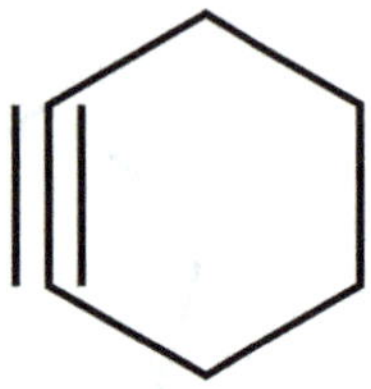

Cyclohexyne

Cyclooctyne

Cyclic Alkynes

Figure 7.15: Alkyne Molecules

## Aromatic Hydrocarbons

The aromatic hydrocarbon functional group (sometimes identified by the term **arene**) is a very large and diverse class of organic compounds that exhibit both unusual stability and, surprisingly, a characteristic and very distinct odor. In fact, the name of the functional group derives from the word *aroma*! The key to the unusual stability of the members of this functional group is the phenomenon of *delocalized electrons,* which we first met during our brief introduction to the valence bond theory in chapter 6. The simplest hydrocarbon that exhibits the properties of the aromatic hydrocarbon functional group (collectively called **aromaticity**) is the molecule benzene. While there are aromatic compounds that include either ring structures built with atoms *other* than carbon (heteroatoms) or non-ring structures, we focus here on those aromatic compounds containing one or more benzene rings. Figure 7.16 displays some representative members of the aromatic functional group.

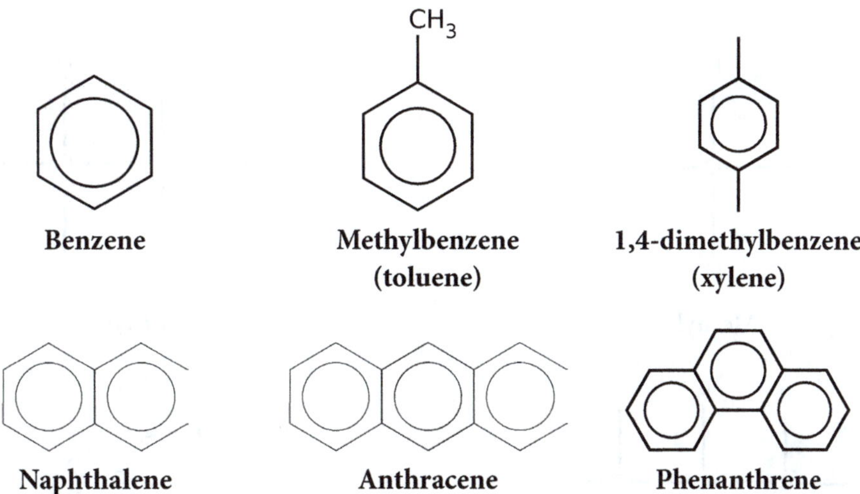


Figure 7.16: Aromatic Compounds

## Naming Organic Molecules

The identification of a systematic structural pattern among the molecules of organic chemistry (the functional groups) might suggest to the reader that there is a methodical and consistent procedure to give *names* to these molecules. In fact, the field of organic chemistry has developed a powerful and very organized approach to give a unique name to each organic molecule. This approach has been codified by the International Union of Pure and Applied Chemistry (IUPAC) and is universally used by chemists around the world. Each organic name consists of three parts:

1. A prefix
2. A root
3. A suffix

In order to understand the meaning of these three principal parts of an organic name, we *imagine* that every organic molecule can be constructed by starting with either an alkane (open-chain or ring) or an aromatic hydrocarbon. We call this imaginary molecule the **parent compound**. Then, the *prefix* in the name of any organic molecule identifies the location of each **substituent** in that molecule. Of course, the question is, what is a **substituent**? A substituent is either a single atom or a group of atoms (other than one of the functional groups noted above) that replaces a hydrogen atom in the molecule's **parent compound**. One of the most significant groups of atoms that plays the role of a substituent is the **alkyl group**. An alkyl group is simply an alkane molecule with one hydrogen atom removed. The removal of the hydrogen atom during the formation of an alkyl group provides an open bonding site, allowing the alkyl group to form a link with a parent compound. The name **alkyl** is derived from the word **alkane** by replacing the *ane* ending with the *yl* ending. Figure 7.17 displays eight alkyl groups that commonly occur as substituents in organic molecules.

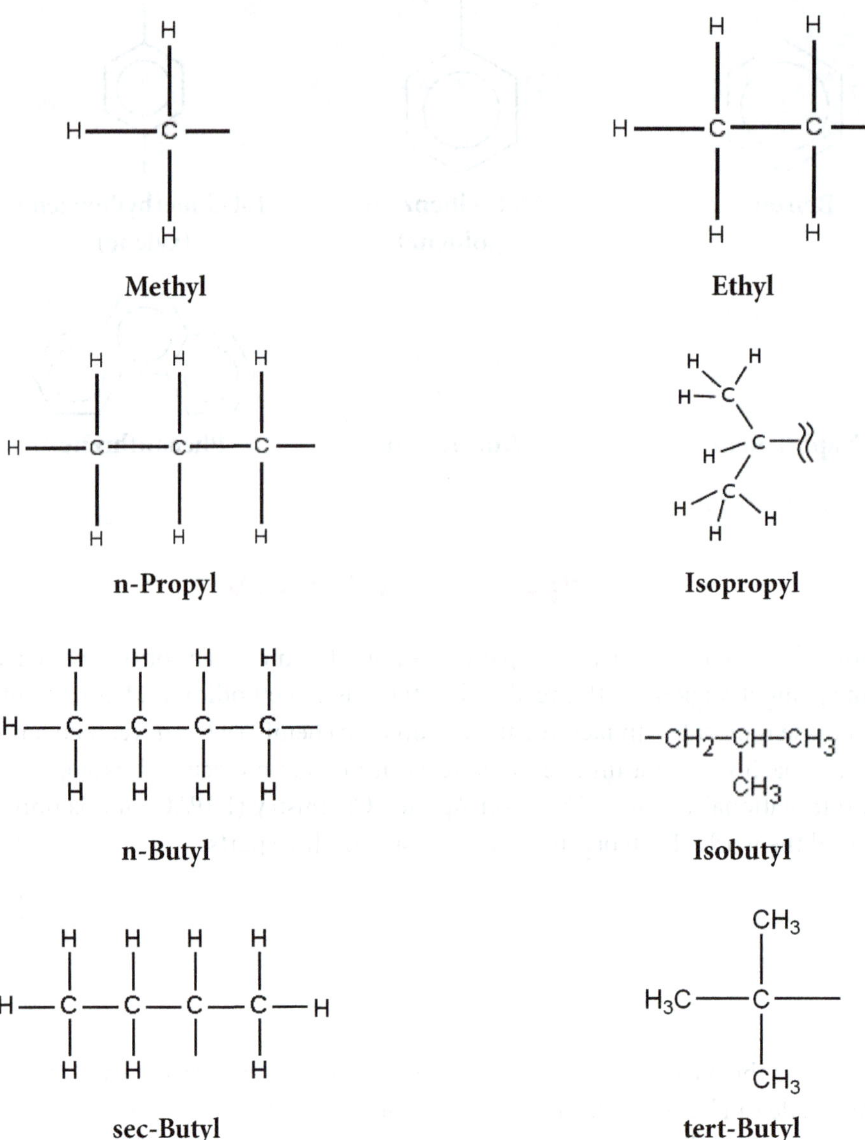


Figure 7.17: Alkyl Groups

Note that the names of the alkyl groups having less than five carbon atoms are derived from the four simplest alkanes: methane, ethane, propane, and butane. Once the number of carbon atoms in the substituent reaches five, the systematic *Greek prefixes* (see table 4.3) are used. Table 7.1 list the first ten *IUPAC prefixes* used to name the alkyl groups. The key to giving our organic molecule a name is to identify the **parent compound**. This is exactly the role of the *root*.

The key characteristic that the root identifies is the *number* of carbon atoms in the aliphatic parent compound. (Because the aromatic molecules that we will meet in this text are all built around benzene, the number of carbon atoms is already known. Benzene has six carbon atoms.) Finally, the *suffix* identifies the **functional group** to which the molecule belongs. In effect, the suffix is the molecule's *family name.*

Table 7.1: IUPAC Prefixes

| Number of Carbon Atoms | Prefix |
|---|---|
| 1 | Meth |
| 2 | Eth |
| 3 | Prop |
| 4 | But |
| 5 | Pent |
| 6 | Hex |
| 7 | Hept |
| 8 | Oct |
| 9 | Non |
| 10 | Dec |

## Naming Alkanes

The assignment of an IUPAC name to any organic molecule follows an orderly and systematic procedure. We'll begin by naming the members of the alkane family, and we will discover that the procedure used for the alkanes is the key to assigning an IUPAC name to **every** organic molecule. The same well-defined steps are followed repeatedly:

1. **Name the root:**
   Select the longest continuous chain of carbon atoms as the *imaginary* **parent compound**. (For the alkanes, the parent compound is also an alkane.) The *imaginary* **parent compound** is always an alkane whose name is assigned using the IUPAC prefixes from table 7.1. Consider any atom or group of atoms that is not part of the *imaginary* parent compound (that is, **not included** in the longest continuous chain of carbon atoms) to be a **substituent** that has replaced a hydrogen atom in the *imaginary* parent compound. The **root** of the molecule to be named will be derived from the name of the *imaginary* parent compound.

2. **Identify the prefix:**
   The root will be then preceded by a **prefix**, which includes the names of **all** the substituents bonded to the *imaginary* parent compound. Identify all the substituents that will be included in the **prefix**, giving each substituent its proper name.

3. **Identify the suffix:**
   The **suffix** is simply the indicator of the *family* or *functional group* to which the molecule belongs. In the case of the alkanes this is simply the *ane* ending of the word *alkane*. (In this case, the suffix of the *imaginary* parent compound and the molecule we are naming is the same. This will change when we name members of other functional groups.)

4. **Number the carbon atoms in the *longest continuous chain*:**
   While we have taken important strides by identifying the root, the prefix, and the suffix of the molecule to be named, the naming procedure requires us to be *much more specific*. The first step in this additional level of detail is to number the carbon atoms in the *longest continuous chain* of carbon atoms in the *imaginary* parent compound. The numbering of the carbon atoms can occur in two ways: either starting at the leftmost carbon atom in the longest continuous chain or starting at the rightmost carbon atom in the longest continuous chain.

5. **Associate each substituent with the number of the carbon atom to which it is bonded:**
   For example, if a *methyl group* is bonded to the *imaginary* parent compound at carbon number 2 (where the number was assigned in step 4), assign the name *2-methyl* to that substituent. Because it is possible to number the carbon atoms in the parent compound **two** different ways, left to right or right to left, each substituent can be assigned **two** numbers! Consequently, the left-to-right numbering will assign a set of numbers (integers) to the substituents, and the right-to-left numbering will assign a second set of numbers (also integers) to the substituents.

6. **Choose *one* of the *two* sets of numbers generated in step 5:**
   We now have two sets of numbers; this seems to be getting complicated! How do we choose? If there is only **one** substituent, the answer is easy: choose the numbering that associates the lower of the two numbers with the sole substituent. If there is more than one substituent, add together the numbers originating from the left-to-right numbering. Call this total $\mathbf{S_1}$ (shorthand for *sum number one*). Then add together the numbers originating from the right-to-left numbering and call this total $\mathbf{S_2}$ (shorthand for *sum number two*). Compare $\mathbf{S_1}$ and $\mathbf{S_2}$; they are simply two numbers. Choose the numbering that gives the **smaller** number. If $\mathbf{S_1}$ and $\mathbf{S_2}$ are **equal**, you can choose either numbering system!

7. **If the same substituent occurs multiple times, use an appropriate *Greek* prefix:**
   It often occurs that the *same* substituent occurs multiple times along the *imaginary* parent compound. In order to identify multiple occurrences of the same substituent, a standard *Greek prefix* (table 4.3) is added to the name of the substituent. If the substituent occurs twice, the prefix is *di*; three times the prefix is *tri*; four times the prefix is *tetra*; and so on.

8. **Write the molecules name as a single word:**
   This the last step! Simply combine the prefix, root, and suffix into a single continuous name. If the prefix contains multiple distinct substituents, organize them in alphabetical order. However, in determining the alphabetical order, **do not** use the *Greek prefixes*.

Example 7.3: What is the IUPAC name of the following molecule?

$CH_3C(CH_3)_2CH(CH_3)$ (Condensed formula)

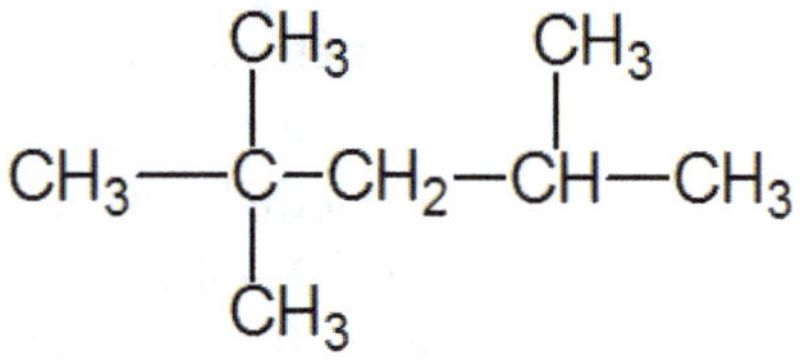

$$CH_3-C(CH_3)_2-CH_2-CH(CH_3)-CH_3$$

(Structural formula)

Answer:

**Organize:** Examine the structural formula. Note that the molecule to be named is an alkane.

**Unknown:** An IUPAC name

**Translate:** Review the eight steps used to name alkanes.

**Solve:** Identify the longest continuous chain of carbon atoms.

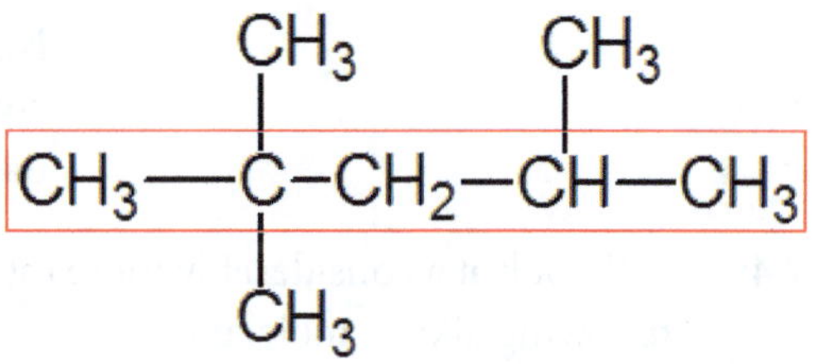

$$CH_3-C(CH_3)_2-CH_2-CH(CH_3)-CH_3$$

The longest continuous chain is a chain of five carbon atoms. Hence, the *imaginary* parent compound is pentane. Note that the name for a five-carbon alkane is *pentane* (following the convention found in table 7.1). This is the root.

The prefix will include *methyl* because there are three methyl groups bonded to the chain of five carbon atoms.

Because the molecule to be named has only single carbon-carbon bonds, the suffix is *ane.*

Now number the carbon atoms in the longest continuous chain:

$$\overset{1}{\underset{5}{CH_3}}-\overset{2}{\underset{4}{C}}(CH_3)_2-\overset{3}{\underset{3}{CH_2}}-\overset{4}{\underset{2}{CH}}(CH_3)-\overset{5}{\underset{1}{CH_3}}$$

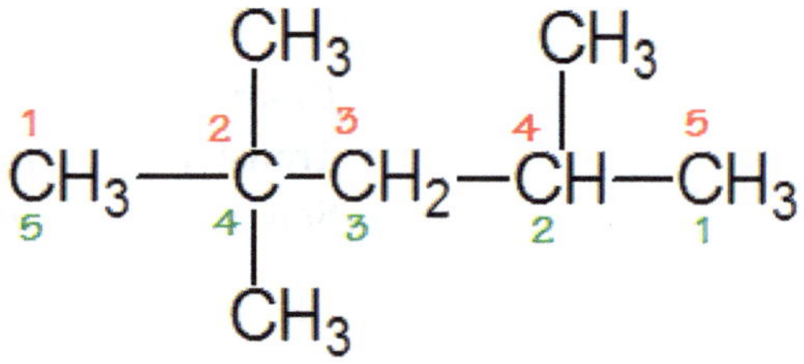

The two possible numbering choices are shown in red and green. For the red numbering (left to right), the substituents are named: 2-methyl, 2-methyl, and 4-methyl. For the green numbering (right to left), the substituents are named: 4-methyl, 4-methyl, and 2-methyl. The left-to-right numbering gives $\mathbf{S}_1 = 8$ $(2 + 2 + 4)$, while the right-to-left numbering gives $\mathbf{S}_2 = 10$ $(4 + 4 + 2)$. Hence, we choose the left-to-right number (red).

Because the *methyl* substituent occurs three times, we use the Greek prefix *tri*.

The full name is

**2,2,4-trimethylpentane**.

Note that numbers are separated from numbers by **commas**, and numbers are separated from words by **dashes**.

Example 7.4: Let's look at a considerably more interesting molecule. What is the IUPAC name of the following alkane molecule?

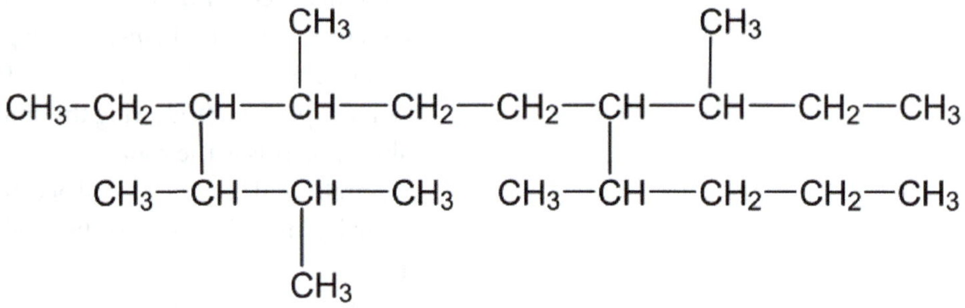

```
                 CH3                                  CH3
                  |                                    |
CH3-CH2-CH-------CH----CH2----CH2----CH-----CH-----CH2----CH3
         |                            |
    CH3-CH-------CH----CH3     CH3----CH-----CH2----CH2----CH3
                  |
                 CH3
```

Answer: **Organize:** As daunting as it may seem, the reader has **all** the tools needed to name this molecule. Begin by examining the structural formula and noting that the molecule is an alkane.

**Unknown:** An IUPAC name

**Translate:** Review the eight steps used to name alkanes.

**Solve:** Identify the longest continuous chain of carbon atoms.

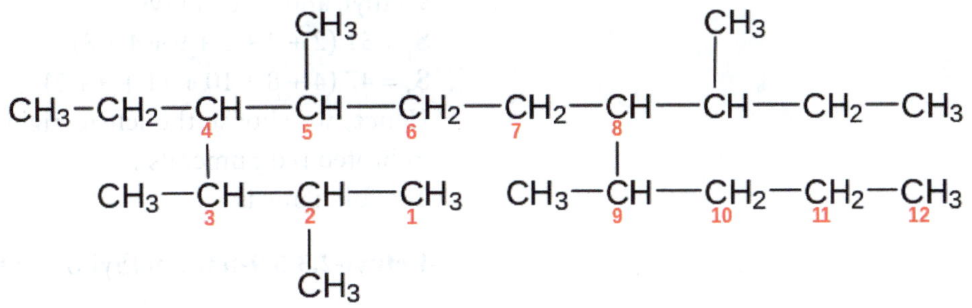

```
                 CH3                                  CH3
                  |                                    |
CH3−CH2−CH──────CH──CH2──CH2──CH─────CH──CH2──CH3
         4        5    6    7    8
         |                       |
  CH3──CH──CH──CH3      CH3──CH──CH2──CH2──CH3
       3    2   1            9   10   11   12
            |
           CH3
```

The longest continuous chain is a chain of twelve carbon atoms. Hence, the *imaginary* parent compound is dodecane. (Here there are several possible choices for a continuous chain of twelve carbon atoms. **All** will of them will yield the **same** name. The numbering is indicated at this step to help with the naming.) To determine the prefix, we look at the following diagram, in which all the substituents are identified by the boxes.

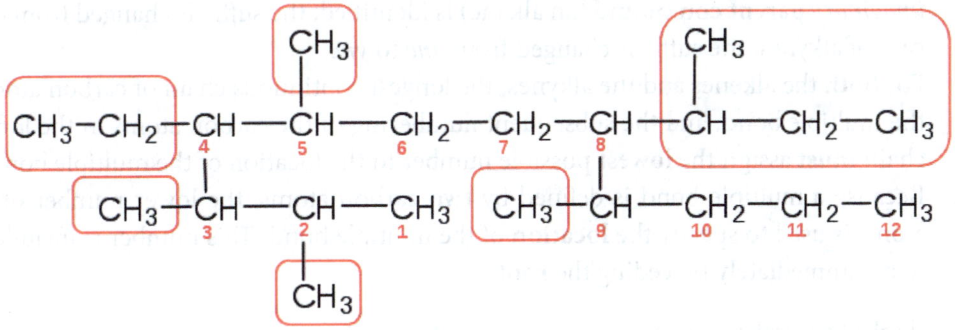

```
                 [CH3]                                [CH3
                  |                                    |
[CH3−CH2]−CH────CH──CH2──CH2──CH─────CH──CH2──CH3]
           4      5    6    7    8
           |                     |
  [CH3]──CH──CH──CH3    [CH3]──CH──CH2──CH2──CH3
         3    2   1             9   10   11   12
              |
            [CH3]
```

The boxes identify *four* methyl groups, *one* ethyl group, and *one* sec-butyl group.

To avoid confusion, only a single set of numbers for the carbon atoms is indicated. We will call it the left-to-right numbering. The **same** chain can be numbered right to left by making carbon 12 carbon number 1 and working in the opposite direction.

The left-to-right numbering gives the following substituents:

2-methyl, 3-methyl, 5-methyl, 9-methyl

4-ethyl, and 8-sec-butyl.

The right-to-left numbering gives the following substituents:

4-methyl, 8-methyl, 10-methyl, 11-methyl,

9-ethyl, and 5-sec-butyl.
$S_1 = 31\ (2 + 3 + 5 + 9 + 4 + 8)$
$S_2 = 47\ (4 + 8 + 10 + 11 + 9 + 5)$
Hence, we choose the left-to-right numbering (the indicated red numerals).
The full name is

**4-ethyl-2,3,5,9-tetramethyl-8-sec-butyldodecane.**

Note that numbers are separated from numbers by **commas**, and numbers are separated from words by **dashes**.

## Naming Alkenes and Alkynes

The procedure for naming alkenes and alkynes is nearly identical to the IUPAC procedure that we followed to name alkanes. There are two differences:

1. The suffix will now change to identify the appropriate family. In the case of the alkenes, when the *imaginary* **parent compound** (an alkane) is identified, the suffix is changed from *ane* to *ene*. In the case of alkynes, the suffix is changed from *ane* to *yne*.
2. For **both** the alkenes and the alkynes, the longest continuous chain of carbon atoms **must** include the multiple bond, and the subsequent numbering of the carbon atoms in the longest continuous chain **must** assign the **lowest** possible number to the location of the multiple bond.
3. Because a multiple bond is defined by **two** carbon atoms, the lower number of the two carbon atoms is used to specify the **location** of the multiple bond. This number is included in the IUPAC name immediately preceding the **root**.

Let's look at several examples.

Example 7.5: What is the IUPAC name of the following molecule?

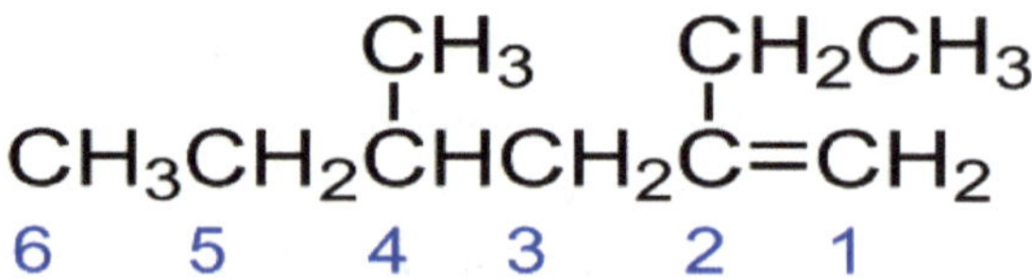


Answer:

**Organize:** Examine the structural formula and note that the molecule contains a double bond.

**Unknown:** An IUPAC name

**Translate:** Review the eight steps used to name alkanes. However, take note that this molecule is an alkene. It has one **double** bond.

**Solve:** Identify the longest continuous chain of carbon atoms that **includes** the double bond.

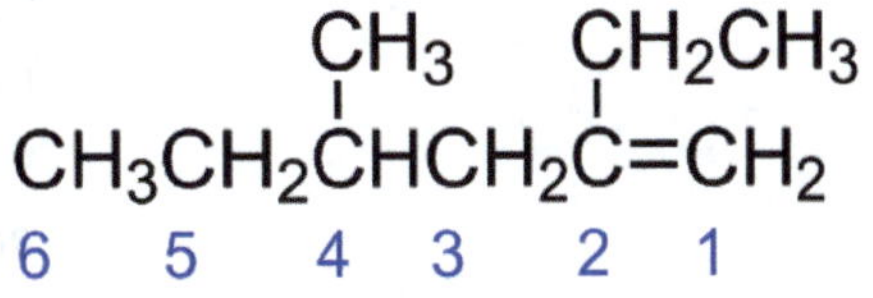

$$\begin{array}{l} \quad\quad\quad\;\; CH_3 \quad\;\; CH_2CH_3 \\ \quad\quad\quad\;\;\; | \quad\quad\quad\; | \\ CH_3CH_2CHCH_2C{=}CH_2 \\ 6 \quad 5 \quad 4 \;\; 3 \quad 2 \;\; 1 \end{array}$$

The longest continuous chain *that includes the double bond* is a chain of six carbon atoms. Notice that there is a continuous chain of seven carbon atoms, **but it does** not **include the double bond**. Hence, the *imaginary* parent compound is hexane. (The numbering is indicated at this step to help with the naming.)
To determine the prefix, we look at the following structural formula:

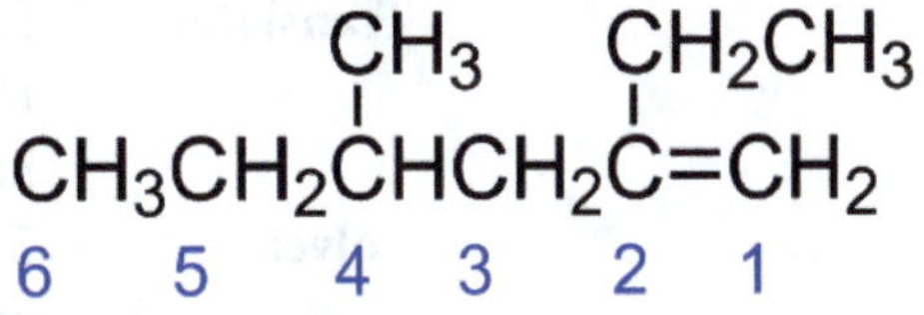

$$\begin{array}{l} \quad\quad\quad\;\; CH_3 \quad\;\; CH_2CH_3 \\ \quad\quad\quad\;\;\; | \quad\quad\quad\; | \\ CH_3CH_2CHCH_2C{=}CH_2 \\ 6 \quad 5 \quad 4 \;\; 3 \quad 2 \;\; 1 \end{array}$$

There is a methyl substituent located at carbon 4 and an ethyl substituent located at carbon 2.
Because the numbering used **must** give the **lowest** possible value to the double bond, we have ***no*** choice. The numbering **must** be done as indicated.
The right-to-left numbering in the diagram gives the following substituents:
4-methyl and 2-ethyl.
The full name is

**2-ethyl-4-methyl-1-hexene.**

Note that the 1 immediately preceding the root identifies the location of the double bond. (The double bond is defined by two carbon atoms. The lower number of the two carbon atoms is used to specify the location of the double bond.) Note that numbers are separated from numbers by **commas,** and numbers are separated from words by **dashes**.

An alternate form of the name is also acceptable:

**2-ethyl-4-methylhex-1-ene.**

The number indicating the location of the double bond is placed immediately adjacent to the *ene* suffix. This convention is very useful in naming *polyfunctional* molecules.

Example 7.6: What is the IUPAC name of the following molecule?

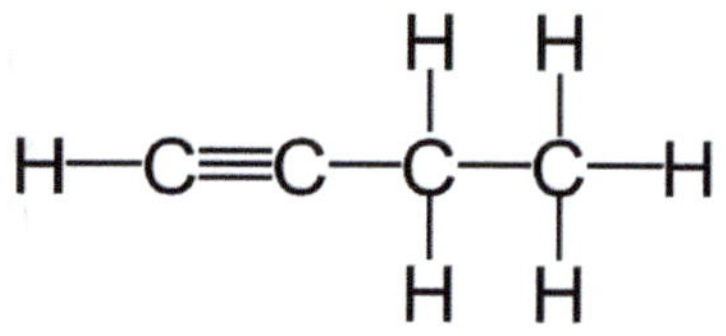

```
            H  H
            |  |
H—C≡C—C—C—H
            |  |
            H  H
```

Answer:

**Organize:** Examine the structural formula and note the triple bond.

**Unknown:** An IUPAC name

**Translate:** Review the eight steps used to name alkanes. However, take note that this molecule is an alkyne. It has one **triple** bond.

**Solve:** Identify the longest continuous chain of carbon atoms that **includes** the triple bond.

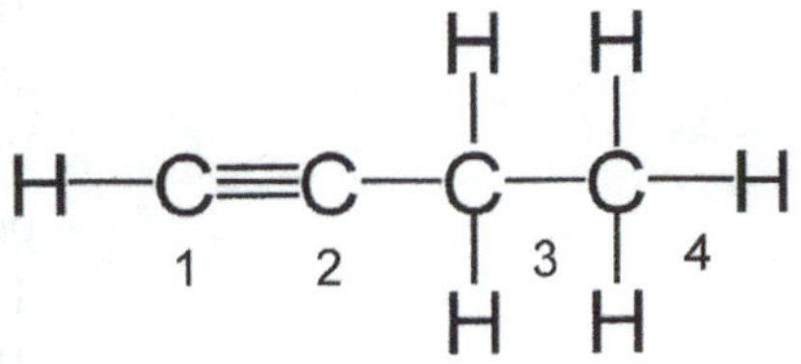

```
            H  H
            |  |
H—C≡C—C—C—H
   1   2  | 3 | 4
            H  H
```

The longest continuous chain *that includes the triple bond* is a chain of four carbon atoms.

The molecule contains **no** substituents. Consequently, there is **no** prefix.

The full name is

**1-butyne.**

Note that the 1 immediately preceding the root identifies the location of the triple bond. (The triple bond is defined by two carbon atoms. The **lower** number of the two carbon atoms is used to specify the location of the triple bond.) Note that numbers are separated from words by **dashes**.

An alternate form of the name is also acceptable:

**but-1-yne.**

The number indicating the location of the triple bond is placed immediately adjacent to the *yne* suffix. This convention is very useful in naming *polyfunctional* molecules.

In the above examples of naming an alkene and an alkyne, the procedure we followed produced the IUPAC name of each molecule. For historical reasons, just as the first four entries of table 7.1 deviate from the systemic *Greek prefixes* (table 4.3), several alkenes and alkynes have **common names** as well as IUPAC names. For example, the IUPAC name for $CH_2CH_2$ is *ethene*, while the common name is *ethylene*; the molecule $CH_3CHCH_2$ has the IUPAC name *propene*, but the common name *propylene*; and the IUPAC name of the molecule $C_2H_2$ is *ethyne*, while its common name is *acetylene*. Finally, the naming of **cyclic** alkanes, alkenes, and alkynes follows exactly the procedures we have followed above with only a single exception: the root of the IUPAC name includes the prefix *cyclo* to distinguish the cyclic compounds from the open-chain compounds.

## Naming Aromatic Hydrocarbons

Because some of the aromatic hydrocarbons include a benzene-like ring structure, they are named as benzene derivatives. That is, the benzene molecule is the *imaginary* **parent compound** that is the root of the IUPAC name for the aromatic hydrocarbon molecules. Because the benzene molecule consists of six identical carbon atoms, the crucial step in naming a ring-based aromatic hydrocarbon molecule is **numbering the carbon atoms** in the benzene ring. The numbering of the carbon atoms in the benzene ring will then allow us to specify the precise locations of any substituents. Let's look at several examples.

Example 7.7: What is the IUPAC name of the following molecule?

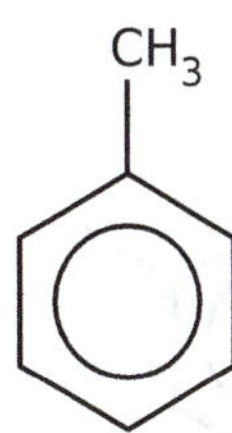


Answer:

**Organize:** Examine the structural formula.

**Unknown:** An IUPAC name

**Translate:** Key observation: the benzene ring

**Solve:** Number the carbon atoms in the benzene ring. Because the numbering can be done either **clockwise** or **counterclockwise**, always choose the order that assigns the **lowest** number to any substituents.

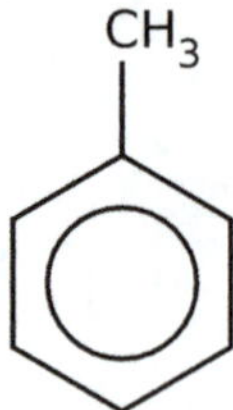


The **root** is benzene, and because some of the aromatic hydrocarbons contain a benzene-like structure, benzene is also the **suffix**. Because the only substituent is the methyl group, the **prefix** is *methyl*. The methyl group would be assigned the number 1, making the prefix *1-methyl*. However, because the methyl group is the only substituent, the conventional agreement (like the one governing molecular formulas) is to omit the 1.
The full name is

**methylbenzene.**

Note that the name is consistent with the idea that aromatic hydrocarbons are named as benzene derivatives. Like several alkenes and alkynes, this molecule also has a **common name**: **toluene**.

Suppose that a benzene ring has two methyl groups as substituents. In this case, there are three possibilities. The next example will demonstrate this.

Example 7.8: What are the IUPAC names of the following molecules?

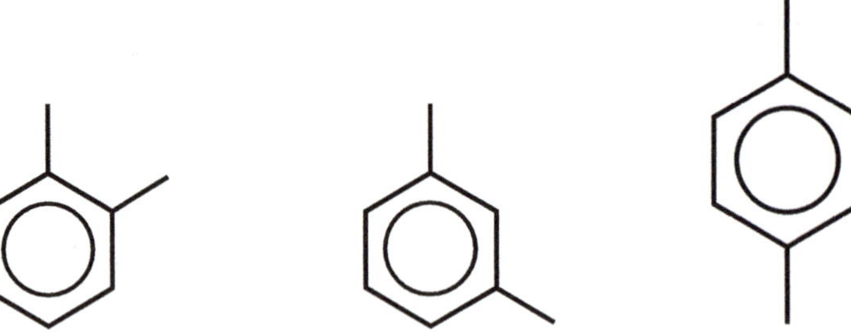

Answer:

**Organize:** Examine each skeletal formula.
**Unknown:** IUPAC names
**Translate:** Key observation: the benzene ring
**Solve:** While the three molecules appear to be very similar, they are quite distinct. In the first molecule, the substituents (methyl groups) are located on **adjacent** carbon atoms. In the second molecule, there

is ***one* carbon atom *between* the carbon atoms to which the methyl groups are bonded.** In the third molecule, there are ***two* carbon atoms *between* the carbon atoms to which the methyl groups are bonded.**

The following diagram, which includes the numbering of the carbon atoms in benzene, emphasizes the distinctions:

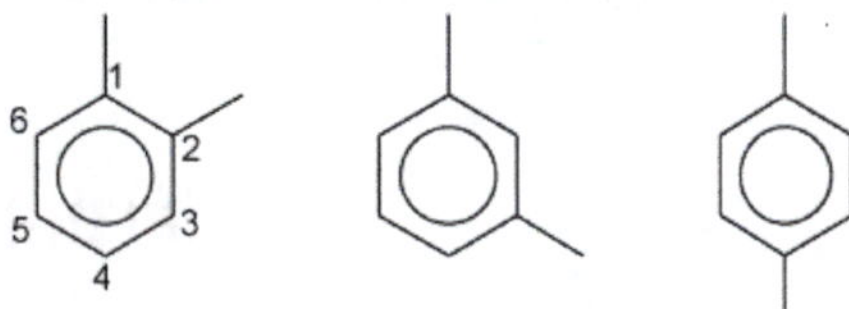


The **root** is benzene, and because some of the aromatic hydrocarbons contain a benzene-like structure, benzene is also the **suffix**. The substituents are methyl groups. Hence the **prefixes** for the three molecules are as follows:

1,2-*dimethyl* (first molecule)
1,3-*dimethyl* (second molecule)
1,4-*dimethyl* (third molecule)

Note the use of the *Greek* prefix *di* because each molecule contains two identical substituents.

The full IUPAC names are as follows:

**1,2-dimethylbenzene**
**1,3-dimethylbenzene**
**1,4-dimethylbenzene**

Note that the name is consistent with the idea that some aromatic hydrocarbons are named as benzene derivatives. As in the case of the aliphatic hydrocarbons, the numbers are separated from numbers by **commas**, and numbers are separated from words by **dashes**.

Because the spatial location of a substituent *directly affects* a molecule's chemical properties, the three locations demonstrated in example 7.8 have been given special designations that can be used to name the molecules. The 1,2 arrangement (adjacent carbon atoms) is called the **ortho** relationship; the 1,3 arrangement (*one* carbon atom *between* the carbon atoms to which the substituents are bonded) is called the **meta** relationship; the 1,4 arrangement (*two* carbon atoms *between* the carbon atoms to which the substituents are bonded) is called the **para** relationship. Figure 7.18 summarizes the use of these special designations to give *common names* to each of the molecules from example 7.8.

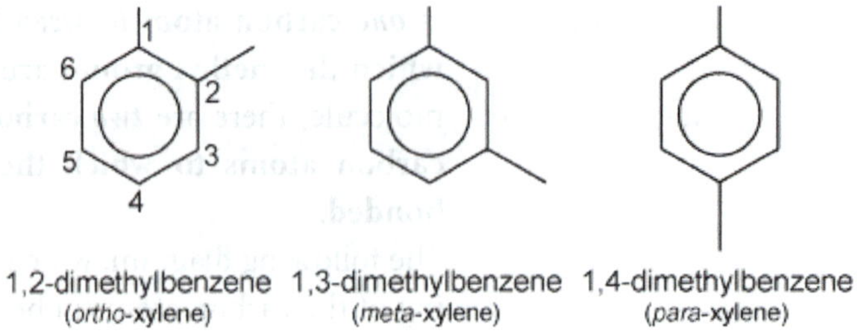

1,2-dimethylbenzene (*ortho*-xylene) 1,3-dimethylbenzene (*meta*-xylene) 1,4-dimethylbenzene (*para*-xylene)

Figure 7.18: Ortho, Meta, and Para Relationships

## Carbohydrates

While the organic molecules which we have already encountered exhibit an amazing diversity and richness (and there are more surprises coming in the later chapters of this book!), the molecules that play a central role in life as we experience it on this planet are often structurally more complex (possessing multiple functional groups), are participants in macromolecular structures involving thousands or tens of thousands of atoms, and finally, possess those characteristics that have allowed the emergence of self-replication. The first class of these molecules that are appropriately designated as *biomolecules* is called **carbohydrate.** Interestingly, the name suggests that the carbohydrates are nothing more than *hydrated carbon atoms.* By examining the molecular formula of *glucose,* one of the simplest carbohydrates, it is easy to understand the origin of the name **carbohydrate.** The modern molecular formula for glucose is $C_6H_{12}O_6$, which was written (before the twentieth century) as $C_6(H_2O)_6$. While the number of each individual element in the two formulas is the same, the pre-twentieth century formula suggests that the six carbon atoms are, in some fashion, associated with six water molecules, making it appear as if the six carbon atoms are being *hydrated* by the water molecules—hence the name **carbohydrate**!

In fact, the members of the carbohydrate class of molecules are now understood to be more complex and much more interesting molecules. **Carbohydrates are polyhydroxy aldehydes, or polyhydroxy ketones, or substances that yield these compounds when hydrolyzed.** Let's examine this definition, piece by piece. The term **polyhydroxy** indicates that each molecule in this class possesses *multiple OH* functional groups. While the *OH* functional group is characteristic of all alcohols (see our earlier discussion above), the carbohydrates are **not** alcohols. The terms **aldehyde** and **ketone** in the definition indicate that the carbohydrates are *polyfunctional* molecules. That is, they are **aldehydes** that also have *multiple OH* groups, or they are **ketones** that also have *multiple OH* groups. In either case, the **aldehyde** or **ketone** functional group takes precedence in determining the chemical properties of the carbohydrates. Finally, the term **hydrolyzed** refers to a critical *chemical reaction* called **hydrolysis** that is central to life processes, so central in fact, that we will examine it in great detail in a later chapter. Suffice to note now that the tem **hydrolysis** is built from two Greek words, *hydro* (water) and *lysis* (to separate) and literally means *to cut or separate with water.* The fact that a carbohydrate can participate in a **hydrolysis reaction** suggests that carbohydrates participate in the formation of macromolecules that can be broken down into much smaller pieces. Indeed, we shall see that this process is one of the crucial steps in the maintenance of life.

The average carbohydrate is approximately 40% carbon and, when oxidized, produces the energy necessary to sustain life processes. Carbohydrate metabolism (which we will study in some detail later in this text) depends on the presence of both the hydroxyl and the carbonyl functional groups. Several terms are routinely used when describing carbohydrate molecules. **Monosaccharides** are carbohydrates that cannot be hydrolyzed into simpler carbohydrate units. The suffix *saccharide* is derived from the Greek *sakkharon*, meaning "sugar" (the Latin word for *sugar* is *saccharum*), and is an indication that the simplest members of the carbohydrate class of molecules are the sugars that are commonly part of our daily experience. The formal definition requires a monosaccharide molecule to contain a minimum of three carbon atoms. **Disaccharides** are carbohydrates that yield two monosaccharides upon hydrolysis. **Oligosaccharides** are carbohydrates composed of three to nine monosaccharide units, while **polysaccharides** are carbohydrates that yield **many** monosaccharide units when hydrolyzed. Polysaccharides range in size from a few tens of monosaccharide units to tens of thousands (or more!) of monosaccharide units, exhibiting the macromolecular structure that is so characteristic of biochemical compounds. These various terms were created as the science of biochemistry (in particular, carbohydrate chemistry) matured over the past century, and the choice of the number *ten* to separate oligosaccharides from polysaccharides is quite arbitrary. Consequently, the most significant terms are *monosaccharide*, *disaccharide*, and *polysaccharide*.

## Carbohydrate Functional Groups

Because the molecules that constitute the carbohydrate class possess the *hydroxyl group*, *OH* (characteristic of the **alcohol functional group**), and either an **aldehyde functional group** or a **ketone functional group**, we take this opportunity to examine these three functional groups in a little more detail. While we met the alcohols, aldehydes, and ketones in our earlier discussion of the three major families of functional groups, we did not attempt *to name* them. As we shall see, the experience we gained by naming the hydrocarbons provides all the tools we shall need to name the alcohols, the aldehydes, and the ketones. The process of assigning names to molecules belonging to these three functional groups will expand our knowledge of each functional group, making them seem less foreign. But our key focus is not on each functional group in isolation but rather the roles played by the alcohol, aldehyde, and ketone functional groups in determining the characteristics of the carbohydrates.

## Naming Alcohols

The procedure for naming alcohol molecules follows closely the procedure we used to name the hydrocarbons:

1. **Name the root:**
   Identify the longest continuous chain of carbon atoms ***that includes the carbon atom to which the hydroxyl group, OH, is bonded*** as the *imaginary* **parent compound**. The *imaginary* **parent compound** is always an alkane whose name is assigned using the IUPAC prefixes from table 7.1. Consider any atom or group of atoms (except the *OH*!) that is not part of the *imaginary* parent compound (that is, **not included** in the longest continuous chain of carbon atoms) to be a

**substituent** that has replaced a hydrogen atom in the *imaginary* parent compound. The *OH* is **not** a substituent; it is the **alcohol functional group**. The **root** of the molecule to be named will be derived from the name of the *imaginary* parent compound.

2. **Identify the prefix:**
   The root will be then preceded by a **prefix**, which includes the names of **all** the substituents bonded to the *imaginary* parent compound. Identify all the substituents that will be included in the **prefix**, giving each substituent its proper name.
3. **Identify the suffix:**
   The **suffix** is simply the indicator of the *family* or *functional group* to which the molecule belongs. In the case of an alcohol, replace the finale *e* in the alkane name of the *imaginary* **parent compound** with the suffix *ol*.
4. **Number the carbon atoms in the *longest continuous chain of carbon atoms that includes the carbon atom to which the hydroxyl group, OH, is bonded*:**
   As with the hydrocarbons, number the carbon atoms in the *longest continuous chain of carbon atoms that includes the carbon atom to which the hydroxyl group, OH, is bonded* in the *imaginary* parent compound. The numbering of the carbon atoms **must** assign the **lowest** possible number to the carbon atom to which the hydroxyl group is bonded.
5. **Associate each substituent with the number of the carbon atom to which it is bonded:**
   For example, if a *methyl group* is bonded to the *imaginary* parent compound at carbon number 2 (where the number was assigned in step 4), assign the name *2-methyl* to that substituent.
6. **If the same substituent occurs multiple times, use an appropriate *Greek* prefix:**
   It is often the case that the *same* substituent occurs multiple times along the parent compound. In order to identify multiple occurrences of the same substituent, a standard *Greek prefix* (table 4.3) is added to the name of the substituent. If the substituent occurs twice, the prefix is *di*; three times the prefix is *tri*; four times the prefix is *tetra*; and so on.
7. **Identify the location of the hydroxyl group:**
   It is often possible for the hydroxyl group to be bonded to *several* different carbon atoms in the *imaginary* parent compound. **If** this is the case, the **number** *of the carbon atom to which the hydroxyl group is bonded* **immediately precedes the root** in the name of the alcohol, thereby specifying the carbon atom to which the hydroxyl group is bonded.
8. **Write the molecules name as a single word:**
   This the last step! Simply combine the prefix, root, and suffix into a single continuous name. If the prefix contains multiple distinct substituents, organize them in alphabetical order. However, in determining the alphabetical order, **ignore** any *Greek prefixes*.

The procedure outlined above defines the naming of *aliphatic* alcohols (both open-chain and cyclic alcohols). Remember that, in the case of cyclic molecules, the prefix *cyclo* **must** precede the root. In the case of an *aromatic alcohol* that involves a ring-based aromatic, the hydroxyl group is bonded to a benzene ring, and the alcohol is named a *phenol*. If there are additional substituents bonded to the benzene ring, the carbon atom to which the hydroxyl group is bonded is assigned the number 1, and all the substituents are identified by the appropriate number. Let's look at some examples of naming alcohols.

Example 7.9: What is the IUPAC name of the following molecule?

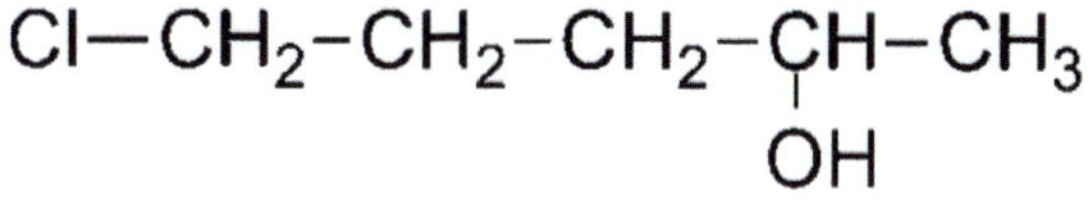

$$Cl-CH_2-CH_2-CH_2-\underset{\displaystyle OH}{\underset{|}{C}}H-CH_3$$

Answer:

**Organize:** Examine the structural formula. Take note of the alcohol functional group.

**Unknown:** An IUPAC name

**Translate:** Review the steps used to name an alcohol.

**Solve:** Identify the longest continuous chain of carbon atoms that includes the carbon atom to which the hydroxyl group, *OH*, is bonded.

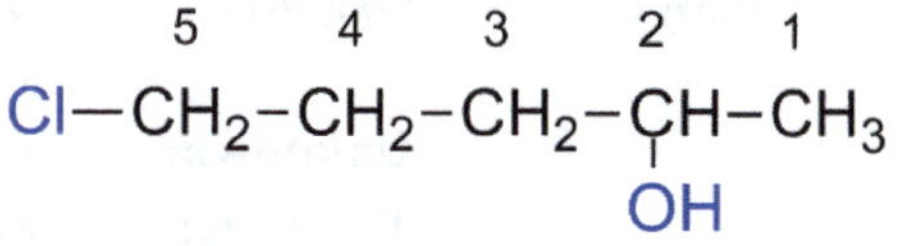

$$\overset{5}{Cl-CH_2}-\overset{4}{CH_2}-\overset{3}{CH_2}-\overset{2}{\underset{\displaystyle OH}{\underset{|}{C}}H}-\overset{1}{CH_3}$$

The longest continuous chain of carbon atoms that includes the carbon atom to which the hydroxyl group, *OH*, is bonded is a chain of five carbon atoms. Hence, the *imaginary* parent compound is pentane. (In this example, there is only **one** choice for the chain of carbon atoms identifying the *imaginary* parent compound.) The root is pentane.

The prefix will include only *chloro*.

The suffix is determined by the presence of the alcohol functional group. The final *e* of the *imaginary* parent compound's name is replaced by *ol*.

The prefix is fully specified by using the above numbering:

*5-chloro*.

The full name is

**5-chloro-2-pentanol**.

Note that the 2 immediately preceding the root identifies the location of the hydroxyl group. It is the number of the carbon atom to which the hydroxyl group is bonded. Note that numbers are separated from numbers by **commas**, and numbers are separated from words by **dashes**.

An alternate form of the name is also acceptable:

**5-chloropentan-2-ol.**

The number indicating the carbon atom to which the hydroxyl group is bonded is placed immediately adjacent to the *ol* suffix.

Example 7.10: What is the IUPAC name of the following molecule?

$$\begin{array}{cccccccccc} CH_3 & - & CH_2 & - & CH & - & CH_2 & - & CH & - CH_3 \\ & & & & | & & & & | & \\ & & & & OH & & & & CH_2 & - CH_3 \end{array}$$

Answer:

**Organize:** Examine the structural formula. Take note of the alcohol functional group.

**Unknown:** An IUPAC name

**Translate:** Review the steps used to name an alcohol.

**Solve:** Identify the longest continuous chain of carbon atoms that includes the carbon atom to which the hydroxyl group, *OH*, is bonded.

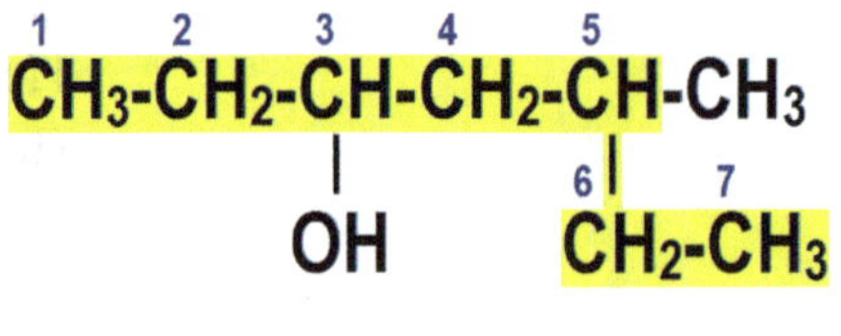

$$\begin{array}{cccccccccc} ^{1}CH_3 & - & ^{2}CH_2 & - & ^{3}CH & - & ^{4}CH_2 & - & ^{5}CH & - CH_3 \\ & & & & | & & & & |\,6 & \quad 7 \\ & & & & OH & & & & CH_2 & - CH_3 \end{array}$$

The longest continuous chain of carbon atoms that includes the carbon atom to which the hydroxyl group, *OH*, is bonded is a chain of seven carbon atoms. Hence, the *imaginary* parent compound is heptane. (In this example, there is only **one** choice for the chain of carbon atoms identifying the *imaginary* parent compound.) The root is heptane.

The prefix will include only *methyl.*

The suffix is determined by the presence of the alcohol functional group. The final *e* of the *imaginary* parent compound's name is replaced by *ol.*

The prefix is fully specified by using the above numbering:

*5-methyl.*

The full name is

**5-methyl-3-heptanol.**

Note that the 3 immediately preceding the root identifies the location of the hydroxyl group. It is the number of the carbon atom to which the hydroxyl group is bonded. Note that numbers are separated from numbers by **commas**, and numbers are separated from words by **dashes**.

An alternate form of the name is also acceptable:

**5-methylheptan-3-ol**.

The number indicating the carbon atom to which the hydroxyl group is bonded is placed immediately adjacent to the *ol* suffix.

## Naming Aldehydes and Ketones

The aldehydes and ketones are both members of the third family of functional groups, whose distinguishing feature is the carbon-oxygen double bond. Because this particular structural configuration is important in a wide variety of chemical contexts, we gave it a special name: the **carbonyl group** (see figure 7.12). By carefully examining figure 7.11, the reader will notice that an aldehyde **always** has one hydrogen atom bonded the carbon atom of the carbonyl group (this carbon atom is often called the **carbonyl carbon**). Because the hydrogen atom is capable of forming **only one** bond, the carbonyl group in an aldehyde **must** occur *at one end of the molecule.* Further note, again, as a result of the fact that the hydrogen atom can form **only one** bond, there can be **no** cyclic aldehydes; the carbonyl carbon of an aldehyde cannot be part of a ring structure. In contrast, the carbonyl group of a **ketone** can **never** occur *at one end of an open-chain molecule* because the carbonyl carbon in acetone is *always bonded to* ***two*** *other carbon atoms.* There **are** cyclic ketones! The naming procedure for either an aldehyde or a ketone again uses the tools we have already developed:

1. **Name the root:**
   Identify the longest continuous chain of carbon atoms ***that includes the carbonyl carbon*** as the *imaginary* **parent compound**. The *imaginary* **parent compound** is always an alkane whose name is assigned using the IUPAC prefixes from table 7.1. Consider any atom or group of atoms that is not part of the *imaginary* parent compound (that is, **not included** in the longest continuous chain of carbon atoms) to be a **substituent** that has replaced a hydrogen atom in the *imaginary* parent compound. The carbonyl group is **not** a substituent; it is a **characteristic functional group**. The **root** of the molecule to be named will be derived from the name of the *imaginary* parent compound.
2. **Identify the prefix:**
   The root will be then preceded by a **prefix**, which includes the names of **all** the substituents bonded to the *imaginary* parent compound. Identify all the substituents that will be included in the **prefix**, giving each substituent its proper name.
3. **Identify the suffix:**
   The **suffix** is simply the indicator of the *family* or *functional group* to which the molecule belongs. In the case of an aldehyde, replace the finale *e* in the alkane name of the *imaginary* **parent compound**

with the suffix *al.* For a ketone, replace the final *e* in the alkane name of the *imaginary* **parent compound** with the suffix *one.*

4. **Number the carbon atoms in the *longest continuous chain of carbon atoms that includes the carbonyl carbon*:**
   As with the hydrocarbons, number the carbon atoms in the *longest continuous chain of carbon atoms that includes the carbonyl carbon* in the *imaginary* parent compound. The numbering of the carbon atoms **must** assign the **lowest** possible number to the carbonyl carbon atom.
5. **Associate each substituent with the number of the carbon atom to which it is bonded:**
   For example, if a *methyl group* is bonded to the *imaginary* parent compound at carbon number 2 (where the number was assigned in step 4), assign the name *2-methyl* to that substituent.
6. **If the same substituent occurs multiple times, use an appropriate *Greek* prefix:**
   It often occurs that the *same* substituent occurs multiple times along the *imaginary* parent compound. In order to identify multiple occurrences of the same substituent, a standard *Greek prefix* (table 4.3) is added to the name of the substituent. If the substituent occurs twice, the prefix is *di*; three times the prefix is *tri*; four times the prefix is *tetra*; and so on.
7. **Identify the location of the carbonyl group:**
   While the carbonyl group in an aldehyde **always** occurs *at one end of an open-chain molecule*, in the case of the ketones, it is possible for the carbonyl group to be located *several* different positions in the *imaginary* parent compound. **If** this is the case, the **number** *of the carbonyl carbon* **immediately precedes the root** in the name of the ketone, thereby specifying the location of the carbonyl group.
8. **Write the molecule's name as a single word:**
   This the last step! Simply combine the prefix, root, and suffix into a single continuous name. If the prefix contains multiple distinct substituents, organize them in alphabetical order. However, in determining the alphabetical order, **ignore** any *Greek prefixes.*

Let's examine two examples:

Example 7.11: What is the IUPAC name of the following molecule?

$$
\begin{array}{cccccccccccccc}
 & & & & & & & & & & & O & & \\
 & & & & & & & & & & & \| & & \\
CH_3 & - & CH_2 & - & CH_2 & - & CH & - & CH_2 & - & CH & - & C & - & H \\
 & & & & & & | & & & & | & & & & \\
 & & & & & & CH_3 & & & & CH_2-CH_3 & & & &
\end{array}
$$

| Answer: | **Organize:** | Examine the structural formula. Take note of the aldehyde functional group. |
|---|---|---|
| | **Unknown:** | An IUPAC name |
| | **Translate:** | Review the steps used to name an alcohol. |
| | **Solve:** | Identify the longest continuous chain of carbon atoms that includes the carbonyl carbon atom. |

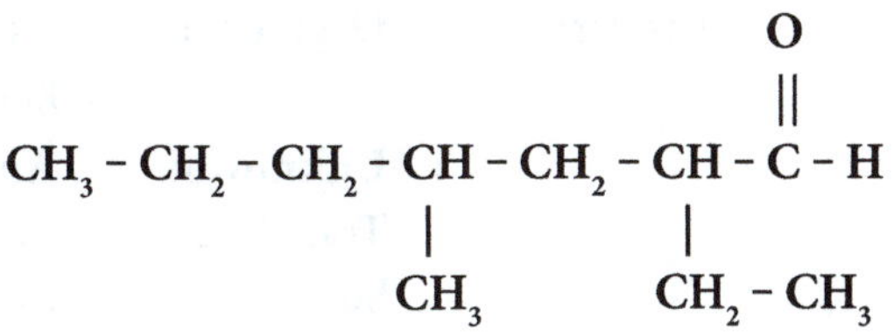

```
                                     O
                                     ||
CH3 - CH2 - CH2 - CH - CH2 - CH - C - H
                  |          |
                  CH3        CH2 - CH3
```

The longest continuous chain of carbon atoms that includes the carbonyl carbon atom is a chain of seven carbon atoms. Hence, the *imaginary* parent compound is heptane. (There is only **one** choice for the chain of carbon atoms identifying the *imaginary* parent compound. The carbonyl carbon atom **must** be assigned the number 1.) The root is heptane.

The prefix will include *methyl* and *ethyl.*

The suffix is determined by the presence of the aldehyde functional group. The final *e* of the *imaginary* parent compound's name is replaced by *al.*

The prefix is fully specified by using the above numbering:

*2-ethyl*
*4-methyl*

The full name is

**2-ethyl-4-methylheptanal.**

Note that because the aldehyde functional group **always** occurs *at one end of the molecule*, there is no need to include the number of the carbonyl carbon. In an aldehyde, the carbonyl carbon is **always** assumed to be the first carbon atom of the *imaginary* parent compound. Note that numbers are separated from numbers by **commas**, and numbers are separated from words by **dashes**.

Example 7.12: What is the IUPAC name of the following molecule?

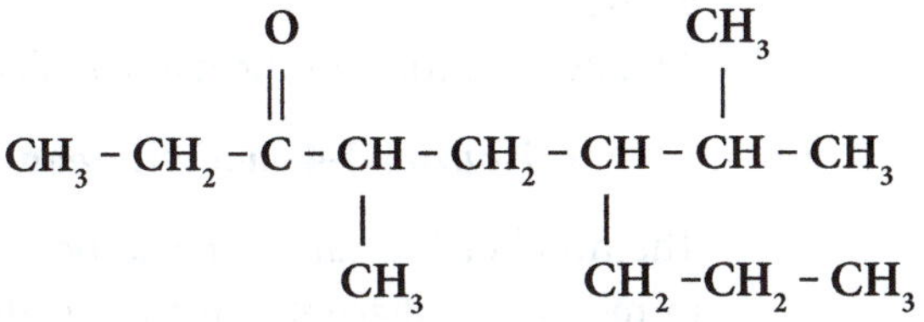

```
              O                 CH3
              ||                |
CH3 - CH2 - C - CH - CH2 - CH - CH - CH3
                |          |
                CH3        CH2 -CH2 - CH3
```

Answer: **Organize:** Examine the structural formula. Take note of the ketone functional group.

**Unknown:** An IUPAC name

**Translate:** Review the steps used to name an alcohol.

**Solve:** Identify the longest continuous chain of carbon atoms that includes the carbonyl carbon atom.

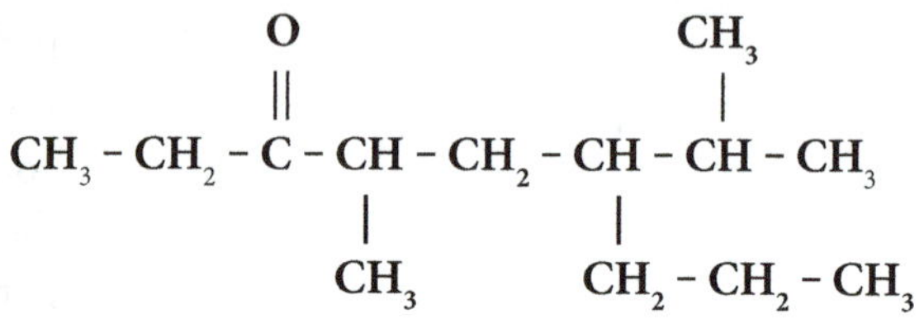

$$
\begin{array}{ccccccccccccccc}
 & & & & \mathrm{O} & & & & & & & & \mathrm{CH_3} & & \\
 & & & & \| & & & & & & & & | & & \\
\mathrm{CH_3} & - & \mathrm{CH_2} & - & \mathrm{C} & - & \mathrm{CH} & - & \mathrm{CH_2} & - & \mathrm{CH} & - & \mathrm{CH} & - & \mathrm{CH_3} \\
 & & & & & & | & & & & | & & & & \\
 & & & & & & \mathrm{CH_3} & & & & \mathrm{CH_2-CH_2-CH_3} & & & &
\end{array}
$$

The longest continuous chain of carbon atoms that includes the carbonyl carbon atom is a chain of nine carbon atoms. Hence, the *imaginary* parent compound is nonane. In order to give the carbonyl carbon atom the **lowest** possible value, number the longest continuous carbon chain that includes the carbonyl carbon from left to right. The root is nonane.

The prefix will include *methyl* and *ethyl.*

The suffix is determined by the presence of the ketone functional group. The final *e* of the *imaginary* parent compound's name is replaced by *one.*

The prefix is fully specified by using the above numbering:

*6-isopropyl*

*4-methyl*

The full name is

**6-isopropyl-4-methyl-3-nonanone.**

Note that numbers are separated from numbers by **commas**, and numbers are separated from words by **dashes**.

An alternate form of the name is also acceptable:

**6-isopropyl-4-methylnonan-3-one.**

The number indicating the carbonyl atom is placed immediately adjacent to the *one* suffix.

## Monosaccharides

As noted earlier, the monosaccharides are the simplest molecules that we commonly identify by the term *sugar*, and several systems of terminology have developed that are routinely applied to the monosaccharides. The suffix *ose* (derived from the last two letters of the Greek word *glukus*, which means "sweet") is routinely combine with the Greek prefixes (table 4.3) to identify monosaccharides containing varying numbers of carbon atoms: for example, *triose* for a sugar containing three carbon atoms, *tetrose* for a sugar containing four carbon atoms, and *pentose* for a sugar containing five carbon atoms. (The molecule glycoaldehyde, containing two carbon atoms, can be called a *diose*, although it is **not** considered to be a sugar. It is the simplest molecule containing both an aldehyde functional group and a hydroxyl functional group. In this sense, it is the simplest possible carbohydrate molecule.)

The suffix *ose* is often preceded by either the prefix *aldo* (from the word *aldehyde*) or by the prefix *keto* (from the word *ketone*), giving rise to the terms *aldose* for a monosaccharide containing an aldehyde functional group and *ketose* for a monosaccharide containing a ketone functional group. (Note that in both aldose and ketose, the letter *o* is not repeated.) In fact, this terminology is given even greater specificity by indicating the **number** of carbon atoms in the monosaccharide: an *aldohexose* is a six-carbon monosaccharide containing the aldehyde functional group; a *ketohexose* is a six-carbon monosaccharide containing the ketone functional group. Finally, we will shortly encounter multiple occurrences of the prefix *glyco*, which is derived from the first syllable of the Greek word *glukus*. The prefix immediately indicates that a saccharide or multiple saccharides are playing a central role in either a biochemical structure or a biologically significant reaction. It is interesting to note that the simple understanding of the origin of the prefix *glyco* is a powerful indicator of the specific biochemistry one expects to observe.

In our later study of metabolism (the biochemical processes that sustain life as we know it), we shall encounter several highly significant five-carbon and six-carbon monosaccharides: glucose, fructose, galactose, ribose, and 2-deoxyribose, to identify only a small number of the key actors. The most abundant monosaccharide found in nature is glucose, and its decomposition in a process called **glycolysis** (note the prefix *glyco*) produces the energy that sustains all life forms. Further, as we shall see shortly, glucose is a component of the disaccharide **sucrose** and the polysaccharides, **starch**, **glycogen**, and **cellulose**. Fructose (also known as **fruit sugar**), along with glucose and galactose, is one of the three dietary monosaccharides that is absorbed directly into the human bloodstream during digestion. Galactose occurs (along with glucose) in the disaccharide lactose and can be chemically converted into glucose by the human body. While glucose and galactose are *aldohexoses*, fructose is a *ketohexose*. Simple diagrams of these three sugars are displayed in figure 7.19. Note that the numbering of the carbon atoms in figure 7.19 assigns the number 1 to the carbonyl carbon of the aldehyde functional groups in glucose and galactose, and number 2 to the carbonyl carbon of the ketone functional group in fructose.

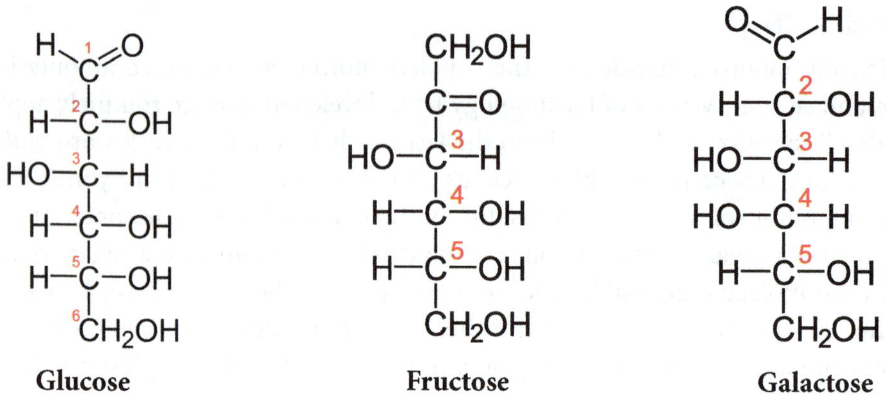

Glucose Fructose Galactose

Figure 7.19: Key Dietary Monosaccharides

In contrast to the key roles played by the dietary monosaccharides as energy sources, the two *aldopentoses*, ribose and 2-deoxyribose, play critical *structural* roles in the **nucleic acids, RNA** and **DNA** (much more about nucleic acids shortly!). The simple structural diagrams of ribose and 2-deoxyribose are depicted in figure 7.20. (The notation "CHO" is another symbol for the aldehyde functional group. Remember that the carbonyl carbon is assigned 1.)

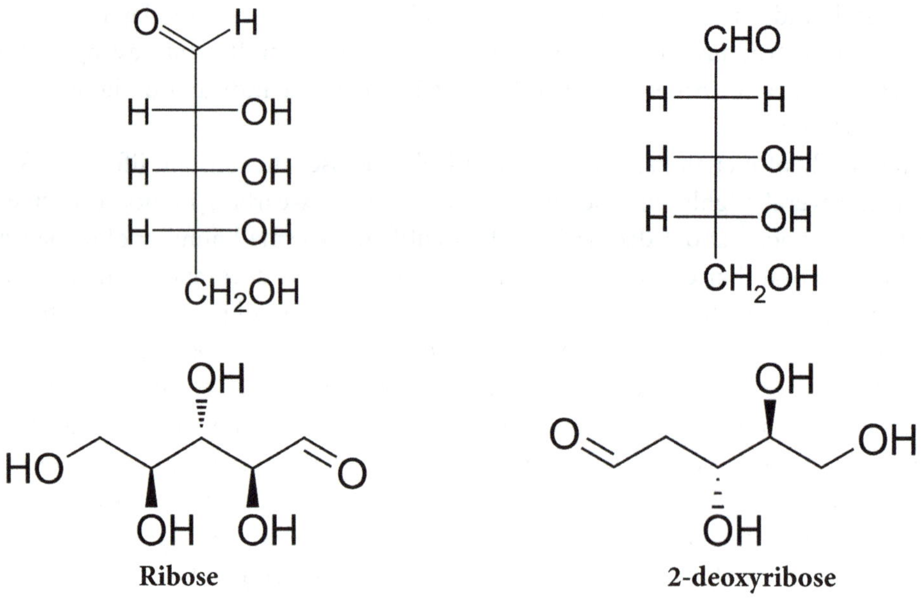

Ribose 2-deoxyribose

Figure 7.20: Aldopentoses

## Two New Biochemically Important Functional Groups: Acetals and Hemiacetals

We have seen numerous examples of the fundamental observation that was made very early in the development of organic chemistry: *the characteristic chemistry of any organic molecule,* ***regardless*** *of its size or complexity, is fundamentally determined by the functional groups found in the molecule.* This very same principle applies to biomolecules, which can be viewed as complex organic molecules that possess the specific characteristics necessary for the maintenance of life. For the carbohydrate family, there are two new *polyfunctional* groups that play significant roles in both the *structure* and *reactivity* of carbohydrate molecules. The first of these two groups is the **acetal**. An acetal is a functional group in which **two ether-like** structures share a common carbon atom. Recall from figure 7.8 that an *ether* is a simple organic functional group in which an oxygen atom is linked to two (possibly different) sets of carbon atoms by single bonds. In the acetal a *common* carbon atom is *shared* by **two** ether functional groups. As in all ethers, all the bonds are single bonds. Figure 7.21 displays an acetal. Note that the *Rs* (*R*, *R'*, and *R"*) in figure 7.21 *may* or *may not* be identical. They simply represent collections of carbon atoms.

OR"
|
R—C—H
|
OR'

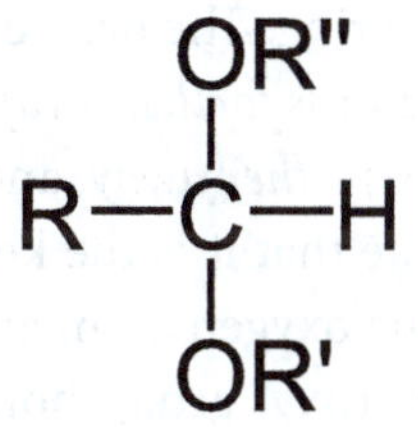

Figure 7.21: Acetal

The second new functional group is the **hemiacetal**. A hemiacetal is a functional group in which **one hydroxyl** group is bonded to a carbon atom that is part of an **ether-like** structure. Recall again that an *ether* is a simple organic functional group in which an oxygen atom is linked to two (possibly different) sets of carbon atoms by single bonds, while the *hydroxyl* group is the characteristic functional group of an alcohol (figure 7.7). In a hemiacetal **one hydroxyl** group is bonded to a carbon atom that is also part of an **ether-like** structure, and all the bonds are single bonds. Figure 7.22 displays a hemiacetal. Note that the *Rs* (*R* and *R'*) in figure 7.22 *may* or *may not* be identical. They simply represent collections of carbon atoms.

OH
|
R—C—H
|
OR'

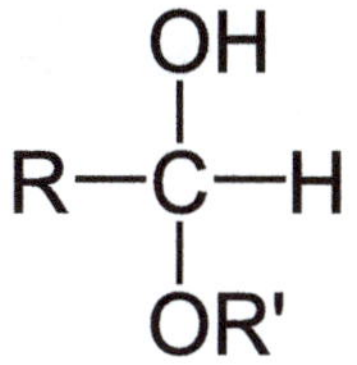

Figure 7.22: Hemiacetal

The formulas displayed in figures 7.19 and 7.20 suggest that the five representative monosaccharides introduced above are *open-chain molecules*. In fact, in a physiological setting, the monosaccharides form cyclic structures that are **not planar** (like benzene) and that exhibit one of the new functional groups just introduced: the hemiacetal. The conventional representation of the cyclic form of the monosaccharides is called a **Haworth projection**. Figure 7.23 shows a Haworth Projection for glucose.

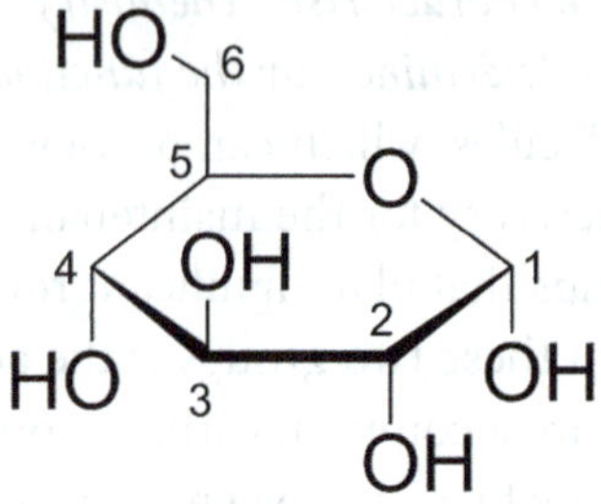

Figure 7.23: A Haworth Projection for Glucose

Like a skeletal diagram, each point in the figure (numbered 1 through 6) represents a carbon atom. The six atoms of the ring (five carbon atoms and one oxygen atom) divide space into a region lying *above* the ring and a region lying *below* the ring. The heavier lines indicate atoms *closer in space* to the observer, while the lighter lines represent atoms *farther away in space* from the observer. Consequently, atoms 2 and 3 are *closest*, atom 4 and 1 are *farther away*, and atoms 5 and 6 are *farthest away*. Carbon atom 1 is called the **anomeric** carbon; note that it is the key carbon atom of a hemiacetal functional group. The anomeric carbon along with the oxygen atom and carbon atom 5 constitute the **ether-like** structure of the hemiacetal, while the hydroxyl group bonded to the anomeric carbon is the other constituent of the hemiacetal.

Then formation of cyclic structures by the monosaccharides actually produces **two** interconvertible forms of the ring molecule that are in equilibrium with one another. These two interconvertible forms are called **anomers** (this is the origin of the name for carbon atom 1 in figure 7.23) and are designated as the **alpha anomer (α)** and the **beta anomer (β)**. Figure 7.24 depicts the equilibrium between the two anomers.

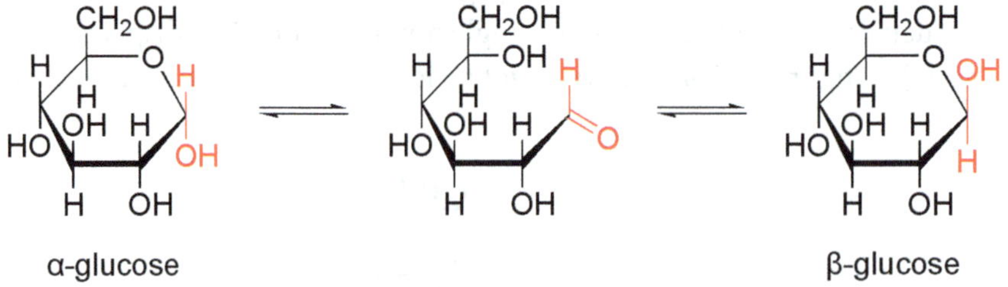

Figure 7.24: Alpha and Beta Anomers of Glucose

The reader should pay close attention to the hydroxyl group and hydrogen atom (colored red in figure 7.24) bonded to the anomeric carbon. In the leftmost panel of figure 7.24 (the α anomer), the hydroxyl group is depicted as lying in the space *below* the ring, while in the rightmost panel of figure 7.24 (the β anomer), the hydroxyl group is depicted as lying in the space *above* the ring.

Under physiological conditions, the six carbon monosaccharides, glucose and fructose, can form either six-membered rings or five-membered rings, both of which include the hemiacetal functional group. The six-member rings are given the generic name **pyranose**, where the reader will recognize the *ose* suffix that is indicative of a sugar. In contrast, the five-member rings are given the name **furanose**. Figure 7.25 shows the six-member and five-member rings formed by glucose, while figure 7.26 displays the corresponding structures for fructose. The physiologically significant form of the ribose sugars, which plays a critical role in the **nucleic acids**, is the five-member ring (a *furanose*) displayed in figure 7.27.

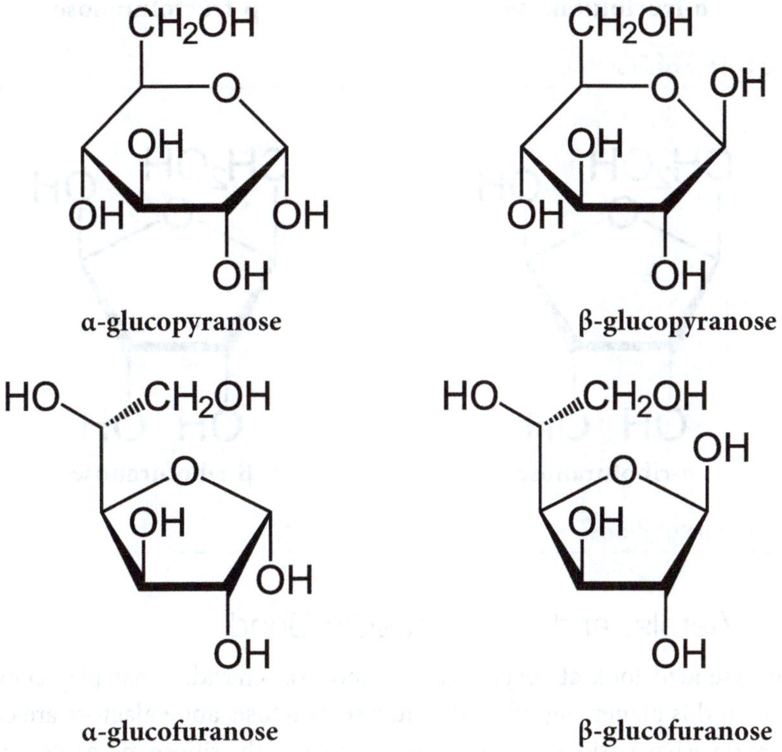

**α-glucopyranose** **β-glucopyranose**

**α-glucofuranose** **β-glucofuranose**

Figure 7.25: Cyclic Forms of Glucose

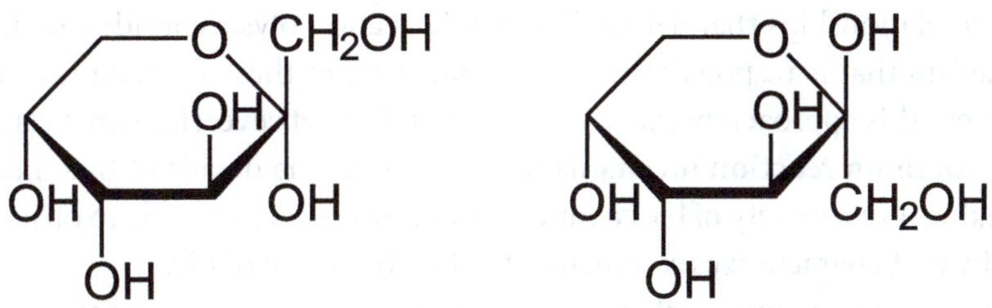

**α-fructopyranose** **β-fructopyranose**

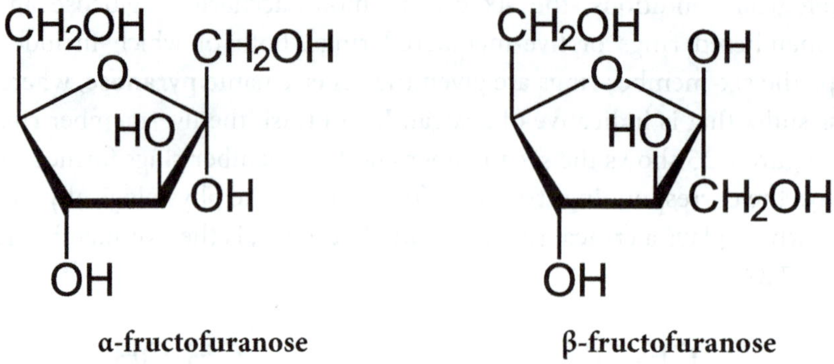

α-fructofuranose β-fructofuranose

Figure 7.26: Cyclic Forms of Fructose

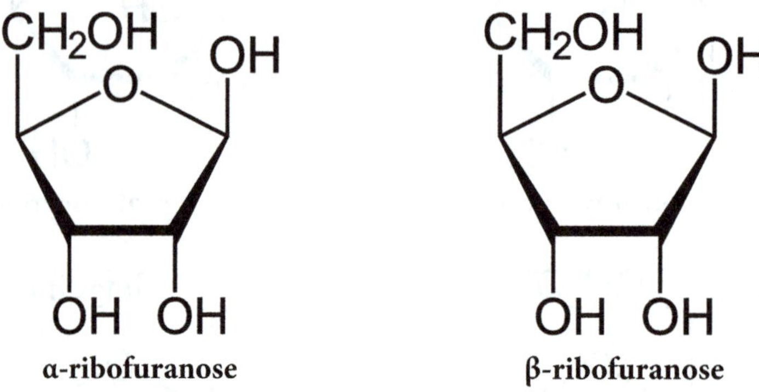


Figure 7.27: Cyclic Forms of Ribose

## Disaccharides, Aetals, and the Glycosidic Bond

We have taken an extended look at several critical monosaccharides that play central roles in the biochemistry of life on this planet. Significantly, glucose, fructose, and galactose are called the *dietary* monosaccharides, and we have briefly noted the importance of the ribose molecules to the structure of the *nucleic acids*, RNA and DNA. However, with the exception of providing simple definitions, we have neither explored the properties of disaccharides and polysaccharides nor examined how these molecules are able to produce monosaccharides when they are *hydrolyzed*. For the present, we will focus on the *structural* relationships that define disaccharides and polysaccharides, saving the details of the reaction mechanism that is responsible for the production of these biochemical structures for a later chapter. We choose this approach because we are about to meet several distinct classes of biomolecules that all share a **common reaction mechanism**. By studying the details of this mechanism in a future chapter and linking it to a variety of biochemical molecules, we will be able to observe one aspect of the remarkable unity that characterizes so much of the biochemistry of life.

Disaccharides, as the name implies, are constructed from two monosaccharides through the formation of a **glycosidic bond**. Figure 7.28 is a schematic highlighting the key features of a glycosidic bond between two glucose molecules. By examining the figure, let's carefully note the important characteristics of the glycosidic bond. Beginning with the upper portion of the figure, note that the two leftmost glucose molecules are designated as α anomers, meaning that the hydroxyl groups bonded to the anomeric carbon of each molecule (carbon atom 1 in the figure) lie *below* the molecular rings. The glycosidic bond forms as a result of a reaction between the hydroxyl group bonded to carbon atom 1 of the first glucose molecule and the hydroxyl group bonded to carbon atom 4 of the second glucose molecule (both are encircled by the dashed region in the figure). The result of this reaction leaves an oxygen atom bonded both to carbon atom 1 (the anomeric carbon) of the first glucose molecule and to carbon atom 4 of the second glucose molecule while producing a disaccharide molecule and a water molecule as the products of the reaction (right half of the upper portion of the figure). The disaccharide (also called a **glycoside** because it is a molecule made up of two simpler components joined by a **glycosidic bond**) also exhibits one of our two new functional groups: the acetal. To identify the acetal structure, examine the disaccharide molecule and note that the anomeric carbon (labeled 1) of the left molecular ring in the disaccharide is *shared by* two ether-like *structures*. This is the defining signature of the acetal functional group. Because the glycosidic bond formed in the upper half of the figure links the anomeric carbon of the first α anomer with carbon atom 4 of a second molecule, the bond is given the formal name **α(1→ 4) glycosidic bond**.

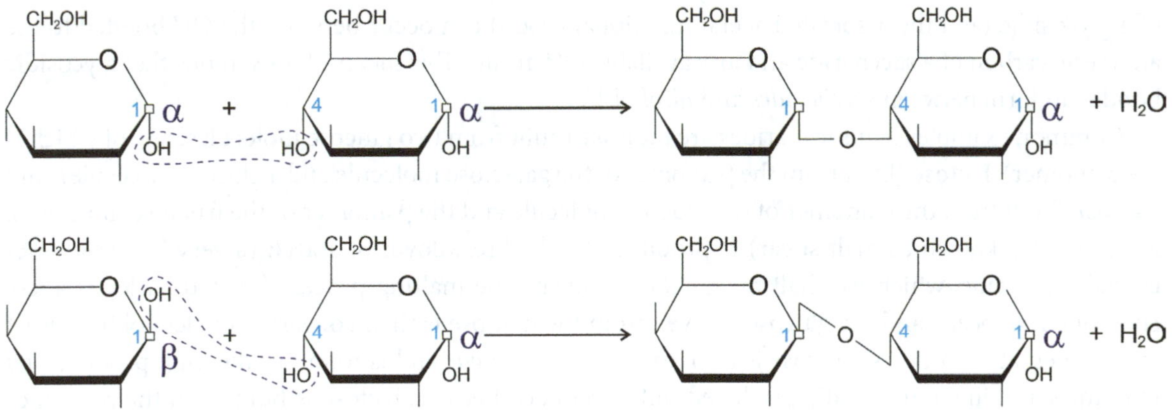

Figure 7.28: Glycosidic Bonds

The lower portion of figure 7.28 depicts the formation of a glycosidic bond between a glucose β anomer and a glucose α anomer. As in the upper portion of the figure, the reaction yields a disaccharide molecule and a water molecule as its products with the glycosidic bond, again exhibiting the acetal functional group. However, because two different glucose anomers, one α anomer and one β anomer, are linked to form the disaccharide in the lower portion of the figure, the resulting disaccharide has a **different geometric arrangement in space**. Consequently, a different symbol is used in the figure to identify the new glycosidic bond, and its formal name is the **β(1→ 4) glycosidic bond**.

The disaccharide sucrose is displayed in figure 7.29. In this case, two different monosaccharides, the α anomer of glucose (the pyranose) and the β anomer of fructose (the furanose), are linked by a glycosidic bond. Unlike our previous examples, the glycosidic bond in sucrose links the two anomeric carbons of the original monosaccharides. In the pyranose ring, the anomeric carbon would be labeled 1, but in the furanose ring the anomeric carbon would be labeled 2. Consequently, the formal name of the glycosidic bond depicted in figure 7.29 is the **α,β(1→ 2) glycosidic bond.**

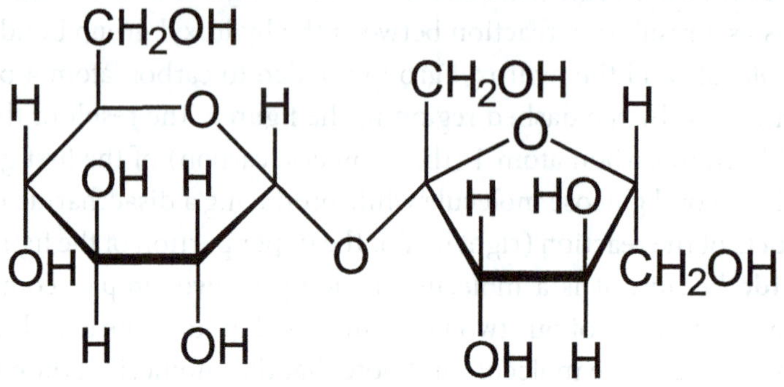

Figure 7.29: Sucrose

While the examples of glycosidic bond formation reviewed previously may suggest that this type of bond is limited to linkages between only two saccharides, this is **not** the case. In fact, the formation of a glycosidic (and the associated acetal functional group) can occur between the *OH* bonded to the anomeric carbon of a saccharide and **any** available *OH* group. This means, for example, that glycosidic bonds can form between *saccharides* and *alcohols*!

Common examples of disaccharides are **maltose** (built from two glucose molecules, one of which is the α anomer), **lactose** (built from the β anomer of the galactose molecule and a glucose molecule), and **sucrose** (built from the α anomer of the glucose molecule and the β anomer of the fructose molecule). Maltose (also known as malt sugar) is produced by the breakdown of **starch** (a very large **polysaccharide** molecule, which we shall meet shortly) during the malting process (malted barley is a key component of beer), and the glucose derived from the maltose is then converted to alcohol by a yeast in a fermentation process. Lactose is also known as *milk sugar* and is found in the milk produced by mammals and in commercially produced milk products. Because lactose is built from the β anomer of the galactose molecule, the glycosidic bond in lactose is a **β(1→ 4) glycosidic bond.** Sucrose is common table sugar and is the most abundant naturally occurring disaccharide. The principal commercial sources of sucrose are sugar cane and sugar beets; with the recent emphasis on renewable energy sources, the sucrose derived from sugar beets is used as the starting point in the production of ethanol. As we noted above, the **α,β(1→ 2) glycosidic bond** links together the anomeric carbon atoms from the glucose and fructose monosaccharide components of sucrose. Because both anomeric carbon atoms are part of the glycosidic bond and are **not** available to participate in other reactions, sucrose **cannot** participate in oxidation-reduction reactions (see chapter 5). In contrast, the monosaccharide glucose **can** participate in oxidation-reduction reactions and is called a **reducing** sugar. **Sucrose is not a reducing sugar.**

## Polysaccharides

As we have seen, the key to the formation of disaccharides from monosaccharide units is the glycosidic bond. However, after two glucose molecules are linked by an α(1→ 4) glycosidic bond, the anomeric carbon of the second glucose unit is still capable of forming **another** glycosidic bond, thereby producing a **trisaccharide**. The right side of figure 7.30 shows the available anomeric carbon atom in the second glucose unit of the disaccharide.

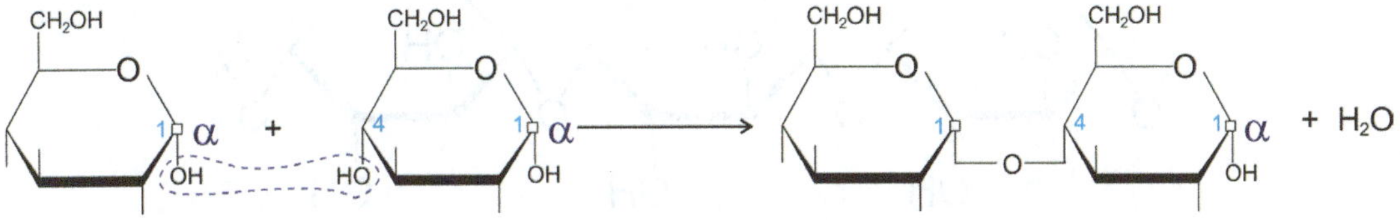

Figure 7.30: Disaccharide Formation

Note that carbon atom 1 of the first glucose unit is linked to carbon atom 4 of the second glucose unit (a glycosidic bond exhibiting the acetal functional group), but the anomeric carbon atom of the second glucose unit (atom 1) is still capable of forming another glycosidic bond. By the successive addition of glucose units, a **polysaccharide** molecule (that is, a molecule containing *many* monosaccharide units) is produced. In effect, polysaccharides are highly effective structures used for **storing** glucose.

The polysaccharide known as **starch** is the glucose storage mechanism used by plants. Starch is made up of two separate carbohydrates: **amylose** and **amylopectin**. Amylose constitutes approximately 20% of starch and is a polysaccharide composed of between 250 and 4,000 glucose units linked together by α(1→ 4) glycosidic bonds (see figure 7.31).

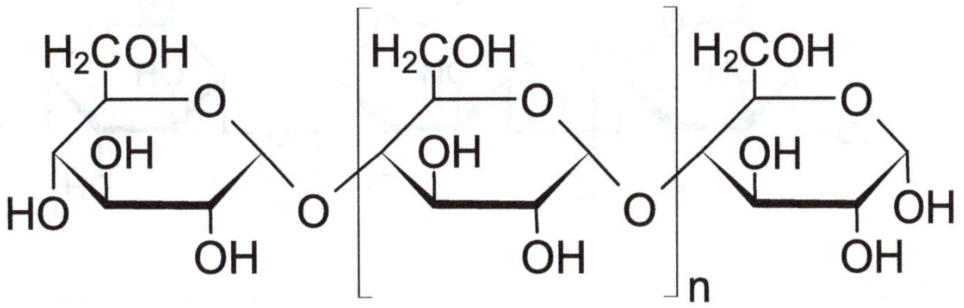

Figure 7.31: Amylose

The remaining component of starch (approximately 80% of the starch polysaccharide) is the carbohydrate **amylopectin**. Amylopectin, like amylose, is composed of glucose units linked by α(1→ 4) glycosidic bonds. However, at approximately every twenty-five glucose units, the linear amylopectin chain **branches** as a result of an α(1→ 6) glycosidic bond (figure 7.32).

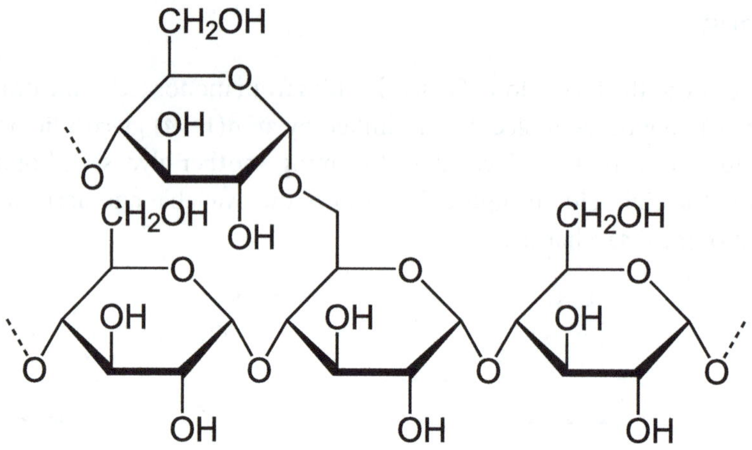

Figure 7.32: Amylopectin

The storage polysaccharide used by animals is called **glycogen**. Glycogen is a polysaccharide that is identical to amylopectin, except that the branching α(1→ 6) glycosidic bonds occur approximately every twelve glucose units (figure 7.33).

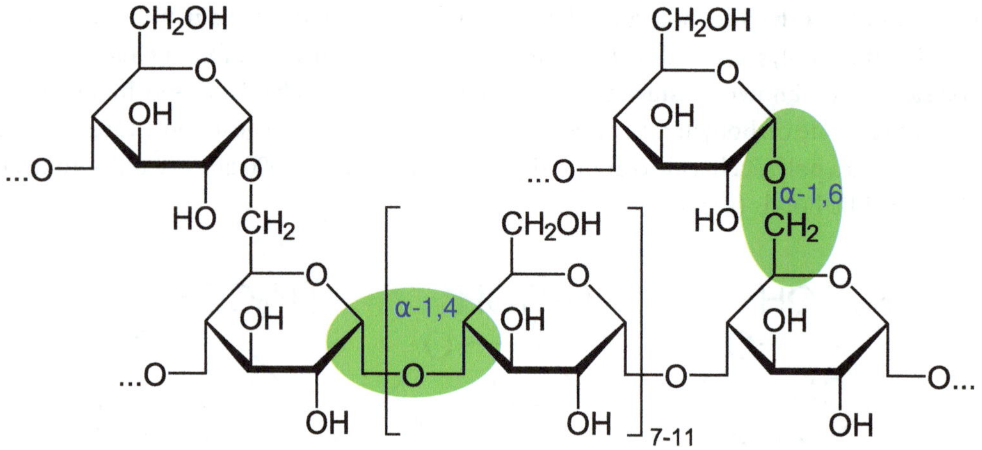

Figure 7.33: Glycogen

In addition to their highly effective **storage** function, polysaccharides also play a critical role in the **structure** of living organism. **Cellulose** is a polysaccharide in which large numbers of glucose units are linked together in very long linear chains by β(1→ 4) glycosidic bonds (figure 7.34).

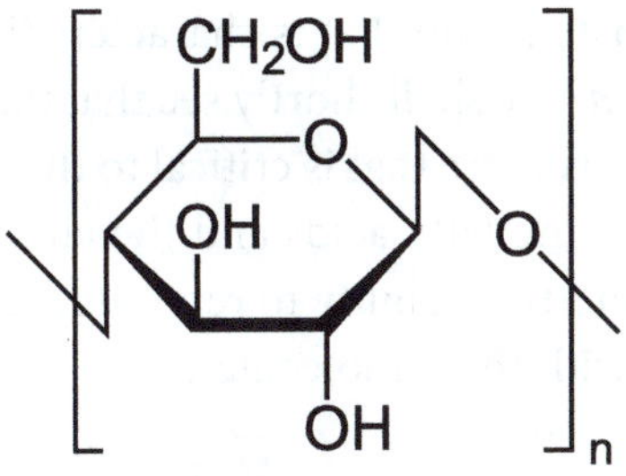

Figure 7.34: β(1→ 4) Glycosidic Bond in Cellulose

Cellulose is a major component of the cell walls of green plants, and its linear alignment allows it to form strong, rigid structures. Cellulose constitutes more than 40% of the wood in trees and is a major component of plant fibers: cotton (~90%), flax (~65%), and hemp (~67%).

Finally, **chitin** is a modified polysaccharide in which one of the hydroxyl groups of glucose is replaced by an acetyl amine group and the glucose units are linked together by β(1→ 4) glycosidic bonds (figure 7.35).

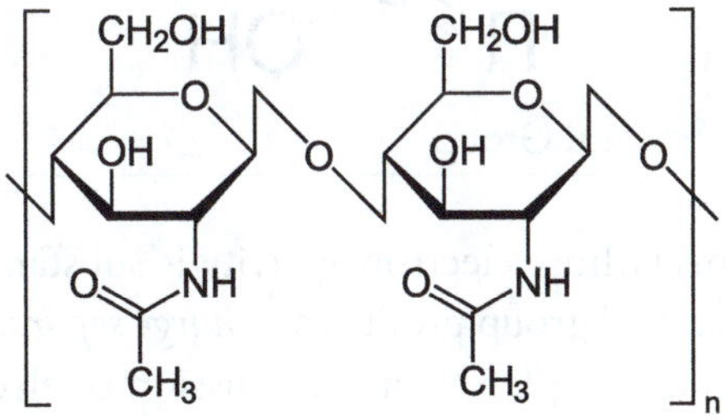

Figure 7.35: Chitin

Chitin is the major component of the cell walls of fungi, the exoskeletons of arthropods, the radulas of mollusks, and the beaks of cephalopods. Interestingly, because of its strength, flexibility, and biodegradability, chitin has been used as surgical thread.

## Biological Acids

In chapter 5 we defined a Brønsted-Lowry acid as a molecule that is capable of *donating a proton*, and we distinguished between *weak* and *strong* acids by observing their behavior as *weak* and *strong electrolytes*. Central to this understanding of *weak acids* is the concept of *chemical equilibrium*, a dynamic state characteristic of most chemical reactions and one that is particularly critical to the biochemistry of life. In fact, two of the categories of biological acids, the **fatty acids** and the **amino acids**, are members of the more general category of organic weak acids known as the **carboxylic acids** (one of the organic functional groups introduced early in this chapter). However, both the reactivity and resulting biochemical structures associated with the fatty acids and the amino acids dramatically overshadow their acid-base chemistry. The **nucleic acids** were identified in the 1870s by Johann Miescher as constituents of the cell nucleus that possess the phosphate group that is characteristic of phosphoric acid. Consequently, they were given the name *nucleic acid*. We shall shortly see that the nucleic acids are, in fact, very large biomolecules possessing a complex structure that is critical to the ability of these molecules to *replicate* themselves. But just as in the case of the fatty acids and the amino acids, these characteristics of the nucleic acids (both their structure and their ability to replicate) are significantly more important than any acid-base chemistry associated with these molecules.

## Fatty Acids

The fatty acids (an abbreviation for *fatty carboxylic acids*) are biologically important molecules that are members of the much larger class of molecules that possess the **carboxylic acid** functional group. This functional group is a member of the third family of functional groups that we met at the beginning of this chapter. Figure 7.36 displays the carboxylic acid functional group.

$$R-\overset{\overset{\displaystyle O}{\|}}{C}-OH$$

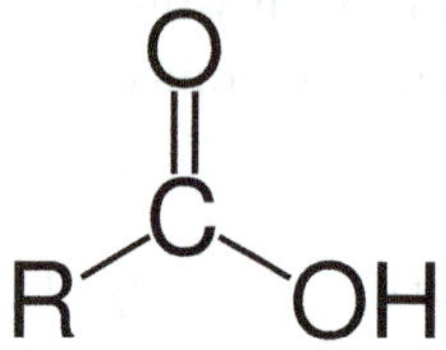

Figure 7.36: Carboxylic Acid Functional Group

The presence of the oxygen atoms (whose electronegativity is substantially greater than that of either carbon or hydrogen) in this functional group produces a *charge separation*, making the carboxylic acid functional group a *polar group* (see chapter 4 for a discussion of electronegativity and polarity). As noted earlier, the *fatty acids* are a subset of the general carboxylic acid family, consisting of molecules containing between twelve and twenty-two carbon atoms in a hydrocarbon chain. A typical saturated fatty acid is shown in figure 7.37.

$$CH_3CH_2CH_2CH_2CH_2CH_2CH_2CH_2CH_2CH_2CH_2CH_2CH_2CH_2CH_2\overset{\overset{\displaystyle O}{\|}}{C}\text{-}OH$$

Figure 7.37: Saturated Fatty Acid with Sixteen Carbon Atoms

Because the long, hydrocarbon chain is *nonpolar*, while the carboxylic acid functional group is *polar*, the fatty acids are called **amphipathic** or **amphiphilic**. (The origin of the terms is the Greek *amphis*, meaning "both," and *philia*, meaning "love"; the molecule is attracted to **both** *polar* and *nonpolar* chemical environments.)

The fatty acids are particularly important because they react chemically with **glycerols** (molecules that are polyhydroxy alcohols) to produce the biologically significant molecules known as **triglycerides**. Figure 7.38 displays the reaction between a glycerol molecule and three fatty acid molecules to form a single triglyceride. The three fatty acid need **not** be identical and, in fact, often are not identical.

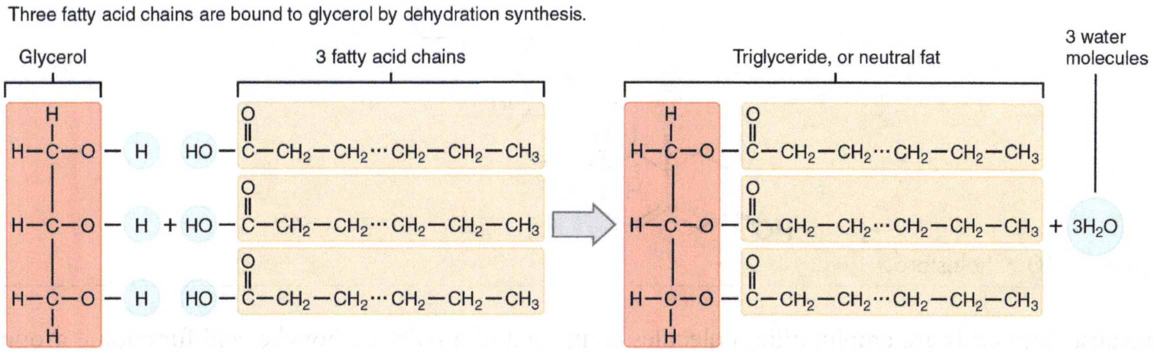

Three fatty acid chains are bound to glycerol by dehydration synthesis.

Glycerol

3 fatty acid chains

Triglyceride, or neutral fat

3 water molecules

Figure 7.38: Formation of a Triglyceride

The chemical bonds that are formed between the glycerol molecule and the three fatty acid molecules (shown in figure 7.38) are called **ester bonds** or **ester linkages**; ester bonds exhibit a chemical structure that is characteristic of another important functional group introduced earlier in this chapter: the **ester functional group**. These structural characteristics of the ester functional group are shown in figure 7.39. As we shall see, the ester functional group plays a role in a number of biologically significant molecules.

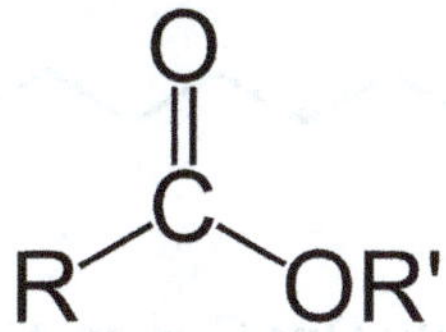

Figure 7.39: Ester Functional Group

The triglycerides are members of a larger and very diverse group of biomolecules known as the **lipids**. Lipids occur in both plants and animals on this planet and are soluble in either nonpolar solvents or solvents of low polarity. Structurally, the lipids are a very diverse group of compounds and include fats (triglycerides), waxes, oils, sterols (a subgroup of the steroid family of molecules), fat-soluble vitamins, other monoglycerides and diglycerides, and lipid molecules that include a phosphate group (phospholipids). A very important sterol is cholesterol (see figure 7.40). Like all other members of the steroid family, cholesterol is built around a core of four fused rings of carbon atoms (a total of seventeen carbon atoms often identified as *four fused carbocyclic rings*). Lipids perform a wide range of biological functions that include energy storage, signaling, and acting as structural components of the cell membrane. (We will discuss the cell membrane at greater length in chapter 11.)

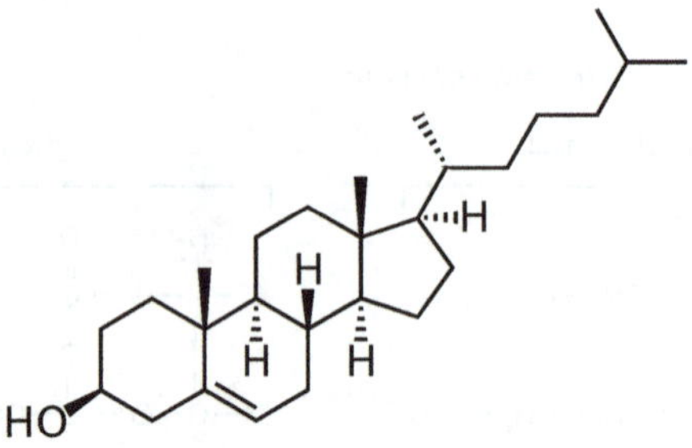

Figure 7.40: Cholesterol

Because fatty acids are amphiphilic molecules composed of a *polar* carboxylic acid functional group and a *nonpolar* long hydrocarbon chain, a perfectly reasonable question to ask is, what are the consequences if the hydrocarbon chain is *unsaturated*? Recall that a *saturated* hydrocarbon possesses **only** single bonds, while an *unsaturated* hydrocarbon possesses **one or more** multiple bonds. Figure 7.37 displayed a typical saturated fatty acid; several more examples are shown in figure 7.41.

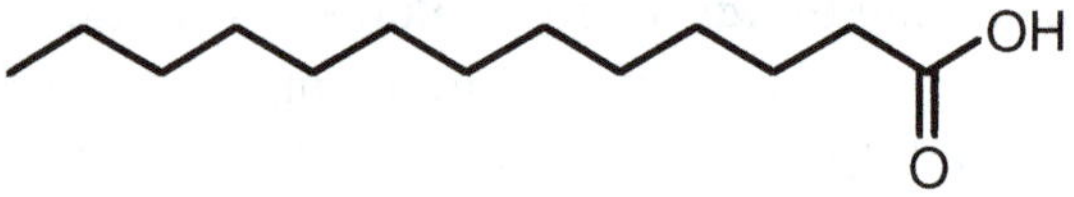

**Thirteen Carbon Atoms**

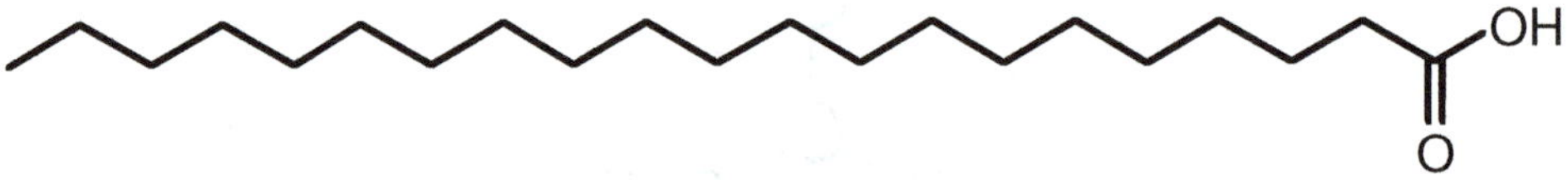

**Twenty-One Carbon Atoms**

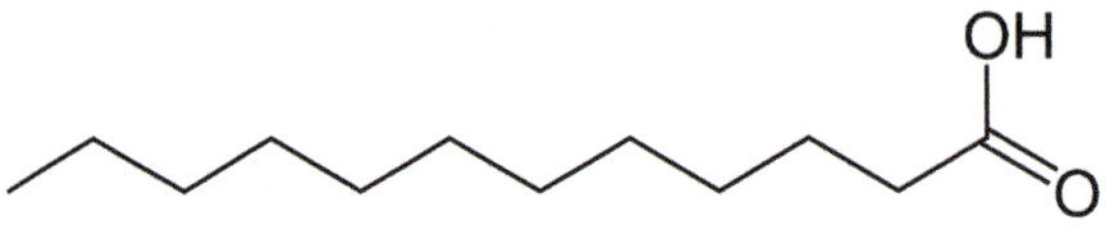

**Twelve Carbon Atoms**

Figure 7.41: Saturated Fatty Acids

In contrast, *unsaturated* fatty acids exhibit a distinctly different geometric arrangement in space as a consequence of the multiple bonds that are part of the molecules. Compare the saturated fatty acids displayed in figure 7.41 with the unsaturated fatty acids shown in figure 7.42.

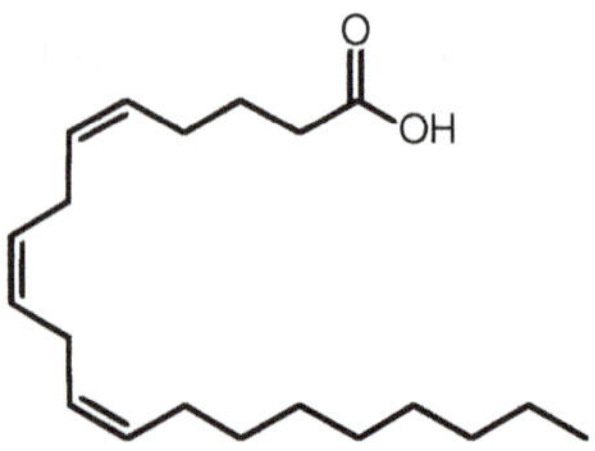

**Twenty Carbon Atoms**

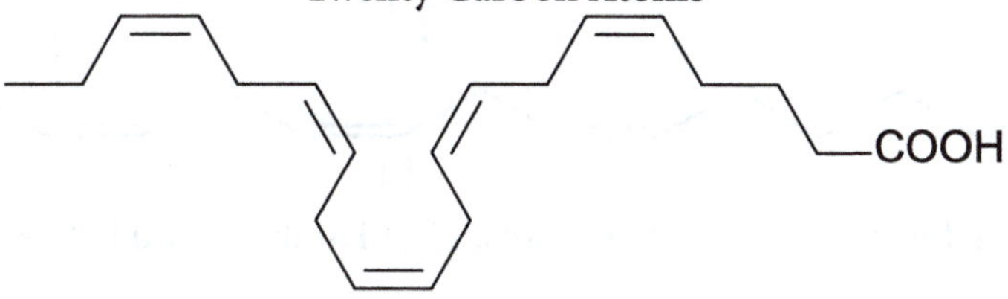


**Twenty Carbon Atoms**

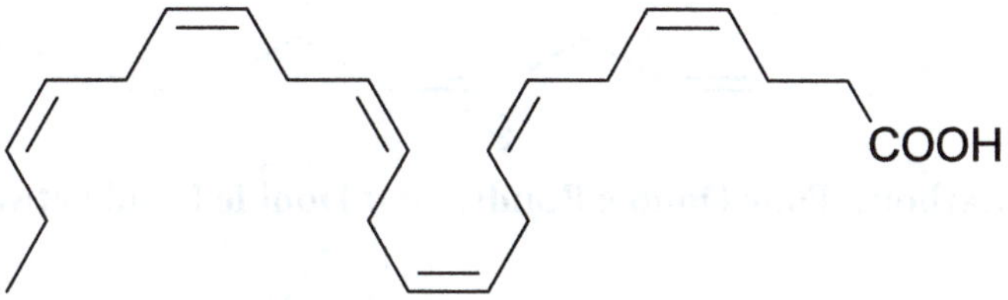


**Twenty-Two Carbon Atoms**

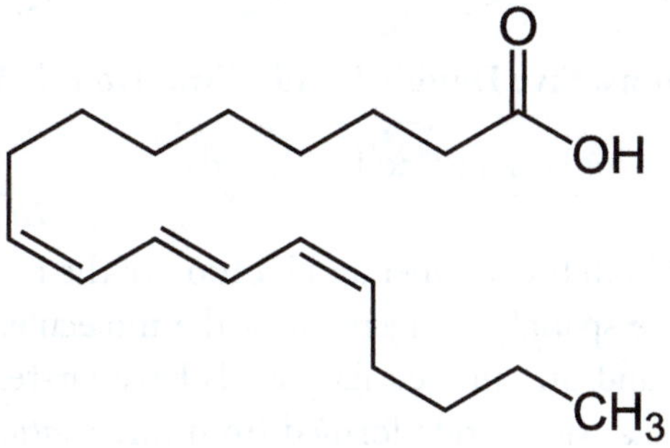


**Eighteen Carbon Atoms**

Figure 7.42: Unsaturated Fatty Acids

Unsaturated fatty acids with double bonds are often identified using a numbering system that originated with the biochemistry community. The carbon atom that is located **farthest** from the carboxylic acid functional group is designated as the ω (**omega**) carbon atom. The carbon atoms of the long hydrocarbon chain are then numbered beginning with the ω **carbon atom**. The number *of the carbon atom at which the first double bond occurs,* along with *the* total number *of double bonds,* is used to generate a name for the fatty acid. Figure 7.43 contains examples of this naming convention (with the omega numbering in red) for unsaturated fatty acids.

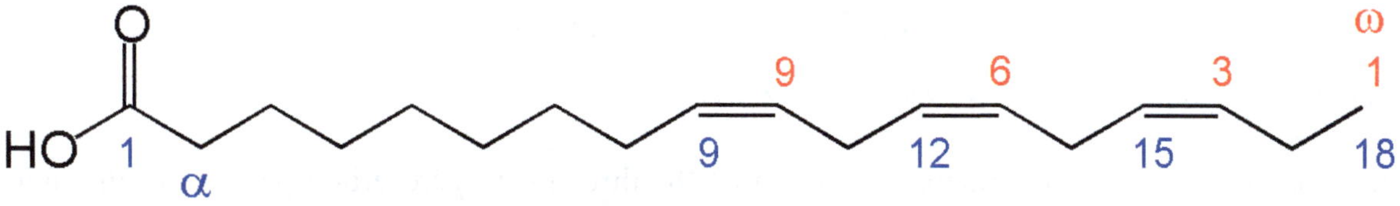


**[18:3], ω-3: Eighteen Carbons, Three Double Bonds, First Double Bond between Atoms 3 and 4**

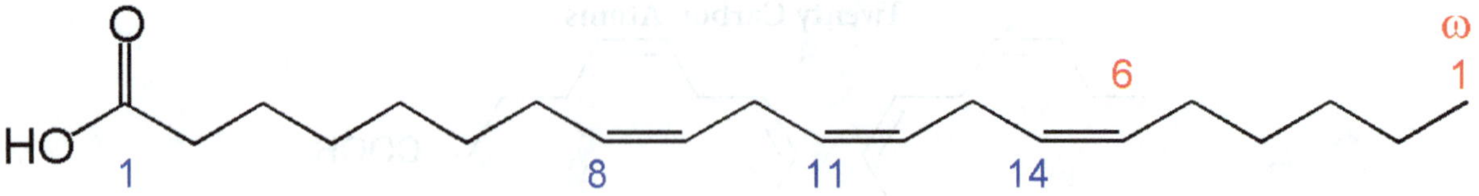

[20:3], ω-6: **Twenty Carbons, Three Double Bonds, First Double Bond between Atoms 6 and 7**

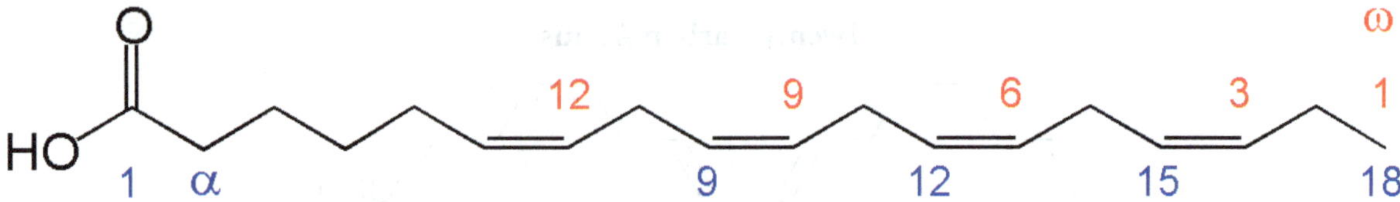


[18:4], ω-3: **Eighteen Carbons, Four Double Bonds, First Double Bond between Atoms 3 and 4**

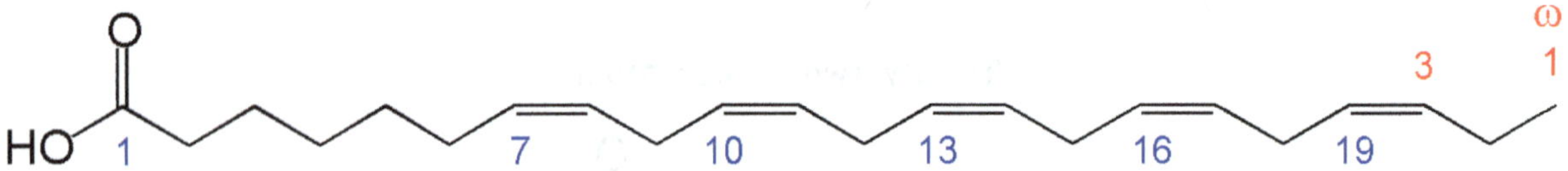

[22:5], ω-3: **Twenty-Two Carbons, Five Double Bonds, First Double Bond between Atoms 3 and 4**

Figure 7.43: Omega Numbering for Unsaturated Fatty Acids

From figure 7.42 it is clear that both the number and location of the multiple bonds in an unsaturated fatty acid dramatically affect the spatial arrangement of the molecule. The differences in the spatial arrangements of the saturated and unsaturated fatty acids have far-reaching effects on the resulting triglycerides. Figure 7.44 shows a triglyceride formed from three *saturated* fatty acids. Note that the skeletal structure implies that the long hydrocarbon portions of the molecule lie **parallel** and **in close proximity** to one another. Compare this spatial arrangement to that of figure 7.45, which shows the skeletal structure of a triglyceride formed from three *unsaturated fatty acids.*

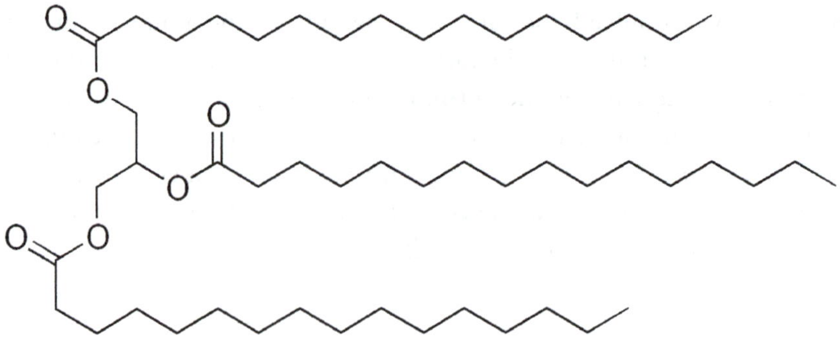

Figure 7.44: Triglyceride with Saturated Fatty Acids

In figure 7.45 the skeletal structure suggests that the three long hydrocarbon portions of the molecule are **neither** parallel **nor** in close proximity to one another. This difference has the important observational consequence that triglycerides formed from saturated fatty acids tend to be **solids** at room temperature and normal atmospheric pressure. The approximately parallel hydrocarbon chains interact more strongly with one another through weak intermolecular forces, resulting in a more closely packed and denser (solid or semisolid) configuration. (We will discuss these intermolecular forces in a later chapter.) In contrast, triglycerides formed from unsaturated fatty acids tend to remain **liquid** at room temperature and normal atmospheric pressure.

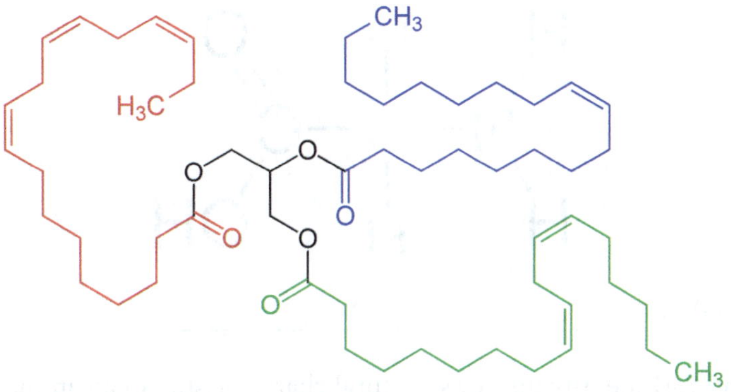

Figure 7.45: Triglyceride with Unsaturated Fatty Acids

## Amino Acids

Like the fatty acids, the **amino acids** are a subset of the carboxylic acid family of molecules and possess the carboxylic functional group (See figure 7.11 and figure 7.36). In addition, the amino acids possess a second functional group, the **amino functional group** (see figure 7.10), which we also met much earlier in this chapter. In addition to the two functional groups, each amino acid is distinguished by a characteristic **side-chain**. The representative structure of an amino acid is show in figure 7.46, where the *R* represents the characteristic side-chain.

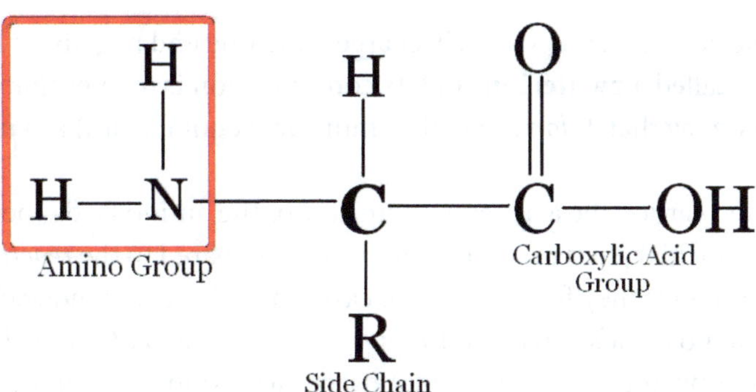


Figure 7.46: Amino Acid

The carbon atom to which the carboxylic acid functional group is bonded in an amino acid is designated as the α **carbon atom**. In the biologically important amino acids both the *amino functional group* and the *side-chain* are bonded to the α *carbon atom*. Consequently, such amino acids are designated as α **amino acids** (see figure 7.47). While a large number of amino acids have been identified on this planet (well in excess of one hundred), there are only twenty-two α amino acids, and only twenty-one of them occur in eukaryotic cells (cells possessing a nucleus and or other structures within a cell membrane).

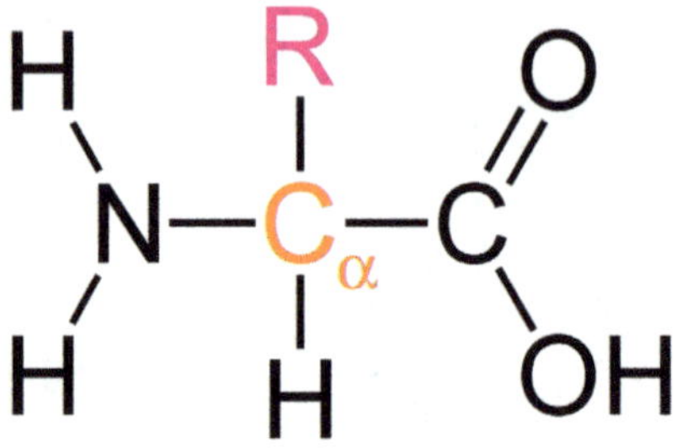

Figure 7.47: α Amino Acid

While figures 7.46 and 7.47 identify the key structural characteristics of an amino acid, the behavior of the amino acids in a physiological setting are, not unsurprisingly, governed by the *chemical equilibrium* depicted in figure 7.48.

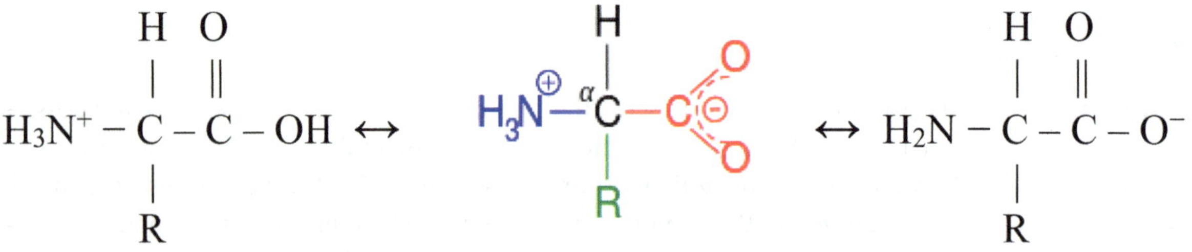

Figure 7.48: Amino Acid Equilibrium

The doubly charged species (one end positively charged and one end negatively negative charged, with a **net** charge of **zero**) is called a **zwitterion**, and its concentration is greater than the concentrations of the other two species (we say that it *dominates* the amino acid equilibrium) at a **unique** pH value called the **isoelectric point**.

However, as we noted earlier, the acid-base chemistry of the amino acids (here we are focusing our attention on the α amino acids) pales in significance when compared to the *reactivity* of the amino acids and the *biochemical structures* they form. As is the case with all organic molecules, it is the *functional group* that determines a molecule's reactivity. Because the α **amino acids** are *polyfunctional* molecules, it is possible (just as we saw in the case of the monosaccharides) for **two** amino acid to react with one another. The reaction takes place between the carboxylic acid functional group (frequently called the **C-terminus**) of one amino acid molecule and the amino functional group (the **N-terminus**) of another amino acid molecule. Figure 7.49 **displays the reaction of two** α **amino acids**.

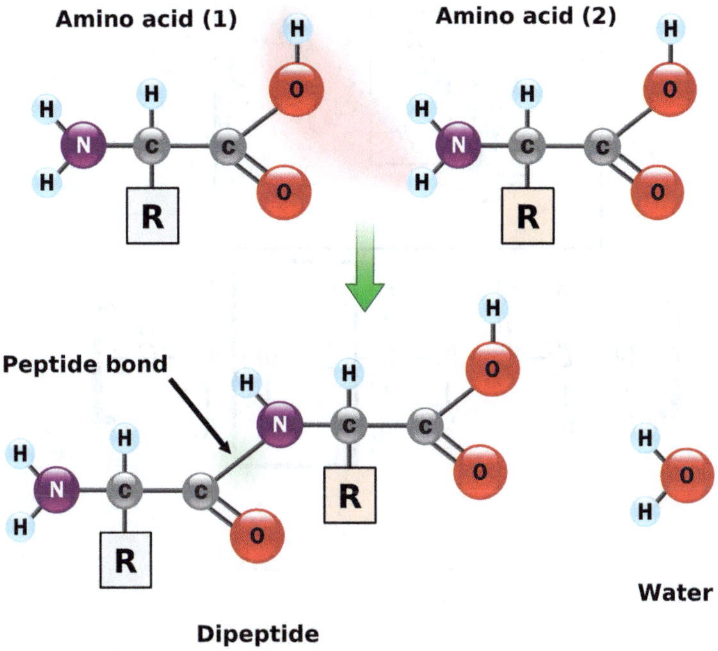

Amino acid (1)

Amino acid (2)

Peptide bond

Dipeptide

Water

Figure 7.49: Peptide Bond Formation

The consequence of the reaction establishes a new bond between the nitrogen of amino acid (2) with the carbonyl carbon of the amino acid (1). This new bond is called a **peptide bond** by the biochemistry community, and the resulting molecule is called a **dipeptide**, because it was formed from **two** amino acid molecules (note the parallel with the naming of a *disaccharide* produced by linking **two** *monosaccharides*). The individual amino acids are often called **amino acid residues.** A careful examination of the peptide bond discloses that the peptide bond is part of the **amide functional group** introduced much earlier in this chapter. Figure 7.50 (as well as the earlier figure 7.11) displays the *amide functional group*. (Note that one of the *R*'s bonded to the nitrogen atom is replaced by a hydrogen atom in the formation of the peptide bond.)

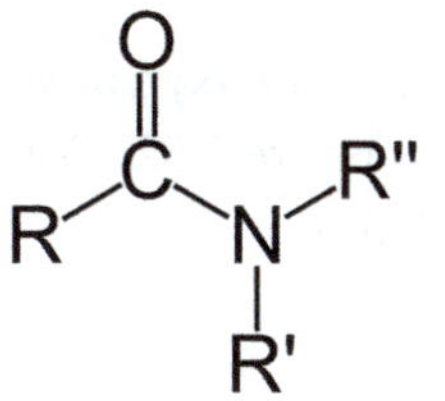

Figure 7.50: Amide Functional Group

Of course, because the *dipeptide* formed from two amino acids possesses **both** a C-terminus and an N-terminus, additional peptide bonds can be created at either end of the molecule, producing a **polypeptide** molecule (exactly analogous to the polysaccharide molecules we observed earlier). Figure 7.51 shows a **tripeptide** molecule (*two* peptide bonds, but *three* amino acid residues).

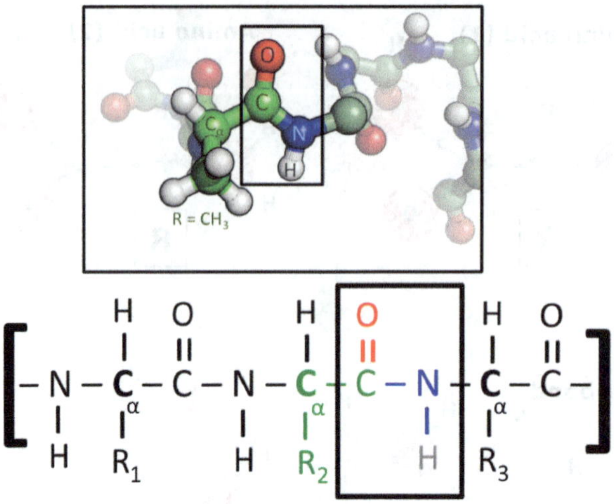

Figure 7.51: Tripeptide

## Nucleic Acids

While the **nucleic acids** were initially identified as constituents of the cell nucleus (thereby associating them with eukaryotic life forms) and exhibited the acidic properties associated with the phosphate group, this simple characterization was subsequently replaced over a period of some seventy-five years by a new understanding that was to place these molecules at the very center of all life on this planet. Further investigations of the nucleic acids revealed both a *range of reactions* and *chemical structures* that rapidly overshadowed the early observation of their acidic properties. Like the polysaccharides and polypeptides, the nucleic acids are *large* molecules composed of many thousands of subunits known as **nucleotides**. In turn, each subunit is made up of three distinct components: a **five-carbon sugar (a pentose)**, a **nitrogenous base (aromatic heterocycles containing carbon and nitrogen atoms)**, and a **phosphate group** (the original reason for designating these compounds as *acids*).

Two different pentose sugars, **ribose** and **2-deoxyribose**, can appear in the nucleic acids. The *beta* anomers of the two furanoses are shown in figure 7.52. Compared to ribose, 2-deoxyribose is missing an oxygen atom (note the arrow in figure 7.52).

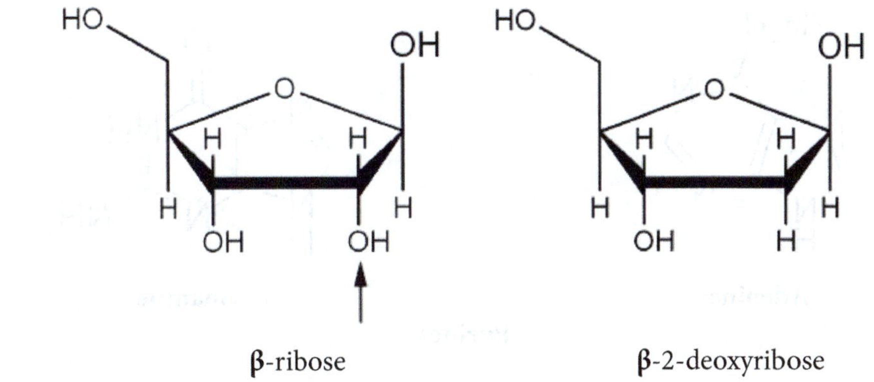

β-ribose β-2-deoxyribose

Figure 7.52: Ribofuranose and 2-Deoxyribofuranose

Because each nucleic acid molecule contains *exclusively* either ribose or 2-deoxyribose (that is, **never a mixture of the two sugars**), the names of the sugars are the origin of the names given to the two classes of nucleic acids found on this planet: **ribonucleic acid (RNA)** and **deoxyribonucleic acid (DNA)**. The **nitrogenous bases** constitute the second component of a **nucleotide** and are members of two large classes of aromatic molecules, the **purines** and the **pyrimidines**. The general structures of these two molecular classes are shown in figure 7.53.

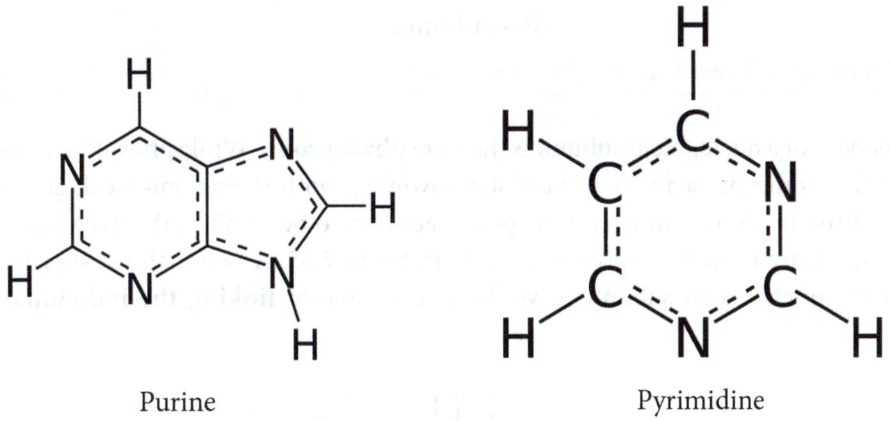

Purine Pyrimidine

Figure 7.53: Nitrogenous Bases

Notice that the purine molecule possesses **two** heterocyclic rings, while the pyrimidine molecule (the molecule with the longer name!) possesses only **one** heterocyclic ring. Only **five** different nitrogenous bases are used in the nucleic acids: adenine, guanine, cytosine, thymine, and uracil. Three of the bases (adenine, guanine, and cytosine) are common to both **RNA** and **DNA**; thymine is found **only** in **DNA**, while uracil is found **only** in **RNA**. The two bases adenine and guanine are *purines*, while cytosine, thymine, and uracil are *pyrimidines*. The chemical structures of all five molecules are depicted in figure 7.54.

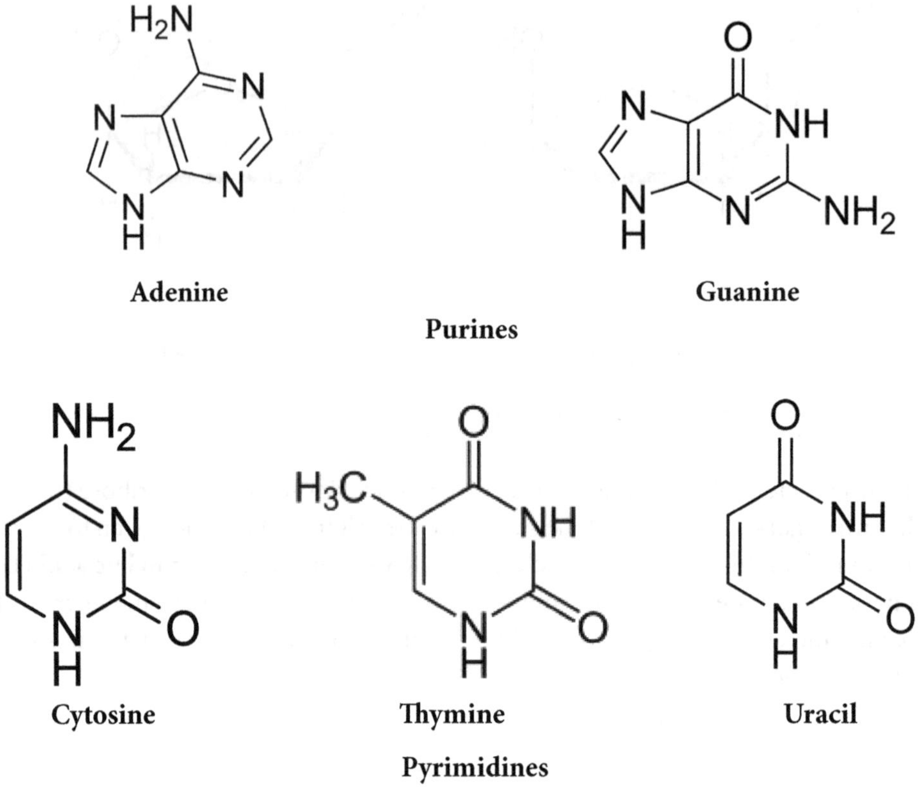

Figure 7.54: Purines and Pyrimidines in DNA and RNA

The final component of a nucleic acid subunit is the **phosphate group**. While phosphoric acid ($H_3PO_4$) exhibits a complex conjugate acid-base chemistry involving both strong and weak acid species, at physiological pH the predominant chemical species appears to be $HPO_4^{2-}$, the hydrogen phosphate species, whose structure is most simply represented in figure 7.55. As we shall see shortly, this third component of the nucleic acid subunit plays the crucial role of **linking the individual subunits together**.

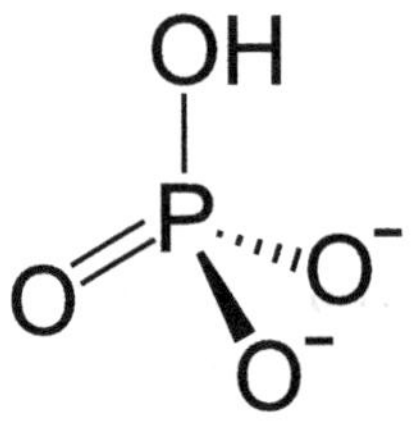

Figure 7.55: Phosphate Group

The first step in building the complete structure of a nucleic acid is the linking together of the three components of a nucleic acid subunit. The process begins by bonding either the ribofuranose or the deoxyribofuranose to one of the five nitrogenous bases to form a structure that is called a **nucleoside**.

The bond linking the two components is between the anomeric carbon atom of the sugar molecule and a nitrogen atom in the purine or pyrimidine ring (see figure 7.56).

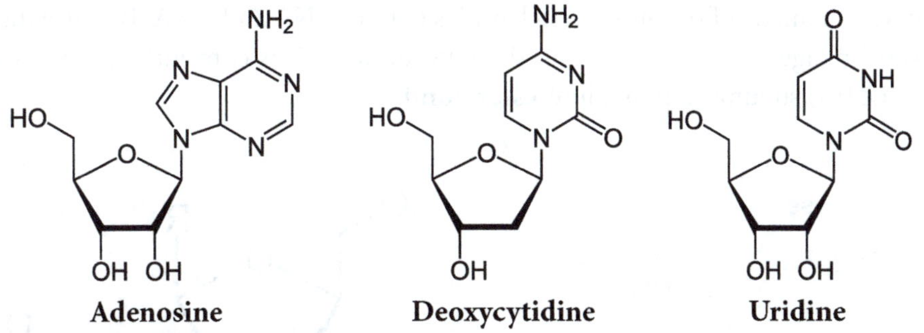

**Adenosine** **Deoxycytidine** **Uridine**

Figure 7.56: Nucleosides

While the bond linking the sugar molecule and the nitrogenous base molecule takes place between a nitrogen atom of the base and the anomeric carbon atom of the sugar, thus resembling a *glycosidic* bond, an actual acetal structure **is not formed**. Consequently, this bond is often called an **N-glycosidic bond** and the resulting structure an N-glycoside. However, IUPAC discourages the use of the term N-glycoside.

Finally, a phosphate group is added to the *nucleoside*, thus forming the complete **nucleotide** subunit. In figure 7.56 a complete *nucleotide* is displayed, showing the furanose, the nitrogenous base cytosine, and the phosphate group.

**Cytidine 5'-monophosphate**

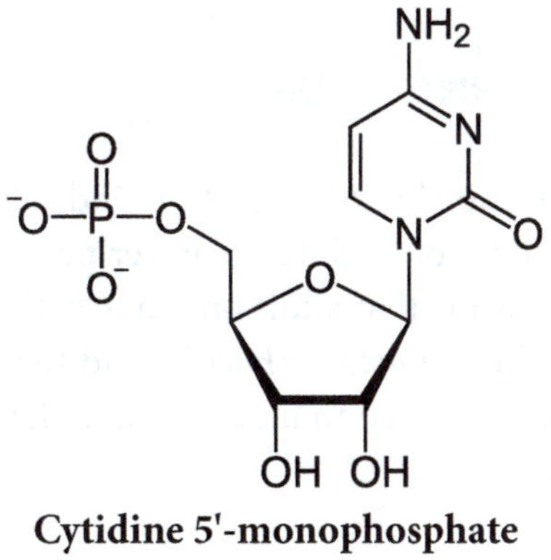

Figure 7.57: Nucleotide

Because the phosphate group is bonded to the 5' carbon of the sugar (the carbon atoms of the sugar molecule are identified by primes; the numbers used for the atoms of the heterocycle are unprimed), the nucleotide is named *5'-monophosphate*. A careful examination of the bond between the sugar and the phosphate group reveals that it appears to be almost identical to an *ester functional group*, with the phosphorus atom replacing the carbon atom of the ordinary ester. Consequently, the linkage is called a **phosphoester**.

The last step in the formation of a complete nucleic acid is to link together many thousands of the nucleotide subunits (remember, nucleic acids are *very large* molecules). This final step requires the formation of a **second phosphoester** bond between a second pentose sugar and the phosphate group. Figure 7.58 shows a schematic of two linked nucleotides in both RNA and DNA. Because there are now **two phosphoester** linkages that, together, link the two subunits, it is conventional to name the total linkage between the two subunits a **phosphodiester bond**.

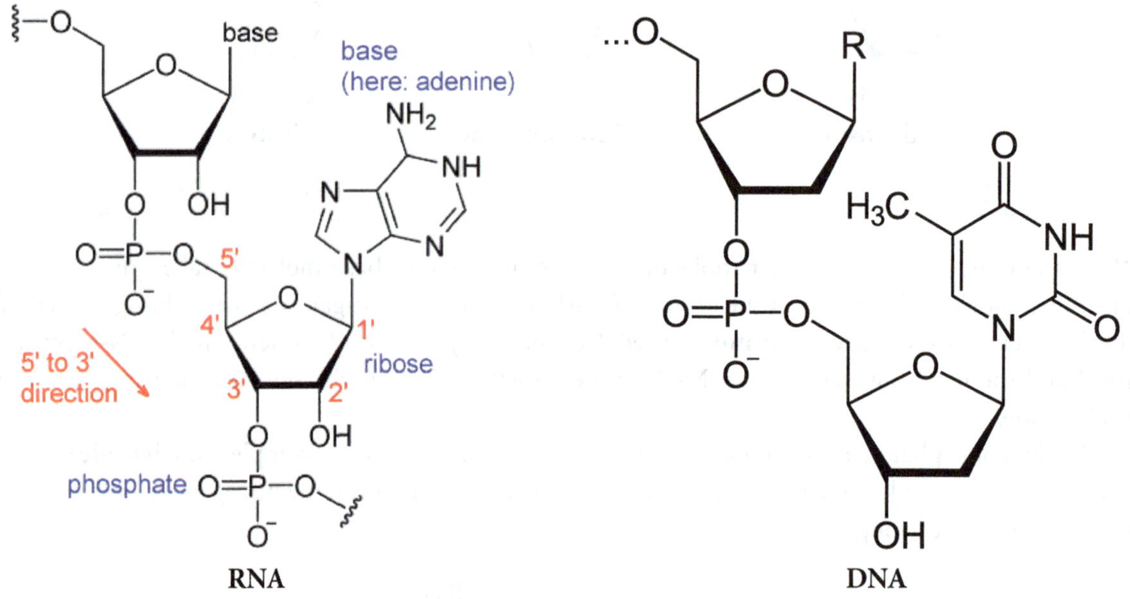

Figure 7.58: Phosphodiester Bonds in RNA and DNA

Note in the RNA panel that the indication "5' to 3' direction" indicates that the upper pentose is bonded to a phosphate group at the 5' carbon atom, while the lower pentose sugar is bonded to a phosphate group at the 3' carbon atom. This means that additional nucleotide subunits can be added **at either end of the molecule**. Hence, just like the *polysaccharides* and the *polypeptides*, it is possible to form **polynucleotides** by continuing the steps used to link two subunits together. As we shall see later, this capability is crucial to life as we understand it.

# Chapter 7 Exercises

1. Which element is at the center of organic chemistry?
2. In stable compounds, how many bonds does the carbon atom form? What is the technical term used to describe this characteristic of the carbon atom?
3. Can the carbon atom form multiple bonds? Explain.
4. Use examples to distinguish the three types of molecular representations: structural, skeletal, and condensed.
5. Define the following terms:
   a. Ball-and-stick model
   b. Wedge-and-dashed model
6. Define the following terms:
   a. Primary carbon atom
   b. Secondary carbon atom
   c. Tertiary carbon atom
   d. Quaternary carbon atom
7. What are saturated and unsaturated compounds?
8. If a carbon atom has undergone $sp^3$ hybridization, how many single bonds can the atom form?
9. Which hybridization scheme must be used by a carbon atom for the atom to form a double bond? Describe the geometry required by a double bond.
10. Which hybridization scheme must be used by a carbon atom for the atom to form a triple bond? Describe the geometry required by a triple bond.
11. What is a functional group?
12. What is the common characteristic of the first family of functional groups? This family is composed of four distinct functional groups. Name the four functional groups in this family.
13. What is a hydrocarbon?
14. What type of bond is characteristic of each of the following functional groups?
    a. Alkanes
    b. Alkenes
    c. Alkynes
15. What is the distinguishing molecular structure found in the aromatic hydrocarbons?
16. The second family of functional groups includes four principal members. Name the four principal functional groups in this family.
17. What is the distinguishing characteristic of the second family of functional groups?
18. Which characteristic molecular structure occurs in all alcohols?
19. What is an ether?
20. What is a thiol?
21. What is a thioether?
22. Define the following members of the amine functional group:
    a. Primary amine
    b. Secondary amine

c. Tertiary amine
d. Quaternary amine

23. What is the characteristic structure present in all five functional groups that constitute the third family of functional groups?
24. Draw the characteristic structure of each of the following functional groups:
    a. Aldehyde
    b. Ketone
    c. Ester
    d. Carboxylic acid
    e. Amide
25. Draw the carbonyl structure.
26. The aliphatic hydrocarbons can exhibit two different molecular structures. What are these structures?
27. List at least four chemical properties of the alkane functional group.
28. True or false? No alkene is a cyclic molecule.
29. What are the three parts that make up the name of every organic molecule?
30. What is the parent compound?
31. What is a substituent?
32. What information does the suffix provide?
33. What is an alkyl group?
34. How many carbon atoms are present in the following alkyl groups?
    a. Methyl
    b. Butyl
    c. n-Propyl
    d. Octyl
    e. Nonyl
    f. Decyl
    g. Isopropyl
35. Give the proper IUPAC name for each of the following molecules:
    a. $CH_3CH_2CH_2CH_2CH_2Cl$
    b. $BrCH_3CHCH_2CHCH_2$
    |          |
    $CH_3$     $CH_2CH_3$
    c. $CH_3CHCH_2CHCH_2Br$
    |          |
    $CH_3$     $CH_2CH_2CHCH_3$
36. Give the proper IUPAC name for each of the following molecules:
    a. $CH_3CH_2CH_2CH{=}CHCl$
    b. $BrCH_2CHCH_2CHCH_3$
    |          |
    $CH_3$     $CH{=}CH_2$

c. $$\begin{array}{l} BrCH_2\overset{}{C}HCH_2CHCH_3 \\ \quad\quad\ \ | \quad\quad\ \ | \\ \quad\ \ CH_3 \quad CH_2CH{=}CHCH_3 \end{array}$$

37. Draw the structures of the following:
    a. 5–isopropyl–4–methyldecane
    b. 2–bromo–4–ethyl–1–hexene
38. Define the following terms:
    a. Ortho
    b. Meta
    c. Para
39. What is a carbohydrate?
40. What is the meaning of the term “hydrolysis”?
41. Define the following terms:
    a. Monosaccharide
    b. Disaccharide
    c. Polysaccharide
    d. Oligosaccharide
42. Give the proper IUPAC name for each of the following molecules:
    a. $CH_3CH_2CH_2CH_2CH_2OH$
    b. $$\begin{array}{l} BrCH_3CHCH_2CHCH_2 \\ \quad\quad\ \ | \quad\quad\ \ | \\ \quad\ \ CH_3 \quad CH_2CH_2OH \end{array}$$
    c. $$\begin{array}{l} CH_3CHCH_2CHCH_2CH_3 \\ \quad\ \ | \quad\quad\ \ | \\ \ \ CH_3 \quad CH_2CH_2CHCH_3 \\ \quad\quad\quad\quad\quad\quad\quad\ \ | \\ \quad\quad\quad\quad\quad\quad\quad OH \end{array}$$
43. Give the proper IUPAC name for each of the following molecules
    a. $$\begin{array}{ccccccccccccc} & & CH_3 & & & & & & & & O & & \\ & & | & & & & & & & & \| & & \\ CH_3 & - & CH & - & CH_2 & - & CH & - & CH_2 & - & CH - C & - & H \\ & & & & & & | & & & & | & & \\ & & & & & & CH_3 & & & & CH_2 - CH_2 - CH_3 & & \end{array}$$
    b. $$\begin{array}{ccccccccccccc} & & & & O & & & & & & CH_2 - CH_3 & & \\ & & & & \| & & & & & & | & & \\ CH_3 & - & CH_2 & - & C & - & CH & - & CH_2 & - & CH - CH & - & CH_3 \\ & & & & & & | & & & & | & & \\ & & & & & & CH_3 & & & & CH_2 - CH_2 - CH_3 & & \end{array}$$

44. Define the terms “aldose” and “ketose.”

45. List five monosaccharides that are biologically important.
46. Identify the following monosaccharides as either aldoses or ketoses:
    a. Fructose
    b. Glucose
    c. Galactose
47. Name the two pentose sugars that play an important structural role in nucleic acids.
48. Define the following terms:
    a. Acetal
    b. Hemiacetal
49. What is the conventional representation of the cyclic form of a monosaccharide called?
50. What is an anomeric carbon atom?
51. What is an anomer?
52. What do the terms "pyranose" and "furanose" mean?
53. What is a glycosidic bond?
54. What is a glycoside?
55. What is the complete name of the glycosidic bond found in sucrose?
56. Is sucrose a reducing sugar? Explain.
57. Explain why a polysaccharide molecule is an effective storage structure for glucose. Which polysaccharide is used by plants to store glucose?
58. What is the polysaccharide used by animals to store glucose?
59. What is cellulose?
60. What is chitin?
61. Name two biologically important carboxylic acids.
62. Define the terms "amphipathic" and "amphiphilic" using the concept of polarity.
63. What are the four molecules that react to form a triglyceride?
64. To which class of biomolecules do the triglycerides belong?
65. What chemical structure characterizes steroid molecules?
66. Explain the "omega nomenclature" used to name fatty acids.
67. What is the difference between a triglycerides constructed from three saturated fatty acids and a triglyceride constructed from the unsaturated fatty acids?
68. Which two functional groups are present in amino acids?
69. What is a zwitterion?
70. What is the isoelectric point?
71. Give the definitions of the C-terminus and the N-terminus of an amino acid.
72. What is a peptide bond? How is it related to the amide functional group?
73. What is an amino acid residue?
74. What is a polypeptide molecule?
75. What is a nucleic acid?
76. What are the two classes of nucleic acids found on Earth?
77. What is a nucleotide?
78. What is a nucleoside?

79. What are the two general classes of molecules that constitute the nitrogenous bases in nucleic acids?
80. Which nitrogenous bases are found in DNA?
81. Which nitrogenous bases are found in RNA?
82. What is a phosphoester?
83. What is a phosphodiester bond?
84. What is a polynucleotide?

# CHAPTER EIGHT

# Mirror, Mirror

## The Role of Isomers

### The Wonder of Geometry

As our story to this point has shown, there is a wonderful and rich diversity associated with chemical *structures*. The structural richness exhibited by the *functional groups* and the *biomolecules* that we surveyed in the previous chapter provides a very clear explanation for chemistry's longstanding preoccupation with the microscopic structural characteristics of our universe. But there are additional subtleties to the structure of molecules that, up to this point, we have ignored. In fact, it was not until the end of the third decade of the nineteenth century that chemists recognized this tremendously important property of our universe's chemical compounds. In 1827 Friedrich Wöhler observed that two compounds possessing the **same** elemental composition (and, consequently, having the **same** *molecular formula*) had distinctly different chemical properties! This same observation was then made several more times until, in 1830, Jöns Jakob Berzelius introduced the term **isomerism** for the phenomenon. The word **isomer** comes from two Greek words, *isos*, meaning "equal," and *méros*, meaning "part," and identifies a compound possessing a **single** *molecular formula* (see chapter 3) but whose atoms are **arranged differently in space**. As we shall soon see, the phrase *"arranged differently in space"* conceals an amazingly rich range of possibilities that makes *isomerism* one of the most important characteristics determining the chemistry of our universe.

The topic of *chemical isomerism* includes a range of complex and often subtle ideas that is responsible for the incredibly rich chemistry we observe in the world around us. The many subtle aspects of this complexity make *chemical isomerism* both a fascinating and a *very large* corner of the science. In fact, the subject is so large that we will be able to address in this text only *some* of its many aspects. Our goal is to gain an appreciation of the amazing richness introduced into chemistry by the possibility of *isomeric structures* without becoming overwhelmed by the massive amount of information that has been collected during the more than two hundred years since the time of Wöhler.

The diagram in figure 8.1 provides an overview of the major categories of isomers and their relationships.

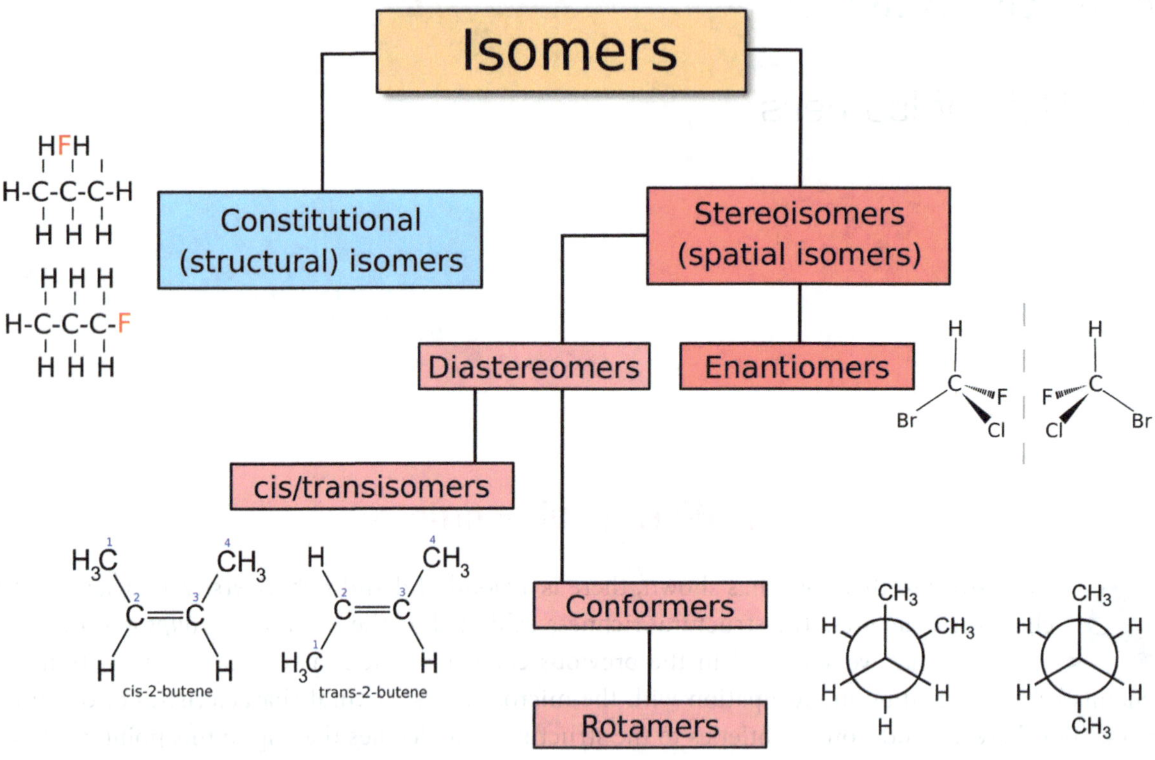


Figure 8.1: Major Categories of Chemical Isomers

While we will examine several major categories of isomers (constitutional isomers, stereoisomers, diastereomers, enantiomers, and cis/trans isomers), the remaining categories and some of the more subtle aspects of the categories we do examine must be left outside the scope of this text. There is simply no space to examine the entirety of this fascinating field, and to even attempt such an ambitious project runs the risk of obscuring key ideas. The reader will be given clear indications of those topics left for future study.

## Constitutional (Structural) Isomers

The **constitutional** (**structural**) **isomers** are molecules that have the **same** molecular formulas but **different** structural formulas; the atoms constituting these isomeric molecules are **arranged differently in space** by simply being *bonded differently to one another*. This emphasis on different molecular structures makes the members of this class of isomers very easy to visualize. Consider the simplest hydrocarbons, the alkanes, and ask about the possibilities. Because the carbon atom can form a maximum of four bonds, the number of distinct isomers increases rapidly as the number of carbon atoms in a molecule increases. Table 8.1 summarizes the number of hydrocarbon isomers for molecules having seven to ten of carbon atoms.

Table 8.1: Alkane Isomers

| Number of Carbon Atoms | Number of Isomers |
|---|---|
| 7 | 9 isomers |
| 8 | 18 isomers |
| 9 | 35 isomers |
| 10 | 75 isomers |

Figure 8.2 shows the simple carbon chains for the cases of four carbon atoms, five carbon atoms, and six carbon atoms.

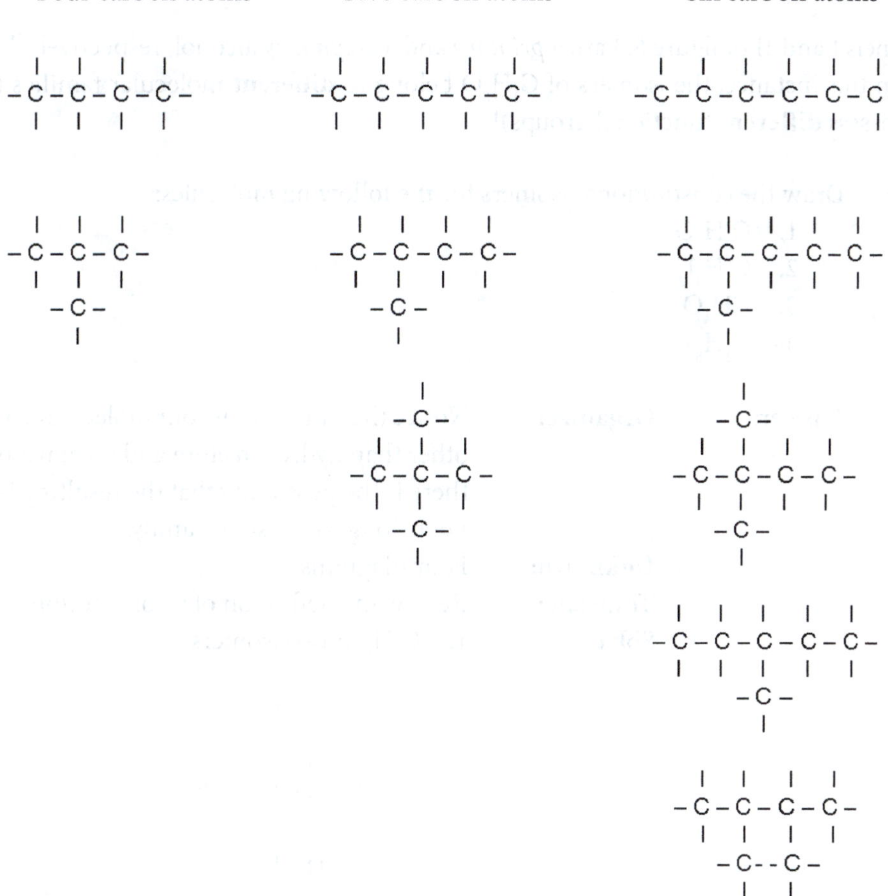


Figure 8.2: Alkane Isomers

In the examples we have considered so far, all the isomers belong to the same family of molecules. The totals in table 8.1 and the images is figure 8.2 are for *alkane* isomers. Let's now look at molecules that contain atoms *in addition to* hydrogen atoms and carbon atoms. Consider the possible isomers of the molecule whose molecular formula is $C_3H_8O$. The three possible isomer of the molecule are depicted in figure 8.3.

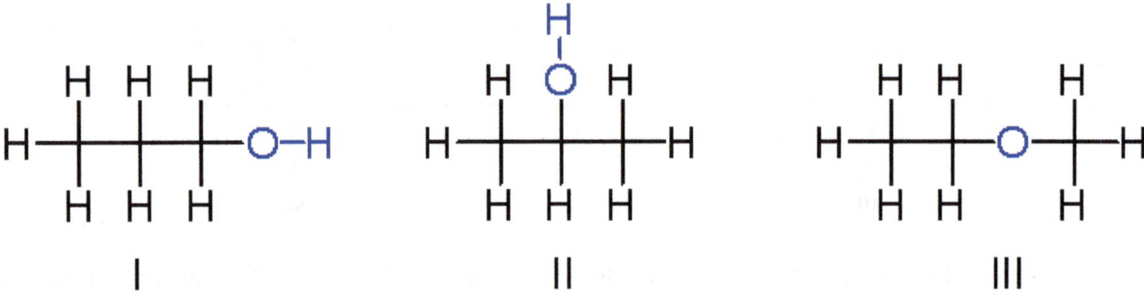


Figure 8.3: Isomers for $C_3H_8O$

Note that panels I and II of figure 8.3 are a *primary* and a *secondary* alcohol, respectively, but panel III is an *ether*. In this instance, the isomers of $C_3H_8O$ belong to **different** molecular families (that is, the molecules possess **different** functional groups)!

Example 8.1: Draw the constitutional isomers for the following molecules:

1. $C_2H_6O$
2. $C_2H_4I_2$
3. $CH_4O$
4. $C_3H_8$

Answer:

**Organize:** Notice that three of the four molecules contain atoms other than hydrogen atom and carbon atoms. Hence, there is the possibility that the resulting isomers *may not* belong to the same family.

**Unknown:** Four diagrams

**Translate:** Review the definition of a constitutional isomer.

**Solve:** 1. $C_2H_6O$; two isomers

```
     H   H
     |   |
 H - C - C - OH
     |   |
     H   H
```

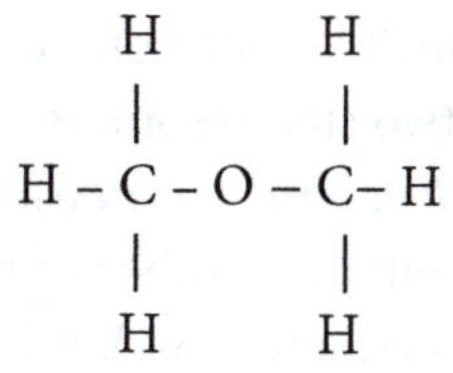

```
   H       H
   |       |
H - C - O - C - H
   |       |
   H       H
```

2. $C_2H_4I_2$; two isomers

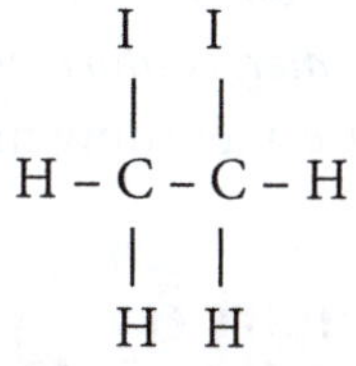

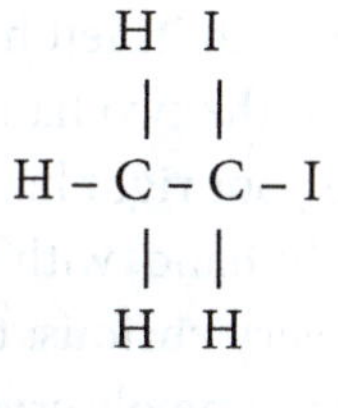

```
    I   I
    |   |
H - C - C - H
    |   |
    H   H
```

```
    H   I
    |   |
H - C - C - I
    |   |
    H   H
```

3. $CH_4O$; only one isomer

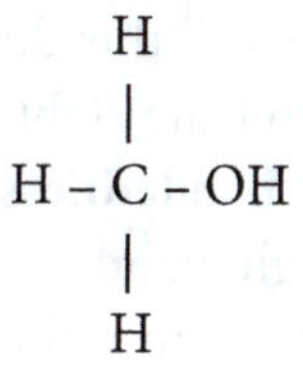

```
    H
    |
H - C - OH
    |
    H
```

4. $C_3H_8$; only one isomer

```
    H   H   H
    |   |   |
H - C - C - C - H
    |   |   |
    H   H   H
```

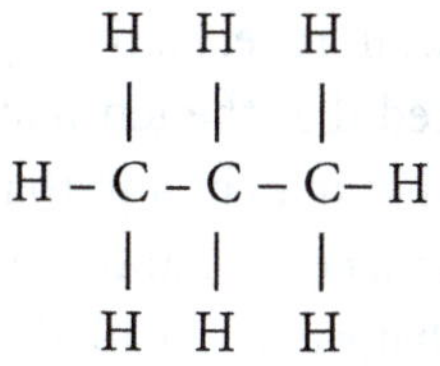

## Stereoisomers

The second major class of isomers is the **stereoisomer**; the word has its origin in the Greek term *stereós*, meaning "solid," and is a direct indication of the importance of a molecule's three-dimensional geometry. Stereoisomers are molecules that have the **same** molecular formula, have the **same** structural formula, but whose atoms are **arranged differently in space**. Note carefully that the *stereoisomers* differer dramatically from the *constitutional* isomers; while their atoms have **different spatial arrangements**, the constituent atoms of two stereoisomers are **connected in exactly the same way** (this is the meaning of the phrase *same structural formula*). In discussing the members of the stereoisomer class of molecules, we shall encounter a number of new terms: **enantiomer**, **diastereomer**, **cis/trans isomer**, **conformer**, and **rotamer** (see figure 8.1). It is important for the reader to note that these new terms are *relational* terms; they draw connections between and among the members of the stereoisomer class of molecules. These terms are not simply categories into which *individual* molecules are placed; this is the role of the two general groups: *constitutional isomers* and *stereoisomers*. In contrast, the terms *enantiomer, diastereomer, cis/trans isomer, conformer, and rotamer* apply to at least **two** and often **more than two** molecules *simultaneously*, establishing important *relationships* between and among stereoisomers.

## Left Hands and Right Hands

By examining the human body with its left hand and its right hand, the reader will note that there is a fundamental difference between the two hands, a difference that has far-reaching consequences at the most fundamental levels. Place your right hand in front of you on a flat surface, palm **down**. Now place your left hand on top of the right hand, with the palm of the left hand resting on top of the back of the right hand. Notice something very curious: the thumb of the left hand lies over the smallest finger of the right hand. Or, making the same observation but using different language to describe it, the two thumbs **do not** lie on top of one another! We have just observed that the left hand is **not superposable** on the right hand. This observation suggests that there is an **essential** or **inherent** difference between a left hand and a right hand. We call this difference **handedness**. This same difference is also evident when comparing the left foot and the right foot. They also possess the property of handedness. The existence of the property called **handedness** is the reason why a right shoe **does not fit** a left foot or why a left glove **does not fit** a right hand.

Now imagine yourself holding your right hand in front of a mirror, with the palm of the right hand **facing** the mirror. (Better yet, actually **do this** with a mirror!) Turn your left hand so that its palm **faces** you. Compare the image seen in the mirror with the left hand facing you. The two are **identical**! This means that the left hand is identical to the mirror image (that is, the image seen in the mirror) of the right hand. This also works in reverse: the right hand is identical to the mirror image of the left hand.

We have now discovered that the left hand is **not superposable** on the right hand (because, as we have seen, the thumbs do not match) and that the left hand **is identical** to the mirror image of the right hand. So we can conclude that each hand is the **nonsuperposable mirror image** of the other hand. This means that the property that we have called *handedness* implies that two objects are **handed if they are nonsuperposable mirror images of one another**!

## Chiral Molecules

Amazingly, the microscopic structures to which we give the name *molecules* also possess the property of *handedness*. A molecule that possesses a **nonsuperposable mirror image** is called **chiral**. The term *chiral* is derived from the Greek word *kheir*, which means hand. If a molecule is **not** *chiral*, we say that it is **achiral**. Consider, for the moment, a molecule with only a single carbon atom. If **four** different groups of atoms are bonded to the carbon atom, it is called an **asymmetric carbon atom**, and the molecule itself is chiral. Figure 8.4 displays a chiral molecule and its nonsuperposable mirror image molecule. The carbon atoms in both molecules are *asymmetric carbon atoms*.

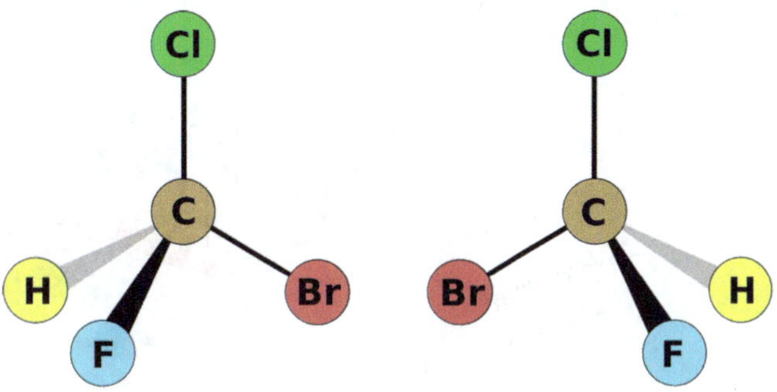


Figure 8.4: Chiral Molecules with Asymmetric Carbon Atoms

## Plane Polarized Light and Optical Activity

The *handedness* of chiral molecules is not simply an abstract geometric property that is a consequence of the arrangement of atoms in space. On the contrary, chiral molecules directly interact with **light** and produce an effect that is measurable in the laboratory. In order to explore this interaction between chiral molecules and light, we must take a short excursion and study some of the characteristics of light itself.

The phenomenon of light has fascinated humankind since antiquity, and the optical investigations of Newton and Huygens in the seventeenth century represent some of the earliest applications of scientific methodology to the natural world. By the nineteenth century, James Clerk Maxwell had achieved a monumental synthesis of electricity, magnetism, and light, providing a rigorous mathematical formulation of the wave interpretation of light that had emerged early in the century from the experiments of Thomas Young. It is important to remember that the wave interpretation of light is a **model**; that is, the wave interpretation is the analogical application of macroscopic categories of thought to a portion of our world that is fundamentally a quantum phenomenon. Consequently, we cannot say that light is a wave; rather, we can only say that the concept of a wave provides **an effective description of many (but not all!) of the observed properties of light**. Further, the description we are about to give is a particularly useful framework with which to understand the observed interaction between light and chiral molecules; it should not be construed as either the only or as the best possible explanation of the marvelous characteristics that make light such a significant part of our world.

From the perspective of Young and Maxwell, light is thought of as a *wave*; this interpretation is an analogy to water waves or the wave of a vibrating string in a string instrument (a violin or guitar, for example). The wave interpretation of light is summarized in figure 8.5.

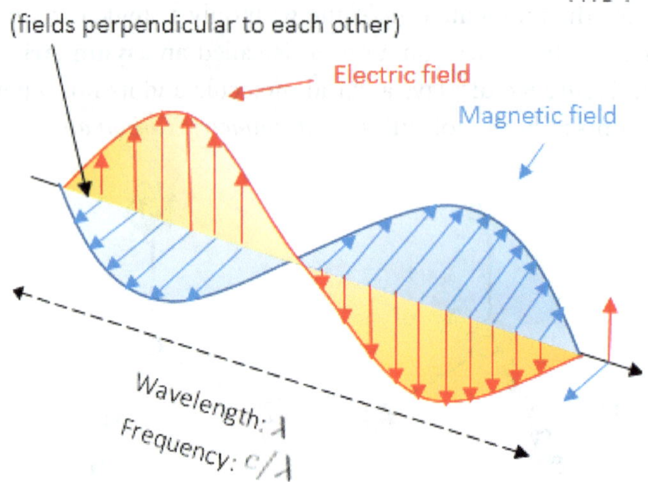


Figure 8.5: Wave Interpretation of Light

It was Maxwell who united the ideas of magnetism (the magnetic field in the figure) and electricity (connected to our earlier concept of *charge* and represented by the electric field in the figure) to explain a *light wave* as a three-dimensional disturbance in space. The direction of propagation is represented by the **black** arrow, while the electric and magnetic properties vary (notice the different heights of the **red** and **blue** arrows in the figure) in directions perpendicular (that is, the angle separating them is 90°) to one another **and** perpendicular to the direction of propagation. One complete cycle of either the electric or magnetic properties is called the **wavelength** (symbolized by the Greek letter *lambda* [λ]), while **c** is the **velocity** of light (the distance light travels per unit time; usually the time unit is *seconds*), and the quantity (**c** ÷ λ) is called the **frequency** (the number of complete cycles completed in one second).

Focusing only on the electric field component of light, figure 8.6 shows that in ordinary light (symbolized by the light bulb, labeled "1" in the figure) the electric field can be oriented in *any* direction that is perpendicular to the direction of propagation (labeled "2" in the figure). However, there are materials called *polarizers* (labeled "3" in figure 8.6), which act as *optical filters* that select one specific direction of the electric field (shown as item "4" in figure 8.6).

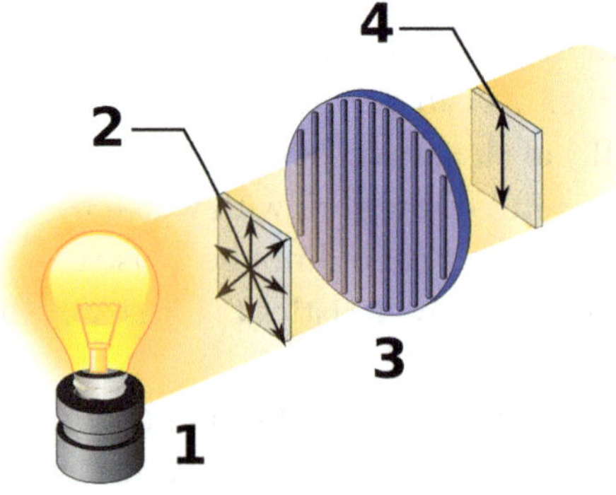


Figure 8.6: Linear Polarization of Light

Each direction is called a specific **linear polarization** of the light. (Another type of polarization, known as **circular polarization**, is also possible. We will not discuss *circular polarization.*) Figure 8.7 depicts another representation of *linear polarization* and emphasizes that the electric field continues to oscillate in the *linearly polarized light.*

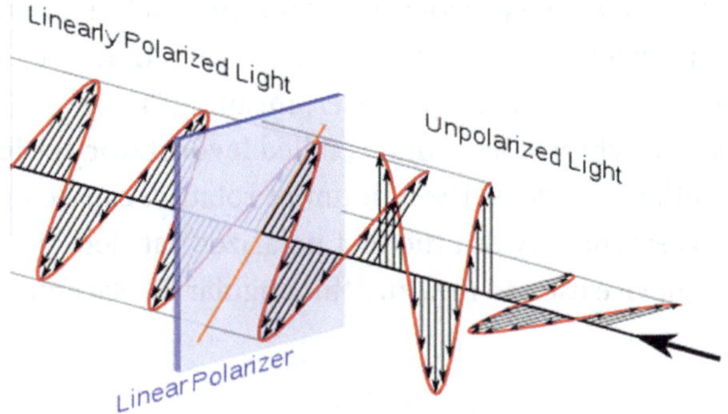


Figure 8.7: Linearly Polarized Light

While the above aspects of the wave model of light are very abstract (Maxwell's theory is highly mathematical), there are simple effects that can be easily observed. Anyone who has used a pair of Polaroid sunglasses has experienced the effect of an optical filter. By only allowing light whose electric field is oriented in a specific direction to reach your eyes, the glare of reflections is eliminated.

But what is the connection between these rather abstract ideas (even if they are observable) and chemistry? The answer to this question is provided by figure 8.8, which is an extension of figure 8.6. In the figure, the chamber labeled "5" contains either a transparent or translucent solution (that is, a solution that transmits light) of a *chiral* compound. Comparing the orientation of the electric field in the chamber (the double-headed arrow labeled "6") with its orientation after the light has passed through the polarizer (labeled "4"), the reader will note that the double-headed arrow has rotated. This is emphasized by noting the orientation of the double-headed arrow *after* the light has exited the chamber (the right side of the figure, immediately before the second polarizer, which is labeled "7"). In order to maximize the intensity of the exiting light, the observer (labeled "8") must rotate the second polarizer to match the new orientation of the light's electric field.

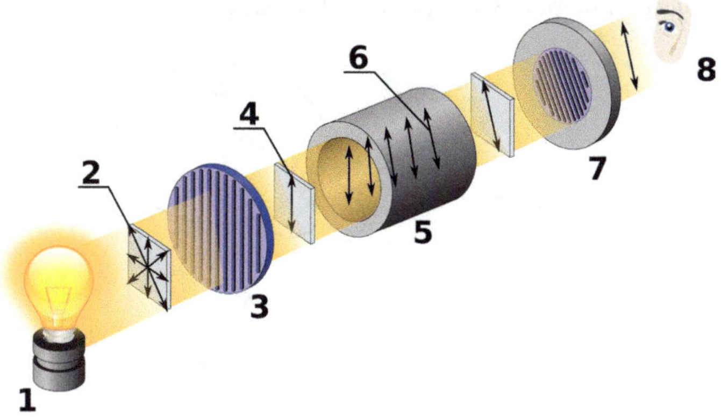


Figure 8.8: Effect on Light of a Chiral Molecule

Amazingly, **every chiral compound will rotate the electric field of plane-polarized light.** Any material that interacts in this way with light is called **optically active.** From the perspective of an observer viewing the light moving *toward* her, the solution of the *chiral* compound is capable of rotating the direction of the light's electric field either **clockwise** or **counterclockwise.** (In figure 8.8, the electric field of the light has been rotated **clockwise,** from the perspective of the observer, by the *chiral* solution in the chamber). If the electric field is rotated in a *clockwise* direction, the chiral compound is called **dextrorotatory** or **dextrorotary** (from the Latin *dexter*, meaning "right," or "on the right-hand side"); for a *counterclockwise* rotation the chiral compound is called **levorotatory** or **levorotary** (from the Latin *laevus*, meaning "left" or "on the left side"). The amount of rotation can be quantitatively measured by comparing the angular displacement between the first polarizer (labeled "4") and the second polarizer (labeled "7"). (The laboratory device used to measure this angular displacement is called a **polarimeter.**)

## Fischer Projections

Substances that are optically active are often given the designation (+) or "d" (lowercase *d*) if they are dextrorotatory or (-) or "l" (lowercase *l*) if they are levorotatory. Another system is used in chemistry (known as the **D/L system**) that is based on the molecule glyceraldehyde and a representation method called a **Fischer projection.** The cartoon presented in figure 8.9 will help us understand the conventions used in a *Fischer projection.* In the figure, the simple molecule is a single carbon atom to which are bound four *different* substituents. Consequently, the carbon atom is an *asymmetric carbon* and the molecule is *chiral* (see figure 8.4). The observer understands that the three-dimensional structure of the molecule corresponds to the diagram shown in the upper right portion of the figure: groups A and B lie **below** the plane of the page (indicated by the dashed lines), while groups C and D lie **above** the plane of the page (indicated by the wedges). This is translated to the *Fischer projection* in the lower right portion of the figure. The intersection point of the two lines (crossing at 90°) is the location of the asymmetric carbon (note that the symbol for carbon is **not** written). The **horizontal** lines passing through the asymmetric carbon correspond to the *wedges* of the upper right portion of the figure, meaning that C and D in the *Fischer projection* lie **above** the plane of the page. The **vertical** lines passing through the asymmetric carbon correspond to the *dashes* of the upper right portion of the figure, meaning that A and B in the *Fischer projection* lie **below** the plane of the page.

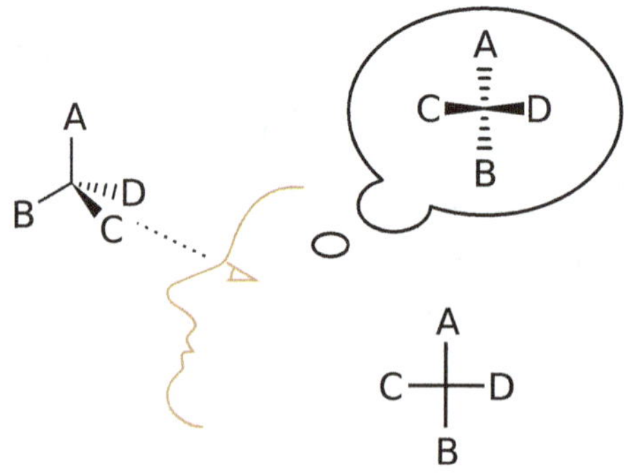

Figure 8.9: Understanding a Fischer Projection

## Glyceraldehyde and the D/L Nomenclature System

Let's now look at a Fischer projection of the glyceraldehyde molecule, a triose monosaccharide, which is the simplest of the aldose carbohydrates. It possesses a single asymmetric carbon atom, and consequently, is a chiral molecule that exhibits optical activity. Figure 8.10 depicts Fischer projections of the two nonsuperposable mirror images of glyceraldehyde. Compare figure 8.11, which shows a three-dimensional representation of D-glyceraldehyde, with its Fischer projection seen in figure 8.10.

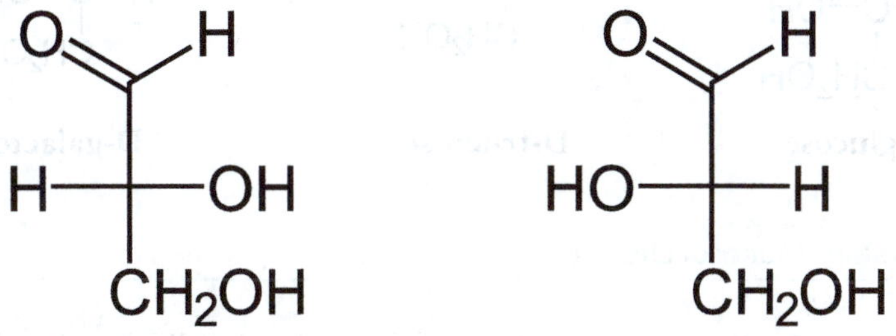

O H
H — OH
$CH_2OH$
**D-glyceraldehyde**

O H
HO — H
$CH_2OH$
**L-glyceraldehyde**

Figure 8.10: Fischer Projections of Glyceraldehyde

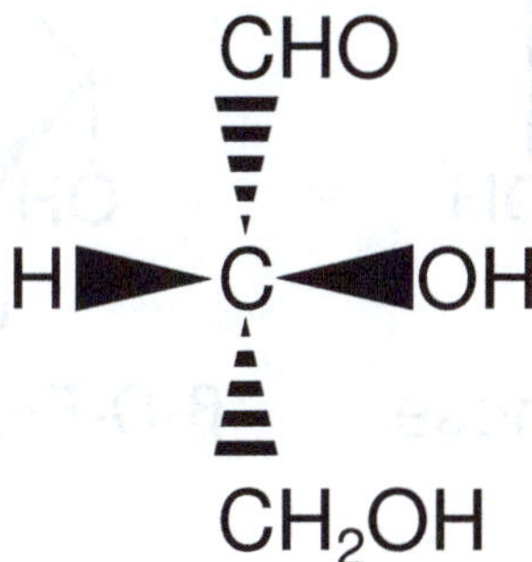


Figure 8.11: D-Glyceraldehyde

Using Fischer projections, chemists in the late nineteenth century defined the D/L nomenclature system. In the Fischer projection of a stereoisomer, first locate the **chiral** carbon atom that is **farthest from the carbonyl group**. If the hydroxyl group (*OH*) that is bonded to that carbon atom lies on the **right side of the vertical line in the Fischer projection**, the stereoisomer is designated as "D." If the hydroxyl group that is bonded to that carbon atom lies on the **left side of the vertical line in the Fischer projection**, the stereoisomer is designated as "L." Compare the designations for glyceraldehyde in figure 8.10. It was fortuitous that the glyceraldehyde geometry assigned the "D" designation in the nineteenth century is indeed the dextrorotatory stereoisomer. This was confirmed in 1951 by structural studies using X-ray crystallography.

Using the D/L system means that the names that we gave to the saccharides in chapter 7 must be amended to identify the specific stereoisomer. Look back to figure 7.19. In order to include the correct stereochemistry, the D/L names appear in figure 8.12.

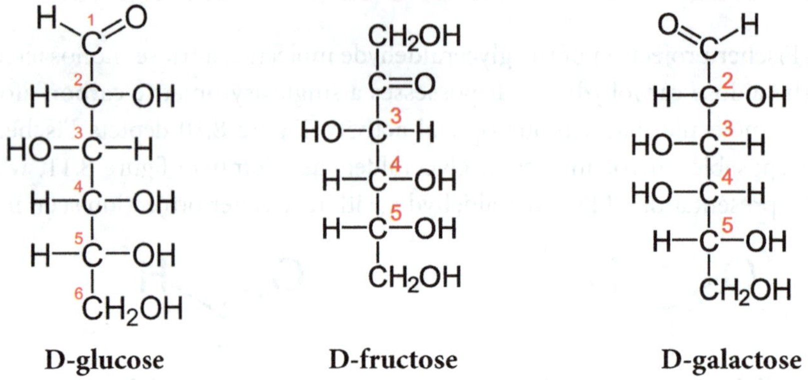

D-glucose
D-fructose
D-galactose

Figure 8.12: Key Dietary Monosaccharides

The identification of the specific stereochemistry of the molecules displayed in figures 7.25, 7.26, and 7.27 produces figures 8.13, 8.14, and 8.15.

α-D-Fructopyranose
β-D-Fructopyranose
α-D-Fructofuranose
β-D-Fructofuranose

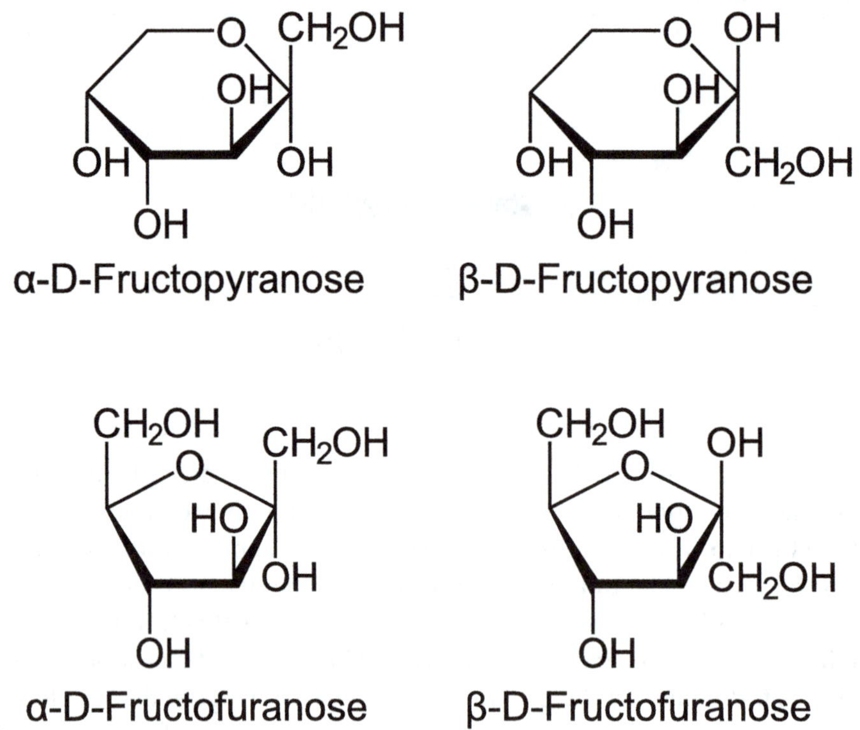

Figure 8.13: Cyclic Forms of D-glucose

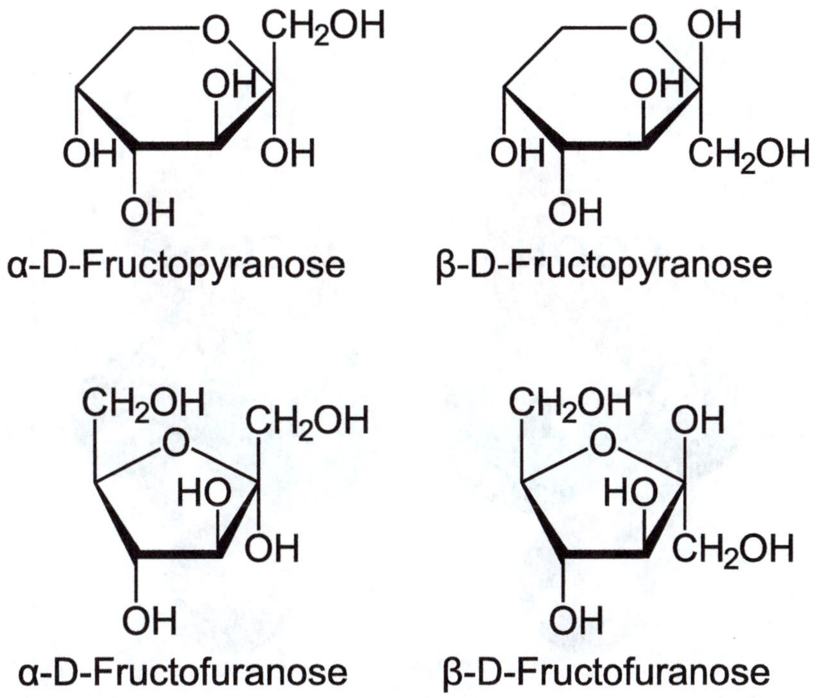

α-D-Fructopyranose β-D-Fructopyranose

α-D-Fructofuranose β-D-Fructofuranose

Figure 8.14: Cyclic Forms of D-fructose

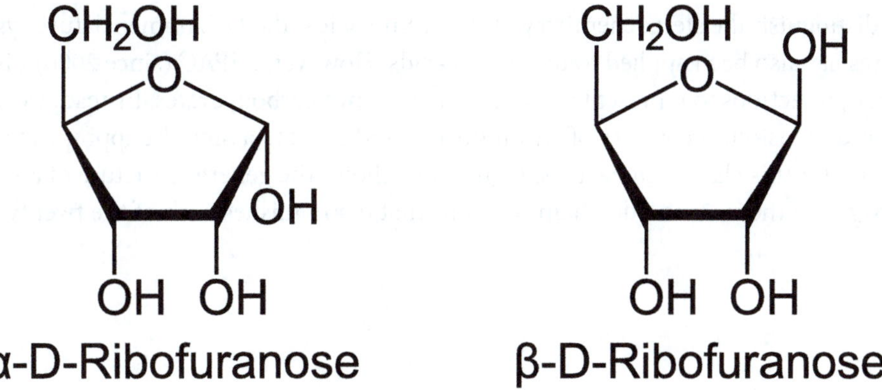


Figure 8.15: Cyclic Forms of D-ribose

## Amino Acids and Chirality

Like the monosaccharides, **α** amino acids (with the exception of glycine) are also *chiral* molecules, and consequently, occur as left-handed and right-handed versions that exhibit optical activity. Figure 8.16 depicts the chiral character of amino acids.

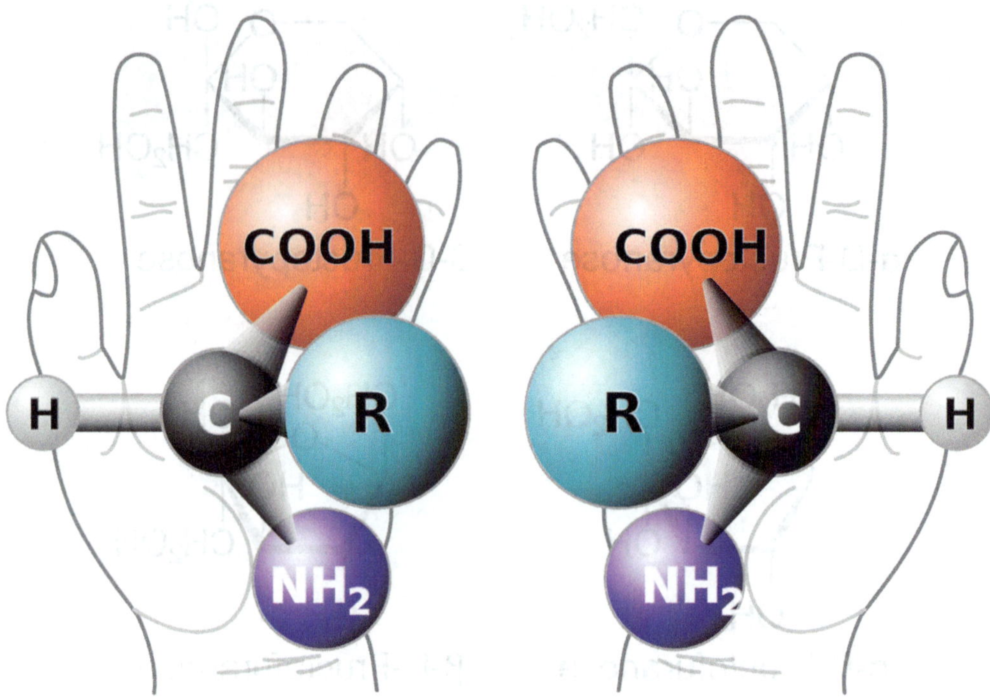


Figure 8.16: Alpha Amino Acids as Chiral Molecules

In order to distinguish the stereochemistry of the amino acids, the D/L nomenclature system used for carbohydrates has also been applied to the amino acids. However, IUPAC (since 2006) discourages the use of Fischer projections to represent molecules that are **not** carbohydrates. Instead, we'll examine the generic three-dimensional geometry of an amino acid and use it to assign the appropriate stereochemical designation for this class of molecules. Figure 8.17 shows the general structure of an α amino acid, where *R* designates the unique side-chain of atoms that distinguishes each of the twenty-two α amino acids.

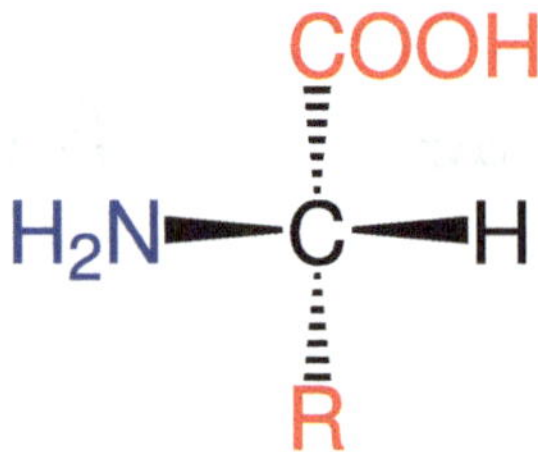


Figure 8.17: α Amino Acid Identified with the "L" Designation

A molecule possessing the stereochemistry shown the figure is identified as an *L amino acid*. In contrast, if the positions of the amino group ($H_2N$) and the hydrogen atom are exchanged, the resulting molecule (the *mirror image* of the molecule represented in figure 8.17) is designated as a *D amino acid*.

Of the twenty-two α amino acids identified on this planet, only one, glycine, is **not** a *chiral* molecule; however, because all the polypeptides (see chapter 7) responsible for life forms on this planet are constructed from *multiple* amino acids, polypeptides **are** *chiral* molecules.

## Enantiomers and Diastereomers

Two *chiral* molecules that are nonsuperposable mirror images of one another are called **enantiomers.** As we noted in our earlier discussion, the term *enantiomer* describes a **relationship between two distinct molecules**; it is a concept that must be applied **simultaneously** to a **pair of molecules.** It is a concept that **cannot** be applied to *single* molecules. Figure 8.18 displays the Fischer projections of two glucose molecules that are related as enantiomers and utilizes the D/L nomenclature system to identify their stereochemistry.

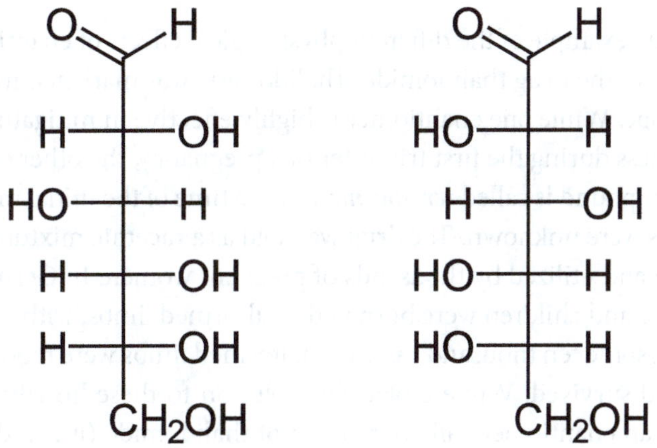

**D-Glucose** **L-Glucose**

Figure 8.18: Glucose Enantiomers

In symmetric environments (environments that are **not** *chiral*, neither right-handed nor left-handed), enantiomers have **identical** physical and chemical properties. A mixture that contains equal numbers of two enantiomers is called a **racemic** mixture and does **not** exhibit optical activity. On the other hand, in environments that **are chiral**, enantiomers exhibit **different** chemical reactivity in their interactions with other enantiomeric compounds. For example, figure 8.19 shows two enantiomers known as **carvones.** The levorotatory enantiomer is perceived by human olfactory receptors as *spearmint*, while the dextrorotatory enantiomer is perceived as *caraway.* This implies that the human olfactory receptors must contain *chiral* components! Similarly, the enantiomers of a pharmaceutical compound behave differently in a *chiral* physiological environment, leading to the use of enantiomerically pure drug formulations.

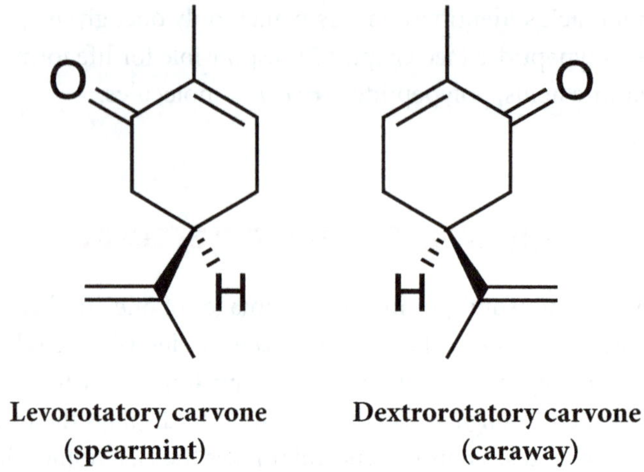

Levorotatory carvone (spearmint)

Dextrorotatory carvone (caraway)

Figure 8.19: Carvone Enantiomers

One particularly stunning example of the different physiological effects of enantiomers in a physiological environment is the story of the drug thalidomide. Thalidomide was marketed in the late 1950s and early 1960s, primarily in Europe. While one enantiomer is highly effective in mitigating the symptoms associated with morning sickness during the first trimester of a pregnancy, the other enantiomer causes severe birth defects. (Such a compound is called *teratogenic.*) At the time of the initial marketing of thalidomide, its teratogenic properties were unknown. The drug was sold as a racemic mixture of the two enantiomers (depicted in figure 8.20) and utilized by thousands of pregnant women. In Germany alone, between five thousand and seven thousand children were born with malformed limbs, with a survival rate of approximately 40%. Worldwide, some ten thousand cases of malformed limbs were reported; only approximately 50% of the ten thousand survived. While a plausible solution to these horrific numbers might appear to be the production of an enantiomerically pure form of thalidomide (it is indeed possible to produce thalidomide so that only one enantiomer is present), unfortunately, under physiological conditions, either isomer will almost immediately **racemize**. That is, any sample containing **only one** of the isomers will very rapidly convert into a 50:50 mixture of **both** isomers (a *racemic mixture*).

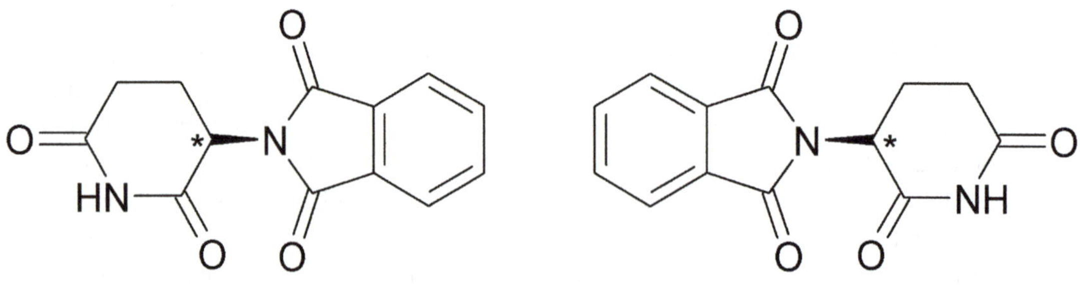


**Morning Sickness Enantiomer**

**Teratogenic Enantiomer**

Figure 8.20: Thalidomide Enantiomers

Most often the presence of an asymmetric carbon atom is sufficient to guarantee the existence of enantiomers. However, this is **not** the case in all situations. Let's look at the tartaric acid molecule. Figure 8.21 displays the Fischer projection of the tartaric acid enantiomers. Note that each enantiomer possesses two symmetric carbon atoms. The tartaric acid enantiomers are *chiral* molecules and are indeed *optically active.*

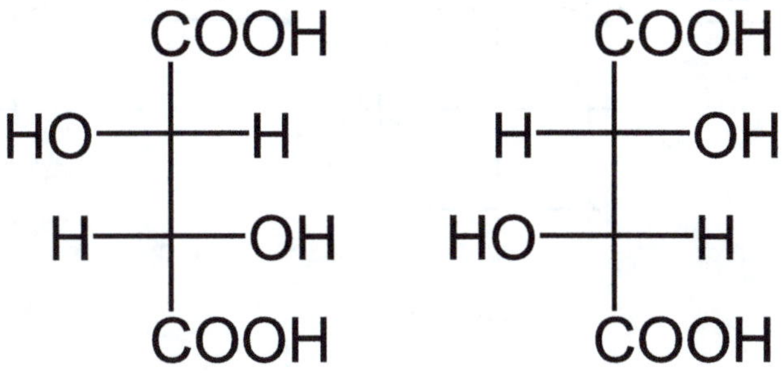

COOH
HO—H
H—OH
COOH

**D-tartaric acid**

COOH
H—OH
HO—H
COOH

**L-tartaric acid**

Figure 8.21: Tartaric Acid Enantiomers

However, there is a **third** possibility. Figure 8.22 depicts the **meso** form of the tartaric acid molecule. A **meso** compound, unlike an enantiomer, **is superposable on its mirror image**. By carefully examining figure 8.22, the reader will observe that the molecule is symmetric with respect to the dashed line. Consequently, although there are two asymmetric carbon atoms in the *meso* form of tartaric acid, the molecule is **not** *chiral* and is **not** *optically active.*

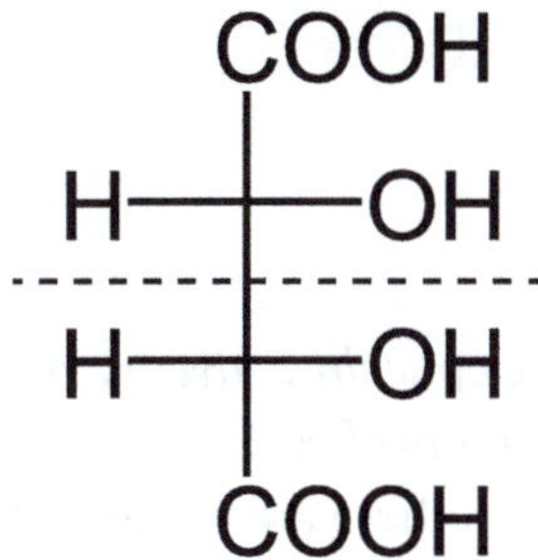


Figure 8.22: Meso Tartaric Acid

In contrast, isomers that are **not** *enantiomers* are called **diastereomers**. Again, the term implies a *relationship*; it **cannot** be applied to a *single* molecule. In figure 8.23 several diastereomers of glucose are displayed. Note that it is possible to have **more than two** diastereomers of a single isomer.

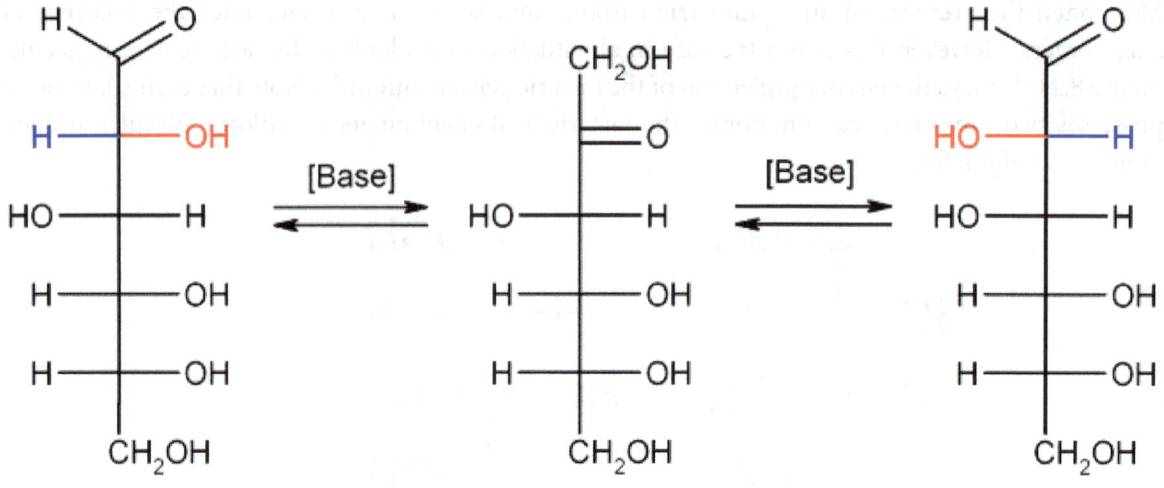

[Base]

[Base]

D-glucose

D-fructose

D-manose

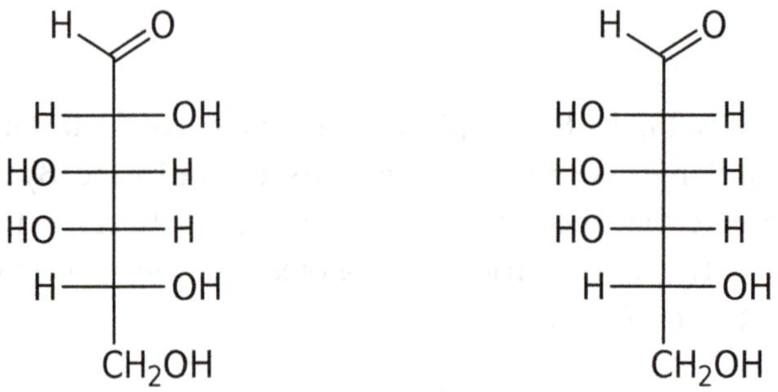

D-galactose

D-talose

Figure 8.23: Diastereomers of Glucose

The molecules displayed in figure 8.23 are *optically active*; each molecule possesses multiple asymmetric carbon atoms. While only the D-enantiomers of each molecule are displayed, there is a corresponding L-enantiomer for each molecule. However, there are many molecules that are related as diastereomers (see figure 8.1) but are **not** *optically active.* The conformational isomers (or conformers) and rotamers are molecules related as diastereomers that differ from one another only as the result of a **simple rotation** around one or more formal carbon-carbon single bonds. Such a rotation around a single bond is energy dependent, and consequently, the stability of *conformational isomers* and *rotamers* depends greatly on the energy available to the molecule. Many *conformational isomers* and *rotamers* rapidly interconvert between various diastereomers and are not considered to be **stable** isomers.

The molecule cyclohexane is an example of a common *conformational isomer* that can exist in a variety of spatial arrangements, with four configurations (see figure 8.24) being most important: **chair (1)**, **half-chair (2)**, **boat (4)**, and **twisted-boat (3 and 5)**.

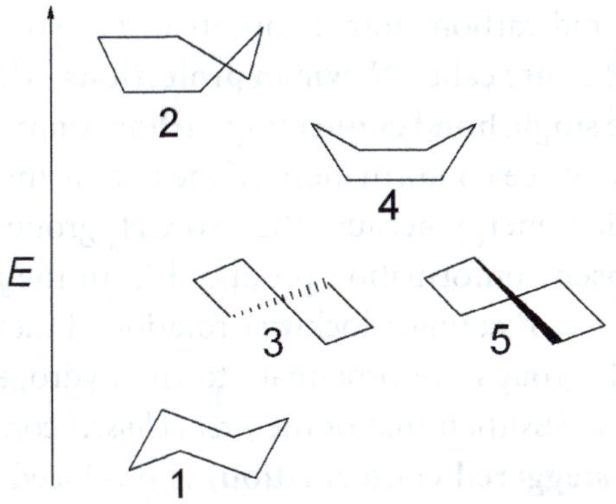

Figure 8.24: Cyclohexane Conformations

At 298.15 K (near room temperature), more than 99% of the molecules in a cyclohexane solution will be in one of two possible chair conformations. Only the *chair* and the *twisted-boat* can be isolated in pure form.

The butane molecule provides a convenient example of various *rotamers*. Figure 8.25 displays the butane molecule from two perspectives.

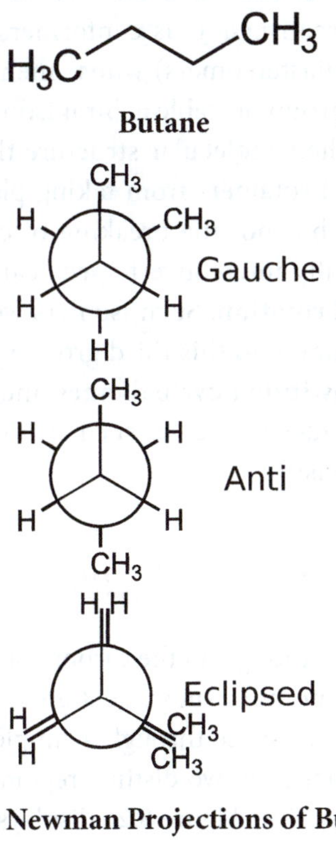

**Butane**

**Three Newman Projections of Butane**

Figure 8.25: Conformations of Butane

The upper panel of the figure displays a modified skeletal structure of butane showing carbon atom 1 with its three hydrogen atoms and carbon atom 4 with its three hydrogen atoms as $CH_3$ groups. The three lower diagrams of figure 8.25 are called **Newman projections**, which assumes that the observer is viewing the molecule **along the single bond connecting carbon atom 2 and carbon atom 3**. From this point of view, the figure displays three conformations of the butane molecule. The **eclipsed** conformation causes the *largest* interaction energy because the two $CH_3$ groups are in their closest proximity. In fact, there is a **second eclipsed** conformation produced from the *first eclipsed* conformation (the one depicted in figure 8.25) by a 120° counterclockwise rotation of the front $CH_3$ group. In this *second eclipsed* configuration, the $CH_3$ groups are proximate to the hydrogen atoms, causing a significant interaction energy, although it is *less* than that of the *first eclipsed* conformation. The **gauche** conformation (sometimes called the **staggered conformation**) is produced by beginning with the *eclipsed* conformation and performing a 60° counterclockwise rotation of the front $CH_3$ group. The interaction energy associated with the *gauche* is *less than* the interaction energies associated with **both** *eclipsed* conformations. Finally, in the **anti** conformation the two $CH_3$ groups are separated by an angle of 180° in the Newman projection. This configuration exhibits the *smallest* interaction energy.

## Cis/Trans Isomers

While a large fraction of the molecules related as diastereomers either are stable and optically active or are capable of isomeric interconversion as conformers or rotamers, there is a third group of molecules (also related as stable diastereomers) whose members are neither optically active nor conformers or rotamers. This third group of stable achiral isomers (also related as diastereomers) is distinguished by characteristics of their molecular structure that prevent even the **hindered rotation** associated with conformers and rotamers from taking place. (The term **hindered rotation** is rotational motion requiring energy but not the breaking of chemical bonds.) In effect, members of this third group **cannot** rotationally interconvert specifically because their molecular structure imposes the constraint of **restricted rotation**. Such isomers were initially called **cis/trans** isomers. While the variety of molecules belonging to this third group is very large, our focus will be on two major groups of molecules: the **substituted cycloalkanes** and the **open-chain alkenes**. The focus on the substituted cycloalkanes excludes molecules such as cyclohexane, which, as we have already seen, exhibits conformational isomerism.

## Substituted Cycloalkanes

As a result of their ring structure, even though all the carbon-carbon bonds in the substituted cycloalkanes are *single* bonds, nonhydrogen atoms or substituent groups bonded to carbon atoms in the ring are **not** free to change their positions in space through a simple bond rotation. In effect, the carbon ring of these cycloalkanes divides space into two distinct regions, one lying **above** the ring of carbon atoms and one lying **below** the carbon ring. Figure 8.26 displays typical examples of the cycloalkanes in which we are interested.

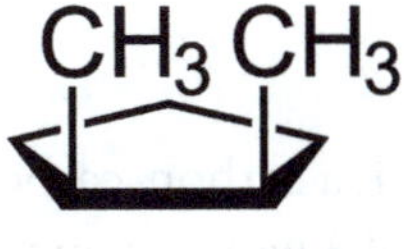


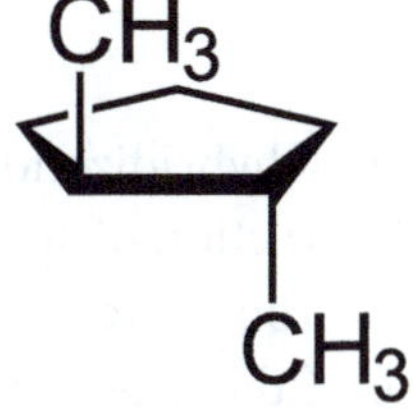


*cis*-1,2-dimethylcyclopentane *trans*-1,2-dimethylcyclopentane

Figure 8.26: Substituted Cycloalkane Isomers

Pay special attention to the $CH_3$ groups bonded as substituents to the cyclopentane rings. In the left panel of the figure, both methyl groups lie in the region of space **above** the cyclopentane ring; more generally, the methyl groups lie on the **same side of the ring**. To designate this arrangement, the prefix cis (from the Latin meaning "on the same side") is added to the complete IUPAC name of the molecule. Note that the prefix is separated from the formal IUPAC name by a dash. In print, the prefix is written in italics. On the other hand, in the right panel of the figure, one methyl group lies in the region of space **above** the cyclopentane ring, while the other methyl group lies in the region of space **below** the cyclopentane ring. The molecule displayed in the right panel of the figure **cannot** be converted into the molecule shown in the left panel **without breaking bonds**, that is, by means of a simple rotation. The two molecules are indeed different, but they are isomers because they possess the same chemical formula. To designate the configuration in the right panel of the figure, the prefix trans (from Latin, meaning "on the other side" or "across") is added to the complete IUPAC name of the molecule. Again, the prefix is separated from the formal IUPAC name by a dash. In print, the prefix is written in italics.

Example 8.2: Write the complete IUPAC names for the following molecules:

1. 2.

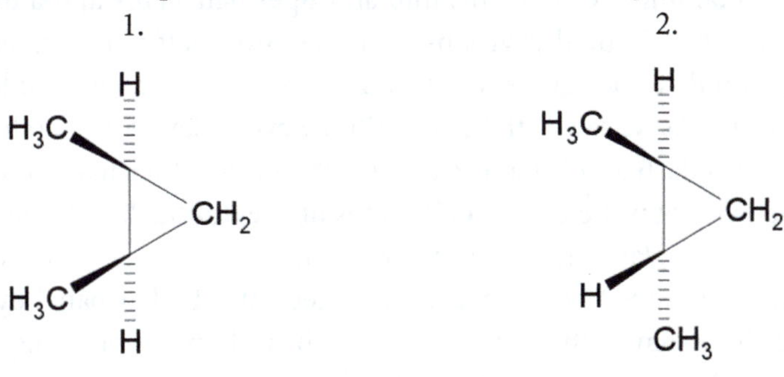


Answer:

**Organize:** Notice that both molecules are cyclic alkanes. Review the procedure for naming cyclic alkanes and pay attention to the stereochemistry given by the dashes and wedges.

**Unknown:** Two names

**Translate:** Review the use of the prefixes *cis* and *trans*.

**Solve:**
1. *cis*-dimethylcyclopropane
2. *trans*-dimethylcyclopropane

## Alkenes

As we noted in our discussion of hybridization in chapter 4, a carbon-carbon *double* bond consists of a *sigma* bond (formed by the interaction of **sp**$^2$ hybrid orbitals) with enhanced probability density **along the line** connecting the two carbon nuclei and a *pi* bond. This is formed by two **p** *orbitals* oriented perpendicularly to the plane of the molecule producing enhanced probability density geometrically **above** and **below** the molecular plane of a planar molecule. Consequently, free rotation around the axis of the *sigma* bond is impossible **without breaking the pi bond**; this means that molecules with *double* bonds (*alkenes*) are subject to the constraints of **restricted rotation**. Figure 8.27 displays two examples of alkene isomers whose chemical characteristics are determined by the restricted rotation imposed by the *double* bond.

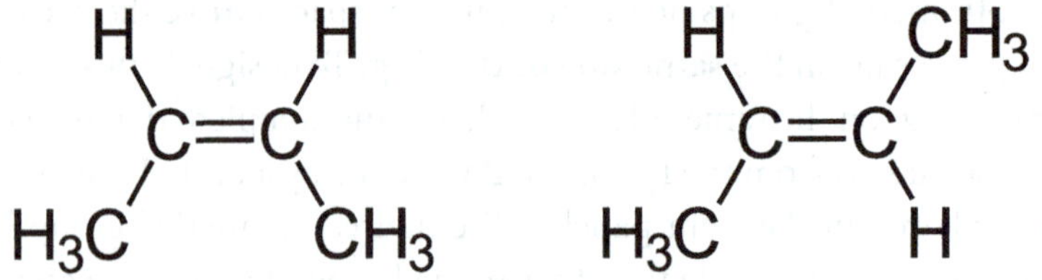

```
   H       H              H        CH3
    \     /                \      /
     C=C                    C=C
    /     \                /      \
 H3C       CH3          H3C        H
```

**cis-2-butene** **trans-2-butene**

Figure 8.27: Alkene Isomers

As noted in chapter 4, the *double* bond in an alkene enforces a **planar geometry** on six atoms: the two carbon atoms linked by the double bond and the four atoms (two at each carbon) bonded to the two double-bonded carbon atoms. Consequently, an imaginary line drawn **along the axis of the double bond** divides the plane containing the six atoms into an **upper** half-plane and a **lower** half-plane. In the left panel of figure 8.27, the two methyl groups lie in the **lower** half-plane, or, equivalently, **on the same side of the double bond**. To designate this arrangement, the prefix *cis* is added to the complete IUPAC name of the molecule. As we saw with the substituted cycloalkanes, the prefix is separated from the formal IUPAC name by a dash and is italicized in print. On the other hand, the in the right panel of the figure, the two methyl groups lie on **opposite sides of the double bond**, one in the **upper** half-plane and one in the **lower** half-plane. The restricted rotation that is characteristic of the double bond prevents the right panel of figure 8.27 from being converted into the left panel by a simple rotation around the double bond. To designate the configuration in the right panel of the figure, the prefix *trans* is added to the complete IUPAC name of the molecule. The prefix is separated from the formal IUPAC name by a dash and is italicized in print.

Example 8.3: Write the complete IUPAC names for the following molecules:

1.

```
          H   H
           \ /
           C=C
           / \
   CH3CH2     CH3
```

2.

```
       H     CH2CH3
         \   /
         C = C
         /   \
   CH3CH2     CH3
```

3.

```
   CH3CH2   CH2CH2CH3
         \   /
         C = C
         /   \
   CH3CH2     CH3
```

Answer: **Organize:** Notice that all the molecules are alkenes. Review the procedure for naming alkenes and pay attention to the stereochemistry.

**Unknown:** Three names

**Translate:** Review the use of the prefixes *cis* and *trans* as applied to alkenes.

**Solve:**
1. *cis*-2-pentene or *cis*-pent-2-ene
2. *trans*-3-methyl-3-hexene
   or
   *trans*-3-methylhex-3-ene
3. 3-ethyl-4-methyl-3-heptene
   or
   3-ethyl-4-methylhept-3-ene
   (Note: No *cis* and *trans* isomers)

Notice that *cis* and *trans* isomers are possible **only** if the carbon atoms linked by the double bond have **two different** atoms or groups of atoms bonded to them. If either carbon has identical atoms or identical groups of atoms bonded to it, *cis* and *trans* isomers are **not** possible. Also note that the **linear geometry** of a *triple* bond (**sp** hybrid orbitals) precludes the existence of *cis* and *trans* isomers. Unlike enantiomers, *cis* and *trans* isomers exhibit **different chemical and physical** properties. Often, the properties of two *cis* and *trans* isomers are very similar, but there are examples for which there are significant difference between the properties of the *cis* isomer and those of the *trans* isomer.

Finally, the use of the *cis* and *trans* designations represents an older system of designating specific diastereomers. If there are **more than two** different substituents bonded to the carbon atoms that are linked by a double bond, confusion can arise with the use of the *cis* and *trans* descriptors. A more general system, known as the E/Z notation (where "E" is from the German *entgegen*, meaning "opposite," and "Z" is from the German *zusammen*, meaning "together"), is recommended by IUPAC to avoid ambiguities in assigning names. The E/Z notational system depends on a set of **priority rules** that assign names to isomers that **may** or **may not** be consistent with the *cis* and *trans* designations. Because our goal is to identify the general principles at work in this particular class of diastereomers, we shall not explore further the details of the E/Z notational system.

# Chapter 8 Exercises

1. What is an isomer?
2. What is a constitutional (structural) isomer?
3. Draw all the isomers of an alkane hydrocarbon made up of seven carbon atoms.
4. What is a stereoisomer?
5. List five types of isomers whose definitions depend on two or more molecules simultaneously. These isomers establish relationships between and among stereoisomers.
6. What is meant by the term "handedness"? Specify the two characteristics that make an object handed.
7. Define the terms "chiral" and "achiral."
8. What is an asymmetric carbon atom?
9. In Maxwell's theory, light is composed of two components. What are they?
10. In the context of the wave theory of light, what are the frequency and the wavelength?
11. What is a polarizer?
12. How does a chiral molecule affect the orientation of the electric field in linearly polarized light?
13. What does the term "optically" active mean?
14. What is meant by the terms "dextrorotatory" and "levorotatory"?
15. What is the name of the experimental device that is used to measure the rotation of the electric field of plane-polarized light by a solution containing a chiral molecule?
16. What is a Fischer projection?
17. Explain the D/L nomenclature system used to identify carbohydrates.
18. True or false? Polypeptides are not chiral molecules.
19. Which α amino acid is not chiral?
20. What are enantiomers?
21. What is a racemic mixture?
22. What is the term that is given to a compound that causes severe birth defects?
23. Is thalidomide a chiral molecule?
24. What is a meso compound?
25. What are diastereomers?
26. How are conformers and rotamers related to one another?
27. List the names of the three rotational isomers of butane depicted in the Newman projections.
28. What are cis/trans isomers? How are they related to the concept of restricted rotation?

# CHAPTER NINE

# Reactivity

## Keys to Putting the "Somethings" Together and Taking Them Apart

### A New Focus: Chemical Reactivity

The topics discussed in chapters 7 and 8 reflect chemistry's longstanding focus on the *structural* aspects of our microscopic model of the universe. We first looked at the rich variety of structures found in both organic chemistry and biochemistry, noting that a crucial key to understanding *chemical reactivity* resides in the characteristic structures of the organic functional groups. But at the same time, we did not concentrate on the *reactive* dimensions of the organic functional groups. Further, we saw how the molecules of life, the lipids, the polysaccharides, the polypeptides, and the nucleic acids, all emerged from characteristic *structural* relationships but again saying very little about the *reactivity* responsible for the emergence of these molecular structures. In many ways this focus on *chemical structures* is the culmination of a story that began in the middle of the nineteenth century.

This *structural* theme continued in our exploration of the major classes of constitutional isomers and stereoisomers; atomic connectivity and the three-dimensional spatial arrangements of atoms determine the myriad of distinct properties of chemical compounds that we observe in our world. The incredibly rich bonding and consequent isomeric capabilities of the carbon atom have led to the awe-some wonder of the multitude of chemical reactions that we call *life*. Yet, in examining these structures, we again have not concentrated of the dynamic aspects of the chemical reactions that are central to life as we know it. It is now time to remedy our relative neglect of *chemical reactivity* by looking at a collection of key reactions that play a central role in our story.

It is important to emphasize that our discussion of *chemical reactivity* will be neither comprehen-sive nor complete. The number and variety of chemical reactions are immense. By being primarily concerned with the carbon atom (as a result of our emphasis on the organic functional groups), we are simply ignoring the rich chemistry that is characteristic of all the other known elements. However, even limiting ourselves to the reactions involving one or more carbon atoms still leaves a formidable number of chemical reactions. We will survey only a very tiny (perhaps the better word is *miniscule*) subset of all the diverse reactions in which the carbon atom is a participant. Our choice will be conditioned by our ultimate goals of understanding the reactions that are absolutely necessary for life to occur. Here, we will discover a major surprise!

In addition to lacking a comprehensive scope, our foray into the world of chemical reactions will necessarily be incomplete in the sense that we will **not** attempt to provide the mechanistic details of the reactions we do choose to study. While understanding each step of a reaction mechanism is the "holy grail" for which molecular dynamicists strive, our goal is much more modest. This viewpoint does not in any way diminish the importance of formulating an effective mechanistic model of chemicals reactions that is both descriptive and predictive. But for the purposes of this text, it is quite sufficient that the reader understand the starting point of each reaction, the central consequence of the reaction (that is, the product molecules and their signature *structures*), and, most importantly, the **common characteristics shared by the reactions we choose to study**. As the reader will soon see, these *common characteristics* provide a powerful organizing principle that connects the many diverse topics we have encountered in the course of our story.

## Condensation

A **condensation reaction** is a chemical *synthesis* in which two or more chemical species (most often, complete molecules) are joined to form a new chemical compound while at the same time producing a small molecule as a byproduct of the reaction. Its name is derived from the Latin *condensare*, meaning "press close together." Among the common small molecules generated as byproducts of a *condensation* reaction are water, ammonia, methanol, hydrogen chloride, and even acetic acid. However, in a biological environment the most common byproduct is water. In this context, *condensation* reactions are given the name **dehydration synthesis reactions** because water is, in effect, "squeezed out" (the reactants are dehydrated).

### Intramolecular Dehydration

Even though the central focus of our story will remain on the dehydration synthesis reactions and their critical role in the biochemistry of life, it is worthwhile to observe that **not all** dehydration **reactions are necessarily** condensations. **That is, there is a class of dehydration** reactions that involve only a **single** molecule (hence, they are **not** *condensations*), and they are called **intramolecular dehydration reactions**. The *intramolecular dehydration* of alcohols is one example of this class of reactions. In the late nineteenth century, the Russian chemist Alexander Zaitsev (also spelled Zaytsev, Saytzeff, or Saytseff) studied these reactions (and a more general class of reactions called **elimination reactions**) and enunciated an empirical "rule of thumb." Figure 9.1 summarizes Zaitsev's rule for two typical alcohols. For the alcohol 2-butanol (see the upper portion of the figure), Zaitsev recognized that the process of eliminating water could produce two possible products, either 2-butene or 1-butene. His experimental observations confirmed that the two products were **not** produced in equal numbers; one isomer (called the **major product**) was favored in that it was produced in larger numbers.

```
 H   OH  H   H                          H               H
 |   |   |   |                          |               |
H – C – C – C – C – H        ➔     H – C – C = C – C – H + H2O
 |   |   |   |                          |   |   |   |
 H   H   H   H                          H   H   H   H
```

**2-butanol** ➔ **2-butene (major product)**

```
                                            H   H
                                            |   |
                             ➔     H – C = C – C – C – H + H2O
                                        |   |   |   |
                                        H   H   H   H
```

**1-butene (minor product)**

```
        OH
        |
CH3CH2CHCHCH3        ➔        CH3CH2C=CHCH3 + H2O
      |                              |
      CH3                            CH3
```

**3-methyl-2-pentanol** ➔ **3-methyl-2-pentene (major product)**

```
                     ➔        CH3CH2CHCH=CH2 + H2O
                                    |
                                    CH3
```

**3-methyl-1-pentene (minor product)**

Figure 9.1: Zaitsev's Rule

Examining many such alcohol reactions (the lower panel of figure 9.1 displays the behavior of 3-methyl-2-pentanol) as well as other eliminations reactions that did not involve alcohols, Zaitsev formulated the **empirical rule** now known as **Zaitsev's rule**: "The alkene that is formed in the largest amount (the major product) is the one that corresponds to the removal of a hydrogen atom from the β carbon (the carbon atom **adjacent to** the carbon atom to which the *OH* group is bonded), which has the fewest number of hydrogen atom substituents."

Let's look more closely at figure 9.1. In the case of the upper panel in the figure (2-butanol), note that there are **two** carbon atoms adjacent to the carbon atom to which the *OH* group is bonded. One of these atoms has **three** hydrogen atoms bonded to it, while the other has only **two** hydrogen atoms bonded to it. Because the hydrogen atom that ultimately combines with the *OH* to form water must come from a β carbon atom, these two carbon atoms are the **only possible sources of a hydrogen atom**. The formation of the **major product** is the result of removing a hydrogen atom from the carbon atom having **only two** hydrogen atoms in the original alcohol. A careful examination of the lower panel in figure 9.1 (3-methyl-2-pentanol) shows that the same behavior is exhibited by this reaction.

There is a simple but highly effective way to remember Zaitsev's rule. Imagine for a moment that a carbon atom's wealth depends on the number of hydrogen atoms bonded to it. In the case of the β carbon atoms in figure 9.1, one of them has **three** hydrogen atoms bonded to it, the other only **two**; we would say that the carbon atom with two hydrogen atoms is *poor* compared to the carbon atom with three hydrogen atoms bonded to it. The carbon with only **two** hydrogen atoms loses one of those hydrogen atoms in the formation of the water molecule. In effect, it becomes *even poorer*. So we can restate Zaitsev's rule as follows: **the poor become poorer**!

Finally, in addition to observing these very specific applications of Zaitsev's rule for *elimination* reactions to alcohols, the previous examples are important because, in addition to being *elimination* reactions, they are most significantly dehydration **reactions**. They are reactions that produce $H_2O$ as a byproduct.

Example 9.1: Use Zaitsev's rule to predict the products of the intramolecular dehydration of the following molecules:

$$\begin{array}{l} \quad\quad\quad OH \\ \quad\quad\quad | \\ 1.\quad CH_3CH_2CHCHCH_2CH_3 \\ \quad\quad\quad\quad | \\ \quad\quad\quad\quad CH_3 \end{array}$$

$$\begin{array}{l} \quad\quad\quad CH_3 \quad OH \\ \quad\quad\quad | \quad\quad\quad | \\ 2.\quad CH_3CH_2CHCH_2CCH_3 \\ \quad\quad\quad\quad\quad\quad | \\ \quad\quad\quad\quad\quad\quad CH_3 \end{array}$$

3.

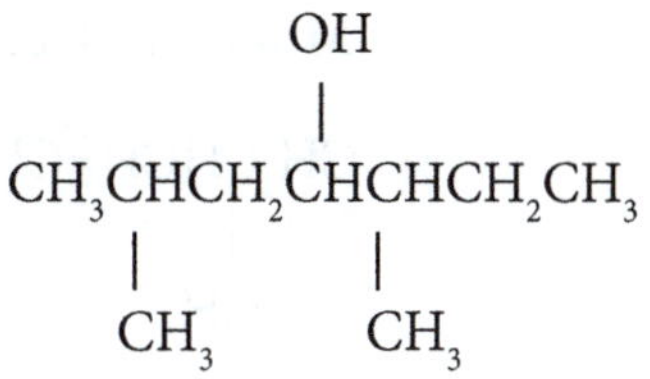

$$\begin{array}{c} \quad\quad\quad\quad OH \\ \quad\quad\quad\quad | \\ CH_3CHCH_2CHCHCH_2CH_3 \\ |\quad\quad\quad\quad | \\ CH_3 \quad\quad CH_3 \end{array}$$

Answer:

**Organize:** The three molecules are all alcohols; two are secondary alcohols, and one is a tertiary alcohol.

**Unknown:** Major and minor products for each dehydration.

**Translate:** Review Zatsev's rule.

**Solve:**

1. **Major** product:

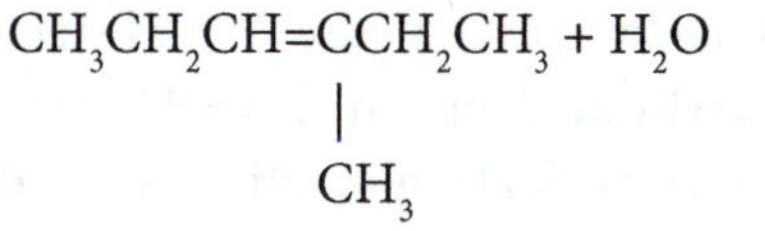

$$\begin{array}{l} CH_3CH_2CH{=}CCH_2CH_3 + H_2O \\ \quad\quad\quad\quad\quad | \\ \quad\quad\quad\quad CH_3 \end{array}$$

**Minor** product:

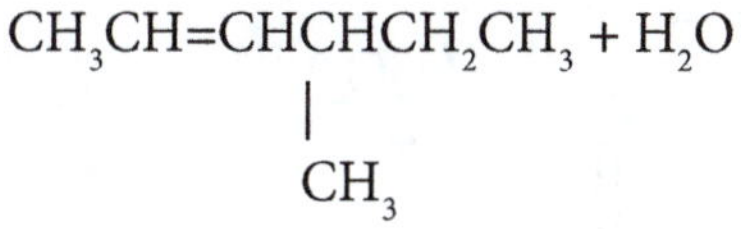

$$\begin{array}{l} CH_3CH{=}CHCHCH_2CH_3 + H_2O \\ \quad\quad\quad\quad | \\ \quad\quad\quad CH_3 \end{array}$$

2. **Major** product:

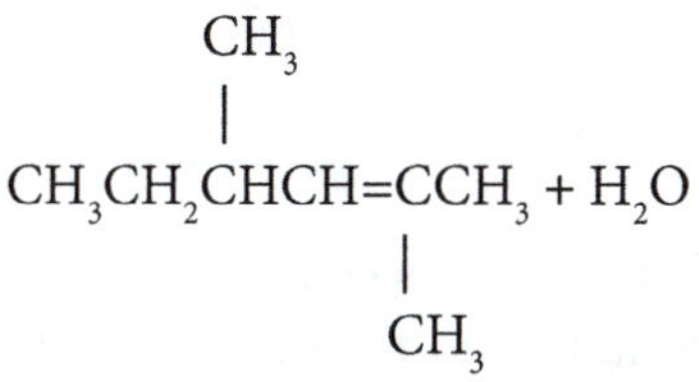

$$\begin{array}{l} \quad\quad\quad CH_3 \\ \quad\quad\quad | \\ CH_3CH_2CHCH{=}CCH_3 + H_2O \\ \quad\quad\quad\quad\quad\quad | \\ \quad\quad\quad\quad\quad CH_3 \end{array}$$

**Minor** product:

$$\begin{array}{l} \quad\quad\quad CH_3 \\ \quad\quad\quad | \\ CH_3CH_2CHCH_2C{=}CH_2 + H_2O \\ \quad\quad\quad\quad\quad\quad | \\ \quad\quad\quad\quad\quad CH_3 \end{array}$$

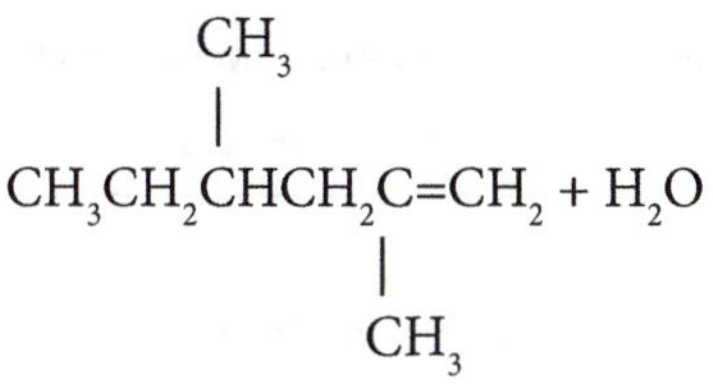

3. **Major** product:

$$\begin{array}{l} CH_3CHCH_2CH{=}CCH_2CH_3 + H_2O \\ \quad | \quad\quad\quad\quad\quad | \\ CH_3 \quad\quad\quad CH_3 \end{array}$$

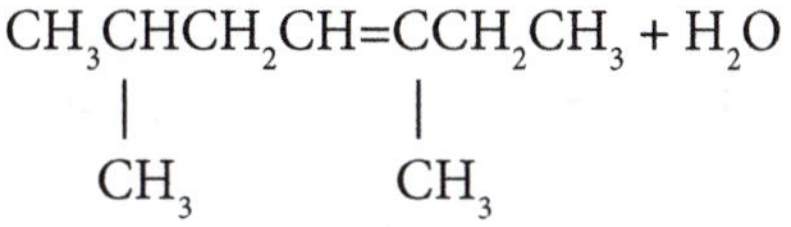

**Minor** product:

$$\begin{array}{l} CH_3\underset{\displaystyle CH_3}{\underset{|}{C}}HCH{=}CH\underset{\displaystyle CH_3}{\underset{|}{C}}HCH_2CH_3 + H_2O \end{array}$$

## Esterification: A Condensation Reaction

We return now to the broad category of *condensation* reactions, in which two or more chemical species are joined to form a new chemical compound while at the same time producing a small molecule as a byproduct of the reaction. We have already met the *ester* functional group (figure 7.39) and have seen the role it plays in several critical biochemical molecules (in particular, in triglycerides, figure 7.38, and nucleic acids, figure 7.58). Esters are formed as the product of a *condensation* reaction between a **carboxylic acid** and an **alcohol** in which a **water molecule** is the small molecule byproduct. Consequently, an **esterification reaction** is certainly a *condensation* reaction, but, more importantly, it is a **dehydration synthesis**! Figure 9.1 summarizes an *esterification* reaction. Note in particular that the reaction is indeed a dehydration; $H_2O$ is the small molecule byproduct.

$$R-\overset{\displaystyle O}{\overset{\|}{C}}-OH\;\;HO-R' \xrightarrow{\;-H_2O\;} R-\overset{\displaystyle O}{\overset{\|}{C}}-O-R'$$

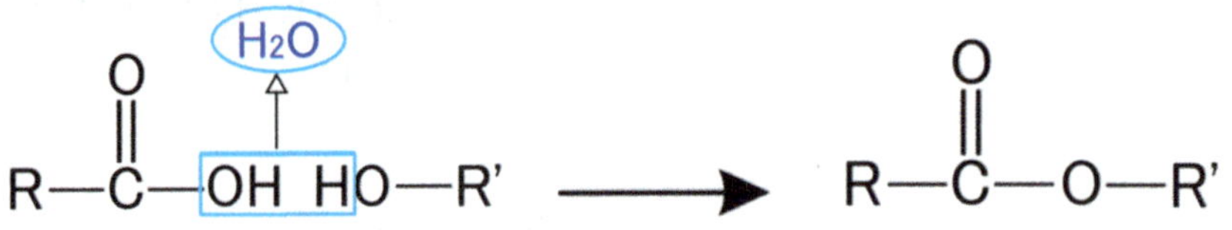

Figure 9.2: Esterification

We saw in figure 7.38 (reproduced here as figure 9.3) the formation of a triglyceride from one glycerol molecule and three fatty acid molecules. It is a dehydration reaction producing a single triglyceride molecule in which three new ester bonds are formed and three water molecules are produced as the small molecule byproducts.

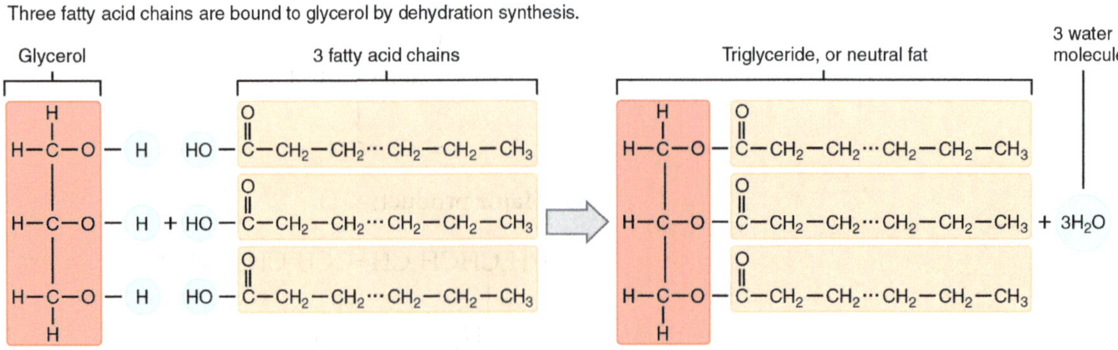

Three fatty acid chains are bound to glycerol by dehydration synthesis.

Glycerol + 3 fatty acid chains → Triglyceride, or neutral fat + 3 water molecules

$$\begin{array}{l} H-C-O-H \quad HO-\overset{O}{\overset{\|}{C}}-CH_2-CH_2\cdots CH_2-CH_2-CH_3 \\ H-C-O-H \;+\; HO-\overset{O}{\overset{\|}{C}}-CH_2-CH_2\cdots CH_2-CH_2-CH_3 \\ H-C-O-H \quad HO-\overset{O}{\overset{\|}{C}}-CH_2-CH_2\cdots CH_2-CH_2-CH_3 \end{array} \Rightarrow \begin{array}{l} H-C-O-\overset{O}{\overset{\|}{C}}-CH_2-CH_2\cdots CH_2-CH_2-CH_3 \\ H-C-O-\overset{O}{\overset{\|}{C}}-CH_2-CH_2\cdots CH_2-CH_2-CH_3 \;+\; 3H_2O \\ H-C-O-\overset{O}{\overset{\|}{C}}-CH_2-CH_2\cdots CH_2-CH_2-CH_3 \end{array}$$

Figure 9.3: Esterification Reaction Producing a Triglyceride

While the lipids in general (and triglycerides in particular) play crucial roles in the biochemistry of a wide variety of life forms on this planet, they are not the only place where the ester bond plays an essential role. We have seen that the *phosphodiester* bond, an analog of the carbon-centered ester functional group, is critical to the nucleic acids RNA and DNA. But, amazingly, esters appear numerous times in the plant kingdom. Many of the attractive flavors and fragrances encountered daily—cherry, strawberry, banana, orange, apricot, and many more—are simple esters synthesized from alcohols and carboxylic acids. Table 9.1 contains a brief list of these esters, their structures, and the associated flavors and fragrances. The list contained in the table is far from being comprehensive.

Table 9.1: Naturally Occurring Esters

| Ester Name | Structural Formula | Flavor or Fragrance |
|---|---|---|
| Butyl acetate | 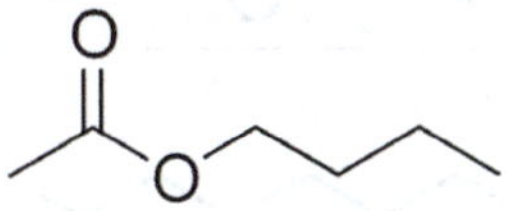 O, O | Apple |
| Ethyl butanoate | 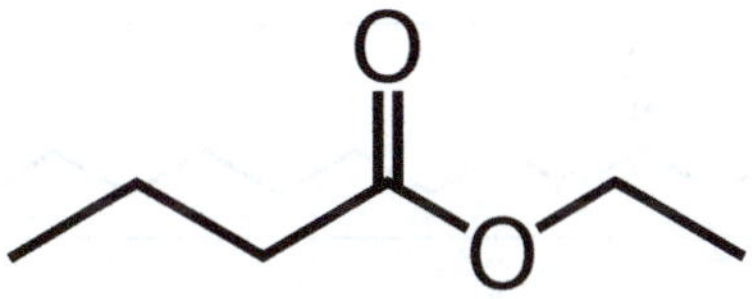 O, O | Banana, pineapple, strawberry |
| Isobutyl acetate | 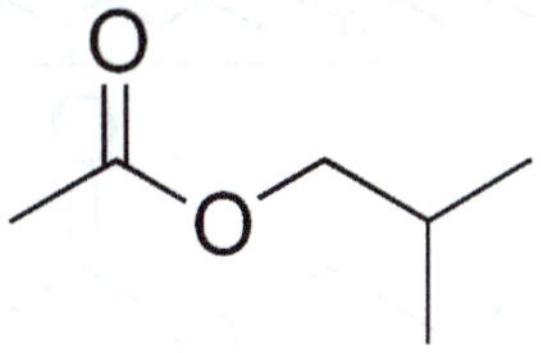 O, O | Cherry, raspberry, strawberry |
| Propyl isobutyrate | 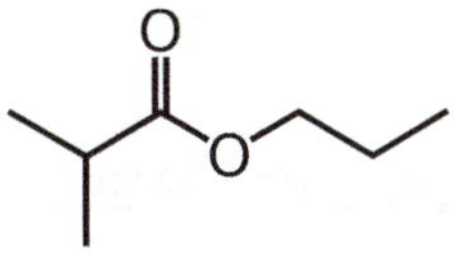 O, O | Rum |
| Propyl hexanoate | 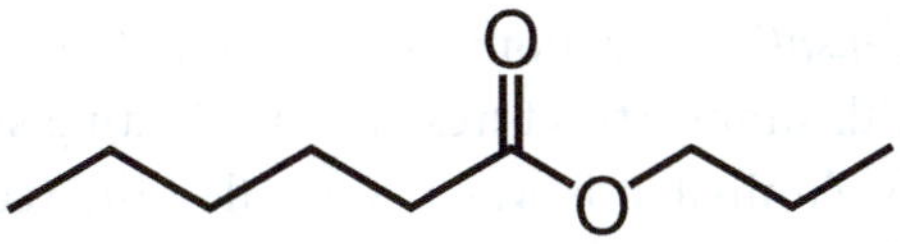 O, O | Blackberry, pineapple, cheese, wine |
| Propyl acetate | 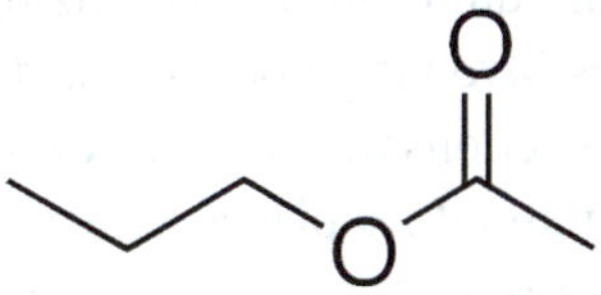 O, O | Pear |
| Methyl butanoate | 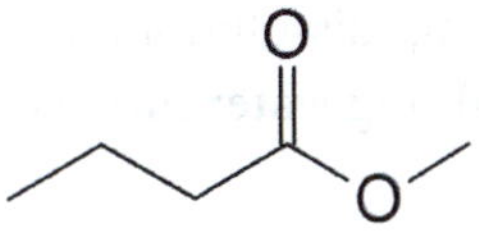 O, O | Pineapple, apple, strawberry |

Table 9.1 (*Continued*)

| | | |
|---|---|---|
| Pentyl acetate | 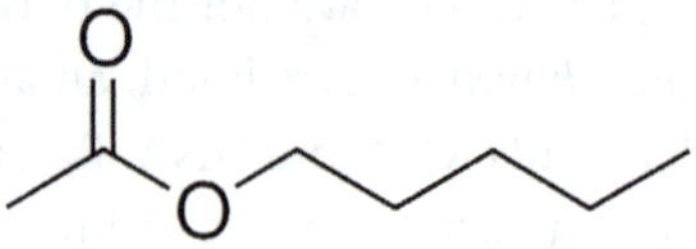 | Pear |
| Methyl salicylate | 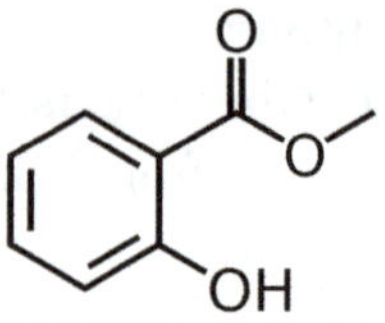 | Root beer, wintergreen |
| Pentyl butanoate | 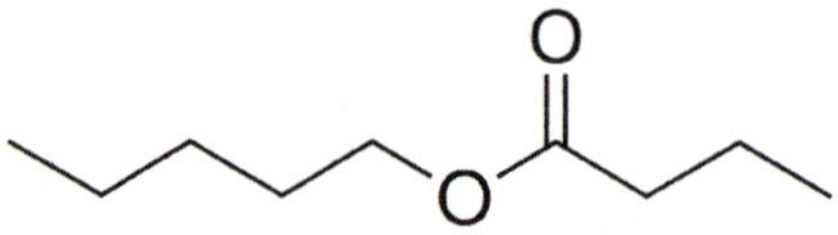 | Apricot, pineapple, pear |
| Octyl acetate | 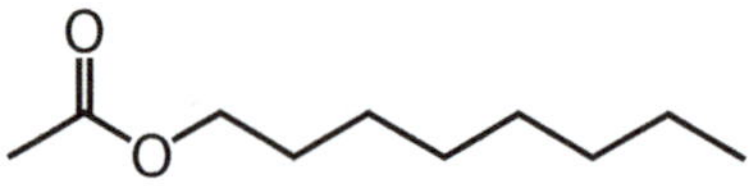 | Orange |
| Octyl butanoate | 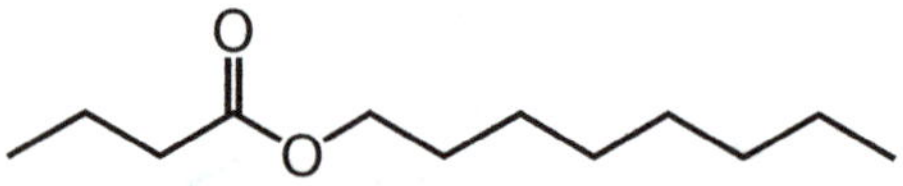 | Parsnip |
| Nonyl caprylate | 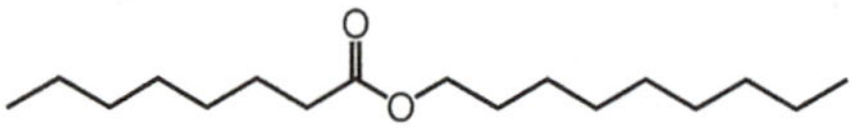 | Orange |
| Isoamyl acetate | 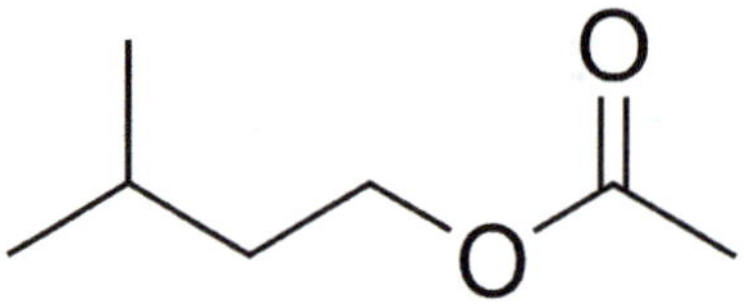 | Pear, banana |

## Hydrolysis Reactions

Our brief introduction to *condensation* reactions demonstrated the chemistry of building more complex structures by starting with simple structures and eliminating small molecules as byproducts. When the small molecule that is eliminated is *water*, we called the reaction a *dehydration* reaction. Based on our experience with chemical equilibrium, recognizing that the vast majority of chemical reactions are indeed two-way processes involving both *forward* and *reverse* reactions, it is perfectly reasonable to expect to find a group of reactions that are the **reverse** of *condensation* reactions. Indeed, such a collection of reactions **does** exist and **does** play a critical role in life processes. These reactions are the **hydrolysis reactions**; the name originates with the Greek word *hydro*, meaning "water," and the Greek word *lysis*, meaning "a loosening, dissolution, decomposition, or untying," and describes chemical reactions that cleave bonds by **adding water molecules**.

## Intramolecular Hydration

We will begin our examination of *hydrolysis* by studying **intramolecular addition** reactions (as opposed to the *elimination* reactions studied by Zaitsev). Again, in the late nineteenth century, another Russian chemist, Vladimir Vasilevich Markovnikov (almost contemporaneously with Zaitsev), studied a group of general *addition* reactions and enunciated an empirical rule summarizing the behavior he observed. **Markovnikov's rule** states, "When an *unsymmetrical* molecule such as HX, where the 'X' can be a single atom, such as *Br*, or a group of atoms, such as *OH*, is inserted at the site of a carbon-carbon double bond, the hydrogen atom from HX becomes bonded to the carbon atom that has the greater number of hydrogen atoms, and the X is bonded to the carbon with the fewer number of hydrogen atoms." As in the case of Zaitsev's rule, Markovnikov actually observed the formation of **two** products, but one of those products (the **major** product) was produced in larger numbers; however, like most empirically based rules, Markovnikov's rule is **not** absolute!

We now focus on a subset of the *addition* reactions Markvnikov studied, namely those reactions that inserted a water molecule (viewed as an *unsymmetrical* molecule, H-*OH*) at the site of a double bond in an alkene (hydration reactions). Figure 9.4 depicts a typical *hydration* that is consistent with Markovnikov's rule. For the alkene, propene (see the upper portion of the figure), Markovnikov recognized that the process of adding water at the site of the double bond could produce two possible products, either 2-propanol or 1-propanol. His experimental observations confirmed that the two

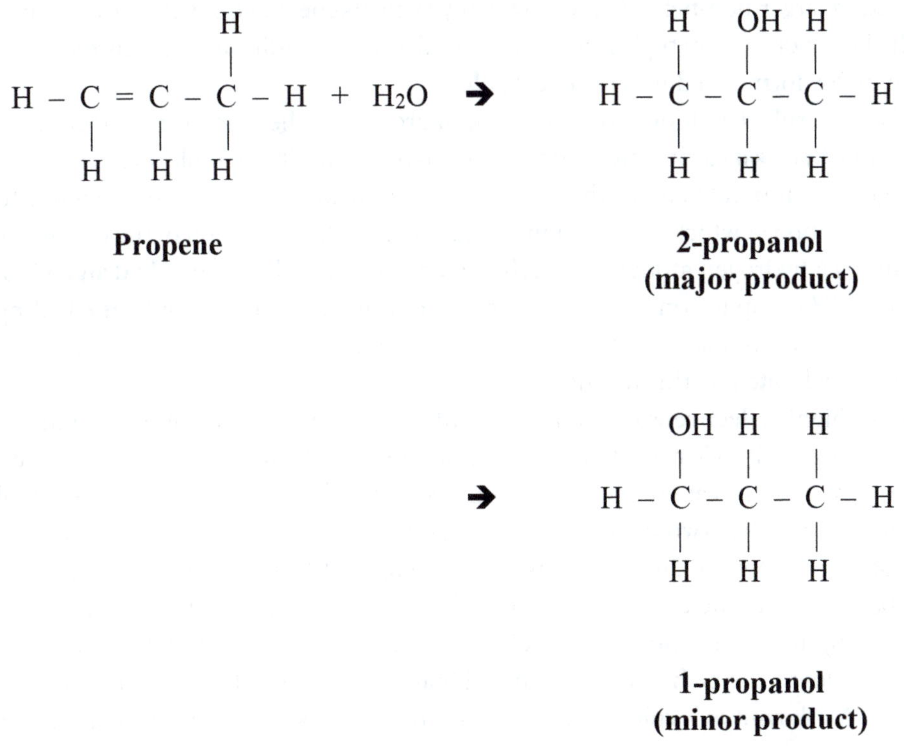

```
                H                                H    OH   H
                |                                |    |    |
H – C = C – C – H  +  H2O   ➔            H – C – C – C – H
    |   |   |                                |    |    |
    H   H   H                                H    H    H

      Propene                                 2-propanol
                                            (major product)


                                             OH    H    H
                                             |     |    |
                            ➔            H – C  –  C  – C – H
                                             |     |    |
                                             H     H    H

                                              1-propanol
                                            (minor product)
```

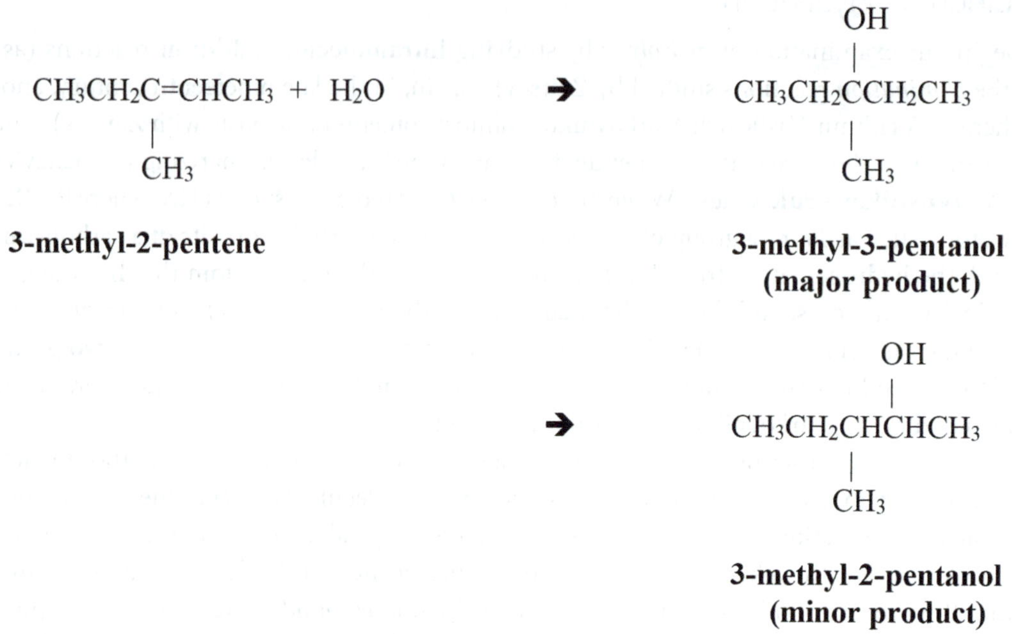

$$CH_3CH_2C{=}CHCH_3 \text{ (with } CH_3 \text{ on C3)} + H_2O \rightarrow CH_3CH_2C(OH)(CH_3)CH_2CH_3$$

**3-methyl-2-pentene** → **3-methyl-3-pentanol (major product)**

$$\rightarrow CH_3CH_2CH(CH_3)CH(OH)CH_3$$

**3-methyl-2-pentanol (minor product)**

Figure 9.4: Markovnikov's Rule

products were **not** produced in equal numbers; one isomer (called the **major product**) was favored in that it was produced in larger numbers. Examining many such alkene reactions (the lower panel of figure 9.4 displays the behavior of 3-methyl-2-pentene) as well as other addition reactions that did not involve water, Markovnikov formulated his **empirical rule**.

As we did with Zaitsev's rule, let's look more closely at figure 9.4. In the case of the upper panel in the figure (propene), note that **two** carbon atoms define the double bond in the molecule. One of these atoms has **two** hydrogen atoms bonded to it, while the other has only **one** hydrogen atom bonded to it. The double bond is the location at which the addition will take place. The formation of the **major product** is the result of adding a hydrogen atom to the carbon atom that **already has** two **hydrogen atoms bonded to it**, while the *OH* group is bonded to the carbon atom that initially had only **one** hydrogen atom bonded to it. A careful examination of the lower panel in figure 9.4 (3-methyl-2-pentene) shows that the same behavior is exhibited by this reaction.

There is a simple but highly effective way to remember Markovnikov's rule. Imagine for a moment (as we did when discussing Zaitsev's rule) that a carbon atom's wealth depends on the number of hydrogen atoms bonded to it. In the case of the two carbon atoms that define the carbon-carbon double bond in figure 9.4, one of them has **two** hydrogen atoms bonded to it, the other only **one**; we would say that the carbon atom with two hydrogen atoms is *rich* compared to the carbon atom with only one hydrogen atom bonded to it. The carbon with **two** hydrogen atoms gains an additional hydrogen atom as a result of adding the water molecule. In effect, it becomes *even richer*. So we can restate Markovnikov's rule as follows: **the rich become richer**! Finally, in addition to observing these very specific applications of Markovnikov's rule for *addition* reactions to alkenes, the previous examples are important because, in addition to being *addition* reactions, they are, most significantly, hydration **reactions**. They are reactions that add $H_2O$ to the starting alkene.

Example 9.2: Use Markovnikov's rule to predict the products of the intramolecular hydration of the following molecules:

1. 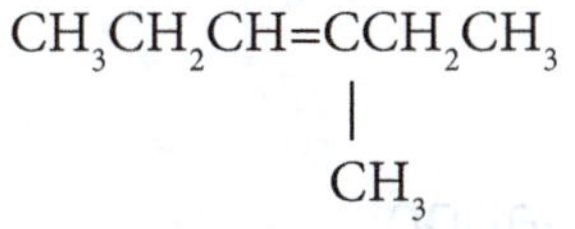

2. 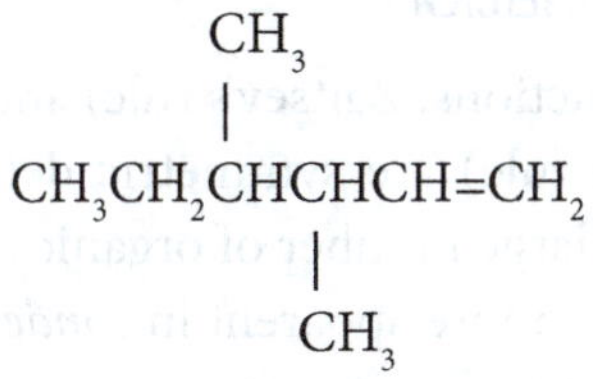

Answer:

**Organize:** The two molecules are alkenes.

**Unknown:** Major and minor products for each dehydration.

**Translate:** Review Markovnikov's rule.

**Solve:** 1. **Major** product:

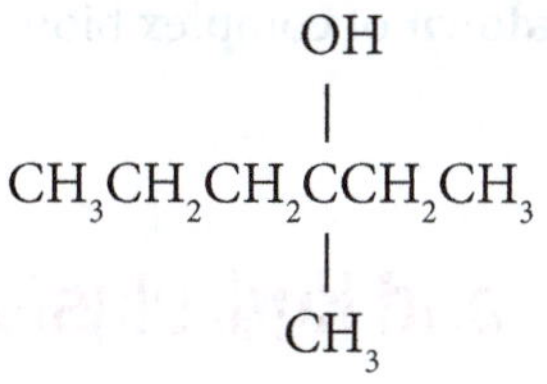

**Minor** product:

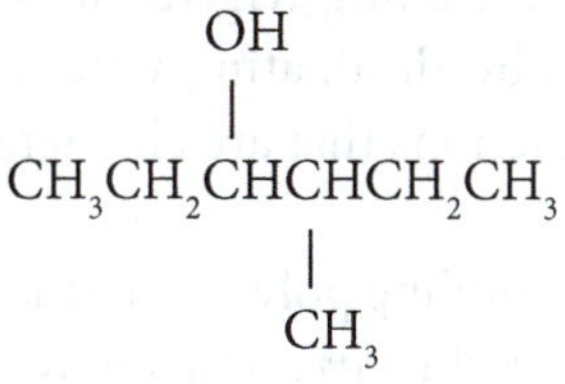

2. **Major** product:

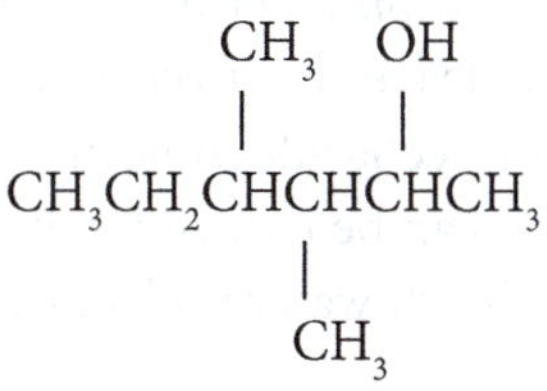

**Minor** product:

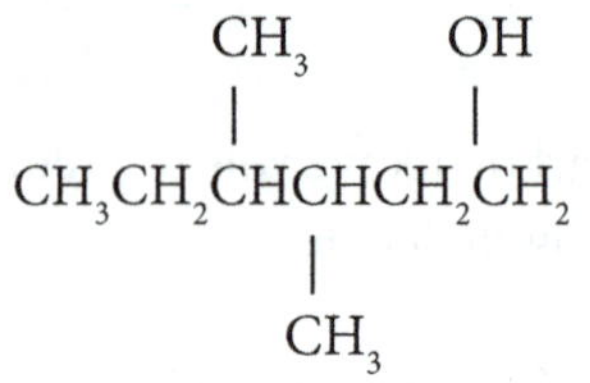

$$\begin{array}{c} \quad CH_3 \quad\quad OH \\ \quad | \quad\quad\quad | \\ CH_3CH_2CHCHCH_2CH_2 \\ | \\ CH_3 \end{array}$$

## Hydrolysis as the Counterpart of Condensation

The *intramolecular dehydration* reactions (*elimination* reactions, Zaitsev's rule) and the *intramolecular hydration* reactions (*addition* reactions, Markovnikov's rule) are symmetric demonstrations of the important role that the small molecule, water, plays in a large number of organic chemistry reactions. However, the critical significance of water becomes much more apparent in *condensation* reactions (a chemical *synthesis* in which two or more chemical species are joined to form a new chemical compound while at the same time **producing** water as a byproduct) and *hydrolysis* reactions (chemical reactions that cleave bonds by **adding** water molecules). Note the symmetric role of water: in a *condensation*, water is **produced**; in a *hydrolysis*, water is **consumed**. In effect, these two classes of reactions are symmetrically responsible for the central biochemical processes that characterize life: the **production** of complex biomolecules meeting the structural, functional, and replication requirements of living organisms and the **breakdown** of complex biomolecules to produce energy and the building blocks of new biomolecules.

# Condensation and Hydrolysis Reactions among Biomolecules

The formation of triglycerides from one glycerol molecule and three fatty acid molecules has already been highlighted as a significant example of an esterification. We now recognize that a hydrolysis is the reverse process, degrading triglycerides to fatty acids and glycerol. The two reactions, one building new molecules by eliminating water and the second decomposing molecules through the addition of water, are reactions that are characteristic of living organisms. But the story does not end here.

Recall the process of building polysaccharides from monosaccharide units. In chapter 7 we focused on the resulting *structure*: the *glycosidic* bond. Figure 7.28 (reproduced here as figure 9.5) depicted the formation of a *glycosidic* bond (and the resulting *acetal* structure) linking two monosaccharide molecules together. The reader should carefully note that this reaction eliminates water as a byproduct. It is, in fact, a *condensation* reaction! But we now know that a reverse chemical process exists; a *hydrolysis* reaction will cleave the glycosidic bond, **add** water, and produce two monosaccharide molecules from a single disaccharide molecule. A simple extension of this process makes it clear that a polysaccharide molecule can be reduced to its monosaccharide constituents by repeated hydrolysis reactions. The addition of each water molecules results in the cleavage of one glycosidic bond.

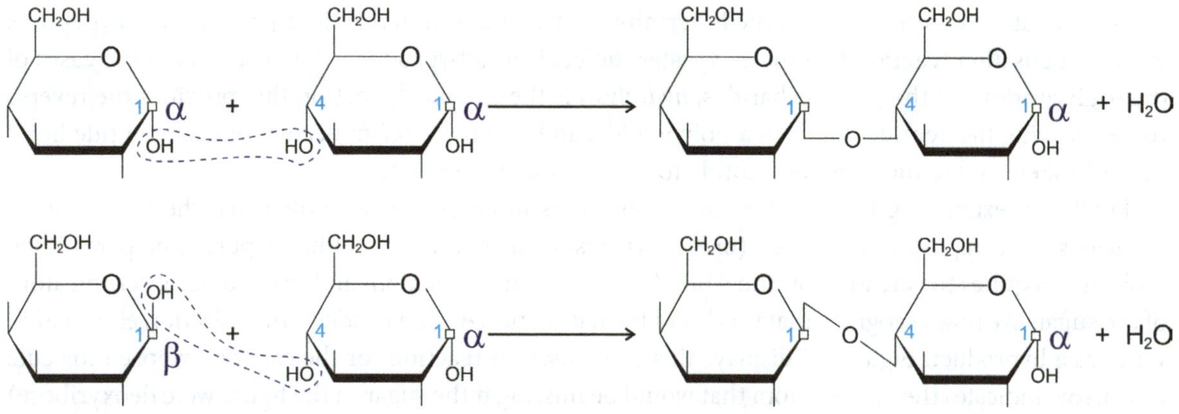

Figure 9.5: Glycosidic Bonds

In our earlier examination of amino acids and the building of *polypeptides*, our attention was focused on another *structural* characteristic of biochemistry: the *peptide* bond. In chapter 7, figure 7.49 (reproduced here as figure 9.6) displayed the formation of a *peptide* bond linking to amino acid residues in a new dipeptide molecule.

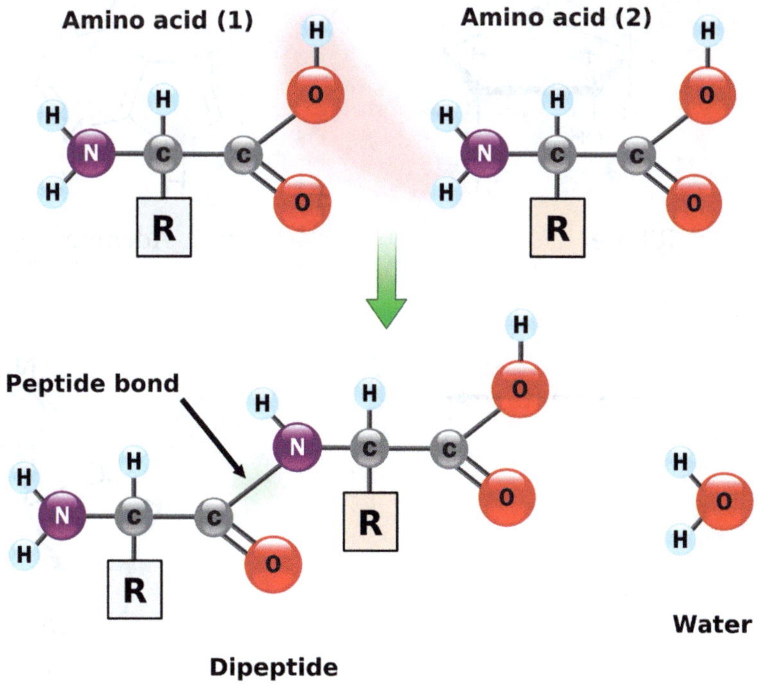


Figure 9.6: Peptide Bond Formation

However, as is clearly seen in the lower right portion of figure 9.6, a water molecule is eliminated with the formation of the peptide bond. Again, the building of a dipeptide from two amino acid residues is a *condensation* reaction whose small molecule byproduct is water! The repeated addition of amino acids either at the C-terminus or at the N-terminus via condensation reactions produces a polypeptide; each condensation reaction eliminates a water molecule as a byproduct. But, again, as in the cases of the triglycerides and the polysaccharides, *hydrolysis* is the chemical reaction that provides the reverse route, cleaving the peptide bonds in a polypeptide, adding one water molecule for each peptide bond that is broken, and reducing a polypeptide to its constituent amino acids.

Finally, in examining the *structure* of nucleic acids in chapter 7, we noted that the formation of a nucleoside, composed of a ribose (or deoxyribose) sugar linked to either a purine or pyrimidine molecule, requires the creation of a new bond between a nitrogen atom and the anomeric carbon atom of the sugar. We now recognize that this bond formation process is a *condensation* reaction eliminating water as a byproduct. Figure 9.7 displays this condensation reaction for the case of a ribose molecule (the arrow indicates the oxygen atom that would be missing if the sugar in the figure were deoxyribose) reacting with an adenine molecule and producing the adenosine nucleoside. Note that during this condensation reaction, a nitrogen atom in adenine loses a hydrogen atom, while the anomeric carbon atom in ribose loses an *OH* group. The hydrogen atom lost by the adenine molecule and the OH group lost by the sugar combine to form the water byproduct.

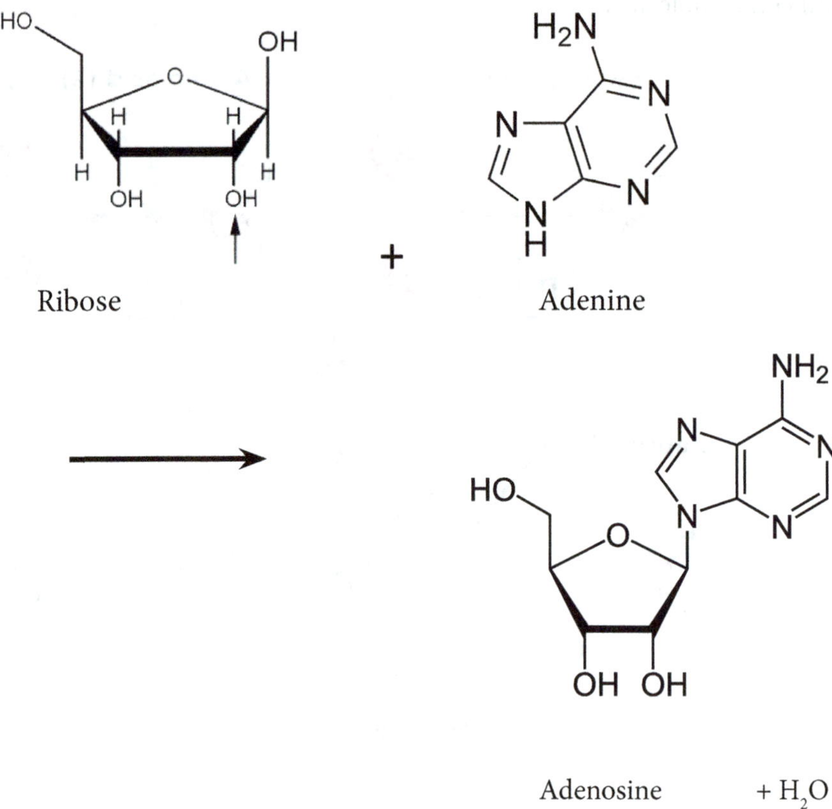


Figure 9.7: Condensation Reaction Forming the Adenosine Nucleoside

The formation of a nucleoside is only one step along the construction path of a complete nucleic acid molecule. Because a nucleic acid is a truly gigantic molecule composed of thousands of *nucleotide* components linked together, the next essential step is to build the *nucleotide* components of the molecule. This is accomplished by linking a phosphate group to a *nucleoside* through the creation of a *phosphoester* bond (an analogue to the ester functional group in which a phosphorus atom has replaced the usual carbon atom). Figure 9.8 summarizes the process for the cytidine nucleoside. Note carefully that the figure depicts another *condensation* reaction! Notice that an OH group is lost by the sugar, and a hydrogen atom is lost from the phosphate group. The phosphate is bonded to the nucleoside, and the hydrogen atom combines with the OH group to form a water molecule as a byproduct.

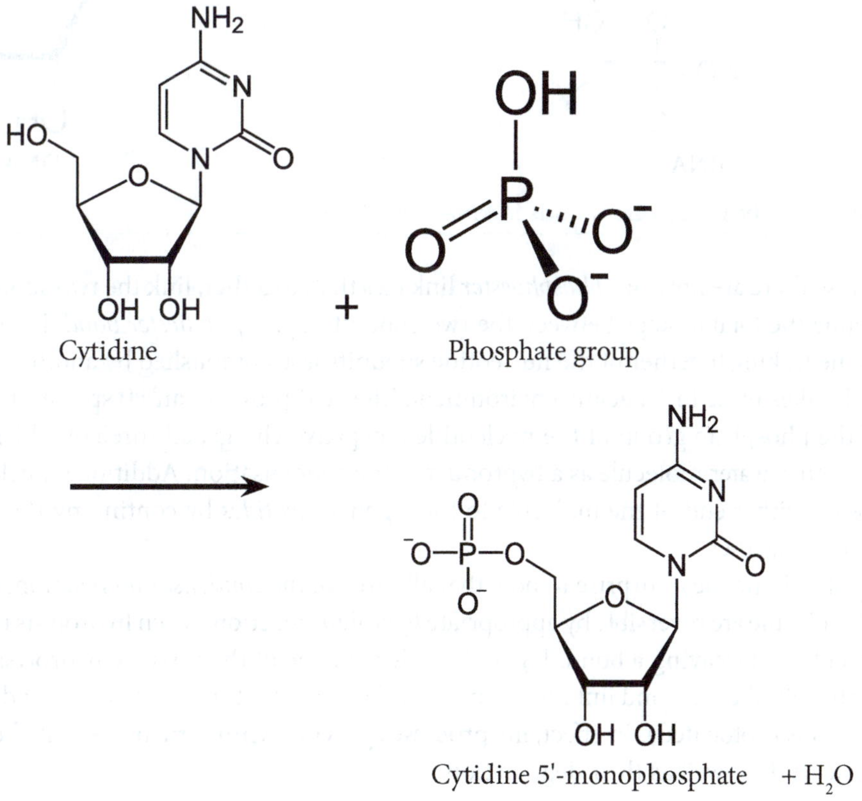

Cytidine + Phosphate group → Cytidine 5'-monophosphate + $H_2O$

Figure 9.8: Condensation Reaction Formation of Cytidine 5'-Monophosphate Nucleotide

The final step in the formation of a complete nucleic acid is the linking together of many thousands of the nucleotide subunits through the formation of a **second phosphoester** bond between a second pentose sugar and the phosphate group. Figure 9.9 (which reproduces figure 7.58) shows a schematic of two linked nucleotides in both RNA and DNA.

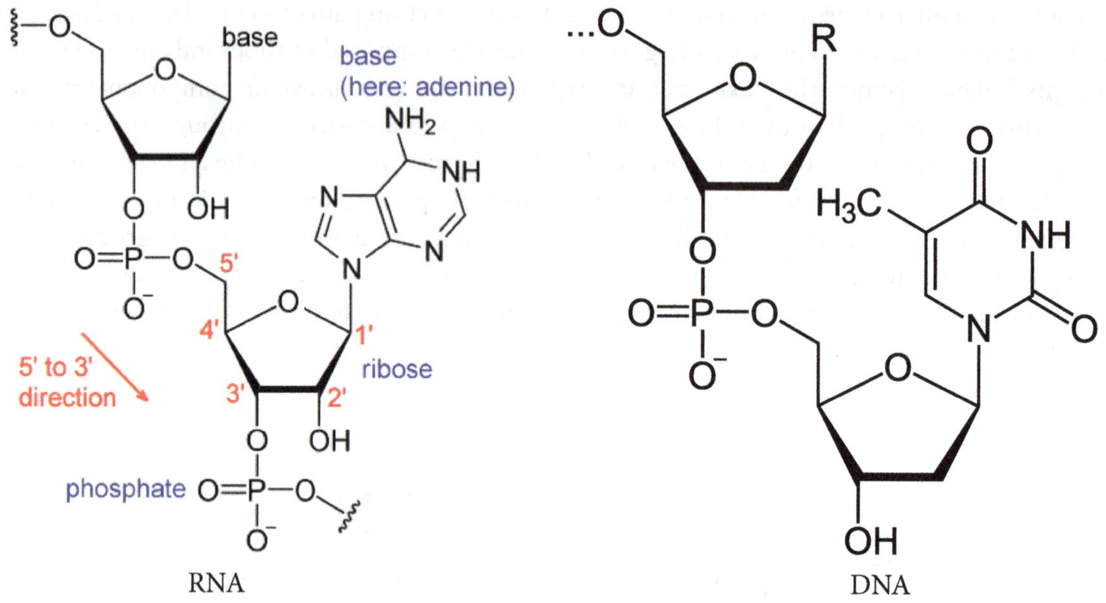

RNA DNA

Figure 9.9: Phosphodiester Bonds in RNA and DNA

Because there are now *two phosphoester* linkages that, together, link the two subunits, it is conventional to name the total linkage between the two subunits a *phosphodiester bond*. It will come as no surprise that the linking together of the nucleotide subunits is accomplished by another *condensation* reaction, which takes place in an acidic environment. The acid provides an $H^+$ species, the sugar an *OH* group, and the phosphate group of the nucleotide a negative charge, all three of which combine to produce yet another water molecule as a byproduct of the condensation. Additional nucleotide subunits can be added at either end of the molecule to form *polynucleotides* by continuing the steps used to link two subunits together.

It should not be a surprise to note that all three of the *condensation* reactions that produce a nucleic acid molecule are reversible by appropriate *hydrolysis* reactions, each hydrolysis reaction adding a water molecule and cleaving a bond. Figure 9.10 depicts one of the possible hydrolysis reactions that cleave the phosphodiester bond linking the nucleotides (note that the nucleoside produced by the hydrolysis reaction is protonated). In effect, life processes provide symmetric paths that allow the construction of nucleic acids as well as their degradation.

## Reactions in Retrospect

The thrust of this chapter has been to link the *structural* focus that dominated the science of chemistry since the beginning of its modern era with an understanding of the fundamentals of *chemical reactivity* that are ultimately responsible for the life processes so characteristic of this planet. While we have explored only a vanishingly small subset of the total collection of chemical reactions, it has become magnificently clear that both *condensation* reactions and *hydrolysis* reactions are central to the biochemistry of life. Central to both of these processes as they unfold in the realm of biochemistry is either the *elimination* or the *addition* of water. (In fact, the name *hydrolysis* implies this.) Consequently, it should not seem remarkable that water is the most abundant molecule in the human body, that water makes up between 50% and 75% of the mass of an average human being, and that approximately 70% of the planet's surface is covered by water. This chapter has demonstrated the crucial links among these reactions, the water molecule, and the fundamental processes that characterize life as we know it.

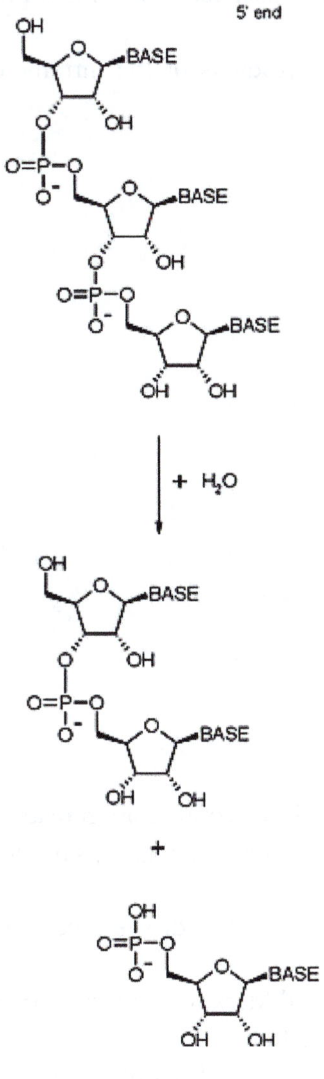

Figure 9.10: Hydrolytic Degradation of an RNA Nucleic Acid

# Chapter 9 Exercises

1. What is a condensation reaction?
2. List five small molecules that are byproducts of condensation reactions.
3. What is the connection between condensation reactions and dehydration reactions?
4. How do intramolecular dehydration reactions differ from condensation reactions?
5. Using Zaitsev's observations, distinguish between a *major* product and a *minor* product.
6. State Zaitsev's rule.
7. Use Zaitsev's rule to predict the products of the intramolecular dehydration of the following molecules:

a.

OH
|
$CH_3CHCH_2CHCH_2CH_3$
|
$CH_3$

b.

$CH_3$ OH
| |
$CH_3CH_2CHCH_2CCH_3$
|
$CH_3$

c.

OH
|
$CH_3CHCHCH_2CH_2CHCH_3$
| |
$CH_3$ $CH_3$

8. What is an esterification reaction? Is it a condensation reaction? If esterification is a condensation reaction, molecules from which two functional groups participate in the reaction?
9. What is a hydrolysis reaction?
10. What small molecule participates in every hydrolysis reaction?
11. State Markovnikov's rule.
12. True or false? Both Zaitsev's rule and Markovnikov's rule are empirical rules.
13. What does the term "hydration reaction" mean?
14. Use Markovnikov's rule to predict the products of the intramolecular hydration of the following molecules:

a. $CH_3CH_2CH_2C{=}CHCH_3$
|
$CH_3$

b.

$$\begin{array}{c} \quad\quad CH_3 \\ \quad\quad | \\ CH_3CH_2CHCHCH{=}CH_2 \\ \quad\quad\quad | \\ \quad\quad\quad CH_3 \end{array}$$

15. Explain the symmetric relationship between condensation reactions and hydrolysis reactions.
16. Explain how a polysaccharide can be decomposed into its constituent monosaccharides by repeated hydrolysis reactions.
17. What type of reaction is responsible for the formation of a polypeptide?
18. What type of reaction cleaves a polypeptide molecule into its constituent amino acid molecules?
19. What type of reaction is responsible for the formation of a nucleoside?
20. What type of reaction is responsible for the formation of a nucleotide? What is the name of the bond that is formed?
21. What type of reaction links nucleotides together to form a nucleic acid molecule (a polynucleotide)?

# CHAPTER TEN

## Energy, Entropy, Gibbs Free Energy, Intermolecular Forces, and Kinetics

### The Links between Structures and Reactivity

In the previous three chapters (chapters 7, 8, and 9) the themes of *structure* and *reactivity* have dominated our discussion; a central conclusion emerging from this discussion was the intimate interconnection between these two themes. Reactivity in chemistry is significantly determined by structural considerations (the *functional groups*), yet the crucial structures of biochemistry are the consequences of key chemical reactions! However, in addition to these dominant thematic considerations, there are several key factors that significantly condition the reactivity of a chemical system and the structures that emerge over time. An exploration of these factors will now lead to brief excursions into thermodynamics, chemical kinetics, and the role of intermolecular forces in chemistry. Each of these branches of science are both very extensive and very complex, making a complete and detailed account of any one of them an impossibility for the purposes of this text. Our goal is to examine several salient concepts in each of these areas that play critical roles in our understanding of chemical *structures* and chemical *reactivity*.

## Thermodynamics

The science of **thermodynamics** traces its origins to the work of von Guericke in the mid-seventeenth century; this was followed by the invention of early steam engines by Newcomen and Watt in the eighteenth century at the beginning of the Industrial Revolution in Europe. In 1824, Sadi Carnot (sometimes called the "father of thermodynamics") published a systematic analysis of the thermal and energetic characteristics of steam engines as well as a determination that the engines of Newcomen and Watt (English inventors) significantly exceeded the efficiency of French designs. The term *thermodynamics* was first used during the nineteenth century and originated from two Greek words: *therme*, meaning "heat," and *dynamis*, meaning "power"; it is indicative of the science's initial preoccupation with the ability of steam engines to produce useful mechanical work as efficiently as possible.

During most of the first 150 years of thermodynamic investigations, attention was focused on the role of *energy* in a **thermodynamic system**. As we noted in chapter 1, there are two types or kinds of energy: *kinetic energy* (energy associated with a mass in motion) and *potential energy* (the energy associated with a mass as a result of the mass's position in space and the action of a conservative force). The central role played by *energy* is reflected in the identification of energy conservation (a principle we also met in chapter 1) as the **first law of thermodynamics**; this conservation principle understood a thermodynamic system's energy to be the sum total of all the *heat* added to or removed from the system and the *work* done by or on the system. While thermodynamic systems were initially identified with specific mechanical devices (such as steam engines), as the science matured, it was realized that a *thermodynamic system* can be any part of the physical universe one chooses to study. In fact, **entirely abstract idealizations** (mathematical models *representing* a portion of the physical universe) can be imagined as *thermodynamic systems* and can be studied with the tools of the science.

It took some time for the science of thermodynamics to understand the two quantities, *heat* and *work*, that constitute the energy of a thermodynamic system. The concept of *work* was defined in chapter 1 as a *force* (defined by Isaac Newton) *acting through or over a distance*. As we saw earlier, this definition has a precise mathematical definition, and the units used to measure a quantity of *work* are *identical* to the units used to measure a quantity of *energy*. On the other hand, the concept of *heat* was initially treated as a **material** substance capable of direct manipulation, much like the pouring of a liquid from one container to another. The quantity of *heat* was defined as the *measurable physical quantity* of a substance called **caloric**. Only after a period of sustained effort by a number of investigators that included the invention of the kinetic molecular theory of gases was the concept of *heat* recognized as an **energy transfer process**. Specifically, when the *energy* of a thermodynamic system **changes as a consequence of a temperature difference between the system and the surroundings** (where the term **surroundings** designates everything in the actual or idealized universe that is **not part** of the *thermodynamic system*), we say that *energy* has been transferred as **heat**. There are two key points to remember here: (1) *heat* is an energy **transfer process**, and (2) there **must** be a **temperature difference** between the thermodynamic system and everything else in the universe that is **not part** of the system (surroundings). Because the kinetic molecular theory of gases identifies *temperature* as a **quantitative measure** of the *average kinetic energy* of molecules, *a temperature difference* between a thermodynamic system and its surroundings means that there is a difference between the *average kinetic energy* of the molecules that belong to the system and the *average kinetic energy* of the molecules that belong to the surroundings. Only when such a difference exists is it possible to observe the phenomenon of *heat*.

## Enthalpy

In order to study the phenomenon of *heat* on our planet, thermodynamicists (the individuals who use the science of thermodynamics to study our world) began by choosing to analyze *idealized* processes (technically termed **reversible processes**) under the assumption that the pressure (see chapter 1) of the thermodynamic system remains **constant**. (The choice of an *idealized* process might seem strange, but in the realm of thermodynamics this approach offered the advantages of both *simplifying* the mathematical effort and providing a model that represented *the best possible result*. The real world could then be compared to *this best possible result.*) While the assumption that the pressure remains **constant** may seem like an unreasonable constraint, it is important to remember that the Earth's atmospheric pressure at any given location changes relatively slowly. To a very reasonable approximation, processes that require only fractions of a second, seconds, or minutes to reach completion (most often, a state of equilibrium) do so under conditions for which the atmospheric pressure is effectively **constant**. This means that the vast array of mechanical and chemical processes that take place on our planet occur under conditions very nearly approximating constant pressure. The *idealizations* created to characterize mechanical and chemical processes that occur under constant pressure conditions thus provide an approach to understand the real processes occurring on our planet. A **change** in the thermodynamic variable (more precisely, the mathematical function) called the **enthalpy** (symbolized by an uppercase **H**) provides a quantitative measure of the *heat* associated with a **constant** pressure physical process.

Just as with mechanical energy, we emphasize that it is **only** possible to measure **differences** in the energy of two different states of a thermodynamic system. Focusing specifically on chemical reactions, the two states we shall consider are the **reactants** (the starting chemical species) and the **products** (the chemical species produced by the reaction). Because the enthalpy change is a quantitative measure of the *heat* associated with a constant pressure physical process, and *heat* itself is an energy transfer process, an *enthalpy change* is measured in energy units. Literally millions of chemical processes on this planet, chemical reactions that are called inorganic (generally not involving the carbon atom), organic, and biochemical, all occur at **constant** pressure. In all of these cases, then, an energy transfer occurs between the reacting system and its surroundings, and the experimental observable associated with this energy transfer is a measured **temperature difference**. The energy transferred between the system and its surroundings is called the **heat of reaction** associated with each chemical process.

For these constant pressure processes, the heat of reaction (the energy that is transferred and identified by a measured temperature difference) is equated with **the enthalpy change**. Equation 10.1 indicates the calculation of an enthalpy change for a thermodynamic system composed of reactants and products. The temperature difference is the experimental observable associated with the enthalpy change:

$$\Delta_R H = H_{\text{products}} - H_{\text{reactants.}} \tag{10.1}$$

The symbol, $\Delta_R H$, which is read as "delta *H* of reaction" (the symbol Δ is the Greek letter *delta*) denotes the enthalpy change that occurs during a chemical reaction as the reactants are transformed into products. The quantity $\Delta_R H$ can be positive, it can be negative, or it can be zero. A chemical reaction is called **exothermic** (more generally, **exoergic**) if $\Delta_R H < 0$; a reaction for which $\Delta_R H > 0$ is called **endothermic** (**endoergic**). An *exothermic* reaction is recognized by an **increase** in the measured temperature, while an *endothermic* reaction is signaled by a **decrease** in the measured temperature. If $\Delta_R H = 0$ (no change in the measured temperature), the reaction is called **thermal neutral**. The energy relationship of reactants and products for an *exothermic* reaction is shown in figure 10.1. The figure plots the relative energy of

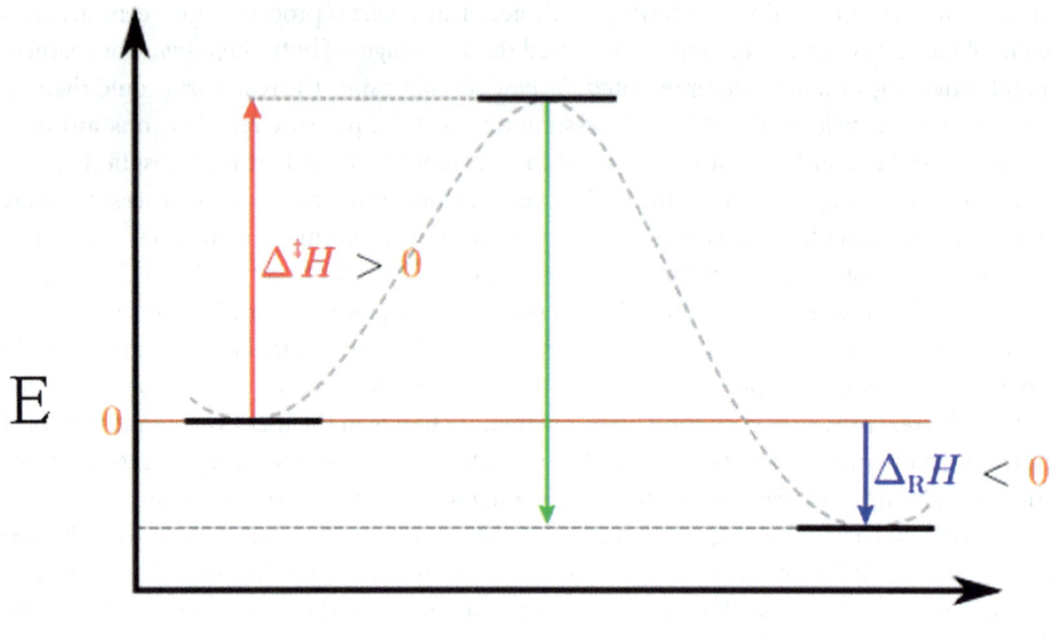


Figure 10.1: Exothermic Reaction Diagram

reactants and products along the vertical axis, while the reaction proceeds from left (reactants) to right (products) along the horizontal axis, which is identified as the reaction coordinate. The figure uses the symbol $\Delta_R H$ to represent the change in enthalpy accompanying a chemical reaction. Figure 10.2 depicts the corresponding energy diagram for an *endothermic* chemical reaction. Note that in figure 10.1 the reactants are in a **higher** energy state than the products, while in figure 10.2 the reactants are in a **lower** energy state than the products.

Both figures include two additional energy quantities: the first is indicated by the symbol $\Delta^{\ddagger}H$ which, identifies the magnitude of an energy barrier separating the reactants from the products; the second is designated by the green arrows, which identify the energy differences between the top of the energy barrier and the energy of the products for the two reactions. The energy barrier is often called the **activation energy** and represents, in the context of the most elementary interpretation, the minimum quantity of energy required for a chemical reaction to occur. (In both figures, the activation energy is formulated in terms of the enthalpy. We shall see shortly that there is a more encompassing definition of the energy that should be used to describe a reaction's activation energy.) In effect, the energetic barrier that separates reactants from products is the most basic physical constraint that prevents reactants from proceeding unimpeded (in the case of an *exothermic* reaction) to products; it is the reason why reactants that are linked exothermically to products continue to exist in our universe.

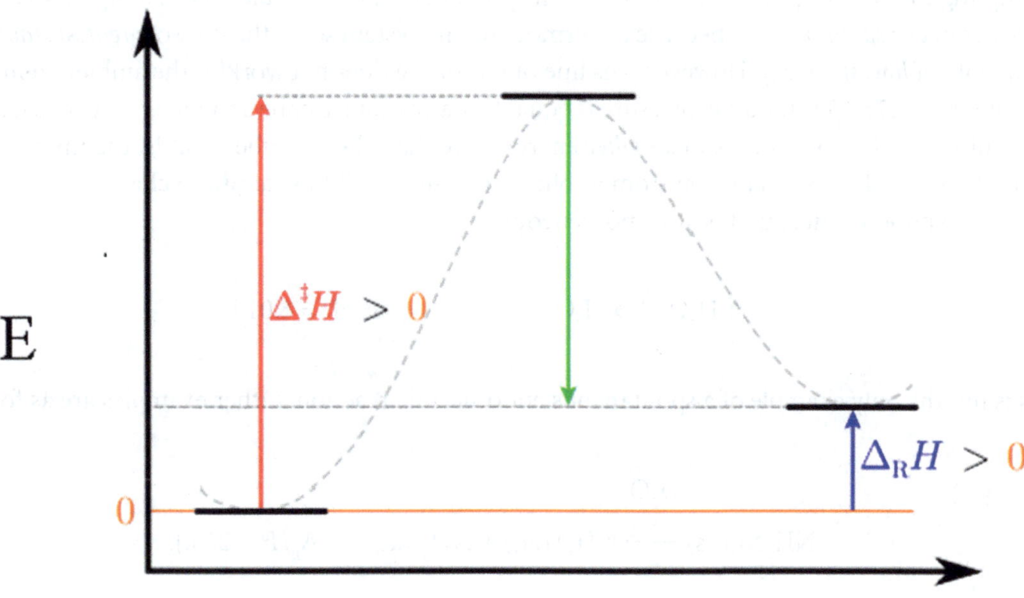


Figure 10.2: Endothermic Reaction Diagram

## Spontaneity

However, despite the apparent existence of *activation energy barriers*, there are processes in our world that seem to require no additional energy, or, equivalently, no additional work (because *energy* and *work* have the same units and, as such, are related aspects of the universe that bring about transformations). Examples are everywhere: water runs downhill, sugar dissolves when mixed with water, water freezes if the ambient temperature is **at or below** 273.15 K, and ice melts if the temperature is **above** 273.15 K (both processes at a pressure of 1 atm). We call such processes **spontaneous processes**. More formally, a **spontaneous process** is a reaction that simply **occurs** or **takes place** under a given set of conditions (i.e., without *useful* work being done, where *useful* work means work for which the forces acting over a distance are **not** the result of pressure and volume changes in the system). The most crucial aspect of a spontaneous process is that it has a *preferred* direction. If, under a set of specified conditions, a reaction **does not occur**, the reaction is said to be **nonspontaneous**.

## Entropy

Curiously, until the middle of the nineteenth century, the science of thermodynamics was silent about *spontaneous processes*. Because the energy conservation principle is at the heart of thermodynamics, it is tempting to use energy as a means to explain spontaneity. For example, one is tempted to argue that processes occur *spontaneously* because a thermodynamic system seeks the state of *greatest stability*, that is, the state of *lowest* energy. However, this line of reasoning does not work! If the ambient temperature is greater than 273.15 K (and the pressure is equal to 1 atm), an ice cube spontaneously melts; however, the resulting puddle of water is in a *higher* energy state than the ice cube. That is, the melting process (a simple physical phase transition from a solid to a liquid; we'll look at **phase changes** in more detail shortly) is endothermic, but it is also spontaneous.

$$H_2O(s) \rightarrow H_2O(l) \qquad \Delta_R H^\circ = 6.01 \text{ kJ.} \tag{10.2}$$

This is not the only example of a spontaneous endothermic reaction. Other examples are as follows:

$$NH_4NO_3(s) \xrightarrow{H_2O} NH_4^+(aq) + NO_3^-(aq) \qquad \Delta_R H^\circ = 25 \text{ kJ.} \tag{10.3}$$

and

$$NH_4Cl(s) \xrightarrow{H_2O} NH_4^+(aq) + Cl^-(aq) \qquad \Delta_R H^\circ = 20 \text{ kJ.} \tag{10.4}$$

(Note: In the symbol $\Delta_R H^\circ$, the superscript indicates that the reactions occur at *standard-state conditions*. This is a technical thermodynamic requirement whose elaboration is beyond the scope of this text.) These examples seem to indicate that while exothermicity (the transition to a *lower* energy state) *favors* the spontaneity of a reaction, it is *not* the sole factor causing spontaneous reactions. Hence, the concept of energy (and the principle of energy conservation) is **not** sufficient to explain all the observed characteristics of the universe. In particular, the central concept of energy conservation leaves the observation of spontaneity unexplained.

The recognition of this glaring inadequacy in the tools of thermodynamics led Rudolf Clausius in 1865 (building on the earlier work of Sadi Carnot) to propose a new thermodynamic variable called **entropy**. The **entropy** of a thermodynamic system (represented by an uppercase **S**) is a **direct measure of the randomness or disorder of a system**; the **greater** the **disorder**, the **greater** the **entropy**. From another point of view, entropy is a description of the number of possible distinct arrangements of the constituents of a system, and, in this sense, it is related to the concept of probability. (Ludwig Boltzmann made this connection explicit with his formulation of the kinetic molecular theory of gases.) The entropy of a system is related to the frequency with which a particular arrangement of the constituents of the system occurs. An ordered system has a **low probability** of occurring, and hence, has a **small entropy value**; a disordered system has a **high probability** of occurring, and hence, has a **large entropy value.** This linking of entropy to the degree of disorder has a very simple and direct interpretation when observing the physical universe. If we select a fixed molar amount of a substance, we can make a very general statement about the entropy associated with the sample, as follows:

$$S_{solid} < S_{liquid} << S_{gas} \tag{10.5}$$

## Entropy and the Second Law of Thermodynamics

The consequence of introducing the new variable, **S**, denoting entropy (it is also a mathematical function, just like enthalpy, ***H***, is a function) was the articulation of the **second law of thermodynamics**: the **total entropy** of an **isolated** system *increases* in the course of a **spontaneous** change. Mathematically, this means that $\Delta_{tot}S > 0$ (where the symbol $\Delta_{tot}S$ is read as "the total entropy change") for **all spontaneous processes taking place in an isolated system**. This statement depends critically on the technical meaning of the term *isolated*; for the purposes of this text, the most important implication of this term (and the only one we shall take note of) is that our **universe is an isolated system**. Consequently, for **all** *spontaneous processes* taking place in our universe, the value of the entropy associated with such a change **must** increase. The change in *entropy* is calculated in exactly the same fashion as the change in enthalpy:

$$\Delta_{tot}S = S_{final} - S_{initial} \tag{10.6}$$

In equation 10.6, $S_{initial}$ is the value of the system's entropy at the beginning of a process, and $S_{final}$ is the value of the system's entropy at the end of a process. If the system is the entire universe and the process is spontaneous, the second law of thermodynamics applies.

The introduction of the concept of entropy represented a significant advance in the science of thermodynamics because, for the first time, the science was able to characterize and quantify spontaneity. The second law of thermodynamics established a direct connection between entropy and spontaneity. However, in order to apply the very general statement of the second law of thermodynamics to any individual process, it is necessary to calculate the entropy change **for the entire universe**! To determine whether or not a single chemical reaction on Earth is spontaneous, the second law of thermodynamics requires the calculation of the change in entropy associated with the reaction **everywhere in the entire universe**, at every star, for every galaxy! Such a task is humanly impossible. Consequently, the invention of the entropy concept with its deep connection to spontaneity seems to be a hollow victory because such an enormous calculation appears hopeless.

## The Gibbs Free Energy

Careful observation of spontaneous processes identified several important phenomena: (1) the transfer of energy from an object at a higher temperature to an object at a lower temperature is **always** spontaneous, (2) the transfer of energy from an object at a lower temperature to an object at a higher temperature is **never** spontaneous, (3) an **increase** in disorder is **always** spontaneous, (4) a **decrease** in disorder is **never** spontaneous, and (5) the temperature of a system seems to be related to spontaneity. All of these observations suggested that there is yet another thermodynamic variable (again, a mathematical function) whose value is related to the *spontaneity* of a process. The **Gibbs free energy** (often simply called the **free energy**) is defined as follows:

$$G = H - TS \tag{10.7}$$

where the symbol $TS$ means the product of $T \times S$. The first important observation is that all the variables in equation 10.7 apply **only** to the thermodynamic system. Second, just as we saw with enthalpy, by restricting ourselves to idealized cases in which the pressure of the thermodynamic system remains **constant**, millions of chemical reactions on this planet can be analyzed with equation 10.7. Finally, we add one additional restriction: we require that the temperature of the thermodynamic system remain **constant**. Again, this is a commonly occurring feature of an immensely large number of chemical reactions on Earth. Imposing the additional condition of **constant** temperature on equation 10.7 gives the following relationship for a chemical reaction (the reaction is the thermodynamic system in which we are interested):

$$\Delta_R G = \Delta_R H - T\Delta_R S\,, \tag{10.8}$$

where the symbol $T\Delta_R S$ means the product of $T \times \Delta_R S$. This equation relates the *change in the Gibbs free energy* for a reaction ($\Delta_R G$) to the *change in the enthalpy* for a reaction ($\Delta_R H$) and the *change in the entropy* for a reaction ($\Delta_R S$), while assuming that both the temperature ($T$) and the pressure ($P$) of the thermodynamic system remain **constant.** The changes in the enthalpy and in the entropy are calculated as before:

$$\Delta_R H = H_{\text{products}} - H_{\text{reactants}} \tag{10.9a}$$

and

$$\Delta_R S = S_{\text{products}} - S_{\text{reactants.}} \tag{10.9b}$$

## Spontaneity and the Gibbs Free Energy

An analysis of the change in the Gibbs free energy for a reaction ($\Delta_R G$) under the **dual** conditions of **constant temperature** and **constant pressure** reveals an amazing conclusion: a chemical reaction taking place under the dual conditions of **constant temperature** and **constant pressure** is spontaneous if and only if $\boldsymbol{\Delta_R G < 0}$. This statement is a shorthand method of stating two separate ideas: (1) if $\boldsymbol{\Delta_R G < 0}$ under the conditions of **constant temperature** and **constant pressure**, a reaction is *spontaneous*;

and (2) if a reaction is spontaneous under the conditions of **constant temperature** and **constant pressure, $\Delta_R G < 0$!** This discovery is tremendously powerful. Because the definition of the Gibbs free energy, G, was done using **only** variables from the thermodynamic system, we now have the means to predict spontaneity of a process by concentrating on **only the system**. Remember, the second law of thermodynamics connected spontaneity with the entropy change of **the entire universe**. Further, because the spontaneity of millions of reactions under the conditions of constant temperature and constant pressure is directly related to the change in the Gibbs free energy that occurs during the course of a reaction, it is now recognized that the change in the Gibbs free energy during a chemical reaction is the fundamental physical characteristic that causes reactions to occur. Looking back to equation 10.7, we note that the definition of the Gibbs free energy is composed of two terms. The first is the enthalpy, $H$; because a change in the enthalpy is equated to the heat of a reaction (an energy transfer process occurring under conditions of constant pressure), the enthalpy can be viewed as an ***energetic*** contribution to the Gibbs free energy. The second term is $TS$, the product of the temperature and the entropy (the product does have units of energy in order to keep the equation consistent); however, the presence of the entropy means that this term can be view as the ***entropic*** contribution to the Gibbs free energy. Consequently, the fundamental physical characteristic causing chemical reactions to occur is composed of two pieces, an *energetic* contribution and an *entropic* contribution. This is an immensely important insight into the fundamental organization of our universe. Displayed graphically, the fundamental role of the Gibbs free energy is emphasized in figures 10.3 and 10.4; these figures should be compared to figures 10.1 and 10.2, which focus on the enthalpy of a chemical reaction. It is now clear

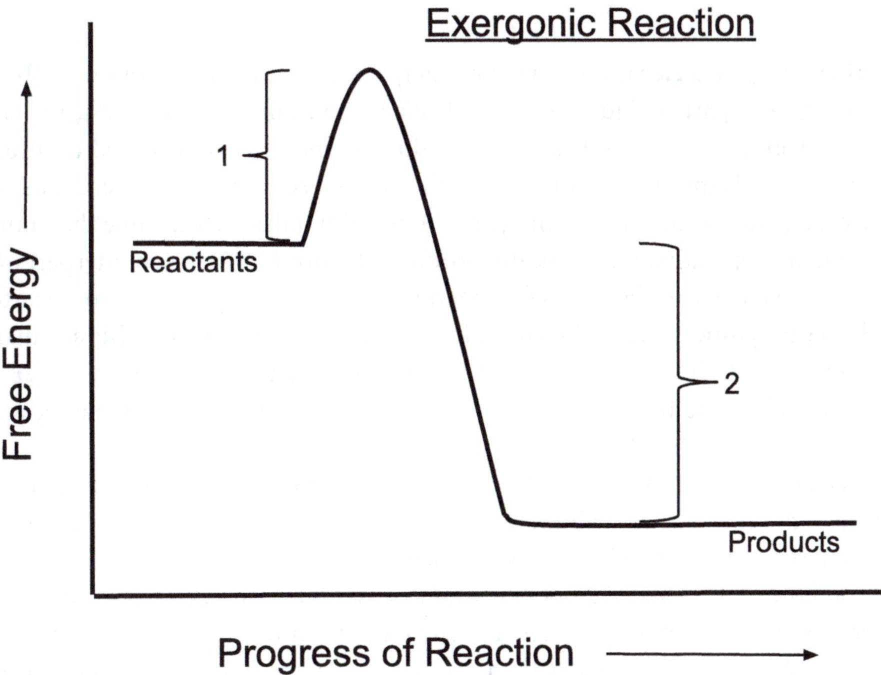


Figure 10.3: Exergonic Reaction Diagram

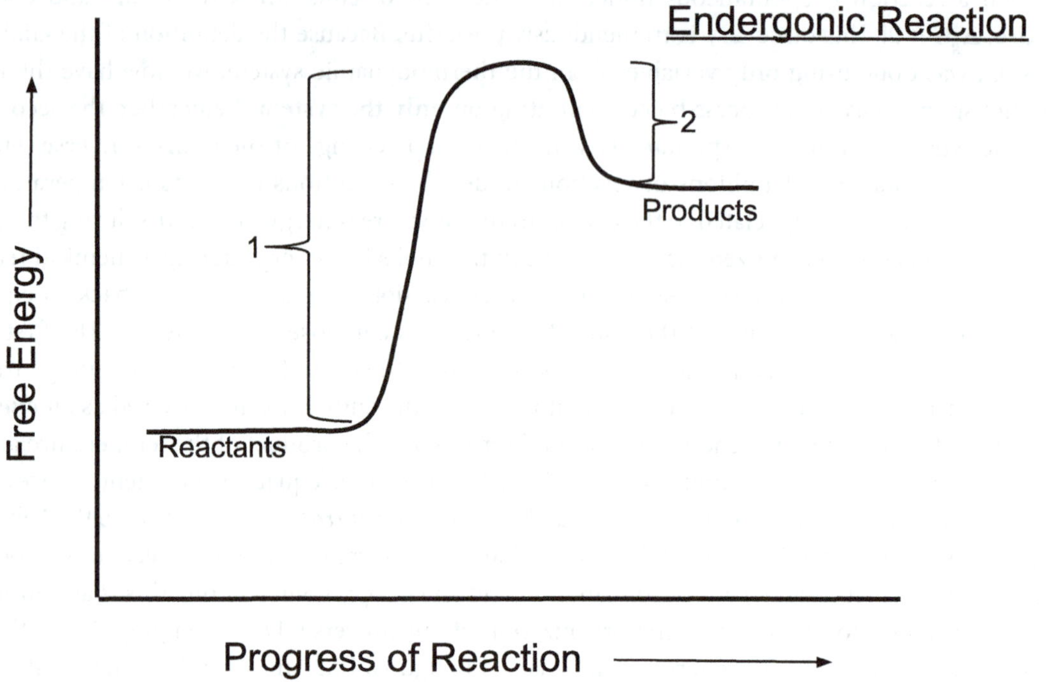


Figure 10.4: Endergonic Reaction Diagram

that the **activation energy barrier** is more appropriately understood in terms of the Gibbs free energy. Figure 10.3 is the counterpart of figure 10.1 and displays the change in the Gibbs free energy as the reaction proceeds from left (reactants) to right (products). The reaction is called **exergonic** because the Gibbs free energy of the products is **lower** than the Gibbs free energy of the reactants. The number 2 identifies the change in the Gibbs free energy as the reaction takes place, while the number 1 labels the free energy barrier separating reactants and products. Figure 10.4 is the counterpart of figure 10.2 and again displays the change in the Gibbs free energy as the reaction proceeds from left to right. The reaction is called **endergonic** because the Gibbs free energy of the products is **higher** than the Gibbs free energy of the reactants. The number 2 identifies the free energy barrier for the **reverse** (products returning to reactants), while the number 1 labels the free energy barrier separating reactants and products in the **forward** direction (reactants to products).

The Gibbs free energy is a powerful tool for exploring the behavior of various chemical reactions using knowledge of the *energetic* contribution ($\Delta_R H$) and the *entropic* contribution ($\Delta_R S$) to a reaction. Figure 10.5 summarizes four possibilities for the values of $\mathbf{\Delta_R H}$ and $\mathbf{\Delta_R S}$.

Under the conditions of constant temperature and constant pressure, equation 10.8 can be used to decide whether or not a reaction is spontaneous. The plus and minus signs in the figure indicate that values for $\mathbf{\Delta_R}$H and $\mathbf{\Delta_R}$S are either greater than zero (plus) or less than zero (minus). If the value of either quantity is exactly equal to zero, the sign of $\Delta_R G$ is determined solely by the nonzero quantity. If the values of both quantities are exactly equal to zero, the system is in a state of equilibrium; there is

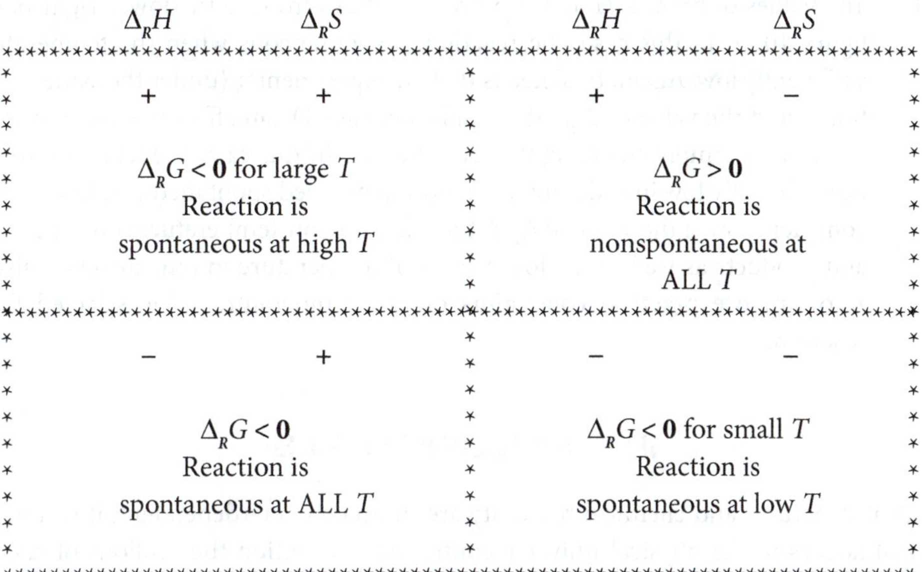


Figure 10.5: Energetic and Entropic Contributions to $\Delta_R G$

**no preferred direction**. As the figure indicates, if the values of neither $\mathbf{\Delta_R H}$ nor $\mathbf{\Delta_R S}$ are equal to zero, there are four cases to consider:

**Case 1:** The values of both $\mathbf{\Delta_R H}$ and $\mathbf{\Delta_R S}$ are greater than zero (see the upper left panel of figure 10.5). In this case, the reaction is spontaneous when the temperature is sufficiently high. This means that an experimenter can affect the spontaneity of a reaction by simply changing the temperature at which a reaction occurs. (This does assume that the value of $\mathbf{\Delta_R H}$ is *unaffected* by a temperature change. This assumption has been observed to be accurate over a very large range of temperature changes.) In this case, the temperature must be sufficiently high in order for a reaction to proceed spontaneously. In addition to consideration of the value of $\mathbf{\Delta_R H}$, the thermal stability of the reactants and products imposes practical constraints on the temperature value selected for the reaction.

**Case 2:** $\mathbf{\Delta_R H} > 0$ and $\mathbf{\Delta_R S} < 0$ (see the upper right panel of figure 10.5). In this case, the reaction is nonspontaneous at **all** temperatures. This means that an experimenter (again, under the same assumption about the value of $\mathbf{\Delta_R H}$ that applies in case 1) **cannot** affect the spontaneity of a reaction by simply changing the temperature.

**Case 3:** $\mathbf{\Delta_R H} < 0$ and $\mathbf{\Delta_R S} > 0$ (see the lower left panel of figure 10.5). In this case, the reaction is spontaneous at **all** temperatures. This means that an experimenter (again, under the same assumption about the value of $\mathbf{\Delta_R H}$ that applies in case 1) **cannot** affect the spontaneity of a reaction by simply changing the temperature.

**Case 4:** The values of both $\Delta_R H$ and $\Delta_R S$ are less than zero (see the lower right panel of figure 10.5). In this case, the reaction is spontaneous when the temperature is sufficiently low. Again, this means that an experimenter (under the same assumption about the value of $\Delta_R H$ that applies in case 1) can affect the spontaneity of a reaction by simply changing the temperature. In this case, the temperature must be sufficiently low in order for a reaction to proceed spontaneously. In addition to consideration of the value of $\Delta_R H$, the effects of low temperatures on the reactants and products as well as the lower limit of temperature in our universe (absolute zero) impose practical constraints on the temperature value selected for the reaction.

## Intermolecular Forces

While chemical structures and chemical reactivity are intimately interdependent, it is now very clear that additional factors in the physical universe control and condition the millions of chemical processes that characterize our world. Our very brief tour of some of the central concepts in the science of thermodynamics highlighted key physical parameters that influence and determine many aspects of chemical reactivity. However, an understanding of the strong intramolecular forces that define chemical structures, of the powerful principle of energy conservation, and of the profound link between entropy and spontaneity, to note only a few of the central concepts we have encountered in the course of this story, are not sufficient if we are to describe successfully some of the very common and most remarkable aspects of our world. We are about to discover that there are additional physical quantities that are responsible for some of the most important macroscopic characteristics that define the world around us. The processes of **melting and freezing** (often described using the single term **fusion**), and **vaporization and condensation**, along with a variety of properties—**friction** (the force that resists relative motion between two bodies when they are in contact), **adhesion** (an attraction originating at the molecular level and exerted between the macroscopic surfaces of objects that are in contact), **cohesion** (a molecular attraction by which the individual particles of a body are bound together in a macroscopic mass), **viscosity** (the property which measures the tendency of a fluid or semifluid to resist flow), and **surface tension** (an attractive force between the molecules at the surface of a fluid and the molecules beneath the surface, which deforms the macroscopic surface to achieve the smallest possible surface area)—are all part of the macroscopic universe. These processes and properties **all** depend on weak interactions either **between** atoms and molecules or **between** entire molecules and are called **intermolecular forces**. (Be careful to make a distinction between the terms *intermolecular* and *intramolecular*. The term *intramolecular* refers to properties **solely inside** or **belonging to** an individual molecule. The parameters that characterize chemical bonds between the atoms of a molecule, for example, are *intramolecular* properties.)

Intermolecular forces may be either **repulsive** or **attractive**; they depend fundamentally on two features of the universe that we encountered much earlier in our story. The first of these characteristics is a purely quantum mechanical phenomenon that was initially recognized during the first quarter of the twentieth century. As we saw in chapter 2, the experimental work of Uhlenbeck and Goudsmit and the theoretical efforts of Pauli led to the articulation of the Pauli exclusion principle, which states that no two electrons in an atom can have the **same** values of the four quantum numbers. This is exactly equivalent to imposing a *limit of two* on the number of electrons associated with any quantum mechanical orbital. In fact, the Pauli exclusion principle is the consequence of the much more general observation that the physical universe is constructed of two fundamentally different types of elementary objects: **fermions** and **bosons.** We met the bosons in our discussion of the interactions of leptons and hadrons. The spin quantum number of all bosons exhibits **only** integer values. In contrast, all objects whose spin quantum number is a half-integer (±½) are called **fermions**; in particular electrons are members of the fermion class. In general, only one fermion can be associated with a particular quantum state. (This is simply another way of requiring that the values of the four quantum numbers of every fermion be unique.) **Consequently, fermions cannot be arbitrarily close to one another in space.** We observe this limitation experimentally as a **repulsive force** between atoms and molecules or between two molecules.

The second characteristic that underpins intermolecular forces is the phenomenon of **charge**, which we also first met in chapter 2. In that chapter we saw that the phenomenon of charge possesses two different and distinct properties that we labeled as *positive* and *negative.* The application of the quantum theory to the behavior of charge at the molecular level provides an explanation of the **attractive** intermolecular forces. While a full quantum mechanical treatment of charge (and, for that matter, of the behavior of fermions) is beyond the scope of this text, we can categorize the various attractive intermolecular forces that have been identified over nearly the past 150 years.

## Van der Waals Forces

In the last quarter of the nineteenth century, Johannes van der Waals used the science of thermodynamics to postulate a rudimentary explanation of attractive intermolecular forces. (While van der Waals also recognized the existence of the weak repulsive forces discussed above, their explanation had to await the development of quantum theory. Van der Waals did not have the theoretical tools as the nineteenth century came to a close to provide a quantitative explanation of the weak repulsive forces.) In recognition of this pioneering work, one complete class of attractive intermolecular forces is now known as **van der Waals forces.** These forces are the interactions of complete, neutral molecules either with one another or with neutral atoms. The van der Waals class of intermolecular forces is divided into two groups; one is called the **dispersion forces** (or **London forces,** in honor of Fritz London, who studied these forces using quantum theory), while the second group is named the **dipole-dipole forces.** An understanding of both groups emerged in the twentieth century after the development of quantum mechanics in the late 1920s, and this understanding fundamentally rests on a detailed description of the concept of **charge separation** that we used to describe molecular polarity in chapter 4. London was able to combine the phenomenon of charge separation and the purely quantum mechanical characteristic that electrons are fermions and to demonstrate that transient (that is, very short-lived) charge separations in atoms and nonpolar molecules accounted for very weak attractive forces (the dispersion forces), which contributed to the macroscopic processes and properties noted previously. The same reasoning can be extended to the case of dipole-dipole forces that occur between polar molecules, again using the insight provided by the new quantum theory. While the term *intermolecular forces* was chosen to reflect the *force* (see chapter 1) exerted by an atom or molecule on another molecule, the strength of these forces is conventionally measured in terms of the quantity of *energy* required to overcome the forces. The strength of the van der Waals forces varies between approximately two and sixty-five kJ $mol^{-1}$ (meaning that between two and sixty-five kilojoules of energy would be required to interrupt a mole of such interactions). This compares with between approximately 125 and 1,100 kJ $mol^{-1}$ required to rupture the chemical bonds linking atoms in a molecule.

## Hydrogen Bonds

Among the wide variety of dipole-dipole interactions that occur between an atom and a molecule or between two molecules, there is one type of dipole-dipole interaction that is of particular significance. This interaction is called a **hydrogen bond** and is formally defined as a special dipole-dipole interaction between a hydrogen atom that is part of a polar bond (e.g., bonds of the type N-H, O-H, and F-H) and an electronegative atom (N, O, and F). Because the hydrogen bond is, most fundamentally, a dipole-dipole interaction, the central characteristic of the hydrogen bond is a *charge separation*. This is schematically indicated in figure 10.6, which is a classical idealization of a hydrogen bond interaction between two water molecules. The symbols $\delta+$ and $\delta-$ are an attempt to represent a charge separation in the molecule, which is best described using the mathematics of the quantum theory. Note carefully that while the water molecule exhibits a charge separation because of the spatial probability distribution of its electrons (as required by quantum mechanics), the molecule itself is *neutral*; **it is not a charged species**. The strength of hydrogen bonds varies from approximately four kJ $mol^{-1}$ to well over 150 kJ $mol^{-1}$. The typical strength of the hydrogen bond interaction in water is approximately 23 kJ $mol^{-1}$.

Figure 10.7 is another depiction of the hydrogen bonds shared by several water molecules, with the symbols $\delta+$ and $\delta-$ again representing a separation of charge in the molecules.

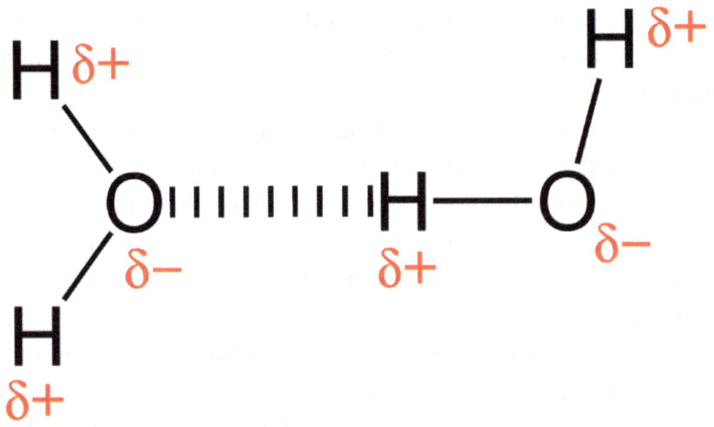

Figure 10.6: Schematic Representation of a Hydrogen Bond

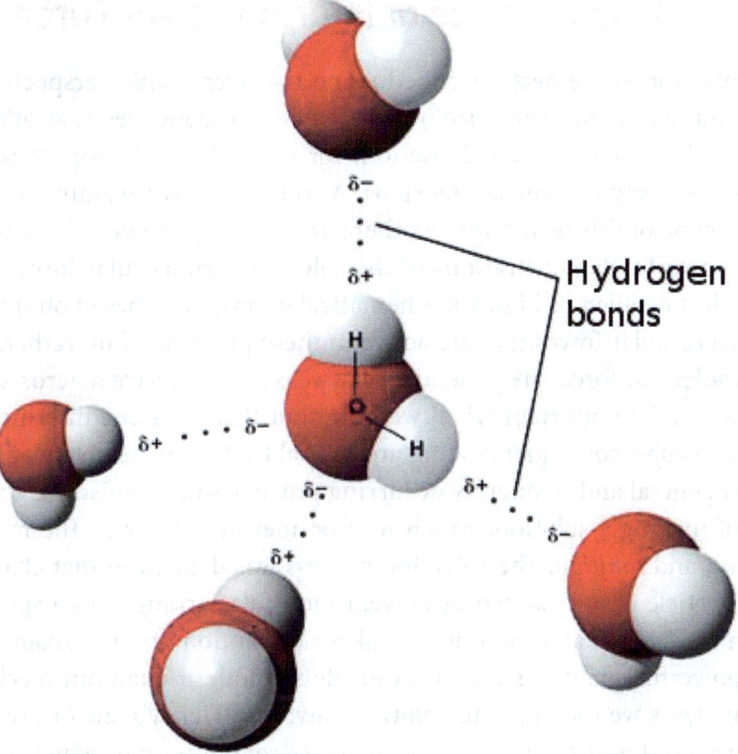


Figure 10.7: Representation of Hydrogen Bonds in Water

## Ion-Dipole Forces

The final category of intermolecular forces consists of interactions between species that are carriers of charge, *ions* (we first met the idea of *ions* in our discussion of Michael Faraday's observations in chapter 4) and dipoles, regions in which there is a spatial separation of charge. Again, the description of ion-dipole forces is done most effectively using the powerful mathematics of the quantum theory to describe the probability distribution associated with charge carriers (the ions) and charge separations (the dipoles). Because of the participation of charge carriers, ion-dipole forces are generally much stronger than the very weak van der Waals forces and are still significantly stronger than a larger majority of hydrogen bond forces. The strength of a typical ion-dipole force varies between approximately 40 and 600 kJ $mol^{-1}$ and depends on the magnitude of the charges associated with the interacting ion.

## A Macroscopic Consequence of Intermolecular Forces: Solvation

While intermolecular forces are best described using the *microscopic* perspective of the quantum theory, they have numerous *macroscopically* observable consequences that affect both processes (melting/freezing and vaporization/condensation, for example) and properties (such as friction, adhesion, cohesion, viscosity, and surface tension). A comprehensive examination of all these effects is well beyond the scope of this text; however, there are several processes that occur frequently and that serve as very powerful demonstrations of the role of intermolecular forces in our world. Our goal is not to provide a detailed or highly mathematical description (based on quantum mechanics) of the critical intermolecular forces that are active in these processes, but, rather, to emphasize that even though intermolecular forces are generally very weak, they do have macroscopic consequences. This understanding will be important when we point out that the same intermolecular forces also have important *microscopic* consequences that are crucial to the biochemistry of life itself.

One of the most general and frequently occurring processes in chemistry is the process of solvation, the process of making a solution, which we first met in chapter 3. The reader will recall the terms *solute*, *solvent*, and *solution*. The solvation process was defined in that chapter as the process in which a solute particle is surrounded by solvent molecules arranged in a specific manner. Note that this definition is a very qualitative one; it makes no mention of microscopic forces that can be described by the powerful quantitative tools available in modern quantum mechanics. In fact, the two quantitative analyses we did apply to solutions involved Henry's law (a pre-twentieth century macroscopic observation) and the calculation of *solute concentrations*, which is based on macroscopic measurements. We now have the possibility to understand the solvation process from the microscopic perspective by recognizing that the manner in which a solute particle is surrounded by solvent molecules depends entirely on the various intermolecular forces at work between the solute and the solvent. Amazingly, a very simple but very powerful generalization that is based on early macroscopic observations is called the **golden rule** and is easily stated as follows: *like dissolves like.* From the perspective of intermolecular forces and the solvation process, this means that solvent molecules can effectively surround a solute particle if **the same intermolecular forces** are shared by the solvent molecules and the solute particle. Consequently, polar solvents easily solvate polar solutes; nonpolar solvents solvate nonpolar solutes. However, polar and nonpolar substances **do not** easily form a solution: note that oil and water do **not** mix!

The golden rule is a very powerful generalization. However, like so many other generalizations, the complexity of the interactions mediated by intermolecular forces can make the application of the golden rule very difficult. This is particularly true if several different intermolecular forces are simultaneously participating in the solvation process. In these cases, the successful application of the golden rule requires a careful analysis of the various intermolecular interactions, the relative strengths of the interacting forces, and the possible secondary chemical reactions (such as the formation of molecular dimers) that may occur. However, the golden rule, when viewed in the light of the totality of intermolecular forces, provides a highly effective framework with which to understand the solvation process.

## A Second Macroscopic Consequence of Intermolecular Forces: Phase Transitions

In addition to the formation of solutions, there is a second process that is a ubiquitous characteristic of our world: **a phase transition**. In chapter 3 we met the three classical states of matter: solids, liquids, and gases. The simplest application of the science of thermodynamics identifies each of these states of matter as a phase and the process of matter changing from one state to another as a phase transition. (You may wonder why there is a need for two different words: *state* and *phase*. As the science of thermodynamics developed, it was recognized that it is possible for several different *phases* to exist in a *single state of matter*. Consequently, the more precise term is a *phase transition*. For our purposes, the three classical states of matter will be treated as only three distinct phases, and the phrase *phase transition* has exactly the same meaning as the term *change of state*. The reader should be aware, however, that the terms *state* and *phase* are not identical.) Figure 10.8 summarizes the variety of phase transitions

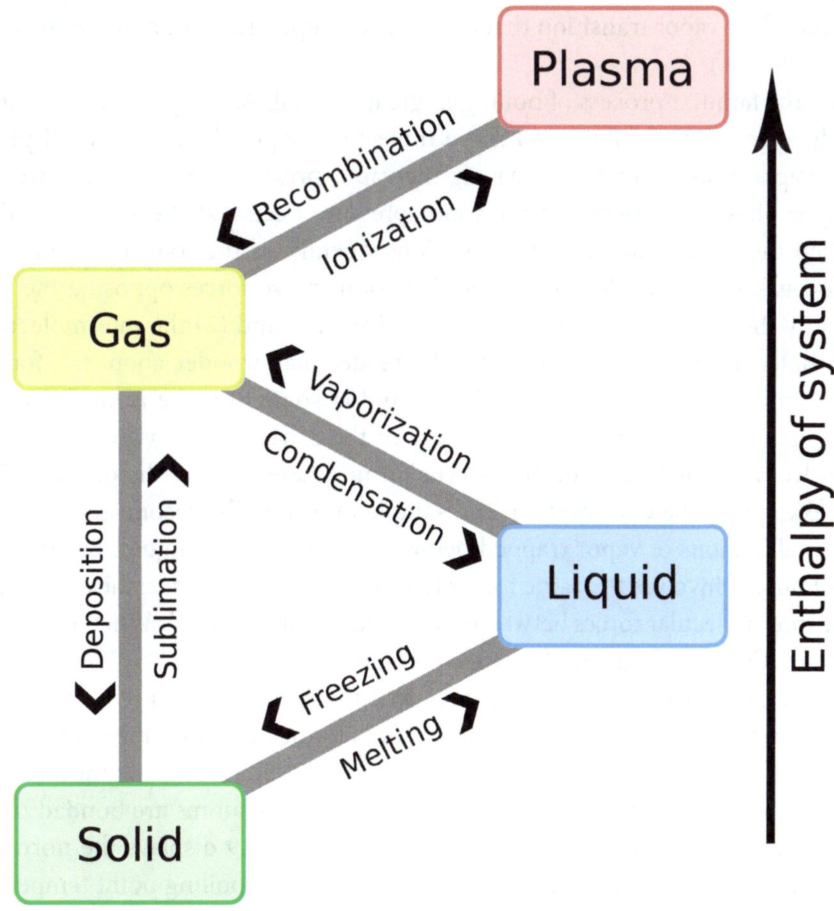


Figure 10.8: Common Phase Transitions

that are observed to occur between the three classical states of matter as well as the special *plasma* state. (A *plasma* is a state of matter in which all the electrons are no longer bound to the nuclei of atoms. The plasma state occurs only under extreme temperature and pressure conditions; the atmosphere of stars is one example of a plasma.)

The vaporization/condensation phase transition is a particularly informative example in which intermolecular forces play a significant role. There are three important terms that we will need in order to explore the role of intermolecular forces in this phase transition. The **vapor pressure** is the pressure exerted by the gaseous molecules that have escaped from the surface of a liquid. In order to define the vapor pressure of a liquid, the liquid and vapor must be in a state of equilibrium (see chapter 5) with each other. The **boiling point** is the temperature at which the vapor pressure of a liquid is **equal** to the external pressure exerted on the liquid. The **normal boiling point** is the temperature when the external pressure is exactly to 1 atm. (Note that the term *vaporization* is most commonly used to identify the phase transition from a liquid to a vapor (gas), with the word *condensation* denoting the reverse process. There are two type of *vaporization*: evaporation and boiling. *Evaporation* refers to the liquid-to-vapor transition that occurs at a temperature **below** the boiling point for a given pressure. *Boiling* refers to the liquid-to-vapor transition that occurs at a temperature **equal to or above** the boiling point for a given pressure.)

Now, let's examine the familiar process of boiling in greater detail. As the previous definitions suggest, as energy is added to a liquid, more and more molecules originally in the liquid phase make the transition to the vapor phase, thereby increasing the vapor pressure; the transition from the liquid phase to the vapor phase that **increases** the vapor pressure occurs at the surface of the liquid, the boundary between the liquid and vapor phases. Consequently, as the transition from the liquid to the vapor phase occurs, the molecules in the liquid encounter two forces opposing the transition: (1) the force exerted by the external pressure on the liquid surface and (2) the intermolecular forces acting between the molecules in the liquid phase. (The reader may wonder about the formation of vapor vesicles, commonly called *bubbles*, in a boiling liquid at some distance from the surface. The bubbles actually originate from two sources. First, early in the process of transferring energy to the liquid, bubbles form from the molecules of the external atmosphere that are dissolved in the liquid. The most common example is the dissolved air in a sample of water. The second source of a bubble is the formation of small regions of vapor trapped below the surface of the liquid. But the formation of this second type of bubble involves the same forces that operate at the surface: the external pressure on the liquid and the intermolecular forces between the molecules of the liquid. When these bubbles of vapor rise to the surface, they contribute to the vapor pressure of the liquid.)

If we examine a series of *different* liquids, the force that the external pressure exerts remains the same for each member of the series; consequently, differences in the behavior of the substances in the phase transition from the liquid phase to the vapor phase may be attributed solely to the intermolecular forces. First, let's consider a set of small molecules in which hydrogen atoms are bonded to elements from the first four rows of the conventional periodic table. Figure 10.9 displays the normal boiling points of the compounds. Note that the figure shows a gradual rise in boiling point temperatures for hydrogen compounds which include elements from rows two, three, and four. This gradual increase is consistent with the increasing mass of the compounds. However, note the sharp **increase** in the boiling point temperatures for HF and $H_2O$, hydrogen compounds of elements from the first row of

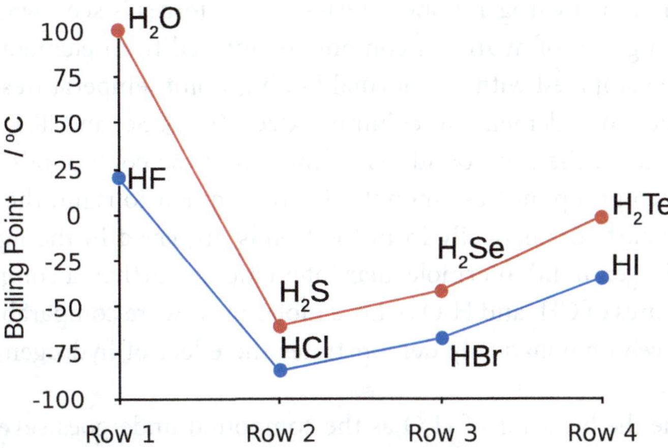


Figure 10.9: Comparison of Boiling Point Temperatures

the periodic table. The masses of both the HF molecule and the $H_2O$ molecule are **less than** the masses of the other molecules whose boiling points are plotted in the figure, so the increase is certainly **not** explained by the mass differences. However, of all the compounds whose boiling points are displayed in figure 10.9, **only** HF and $H_2O$ exhibit hydrogen bond intermolecular forces! Consequently, we can understand the dramatic **increase** in the boiling point temperatures of HF and $H_2O$ to be the result of the energy needed to break the hydrogen bond interactions.

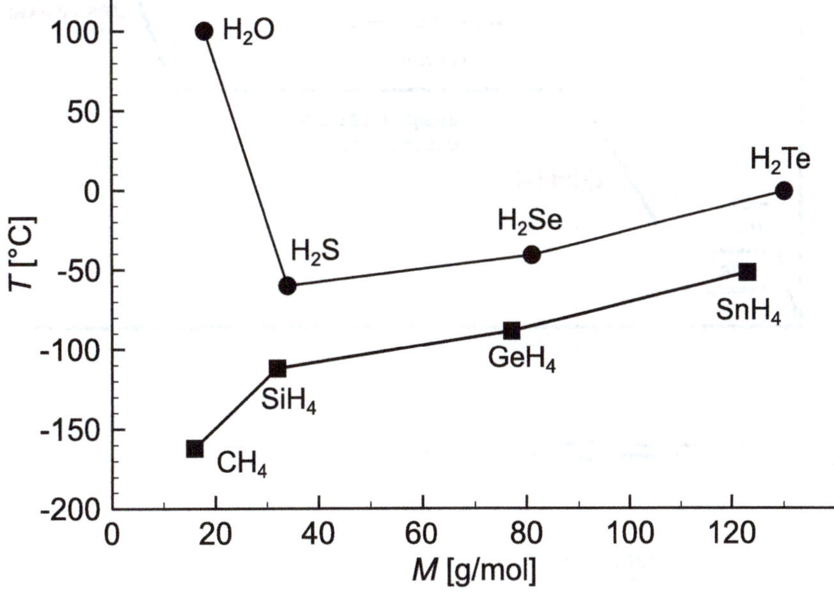


Figure 10.10: Boiling-Point Temperatures as a Function of Molar Masses

A similar effect caused by hydrogen bond intermolecular forces is seen when the normal boiling point temperatures of a group of hydrogen compounds formed from elements in column fourteen (C, Si, Ge, and Sn) are compared with the normal boiling-point temperatures of a group of hydrogen compounds formed from elements in column sixteen (O, S, Se, and Te). Figure 10.10 displays the normal boiling points of the compounds as a function of the compounds' molar masses. While the molar masses of each group increase monotonically from left to right, the normal boiling point temperature of $H_2O$ departs dramatically from the trends displayed in the figure. Again, note that **only** $H_2O$ exhibits hydrogen bond intermolecular interactions. Further, a comparison of the normal boiling-point temperatures of $CH_4$ and $H_2O$ (whose molar masses are comparable at 16.03 g $mol^{-1}$ and 18.02 g $mol^{-1}$, respectively) emphatically demonstrates the effect of hydrogen bond intermolecular forces in $H_2O$.

Finally, let's examine the behavior of $H_2O$ as the compound undergoes **several** phase transitions, the first phase transition (melting/freezing) from the solid phase to the liquid phase and a second phase transition (vaporization/condensation) from the liquid phase to the vapor phase. Figure 10.11 displays the phase transitions as a function of increasing energy (plotted along the horizontal axis

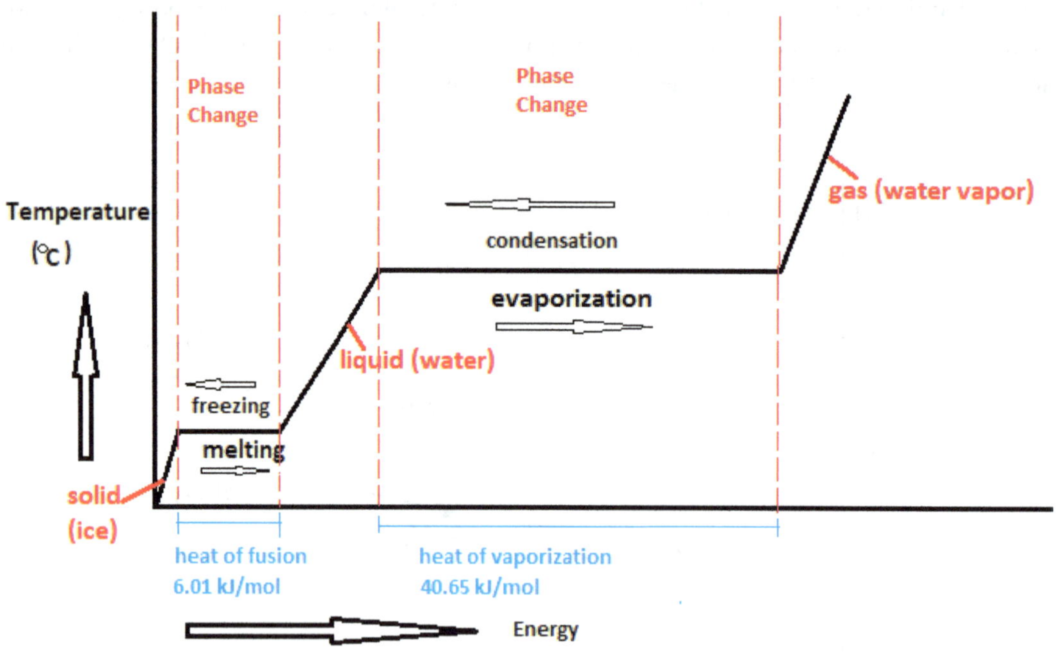


Figure 10.11: Phase Transitions of Water

from left to right), while the vertical axis plots the increasing temperature. Note that during the two phase transitions, the plotted curve is **horizontal**, meaning that the temperature is **constant during the phase transition**. Even though energy is increasing (left to right), the temperature of the $H_2O$ does **not** increase during the phase transitions. The energy that is transferred to the $H_2O$ is utilized in overcoming the intermolecular forces (principally, hydrogen bond forces) as the water first makes the phase transition from the solid phase to the liquid phase and, second, from the liquid phase to the vapor phase. The temperature does **not** increase until each phase transition is complete.

## Chemical Kinetics: The Role of Time in Chemistry

As we began this chapter, our brief exploration of the science of thermodynamics introduced fundamental concepts (heat and work, enthalpy, entropy, spontaneity, and free energy) that provide a critical framework with which to understand key aspects of chemical reactivity. The recognition that a large number of macroscopic observations can be described and interpreted in terms of microscopic intermolecular forces further enhanced and deepened our appreciation of the many subtle factors affecting the transformations we observe in our world. The specific example of phase transitions clearly demonstrated the important role that intermolecular forces play in macroscopically observed transformations. However, there is one critical parameter about which all of these previous investigations have remained essentially silent: **time**. We first encountered the concept of time very early in our story (chapter 1) and have seen that it plays a role in the decay of radioactive elements (half-life). While the statement of the most general form of the Schrödinger equation and its wave function solutions (chapter 2) also contain time as a parameter, we did not explore the implications of this time parameter for chemistry. Finally, our initial discussion of chemical equilibrium (chapter 5) used time only as a marker to identify the point at which the reactant concentrations and product concentrations no longer changed.

The science of **chemical kinetics** is the branch of modern chemistry that studies the temporal characteristics of chemical reactions. Beginning near the middle of the nineteenth century, chemical kinetics takes its name from the Greek word *kinesis*, meaning "motion," and studies the **rates** of chemical processes, attempting to understand the fundamental mechanisms and physical factors that determine the course of a chemical reaction as a function of **time**. While thermodynamics can predict whether or not a chemical reaction will occur (its *spontaneity*), it is completely silent about the **rate** of a chemical reaction. The **rate of reaction** is determined by measuring either the amount of a product generated or the amount of a reactant consumed by a chemical reaction during a specified time period. The amount of either product or reactant is most commonly measured by the **molar concentration** (review chapter 3 for a definition of *molarity*), monitoring either the **increase** or **decrease** of the molar concentration of a particular species. As one would expect, the rate of reaction can be influenced (either increased or decreased) by adjusting the **temperature** at which the reaction occurs, by the **concentrations** of the reacting species, and by the **pH** of the reacting solution. (This is actually a special case of reactant concentrations when either hydronium ions or hydroxide ions are reactants.)

Chemical kinetics is a rich and very extensive branch of the chemical sciences; a very large body of both experimental observations and theoretical analyses has been completed since the inception of the science. Today, a truly impressive amount of experimental data about chemical reactions occurring in the vapor phase, about reactions that depend on the interaction between light and a chemical species (*photochemistry*), and even about reactions occurring in condensed phases (despite the complexity of describing liquids and solids) have been cataloged and analyzed by a succession of powerful theoretical models. It is well beyond the scope of this text to explore the intricacies of this field; numerous texts devoted entirely to the subject are readily available. Our goal here is only to highlight several key insights.

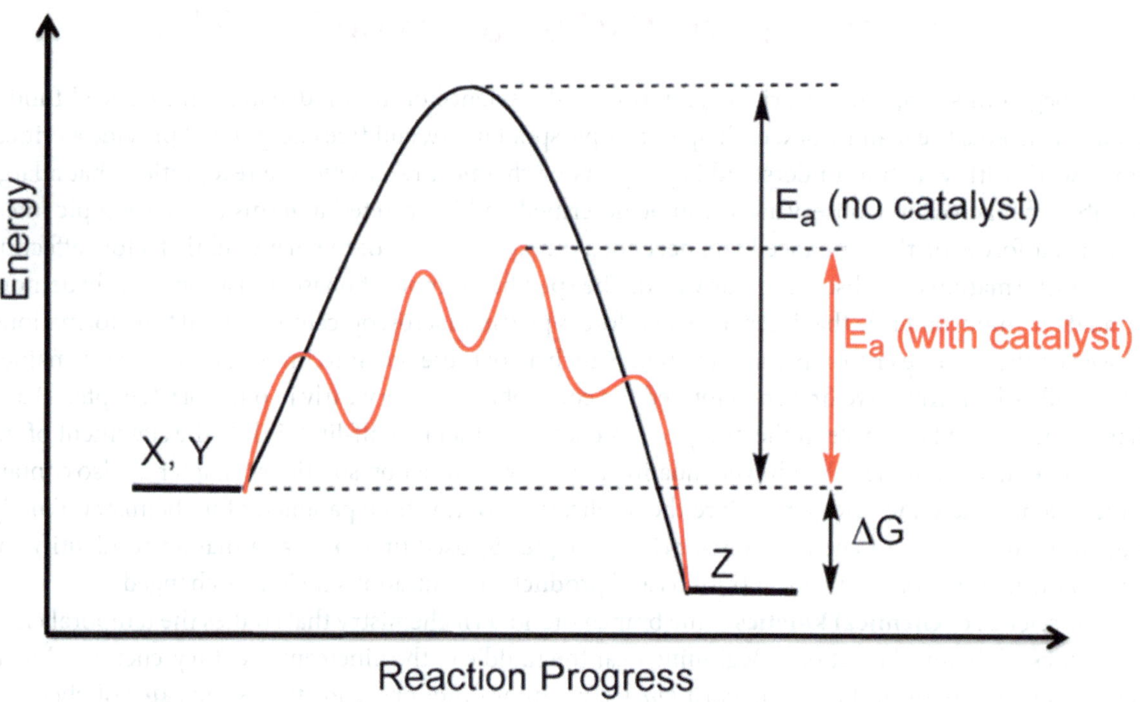


Figure 10.12: Effect of a Catalyst on Barrier to Reaction

As the nineteenth century ended, the Swedish scientist Svante Arrhenius (a physicist who was one of the founders of physical chemistry and who was awarded the 1903 Nobel Prize for chemistry) proposed a mathematical relation between a key parameter characterizing the rates of chemical reactions and the activation energy barrier separating reactants and products. (Arrhenius used the enthalpy to define the energy barrier. See figures 10.1 and 10.2.) It is now clear that the activation energy barrier is more appropriately understood in terms of the Gibbs free energy, and figure 10.12 displays the activation energy barrier in terms of this new understanding. While the mathematical relationship proposed by Arrhenius has proven to be a highly useful insight and has contributed greatly to humankind's understanding of chemical kinetics, we choose not to examine it in greater detail here. The figure, however, does show an important property of our physical universe.

Let's examine figure 10.12 more closely. The energy is plotted along the vertical axis, with *X* and *Y* representing the reactants of the exoergic reaction, and with *Z* representing the products. The free energy change resulting from the chemical reaction is represented by the symbol $\Delta G$. The two energy magnitudes, $E_a$ (no catalyst) and $E_a$ (with catalyst) represent the **two different** energy barriers separating the reactants and products. The reader may be puzzled by the existence of two energy barriers. However, the key to understanding these barriers is the term **catalyst**. A **catalyst** is a substance that **increases the rate of reaction**, even though it is **not consumed** during the course of the reaction; while it participates in the reaction, it is **not** "used up," degraded, or changed. The central effect of a catalyst is to **lower** the activation energy barrier (and perhaps change the overall shape of the barrier as shown in the figure, implying that there may be additional paths from reactants to products as a result of the action of a catalyst.) and thereby **increasing the rate of reaction.**

Note that a catalyst is a fourth way to influence the rate of reaction (as noted previously, the other three are temperature, reactant concentrations, and pH); however, unlike the other three factors, a catalyst will **only increase** the rate of reaction. Further, it should be noted that a catalyst **will** not **change** the chemical equilibrium of a reaction, because a lower activation energy barrier **increases both** the forward reaction rate and the reverse reaction rate. Consequently, the point at which the concentrations of the reactants and the products no longer change with time remains **unchanged.** Finally, the presence of a catalyst does **not** change the energy relationship between the reactants and the product (represented by the symbol $\Delta G$ in figure 10.12).

## Enzymes as Biological Catalysts

While catalytic activity is crucial to a very wide range of chemical reactions in the petroleum, pharmaceutical, and consumer products industries as well as in a number of environmental applications (catalytic converters in motor vehicles, for example), our primary interest is the role of catalysts in biochemical systems. Biochemical catalysts are large polypeptides ranging in size from just over fifty to well over two thousand amino acid residues and are given the name **enzyme**, from the Greek *enzymon*, meaning "yeast" or "leaven." Every enzyme molecule possesses a specialized region with a highly

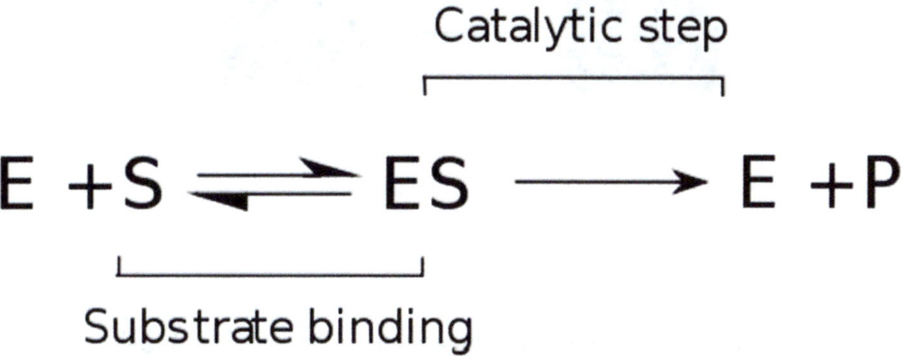


Figure 10.13: Simple Schematic of Enzyme-Substrate Binding

specific three-dimensional geometry that is called the **active site**. (We shall explore the formation of the specialized geometry of active sites when we discuss the folding process by which the large polypeptide molecules, known as **proteins** by biochemists, are formed.) The active site is the region of the enzyme that binds to the reactant to form an enzyme-substrate complex (ES-complex). (The reactant is known as the **substrate**; typically, the name of an enzyme is formed by adding the suffix *-ase* to the name of the substrate.) The highly specific geometry of an enzyme's active site limits the number of substrates that will form an ES-complex to only a few and, for many enzymes, to only one. This characteristic is called **substrate specificity**. The formation of the ES-complex is the crucial step that lowers the activation energy barrier, permitting the reaction to form products and freeing the enzyme to catalyze another reaction with the substrate. Figure 10.13 provides a schematic summary of the process, where *E* represents the enzyme, *S* the substrate, *ES* the ES-complex, and *P* the products.

The complete catalytic action of enzymes sometimes requires the presence of additional components that also bind to the enzyme. There are **two** types of complementary components that participate in enzyme activity. **Cofactors** are inorganic substances (some as simple as metallic ions) often located at the active sites of enzymes that are not part of the molecular structure of the enzyme but are held tightly to the enzyme by intermolecular forces. **Coenzymes** are small *organic* molecules that can be either tightly or loosely bound to an enzyme itself. Because the coenzymes are often modified during the course of an enzyme-catalyzed reaction, they are often considered to be a special class of substrates needed to complete a reaction.

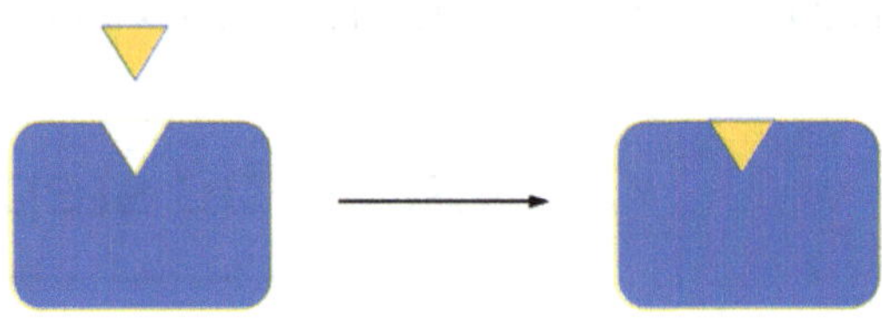

Figure 10.14: Simple Schematic of Enzyme-Substrate Binding (Lock-in-Key Model)

## Models of Enzyme Activity

Several different models have been proposed in an effort to understand the activity of enzymes. Because the active sites of enzymes exhibit a highly specific three-dimensional geometry, the earliest proposed model was the **lock-in-key model**, which suggests that the geometry of a substrate exactly matches the geometry of the associated enzyme as a key fits into a lock. This model very easily explains the **high specificity** of enzyme catalysis: the geometries of both the substrate and the enzyme must **exactly** match for a reaction to proceed. Figure 10.14 displays a simple representation of the lock-in-key model.

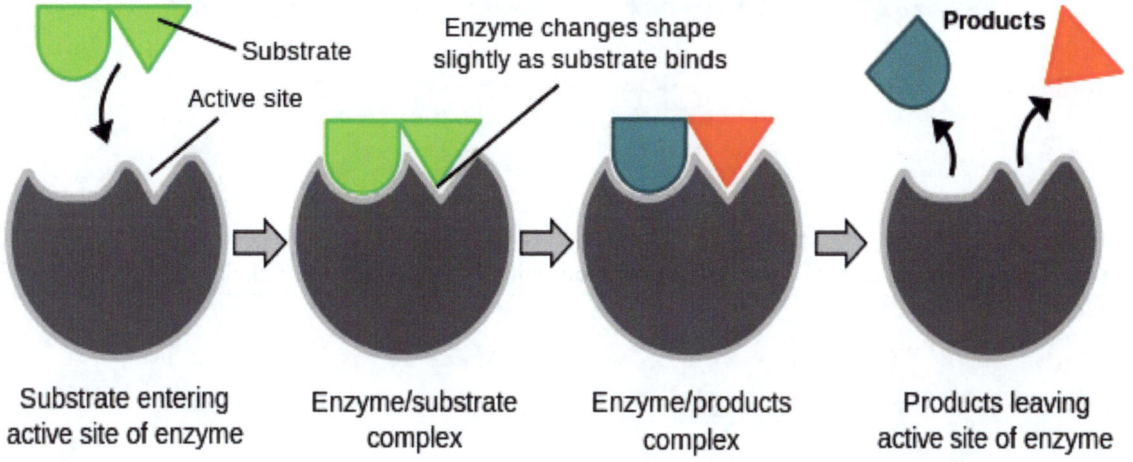


Figure 10.15: Simple Schematic of Enzyme-Substrate Binding (Induced-Fit Model)

In contrast, the **induced-fit** model (proposed in the late 1950s) suggests a more *dynamic* interpretation of the enzyme-substrate interaction. This model proposes that the geometry of the enzyme's active site actually **changes shape** as the enzyme-substrate complex forms. In fact, the geometry of the substrate itself may also change during its interaction with the flexible geometry of the active site. Once the geometric adaptation is complete, the products form and leave the enzyme unaltered, ready to accept another substrate. Figure 10.15 is a schematic representation of the induced-fit model, showing the adaptation of the enzyme's active site geometry (however, the figure does **not** show any change in the geometry of the substrate itself).

Finally, as the twentieth century drew to a close, a third model of enzyme activity was proposed; it is called the **conformational selection/population shift model**. This model suggests that the active site of an enzyme exists in multiple conformational states, and the substrate geometry has a high affinity of bonding with one of these states. In effect, a population of active site geometries is present, and one of these geometries exhibits a high probability of matching the substrate's geometry. This maximum probability determines the binding of the substrate to the enzyme. Figure 10.16 is a simple representation of the conformational selection/population shift model.

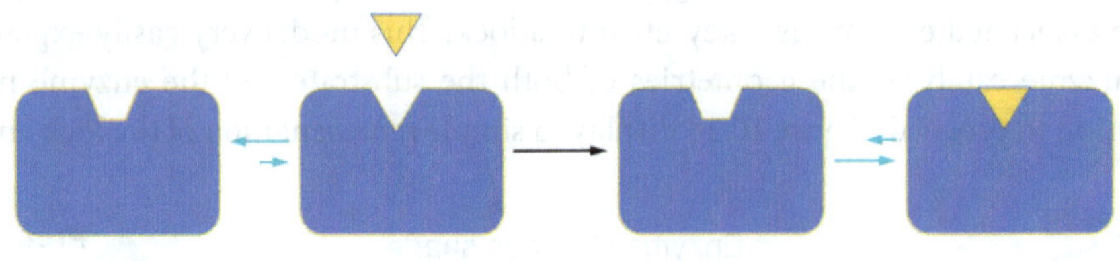

Figure 10.16: Conformational Selection/Population Shift Model of Enzyme Activity

## Enzyme Effects

Each of the proposed models of enzyme activity is designed to account for the important factors that control the effectiveness of enzymes in promoting biochemical reactions. While figures 10.14 and 10.16 might suggest that only a single substrate is involved in enzyme catalyzed reactions, complex biochemical systems often involve *multiple* substrates (in figure 10.15, the case of two substrates is depicted); consequently, one key role of the enzyme's active site is to maintain an optimal spatial separation among the enzyme and all the interacting substrates that effectively promotes a chemical reaction. In the case of a single substrate, the creation of the most favorable spatial separation between the substrate and the enzyme (along with any necessary coenzyme and cofactor) is critical for a unimolecular process to occur; for multiple substrates, the active site of the enzyme maintains the most advantageous spatial separation among the enzyme and all the substrates to catalyze the reaction. The ability of the enzyme's active site to maintain this optimal spatial separation is called the **proximity effect** of an enzyme.

However, it is not only the proximity of the enzymes and the substrate(s) that allows a reaction to proceed more rapidly. In addition to the simple spatial separation, the complete geometric arrangement of the enzyme (again, along with any necessary coenzyme and cofactor) and the substrate(s) is critical for a chemical reaction to occur. The active site of the enzyme, because it possesses a specific geometric shape, imposes a very particular three-dimensional arrangement on the enzymes and the substrate(s). The fact that the active site of an enzyme determines a specific three-dimensional geometric arrangement of the enzyme and the substrate(s) is called the enzyme's **orientation effect**.

Finally, the formation of the ES-complex **reduces** the magnitude of the activation energy barrier separating the reactants (the substrate[s]) from the products of the reaction, thereby **increasing** the rate of reaction. In effect, the strength of the chemicals bonds that maintain the chemical structure of the substrate(s) is **reduced**, and these reduced energetics promote the formation of the products. This energetic change in the intramolecular bonds of the substrate(s) as a result of the enzyme's participation in the reaction is called the **bond energy effect**.

## Physical Factors Affecting Enzyme Activity

Because enzymes are *biological catalysts*, the same three factors we met earlier (temperature, reactant concentrations, and pH) will affect enzyme activity in a biological environment. Enzymes operating in biological organisms exhibit their optimal activity at the characteristic temperature of the organism. In the case of the human body, this temperature is approximately 310.15 K. The impact of reactant concentration and pH on enzyme activity underscores the importance of both hydration and electrolyte balance for the human body. However, there are two additional factors that play critical roles in the activity of enzymes: **inhibitory control** and **allosteric control**.

### Inhibitory Control

An **inhibitor** is any substance that reduces an enzyme's activity; there are several type of enzyme inhibition. **Competitive inhibitors** directly contend with an enzyme's substrate for access to the active site of the enzyme. These substances possess a molecular geometry very similar to the geometry of the substrate, making it possible for them to interact at the active site. In effect, the close correspondence between the geometry of the actual substrate and that of the inhibitor permits the inhibitor to occupy the active site, denying the substrate access to the enzyme. Figure 10.17 presents a graphical depiction of competitive inhibition.

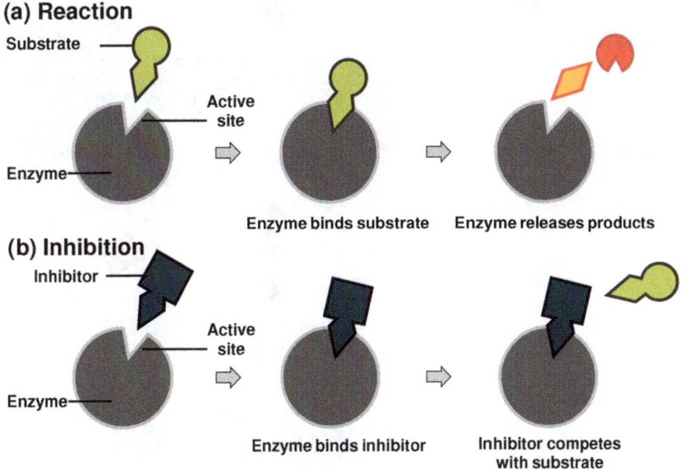


Figure 10.17: Competitive Inhibition

In contrast to the process of competitive inhibition, noncompetitive inhibitors do not approximate the geometry of the actual substrate and do not bind to the enzyme at the active site. Instead, a noncompetitive inhibitor binds at a site remote from the enzyme's active site and prevents the enzyme from completing a conformational change necessary to complete the chemical reaction. In noncompetitive inhibition the inhibitor and the substrate do not compete for access to the enzyme's active site. However, the binding of the inhibitor prevents the enzyme from catalyzing the reaction of the substrate, even though the substrate does succeed in binding at the active site. Figure 10.18 presents a diagrammatic representation of noncompetitive inhibition.

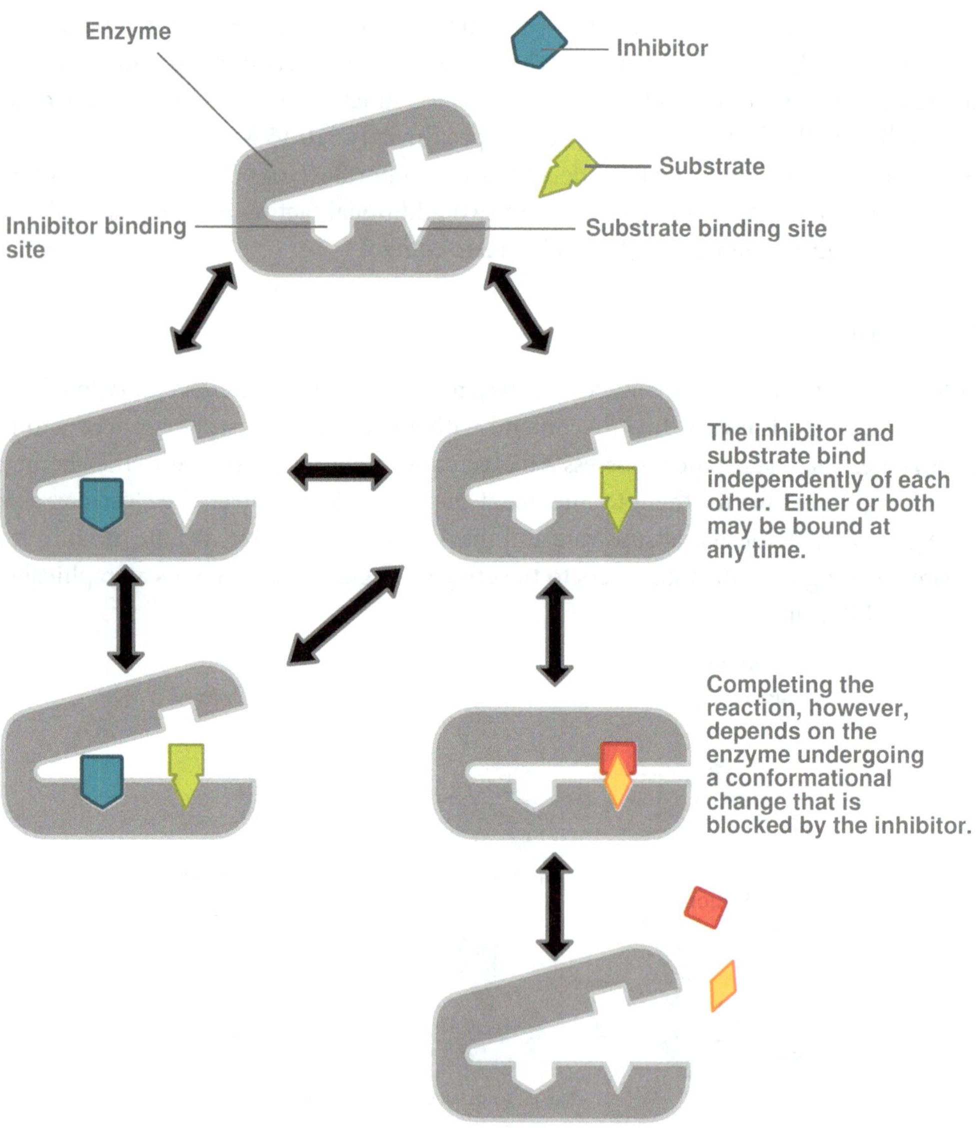


Figure 10.18: Noncompetitive Inhibition

While the competitive and noncompetitive inhibition processes are often reversible, in that neither process permanently deactivates an enzyme, there is a class of inhibitors, called irreversible inhibitors, that permanently changes an enzyme so that it can no longer catalyze a specific chemical reaction. An irreversible inhibitor permanently binds at the active site, structurally modifying the active site so that its geometry and the geometry of the substrate are no longer compatible. Consequently, the enzyme is permanently deactivated.

## Allosteric Control

An enzyme's activity can also be affected by the binding of an **effector molecule** at the enzyme's **allosteric site**, which is a region of the enzyme molecule that is distinct from the active site. Interestingly, the allosteric control of an enzyme's activity can either **enhance** an enzyme's activity or **inhibit** an enzyme's activity. The binding of the effector molecule changes the geometry of the active site, either making it possible for a substrate molecule to bind at the enzyme's active site or changing the geometry so a substrate **cannot** bind at the active site. If the ES-complex forms, the catalytic effect of the enzymes promotes the chemical reaction; if the allosteric control prevents the formation of the ES-complex, no reaction takes place. In Figure 10.19 a diagrammatic representation of **allosteric inhibition is presented.**

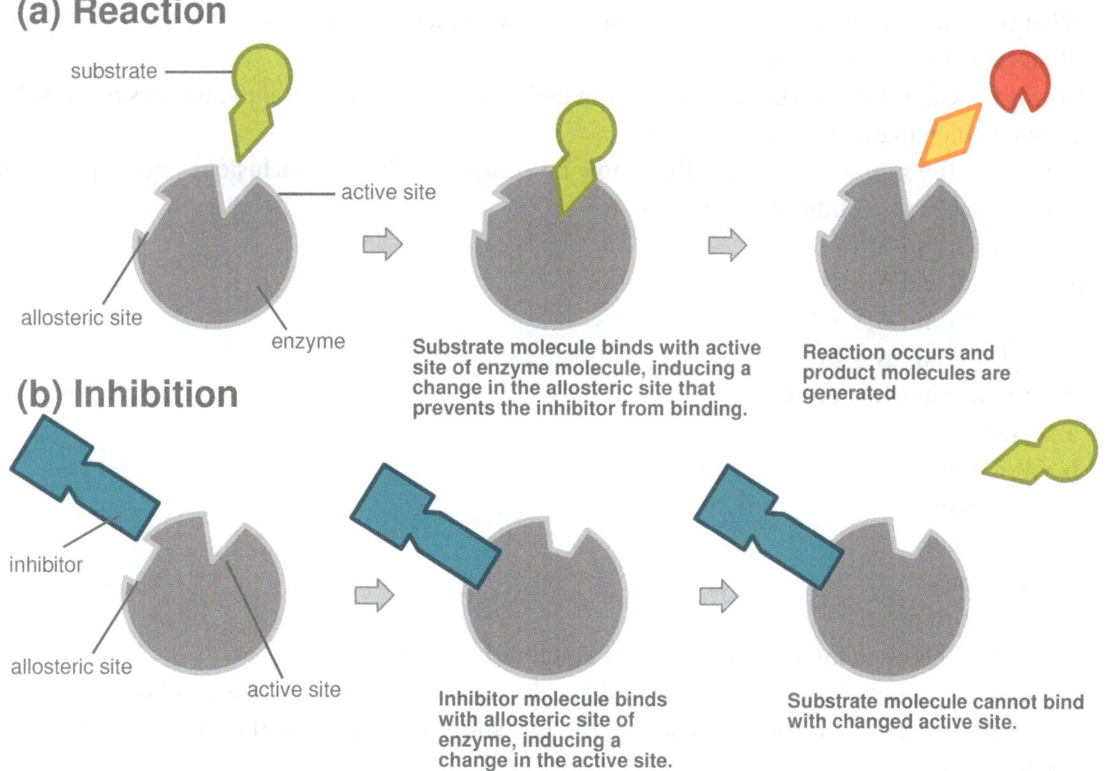


Figure 10.19: Allosteric Inhibition

# Chapter 10 Exercises

1. What is the first law of thermodynamics?
2. Which two physical quantities contribute to a thermodynamic system's total energy?
3. True or false? Heat is a physical substance.
4. True or false? A temperature difference between the thermodynamic system and its surroundings is not necessary for the energy transfer process known as heat.
5. In the context of the kinetic molecular theory of gases, what is temperature?
6. What is enthalpy? What units are used to measure enthalpy?
7. True or false? Only energy differences are measurable.
8. What does a measured temperature difference identify?
9. Using the enthalpy, define an endothermic (endoergic) reaction. Sketch a simple energy diagram.
10. Using the enthalpy, define an exothermic (exoergic) reaction. Sketch a simple energy diagram.
11. What is a reaction for which $\Delta_R H = 0$ called?
12. What is a spontaneous process?
13. What is entropy?
14. Does energy conservation provide an adequate explanation of spontaneity? Explain your answer.
15. How are the values of the entropy assigned to a solid, liquid, and gas related? Using the definition of entropy, explain this relationship.
16. What two concepts are connected by the second law of thermodynamics?
17. What is the Gibbs free energy?
18. How is the Gibbs free energy related to spontaneity? Which two physical parameters must be held constant when making this connection?
19. Determine the sign of $\Delta_R G$ under the following conditions. What conclusions about spontaneity can be reached for conditions a through d?
    a. $\Delta_R H > 0$ and $\Delta_R S > 0$
    b. $\Delta_R H > 0$ and $\Delta_R S < 0$
    c. $\Delta_R H < 0$ and $\Delta_R S > 0$
    d. $\Delta_R H < 0$ and $\Delta_R S < 0$
20. Define the following terms:
    a. Fusion
    b. Friction
    c. Adhesion
    d. Cohesion
    e. Viscosity
    f. Surface tension
21. Distinguish between intermolecular forces and intramolecular forces.
22. By using the values of their quantum numbers, distinguish between fermions and bosons.
23. What are van der Waals forces? What are the two subgroups of forces that make up the van der Waals forces?
24. What is a hydrogen bond?
25. In the context of the solvation process, what is the golden rule?

26. What is meant by the term "phase transition"?
27. What is the vapor pressure of a liquid?
28. Define the term "boiling point." What is the normal boiling point?
29. Using water as an example, how does the temperature of water behave during a phase transition?
30. What is the rate of reaction?
31. What is a catalyst? What appears to be the principal effect of a catalyst?
32. Does a catalyst change the chemical equilibrium of a reaction? Explain.
33. What is a biological catalyst called?
34. What is the active site of an enzyme?
35. Explain the term "substrate specificity."
36. Define the terms "enzyme cofactors" and "coenzymes."
37. Explain the lock-in-key model of enzyme activity.
38. Discuss the induced-fit model of enzyme activity.
39. What is the conformational selection/population shift model of enzyme activity?
40. In the context of enzyme activity, what are the proximity effect, the orientation effect, and the bond energy effect?
41. For enzymes, discuss inhibitory control and allosteric control.
42. What are competitive inhibitors and noncompetitive inhibitors of enzyme activity?
43. Distinguish between reversible and irreversible inhibition.

# CHAPTER ELEVEN
## Building Complex Structures

Our long journey exploring all the preliminary steps that have been required to make *life* possible (at least, as we experience *life* on this planet), has led us on a circuitous route, traversing enormous distances in space and an amazing span of time. If we pause here a moment and look back over this vast landscape of topics, it is impossible to ignore the impressive array of the physical processes that are at work in our world, the sheer number of constituent components that define our universe, and the amazing complementary and coordinated interactions that sustain the integrity of our existence. Beginning with the dawn of time and space, atoms possessing an astounding array of key chemical properties have formed, leading first to simple molecules and, subsequently, to the large biomolecules of life. The energetic and entropic imperatives of this universe have driven and directed the delicate interplay of structure and reactivity, bringing into existence the world we see today. Together, this entire panoply of events would be viewed by some as awesomely beautiful!

As this chapter begins, we are interested in understanding how all these microscopic events (which have macroscopic consequences) have successfully built the **complex structures** of our world. These are structures that depend not simply on a single molecular type or functional group or on a single specific physical interaction. Rather, these complex structures depend on the coordinated interaction of functions and properties that together produce the macroscopic conditions that are critical to life. In particular, this chapter focuses its attention on three key levels of complexity (our complex structures) that are central to life on this planet: **membranes**, **hereditary repositories**, and **proteins**. We shall see how each of these complex structures is built from the biomolecules we met earlier in our story and how the resulting complex structures define life processes. The question of how the majestic edifice built of these complex structures is supplied with the energy needed to sustain life will be reserved for the final chapter of our story.

## The Cell

We begin by examining the structure of a typical **eukaryotic** cell (from the Greek *eu*, meaning "well," and *karyon*, meaning "nut" or "kernel"), which is a cell possessing a nucleus and other organelles enclosed in membranes. Figure 11.1 is a cartoon highlighting some of the key cell structures.

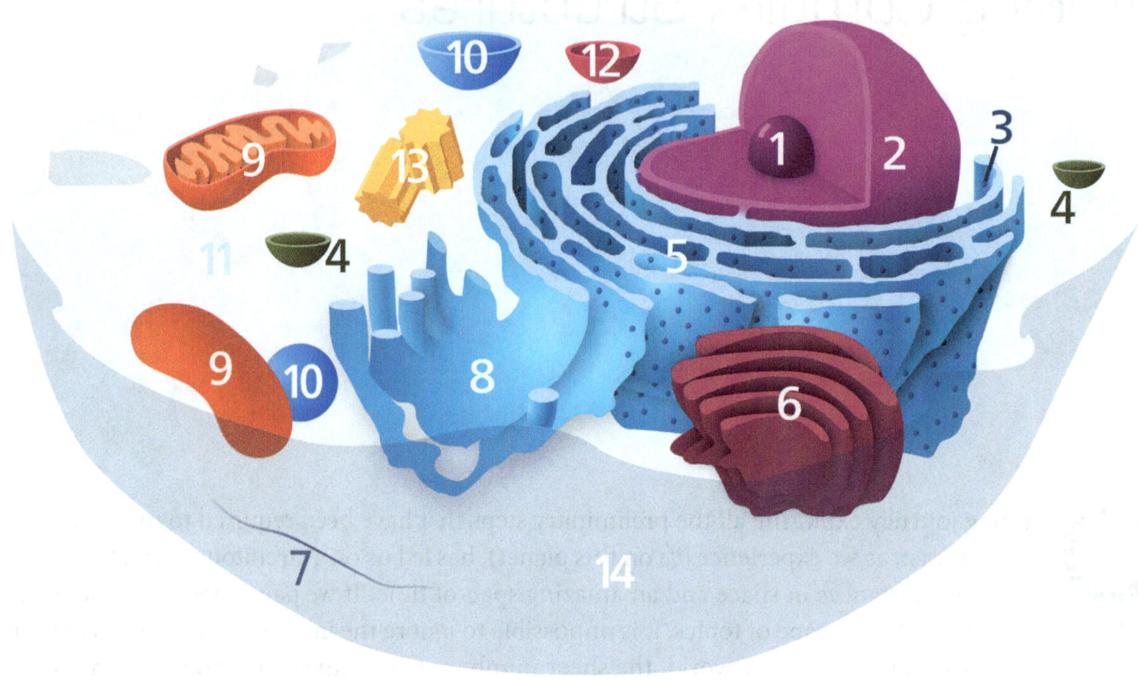


Figure 11.1: Eukaryotic Cell

A number of cell organelles are labeled in figure 11.1:

1. **Nucleolus:** This is a region in the nucleus of the cell that plays a central role in transferring hereditary information to other parts of the cell.
2. **Nucleus:** This is the part of the cell that contains most of the cell's hereditary information, organized in structures called **chromosomes**. (We will examine the chromosome shortly in much greater detail.)
3. **Ribosome (the dots):** The ribosome is the site of **protein synthesis**. Proteins are one of the three complex structures we will examine later in this chapter. They are highly specialized **polypeptides**, our old friends from an earlier chapter.
4. **Vesicle:** The vesicle is an organelle (an organelle is a specialized structure within a cell) located in the **cytosol** of the cell and separated from the cytosol by a lipid membrane. (The **cytosol** is the gel-like intracellular fluid contained within the cell membrane; it is approximately 80% water and is usually transparent.)

5. **Rough endoplasmic reticulum:** This is an interconnected network of flattened, membrane-enclosed sacs or tubes. The outer face of the endoplasmic reticulum is studded with **ribosomes** and is the site of **protein synthesis** in the cell. (We will discuss both **ribosomes** and **protein synthesis** in more detail later in this chapter.)
6. **Golgi apparatus (sometimes called a Golgi body):** This cell organelle was first identified in the late nineteenth century and plays a role in modifying, sorting, and preparing large macromolecules that are synthesized in the cell for secretion by the cell.
7. **Cytoskeleton:** The cytoskeleton is a protein mosaic that forms a skeleton within the cytosol and participates in intracellular transport and cellular division.
8. **Smooth endoplasmic reticulum:** Like the rough endoplasmic reticulum, the smooth endoplasmic reticulum is an interconnected network of flattened, membrane-enclosed sacs or tubes. It lacks ribosomes but participates in **lipid metabolism**, **carbohydrate metabolism**, and the removal of toxins from the cell. (**Metabolism** will be the central topic of the next chapter.)
9. **Mitochondrion:** The mitochondria are cell organelles that generate **most** of the cell's chemical energy. The mitochondria possess **hereditary repositories** that are **independent** of the cell's hereditary molecules.
10. **Vacuole:** Vacuoles are simply enclosed compartments within the cytosol, separated from the cytosol by a **membrane**. They participate in a number of distinct activities within the cell.
11. **Cytosol:** The cytosol is the gel-like intracellular fluid contained within the cell membrane; it is approximately 80% water and is usually transparent. It is the location of multiple processes that occur within a cell: the transmission of molecular signals from the cell's outer membrane to other regions within the cell and the transport of the products of metabolism to other sites within the cell.
12. **Lysosome:** Lysosomes are spherical organelles within the cell, defined by a **membrane** and containing substances capable of breaking down large biomolecules.
13. **Centriole:** The centrioles are cylindrical cell organelles that appear to play a role in cell mitosis (the process of cell division).
14. **Cell membrane:** This is the biological **membrane** that separates a cell's interior from the external environment. It is selectively permeable, and its central function is to protect the cell multiple organelles from the cell's surroundings.

As this brief summary indicates, a eukaryotic cell is a fascinating place filled with complex structures. The above cartoon diagram and brief descriptions of the cell's components and their activities clearly demonstrate the marvelous interacting dynamics that characterize the functioning of a cell.

## Mitochondria and Ribosomes

At this point there are two very important cellular structures that deserve our greater attention: the **mitochondria** and the **ribosomes**. Let's take a closer look at a mitochondrion. (The word *mitochondria* is simply the plural form of *mitochondrion*.) Figure 11.2 is a simple diagrammatic representation of a mitochondrion, highlighting several of its important structures.

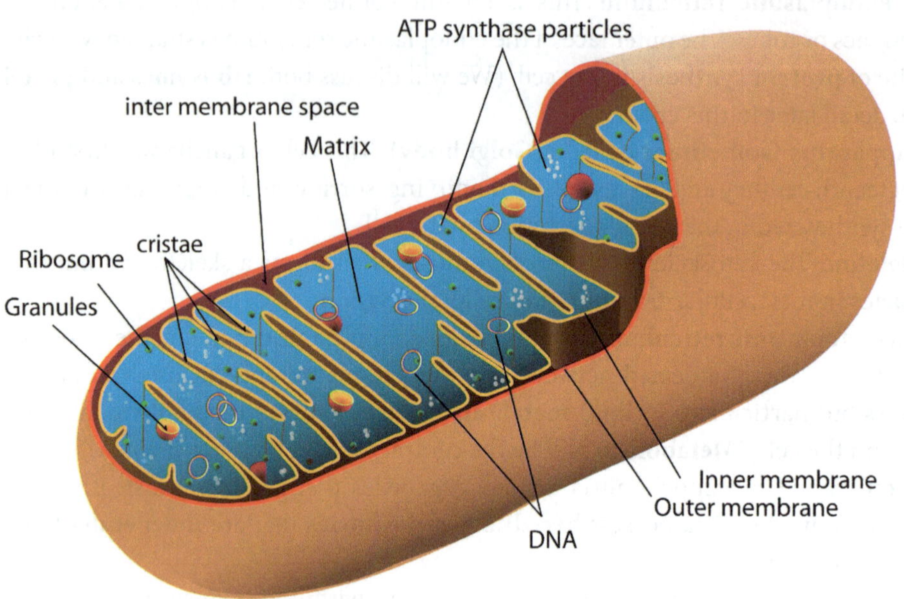


Figure 11.2: Mitochondrion

**Outer membrane:** The outer membrane encloses the entire organelle and is composed of **phospholipids** (a lipid molecule incorporating fatty acids, glycerol, and the phosphate group) as well as proteins. The outer membrane also contains a number of enzymes involved in a wide variety of cellular functions.

**Inner membrane:** The inner membrane of the mitochondrion forms the internal compartments known as **cristae** (see below). The inner mitochondrial membrane is also the site at which a key component of metabolism takes place. (This metabolic process is known as the **electron transport chain** and **oxidative phosphorylation**, and it will be discussed in the next chapter.)

**Inter membrane space:** This is the space between the outer and inner mitochondrial membranes.

**Cristae:** These are the compartments formed by the folding of the inner mitochondrial membrane. As a result of the folding, the mitochondria have a very large surface area at which metabolic processes occur.

**Matrix:** This is the actual space bounded by the inner mitochondrial membrane, and it contains the mitochondrion's ribosome and hereditary material (DNA). The fluid of the matrix is more viscous than the relatively aqueous cytosol.

**Ribosome:** The ribosomes are the structures that synthesize the specific mitochondrial proteins using information derived from the mitochondrial DNA.

**Granules:** While not well understood, the mitochondrial granules located in the matrix appear to play a role in regulating the internal environment of the mitochondrion.

**DNA:** The mitochondrion possesses hereditary information that is **independent** of the cell's hereditary information that is located in the cell's nucleus.

**ATP synthase particles:** These are sites located on the inner mitochondrial membrane at which the final metabolic step (oxidative phosphorylation) occurs.

While the general structures depicted in figure 11.2 and described briefly above provide an overall outline of the mitochondria, they give only the barest hint of the critically important role played by the mitochondria in the amazing symphony we call *life*. Even though the mitochondria are one of the many components found in a eukaryotic cell, they are the essential sources of cellular energy. In the following chapter we shall discover that the most fundamental levels of catabolism (the degradation of chemical compounds to produce energy, a process often referred to as **cellular respiration**) occur **only** in the mitochondria. The mitochondria are the sites of a redox process that provides **most** of the energy required by every aspect of life. Interestingly, because the mitochondria possess independent hereditary material (independent of the cell's hereditary information concentrated in the nucleus), it has been suggested (early in the twenty-first century) that the mitochondria are in fact **endosymbionts**. That is, the mitochondria were originally independent **prokaryotic** organisms capable of performing the redox processes that eukaryotic cells cannot perform by themselves. These prokaryotic cells established a symbiotic relationship, living inside the eukaryotic cells. This hypothesis has been supported (as recently as 2011) by noting that a group of bacteria appear to share a relatively recent common ancestor with the mitochondria found in eukaryotic cells.

The second cellular structure that plays an indispensable part in the process of life as we understand it is the **ribosome**. The above discussion has already indicated that the eukaryotic cell itself and the mitochondria **both** contain ribosomes, suggesting that the ribosome must perform a centrally important function for both the cell as a whole and as well as for the mitochondria. Indeed, the two words we used earlier, **protein synthesis**, are the keys to understanding the critical function of the ribosome. Figure 11.3 is a simple cartoon that highlights several key parts of the ribosome. Note that there are two subunits, a **large subunit** and a **small subunit**. The two regions of the large subunit identified as the **A site** and the **P site** are the two key locations in the process of protein synthesis that we shall examine in some detail at the end of this chapter.

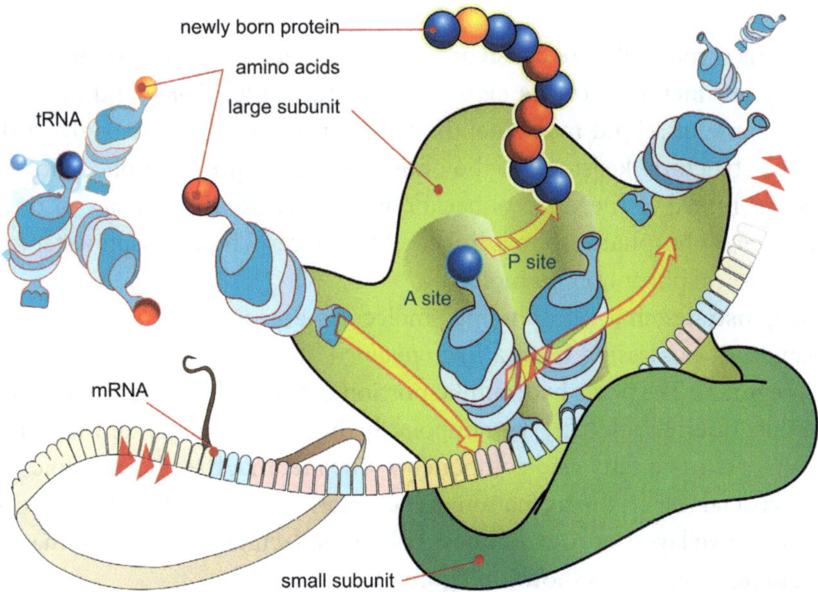


Figure 11.3: Ribosome

The reader may wonder why we have increasingly used the term *proteins*; after all, they are *only* specialized polypeptides. Why not simply use the single term *polypeptide*? Part of the answer to this question lies within the discipline of biochemistry itself, where the term *protein* is ubiquitous. The word itself comes from the Greek *proteios*, meaning "primary," "in the lead," or "standing in front," which is indicative of the more profound importance of proteins. We will soon find that proteins are essential to almost every facet of life. The structural proteins are the basis of every muscle and are essential to the cytoskeleton of every eukaryotic cell. Proteins are the enzymes that catalyze biochemical reactions and are critical participants in metabolic processes; without the tremendous increase in the rate of reaction achieved by these biochemical catalysts, life processes would occur millions of times more slowly. Proteins play crucial roles in **intercellular signaling**, the **immune response**, and the ability of cells to **adhere** to one another (essential for wound healing, for example). It is not an exaggeration to say that the story of life on this planet is really the story of proteins.

## Membranes

Our earlier examination of the eukaryotic cell and several of its most important organelles implicitly highlighted the importance of the many diverse **membrane** structures that are present in a cell. Membranes are the boundaries that separate individual cells from one another. Further, they form the boundaries of the organelles inside each cell. Consequently, the membranes serve several critical functions for both the cell and the various cell organelles: (1) they separate cells and organelles from their external environment, thereby defining a cell and each organelle within a cell; (2) they provide selective transport of nutrients into cells and organelles and transport the waste products generated by cells and organelles to the external environment; and (3) they selectively control the transfer of the specialized product molecules generated in each organelle, making them available to other regions in the cell.

Membranes in eukaryotic cells are constructed from molecules known as **phospholipids**. As their name implies, they are members of the class of molecules called *lipids* and are constructed much like our earlier example of a lipid molecule, the triglyceride. As with the triglycerides, the glycerol molecule forms the structural keystone of the phospholipids; however, a phospholipid has only **two fatty acid** molecules linked to the glycerol backbone by ester bonds. In place of the third fatty acid found in triglycerides, a phosphate group is linked to the glycerol by a phosphoester bond. Figure 11.4 is a schematic of a typical phospholipid molecule.

The figure demonstrates that phospholipid molecules are *amphipathic* molecules (we first met this term in chapter 7), possessing **both** a *hydrophobic* ("water-fearing") segment and a *hydrophilic* ("water-loving") segment. The polar hydrophilic portion of the molecule is made up of the phosphate group bound to the glycerol molecule; the nonpolar hydrophobic portion of the molecule is the long hydrocarbon chain of the two fatty acid components. The fact that the phospholipids are amphipathic molecules plays a crucial role in the formation of membranes; membranes are composed of **phospholipid bilayers**, that is, two layers or rows of phospholipid molecules organized so that the hydrocarbon tails are adjacent to one another, as shown in figure 11.5.

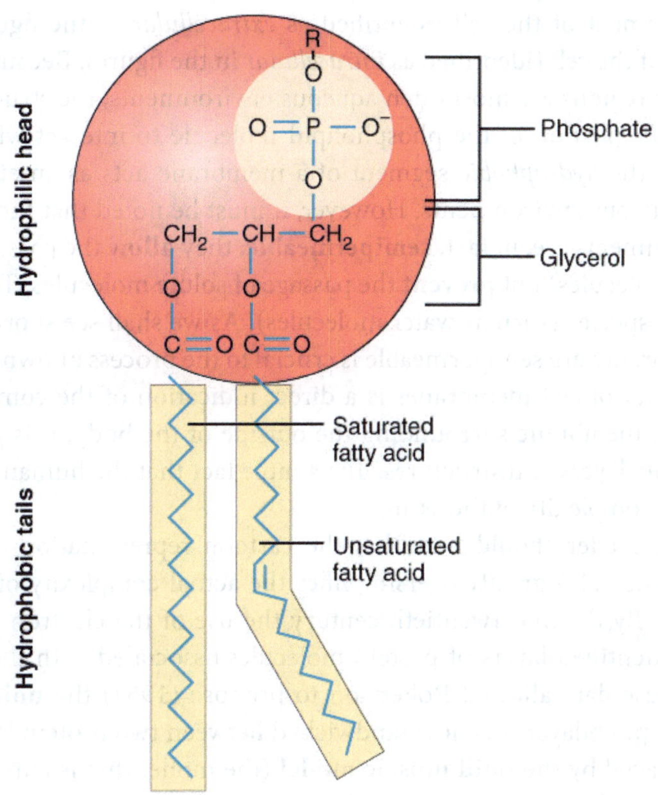


Figure 11.4: Phospholipid

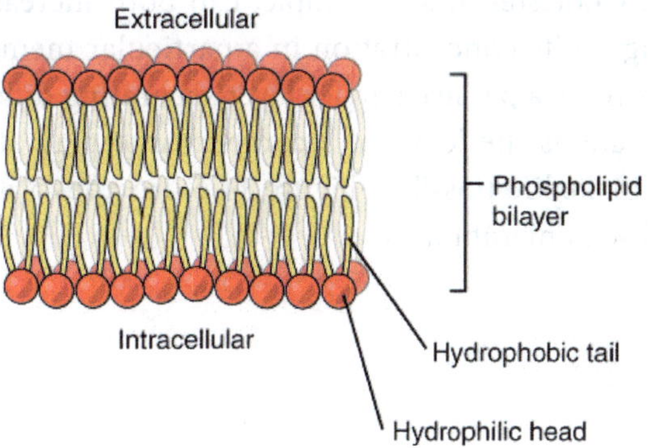


Figure 11.5: Membrane Structure

The polar portion of the phospholipid molecules form the outer layers of the membrane, one facing the external environment of the cell (identified as *extracellular* in the figure), the other oriented toward the interior of the cell (identified as *intracellular* in the figure). Because both the *extracellular* and the *intracellular* regions are most often aqueous environments, the structure of the membranes allows the *hydrophilic* portion of the phospholipid molecule to interact with these environments. On the other hand, the *hydrophobic* segment of a membrane acts as an effective barrier between the two different aqueous environments. However, it must be noted that the barriers separating the two aqueous environments are, in fact, **semipermeable**; they **allow** the passage of solvent molecules (principally water molecules) but **prevent** the passage of solute molecules. They are most permeable to small, uncharged species (such as water molecules). As we shall see shortly, the fact that cellular and organelle membranes are semipermeable is crucial to the process known as **osmosis** (see below). (The semipermeability of cell membranes is a direct indication of the complexity of human **skin**. Skin is **not** simply a membrane surrounding the outside of the body; it is a complex organ that is composed of multiple layers and structures. The simple fact that the human body does not "leak" is an indication of the complexity of the skin.)

Additionally, the reader should note that the cartoon representation of the phospholipid bilayer depicted in figure 11.5 greatly oversimplifies the actual complexity of the membranes found in living organisms. By the mid-twentieth century the use of the electron microscope (developed during the 1930s) identified layers of protein molecules associated with the two hydrophilic sides of a membrane. These data allowed Robertson to propose (1957) the **unit membrane** model, in which the phospholipid bilayer was now sandwiched between two protein layers. In the early 1970s this model was replaced by the **fluid mosaic model** (the model that is currently used), which suggests that a number of distinct molecules and structures dynamically "float" in the lipid bilayer of the membrane: carbohydrates, glycoproteins (molecules formed from a sugar and a polypeptide), cholesterol, protein channels (important for transmembrane transport, as we shall shortly see), and the cytoskeleton.

Each of these molecules and structures contributes to the dynamic functioning of a cell and the organelles within a cell. Cholesterol, for example, can both increase and decrease the fluidity of the membrane, depending on its concentration in a particular membrane region. In a very real sense, the membrane is not simply a passive barrier layer separating different cellular environments. On the contrary, the membrane is the locus of a complex interplay of molecules, structures, and molecular interactions that make life possible. Figure 11.6 provides a more realistic representation of the complexity associated with membranes.

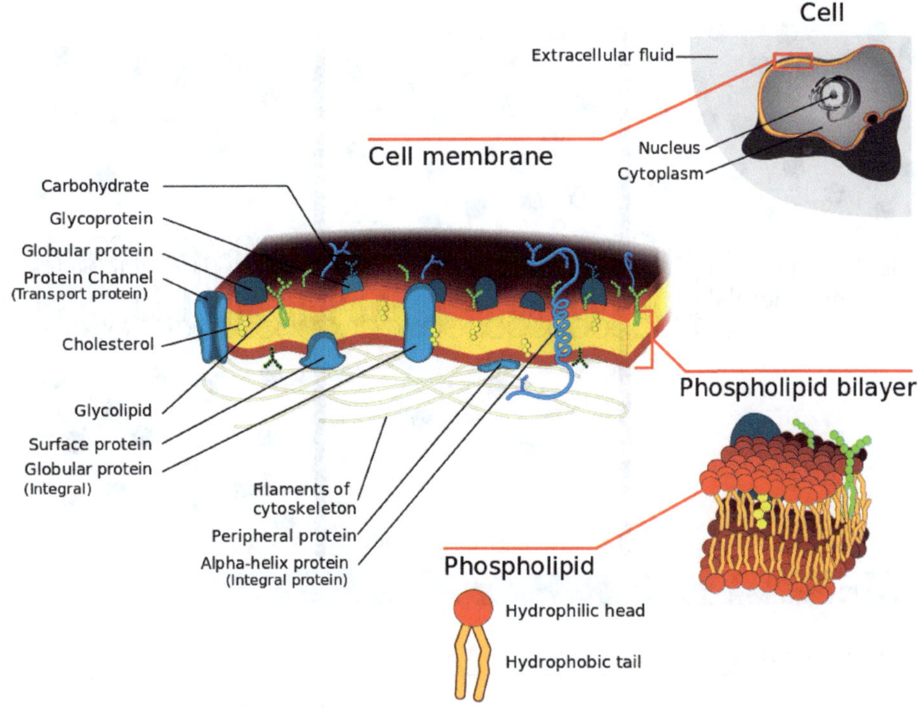


Figure 11.6: Membrane Complexity

## Transport within Membranes

Because membranes are much more than simple passive barriers separating two different environments, it is not surprising to discover the **active** role played by the membrane (more specifically, the complex molecules and structures that make up a membrane) in both controlling and facilitating the movement of solute molecules from one side of the membrane to the other side. In fact, the complex structure of membranes enables them to be **selectively permeable with respect to the solute**, exercising direct control on the solute molecules that pass between the two regions bounding a membrane. We shall consider three processes: (1) **simple diffusion**, (2) **facilitated transport**, and (3) **active transport**.

Simple diffusion (from the Latin *diffudere*, meaning "to spread out") is best understood from a particulate point of view and is more precisely termed *molecular diffusion*. It is a microscopic phenomenon that occurs spontaneously in our world and is caused by the **randomly distributed kinetic energy** associated with the particles that are spreading out. Diffusion effectively explains the movement of particles (molecules) from a region of **higher** concentration to a region of **lower** concentration, although a concentration gradient (a difference in the concentration between two regions) is not required by the diffusion process. In the context of a membrane separating regions of two **different** concentrations, simple diffusion describes the observed movement of particles across the membrane.

Figure 11.7 displays graphically the diffusive process. The three panels of the figure display diffusion as a function of time. The time increases from left to right in the figure, and the concentration disparity is greatest in the leftmost panel of the figure and is the smallest in the rightmost panel.

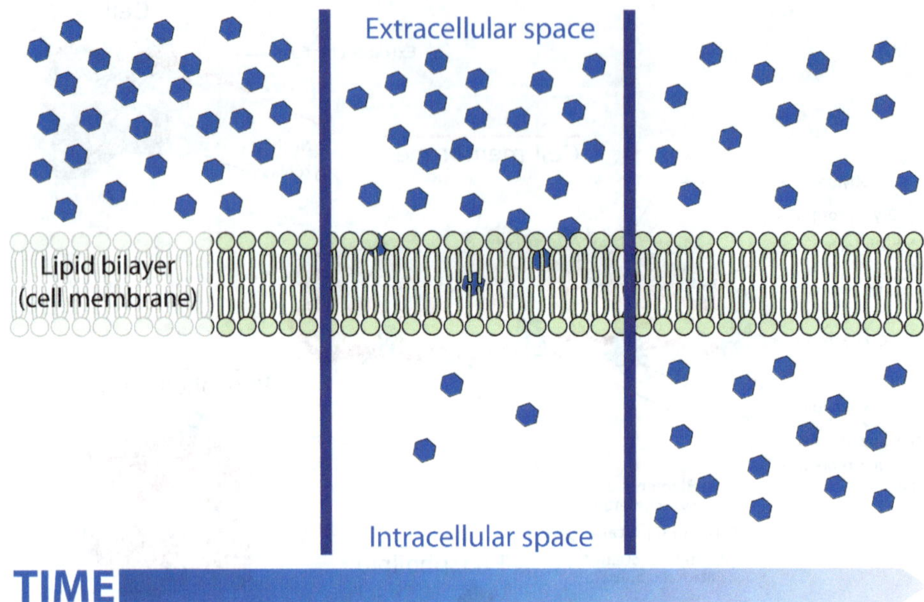


Figure 11.7: Simple Diffusion

Because of the bilayer structure of membranes, polar molecules and charged species diffuse very slowly across the membranes of the cell. Consequently, to make it possible for polar molecules and charged species to traverse a membrane in a more timely fashion, polar protein structures (ionophores and aquaporins are two examples) are integrated directly into the membrane itself (see figure 11.6) and provide a channel through the hydrophobic portion of the membrane. This facilitated transport still relies on the randomly distributed kinetic energy associated with the molecules crossing the membrane and, like simple diffusion, involves movement from a region of higher concentration to a region of lower concentration. Figure 11.8 provides a cartoon showing the facilitated transport process.

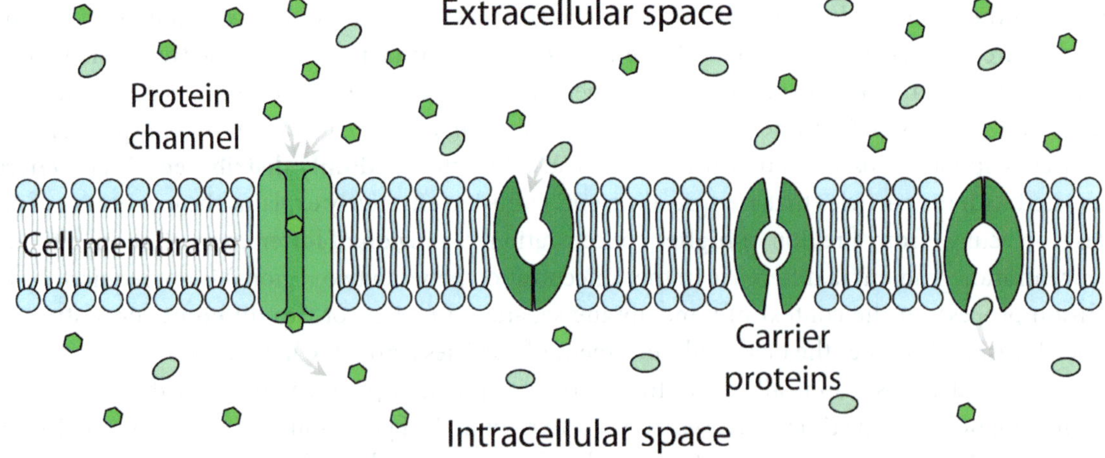


Figure 11.8: Facilitated Transport

The third commonly occurring membrane transport mechanism is called active transport. Unlike the two previous transport processes that we have discussed, active transport requires the expenditure of cellular energy. (In chapter 12, we will discover that organisms store cellular energy in the ATP molecule.) This process moves molecules from a region of lower concentration to a region of higher concentration and must utilize stored energy to complete the process. There are many specific occurrences of the active transport in cellular systems. Figure 11.9 depicts the mechanism used by cells to transport sodium ions and potassium ions across the cell membrane. Notice that the concentration of sodium ions is larger in the upper half of the diagram and that sodium ions are transported from a region of lower concentration to a region of higher concentration (from the bottom of the figure to the top of the figure). The situation for the potassium ions in the figure is exactly reversed: the concentration of potassium ions is larger in the lower half of the figure. Again, potassium ions are transported from a region of lower concentration to a region of higher concentration (from the top of the figure to the bottom of the figure). The energy used by the cell in the active transport process is liberated by converting an ATP molecule to an ADP molecule and a free phosphate group (symbolized by the $P_i$ symbol in the figure).

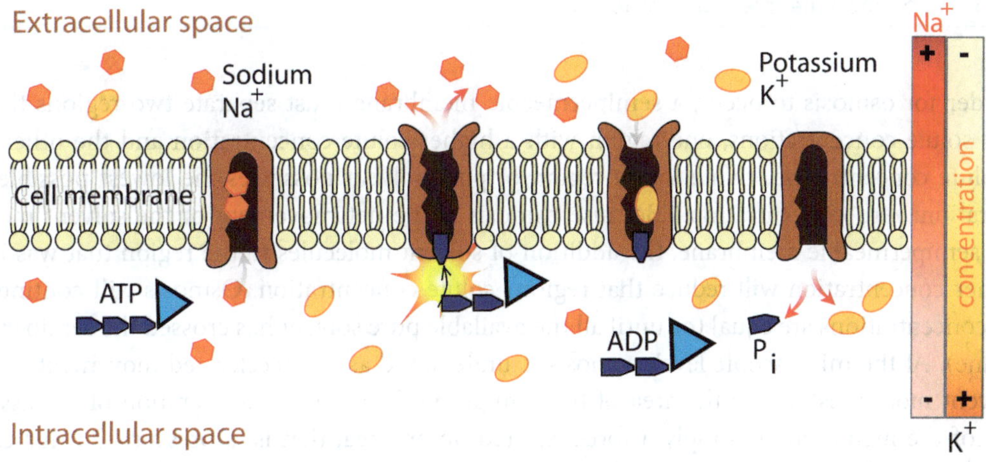


Figure 11.9: Active Transport

## Osmosis

Another experimentally observed phenomenon associated with cell membranes is called **osmosis**. Up to this point in our discussion membrane transport, we have focused our attention on the solute molecules present in either the external or the internal environments. **Osmosis**, on the other hand, is the net movement of solvent molecules through a **semipermeable membrane**, either from a **pure solvent** or from a **dilute solution** into a region with a **higher** solute concentration. The key to osmosis is the semipermeable membrane, that is, a membrane that **allows** the passage of solvent molecules but **prevents** the passage of solute molecules. As we noted earlier, the membranes that are found in living organisms (both cell and organelle membranes) are essentially semipermeable with respect to the water solvent. Figure 11.10 is a schematic depiction of a semipermeable membrane.

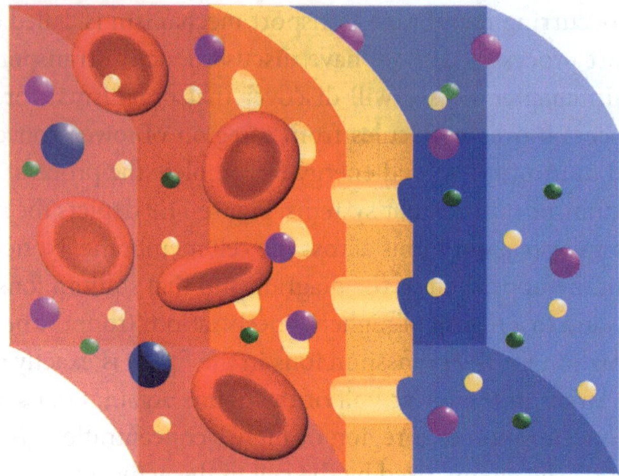

Note that the membrane allows the passage of some constituents in the solution (the spherical objects) but prevents the passage of other objects (the large disks).

Figure 11.10: Semipermeable Membrane

In order for osmosis to occur, a semipermeable membrane must separate two regions that have different solute concentrations, one region with a higher solute concentration and the other with a lower solute concentration. (Note that a region of pure solvent possesses the lowest possible solute concentration: zero concentration.) Because osmosis is the net movement of the solvent molecules across a semipermeable membrane, the addition of solvent molecules to the region that was initially at a higher concentration will reduce that region's solute concentration. Osmosis will continue until the two concentrations are equal (or until all the available pure solvent has crossed the semipermeable membrane). At the microscopic level, osmosis is understood as the accelerated movement of a mass (the solvent molecules) across the area of the semipermeable; but the acceleration of a mass across the area of the membrane is simply a force exerted on an area, that is, a pressure! In this case the pressure is called the osmotic pressure. The osmotic pressure of a solution is measured by determining the opposing pressure required to stop the osmosis. (Practically, this means applying an opposing pressure until the concentration in the region that was initially at a higher concentration stops changing.)

There are three important terms associated with osmosis:

1. **Isotonic:** Two solutions of **equal** concentrations and, hence, exhibiting the **same** osmotic pressure are called isotonic.
2. **Hypertonic:** For two solutions having **different** concentrations and, hence, **unequal** osmotic pressures, the **more concentrated** one (the one with a higher concentration of solute) is called hypertonic.
3. **Hypotonic:** For two solutions having **different** concentrations and, hence, **unequal** osmotic pressures, the **more dilute** one (the one with a lower concentration of solute) is called hypotonic.

# Hereditary Repositories

Over a period that extends through thousands of years of human history, humankind has recognized identifiable patterns in the physical characteristics, commonly called *traits*, which are found within the human population. Eye color, hair color, height, skin color, and facial shapes, to name only a few of the most obvious traits, have been recognized to occur in identifiable patterns within families, clans, tribes, and races. The total collection of these traits or characteristics is known as a **phenotype**. The human recognition of and concern for a species' phenotype is even more evident in the history of both horse and dog breeding. It appears that humankind has, through selective breeding, enhanced the expression of specific traits in horses since the second century BCE. In the case of dogs, selective breeding by humankind may extend over more than fifteen thousand years, culminating in the wide diversity of modern dog breeds.

In the mid-nineteenth century the Augustian monk Gregor Mendel initiated a careful study of the heritable characteristics of pea plants. Published in an obscure journal, *Verhandlungen des naturforschenden Vereins Brünn* (*Proceedings of the Natural History Society of Brno*), his paper went unrecognized until the beginning of the twentieth century, when independent experimental work and a rediscovery of Mendel's original paper initiated the modern science of genetics. Since the work of Watson and Crick in the early 1950s, our understanding of heredity and the science of genetics has been deeply rooted in our understanding of the biochemistry of living organisms.

## Nucleic Acids: The Keys to Heredity

In chapter 7 we met the nucleic acids and discovered that **each** nucleotide is composed of a sugar molecule (either ribose or deoxyribose), a nitrogenous base (either a purine or a pyrimidine), and a phosphate group. The bases are bonded to the sugars by a bond resembling the *glycosidic* bonds of carbohydrates to form a **nucleoside**. The nucleosides are bonded to a phosphate group via a phosphoester bond to form a **nucleotide**, and, finally, the multiple nucleotide units can be linked together by phosphodiester bonds to form a **polynucleotide**. In figure 11.11 we reproduce figure 7.58, showing the linking of two nucleotides by a phosphodiester bond.

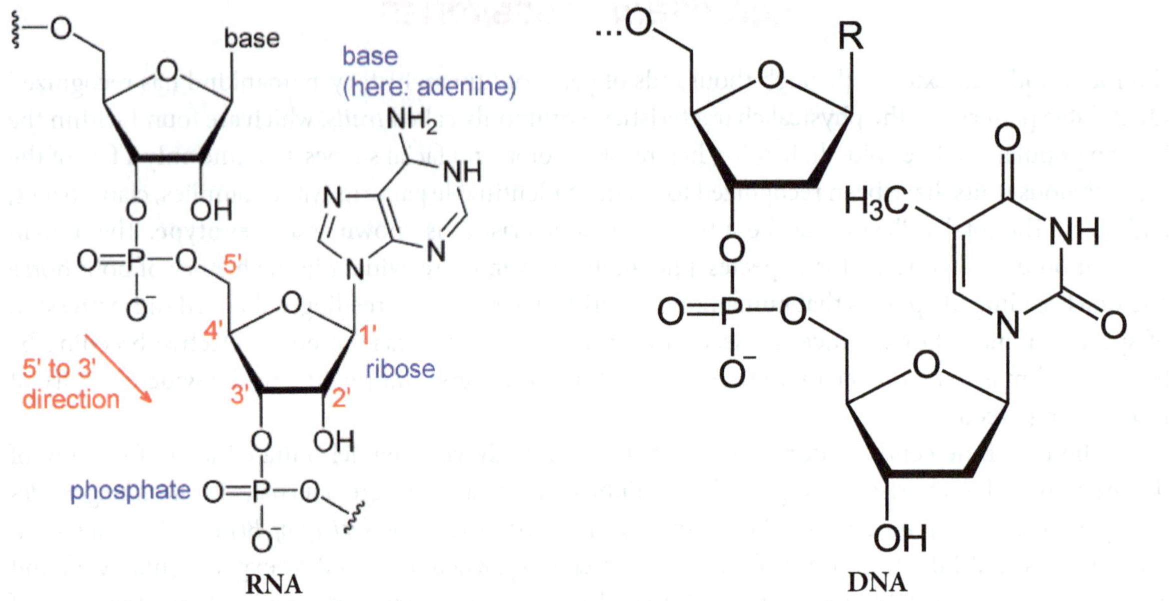

Figure 11.11: Phosphodiester Bonds in RNA and DNA

By the early 1950s, particularly with the work of Watson and Crick, it became clear that the hereditary information that is passed from one generation of living organism to the next is encoded in the deoxyribonucleic acid (DNA) molecules first discovered in the cell's nucleus in the nineteenth century. While it was the observation of the acidic properties of the nucleic acids that initially drew the attention of chemists, this early observation has now paled in significance compared to the critical role that these molecules play in the preservation, transmission, and utilization of the hereditary information that is essential to life.

We now understand that the hereditary biomolecules exhibit **three** levels of structure. Each nucleotide in a DNA polynucleotide includes one of the four nitrogenous base molecules, adenine, thymine, cytosine, or guanine (often abbreviated as A, T, C, and G). Because a polynucleotide molecule is made up of many thousands of nucleotides sequentially linked, the four bases, A, T, C, and G, occur in a specific pattern or sequence. This nucleotide sequence is called the **primary structure** of a nucleic acid. However, because of the different molecular configurations of the purine and pyrimidine molecules (see chapter 7), it is possible to form only adenine-thymine (A-T) and cytosine-guanine (C-G) **complementary pairs** through a hydrogen bond interaction. Figure 11.12 shows the complementary interactions.

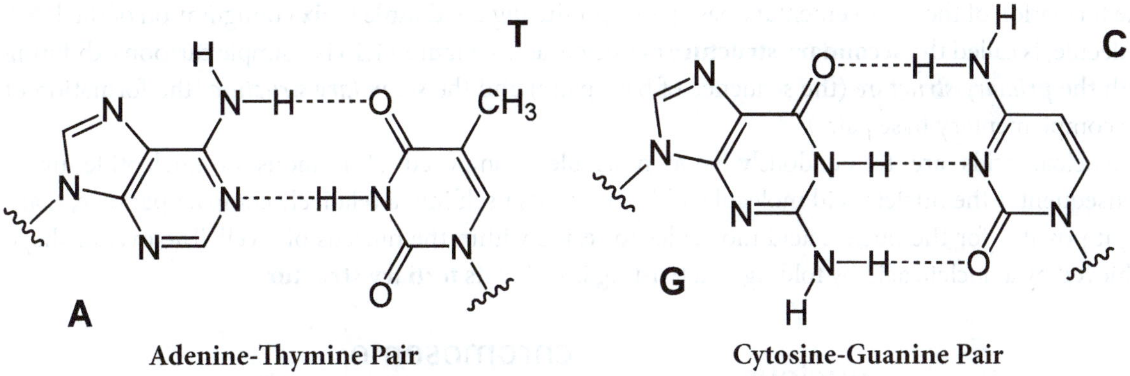

**Adenine-Thymine Pair** **Cytosine-Guanine Pair**

Figure 11.12: Complementary Base Pairs

We first met the hydrogen bond interaction in chapter 10 and noted that the hydrogen bond in water (approximately 23 kJ $mol^{-1}$) is considerably weaker than the strong *intramolecular* bonds that stabilize atoms in molecules. While the hydrogen bonds that are responsible for the formation of complementary base pairs (adenine with thymine and cytosine with guanine) are also much weaker than the intramolecular bonds found in either the purine or pyrimidine molecules, they are sufficiently strong to guarantee the relative stability of the DNA molecule. However, as we shall shortly see, they are not too strong!

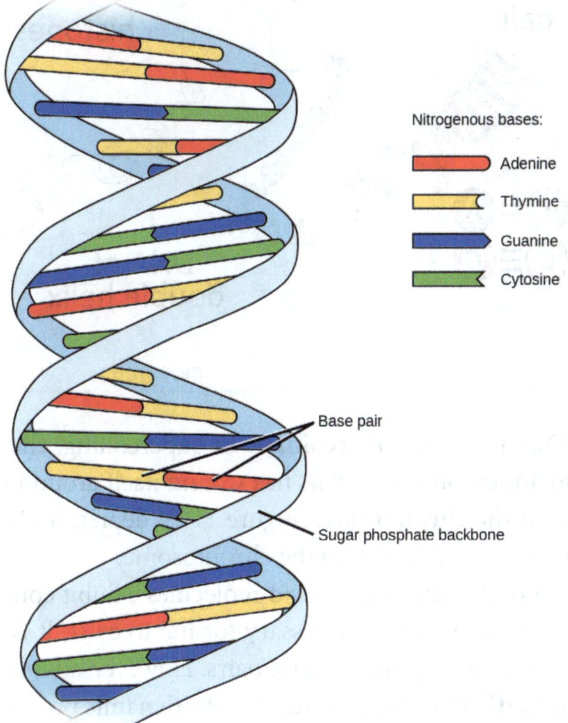


Figure 11.13: DNA Double Helix

The formation of the complementary base pairs, producing the **double helix** configuration of the DNA molecule, is called the **secondary structure** of nucleic acids. Figure 11.13 is a simple cartoon exhibiting both the *primary structure* (the sequence of base pairs) and the *secondary structure* (the formation of the complementary base pairs).

Nucleic acids are tremendously large molecules composed of millions of nucleotide units. Consequently, the nucleic acid molecule folds and twists itself into a relatively compact package, making it possible for the nucleic acid molecules to reside within the nucleus of a cell. The overall shape achieved by a nucleic acid by folding and twisting is called its **tertiary structure**.

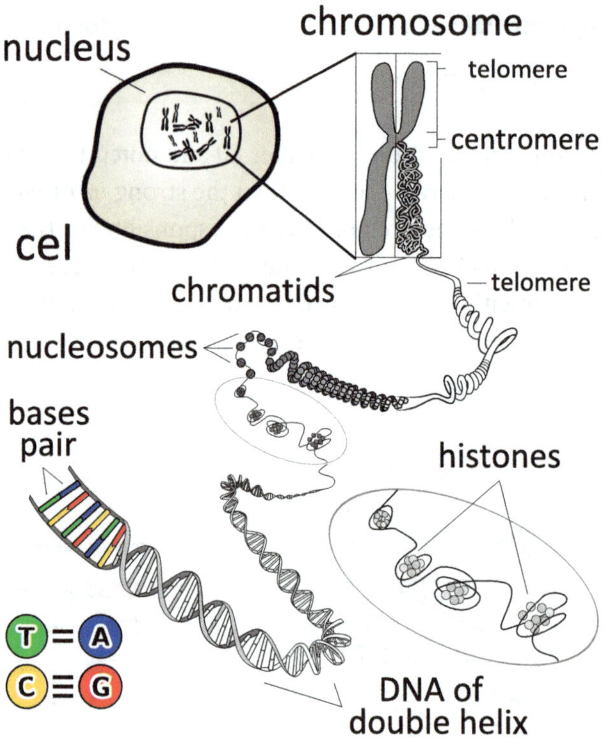


Figure 11.14: Nucleic Acid Supercoiling to Form the Chromosome

The further twisting of DNA is called **supercoiling**. In supercoiling, the DNA is wrapped around protein molecules called **histones**; once the DNA has wound itself around the histone molecules, it is packed into a structure called the **chromosome**. Figure 11.14 depicts a diagram of the process that a nucleic acid molecule follows as it is folded into the chromosome.

The three levels of structure that the nucleic acid molecules exhibit constitute the essential storage structure of the key hereditary information necessary for life to exist. It is the *primary structure* that carries the central information: the sequence of base pairs. However, simply encoding the information that makes life *possible* is not **sufficient** to guarantee that the dynamic processes defining life do, in fact, take place. There are two absolutely critical steps about which we have been silent: **nucleic acid replication** and **protein synthesis**. Let's begin by examining the manner by which nucleic acids replicate.

## Nucleic Acid Replication

The key to understanding the nucleic acid replication process is to remember that the formation of a **double-stranded** molecule (that is, a molecule made up of two long chains; see figure 11.13) took place by the procedure of forming **complementary base pairs**; adenine is **always** paired with thymine, and cytosine is **always** paired with guanine. This means that the two strands contain the **identical sequence** of base pairs; starting at one end of the molecule (let's call it the "top" end), the sequence of base pairs in one strand running from "top" to "bottom" (we'll call the other end of the molecule the "bottom" end) is **identical** to the sequence of base pairs running from "bottom" to "top" (that is, in reverse order). Consequently, if it were possible to separate the two strands and allow complementary base pairing to occur again with each strand, the result is **two identical molecules**! But this is exactly the step at which the crucial hydrogen bond between the base pairs (see figure 11.12) enters the picture. Remember, the hydrogen bonds between the complementary purine and pyrimidine pairs are **strong enough** to provide the nucleic acid molecule with moderate stability. However, the hydrogen bonds **are not so strong** as to prevent the formation of two **single-stranded** polynucleotide molecules from which, through the complementary base pairing process, two identical molecules are formed. Figure 11.15 shows this crucial step in the replication process.

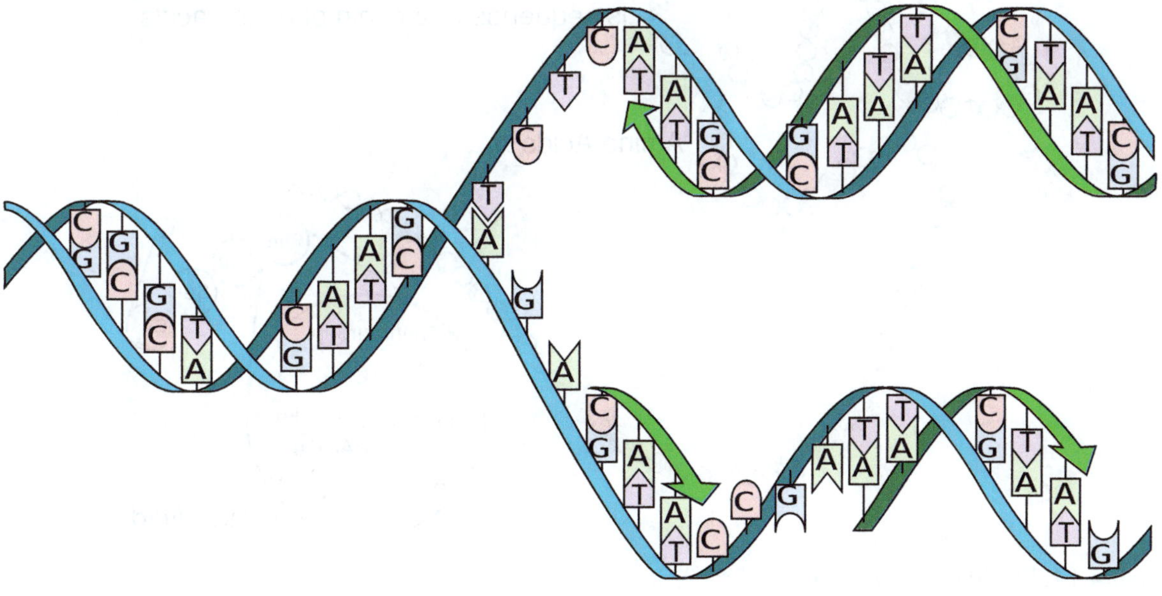

Figure 11.15: Nucleic Acid Replication

# Protein Structure

Since our first encounter with *polypeptide* molecules in chapter 7, we have increasingly used the term **protein** (the term most commonly used in biochemistry for this diverse group of molecules) to describe the polypeptide molecules, noting their importance as enzymes, as structural components of cells, and as integral components of cell membranes. Further, proteins play critical roles in intercellular signaling, the immune response, and the ability of cells to adhere to one another. As we noted earlier in this chapter, it is not an exaggeration to say that the story of life on this planet is really the story of proteins.

Proteins have been studied intensely since the middle of the nineteenth century, and one consequence of this long period of study has been the recognition that protein molecules exhibit characteristic structures. In particular we will examine **four levels** of protein structure. The proteins of eukaryotic cells are polypeptides, and as such, are built from less than one hundred to many thousands of the twenty-one α amino acids that occur in eukaryotic cells. The **primary structure** of a protein is determined by the **number, kind, and sequence** of the amino acid units comprising the polypeptide chain or chains making up the molecule. Because each amino acid is characterized by its side-chain, the primary structure determines the **alignment of side-chains** in a protein, which, in turn, determines the **three-dimensional shape** into which a protein folds. **Hence, the amino acid sequence is the most important factor determining a protein's three-dimensional shape.** Figure 11.16 emphasizes the number, kind, and sequence of the amino acid units in a protein. Figuratively, the amino acid units behave like the individual pearls in a long pearl necklace; however, it is critically important to remember that all the pearls in the necklace are **not identical.**

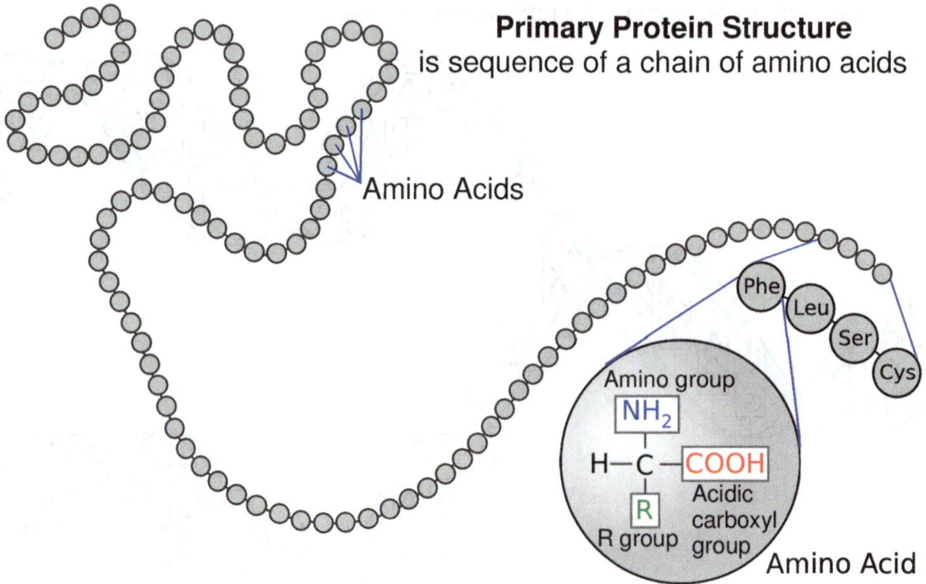


Figure 11.16: Primary Structure of a Protein

One of the initial observations made in the study of proteins was that large segments of the molecules exhibit a particular characteristic regularity. Two large structures were easily identified: the **α-helix** and the **β-pleated-sheet**; they are shown in figure 11.17. The helix is the familiar structure that we have already encountered in the helical arrangement of nucleotides in nucleic acids; the β-pleated-sheet structure resembles the planar pattern that is used in corrugated cardboard. These structures can occur multiple times within a single protein molecule and clearly determine a very characteristic three-dimensional spatial arrangement of segments of a protein. These regular three-dimensional spatial arrangements constitute the **secondary structure** of proteins.

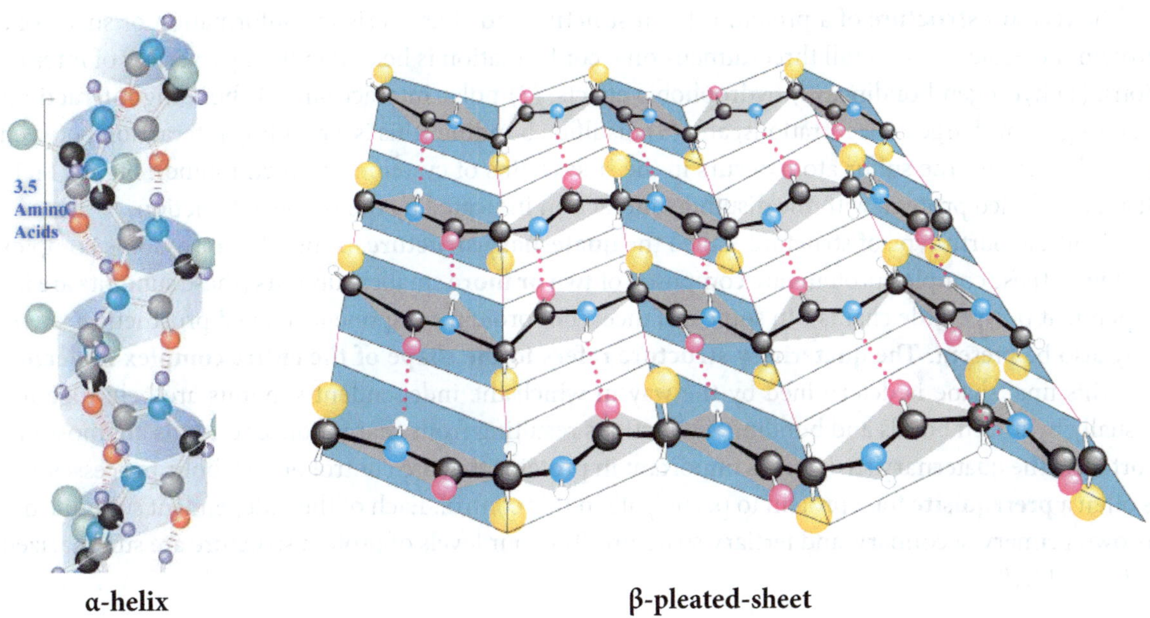


Figure 11.17: Secondary Structure of a Protein

Formally, the **secondary structure** of proteins can be characterized as a regular dimensional structure held together by **hydrogen bonds** between the oxygen atom of a carbonyl group and the hydrogen atom bound to nitrogen atoms of the amino groups in polypeptide chains (note the dashed lines in figure 11.17, indicating hydrogen bonds). A protein's secondary structure is particularly important in determining the strength of a protein.

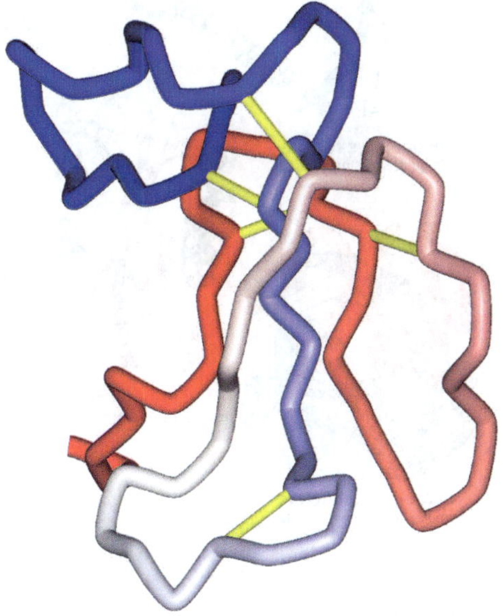

Figure 11.18: Folded Protein Showing Disulfide Bonds

The **tertiary structure** of a protein is the distinctive and characteristic conformation or shape of a protein molecule. This overall three-dimensional conformation is held together by a variety of interactions: (1) hydrogen bonding, (2) hydrophobic effects, (3) polar interactions, (4) bonding interactions resulting from charge concentrations, and (5) disulfide bonding (this is a bonding interaction between two sulfur atoms; the sulfur atom occurs in the side-chains of cysteine and methionine). Figure 11.18 displays a folded protein, with the disulfide interaction indicated by the yellow interaction.

Finally, a fourth type of structure, called the **quaternary structure**, is found in **only** some complex proteins. These complex proteins are composed of **two or more** smaller subunits (these subunits are independent polypeptide chains). In some instances, nonprotein components (called prosthetic groups) may also be present. **The quaternary structure refers to the shape of the entire complex molecule**, and this final shape is determined by the way in which the independent subunits are held together (usually hydrogen bonds and bonding interactions resulting from charge concentrations are most important). The quaternary structure is important in proteins that are control of metabolic processes and is often a **prerequisite** for a protein to participate in this control. Each of the independent subunits has its own primary, secondary, and tertiary structure. The four levels of protein structure are summarized in figure 11.19.

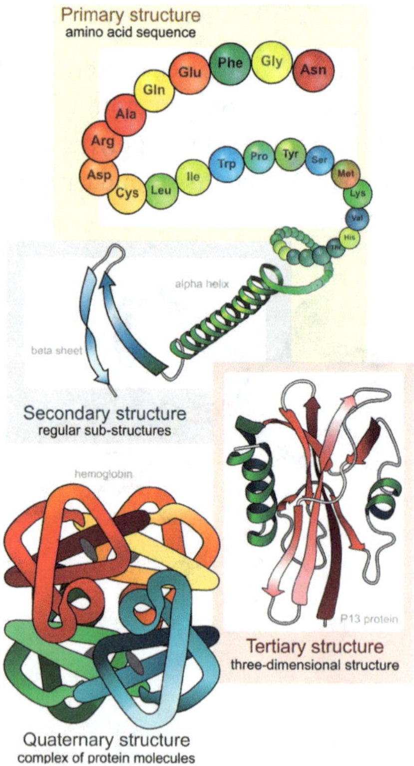


Figure 11.19: Summary of Protein Structures

If the interactions that maintain a protein's secondary, tertiary, and quaternary structure are disrupted by either physical or chemical means, the protein is said to be **denatured**. The denaturation process either completely destroys or significantly diminishes those interactions that maintain a protein's characteristic three-dimensional arrangement. Because it is this characteristic three-dimensional arrangement that determines the biological activity of a protein, a **denatured protein** loses its ability to participate in biochemical processes. The denaturation process **does not**, however, change a protein's primary structure.

## Protein Synthesis

The fact that proteins are ubiquitous in living organisms underscores their importance to the life process. We have already seen that they play a number of critical functions at multiple sites in a cell's biochemistry and that these functions depend crucially on the three-dimensional structure of each protein. However, there is another level of specificity that we have, up to this point, ignored. In addition to their highly specific geometry, proteins are also **individual/species specific**; that is, the proteins of each individual in each separate class of living organisms (by the phrase "class of living organisms" we mean *each species*) are **unique** to that individual/species. (The individual/species specificity is a result of **sexual** reproduction. In the case of **asexual** reproduction, the individual/species specificity is reduced to only **species specificity**. This observation also applies to the phenomenon of multiple individual organisms arising from a single fertilization; among human beings, this phenomenon often, but not always, characterizes multiple births.) Consequently, for either **individual/species specificity** or **species specificity** to occur, it is not surprising that the process of **protein synthesis** must be intimately connected to the hereditary information of each species. In fact, the process of protein synthesis is a multistep process that begins at the hereditary repository; in eukaryotic cells, this is in the cell nucleus.

### Step One: Transcription

The first step in protein synthesis is called **transcription**. In this step, a new molecule called **messenger ribonucleic acid** (messenger RNA, or often abbreviated as **mRNA**) is synthesized using the cell's DNA as a template and controlled by the enzyme RNA polymerase.

This process occurs by the separation of a portion of the DNA double helix into two independent strands (note again the importance of the intermediate strength of the hydrogen bonds that maintain the double helical structure) and then using the base pair sequence of the DNA to produce a complementary base pair sequence in the newly formed mRNA molecule. In figure 11.20 the transcription process is summarized. The single-stranded molecule is the newly synthesized mRNA.

In order to better understand why this process is called *transcription*, let's compare this process to the conversion of an oral presentation (a speech) to a written document (an essay). In making such conversion, the **language used in both the speech and the essay is the** same. The information is simply *transcribed* from the oral form to the written form. It is important to notice that both the **information** and the **language** remain unchanged. Let the DNA molecule correspond to the speech and the mRNA molecule correspond to the essay. If we now consider nucleic acid molecules to be analogous to a language, the *transcription* process in the cell **does** not **change the language**; both DNA and mRNA are nucleic acids! Further, because **complementary base pairing** is used during transcription, **the information does not change.**

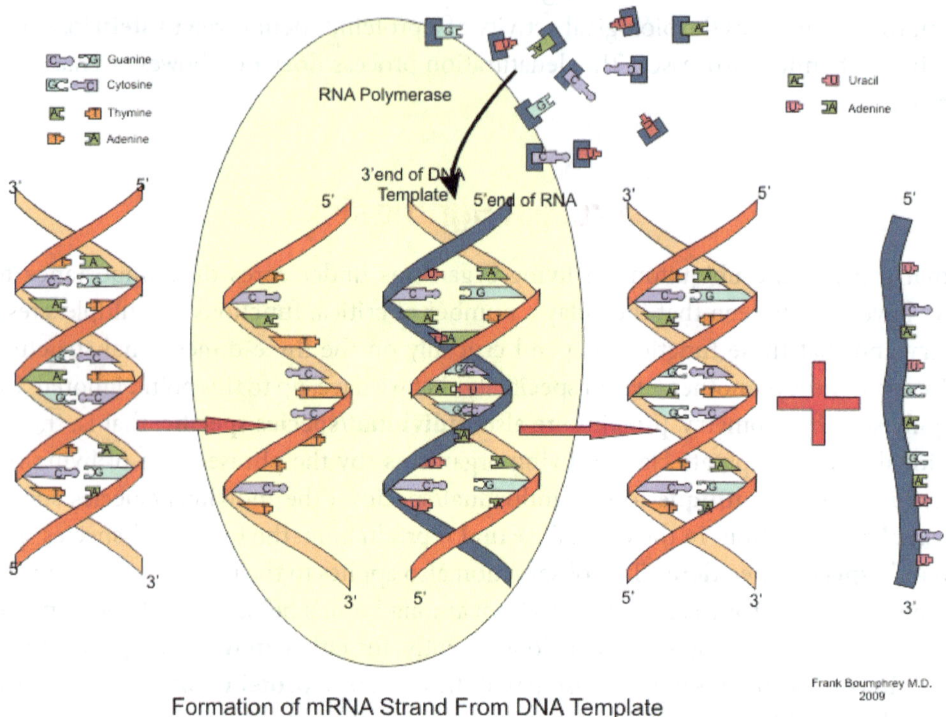


Figure 11.20: Transcription

## Step Two: Translation

The second step of the synthesis occurs **outside** the nucleus at the ribosome and is called **translation**. (Recall that ribosomes are specialized structures located on the cell's rough endoplasmic reticulum. Do not confuse these ribosomes with the specialized ribosomes located in the mitochondria.) This step requires a second molecule found in the cell's cytosol: **transfer ribonucleic acid** (transfer RNA, often abbreviated **tRNA**). As its name implies, tRNA is a nucleic acid that is responsible for *transferring* or *transporting* "something" to a specific location in the cell. The "something" that is transported is an amino acid! Let's look at the structure of tRNA. Figure 11.21 displays a schematic representation of a tRNA molecule in which the colors identify specific regions of the molecule. The violet region is called the **acceptor stem**, to which is attached an **amino acid** (the yellow region) via an ester bond. This formation of an ester bond between an amino acid and a tRNA molecule is called **tRNA activation**. Each of the amino acids has one or more tRNA molecules to which it can attach. The wine-colored region is called the **D-loop**, the blue region the **anticodon loop**, the orange region the **variable loop**, and the green region the **TPsiC-loop** (the name comes from the three nucleotides, thymidine, pseudouridine, and cytidine). The grey region is the **anticodon**. As we shall see shortly, the anticodon distinguishes each tRNA molecule. Notice that the tRNA molecule is a single-stranded nucleic acid molecule; however, its distinctive shape is the result of the complementary base pairing that occurs in the four regions in which the strands of the molecule are parallel to themselves.

Figure 11.21: Transfer RNA (tRNA)

Both the mRNA molecule (synthesized in the cell nucleus) and the tRNA molecule converge on the ribosome that is located on the cell's rough endoplasmic reticulum. The ribosome is composed of two protein/ribosomal RNA (abbreviated rRNA) subunits; the small subunit possesses a grove into which the mRNA fits with its nucleotide bases pointing toward the large subunit. Figure 11.22 depicts the relationship among mRNA, tRNA and the ribosome.

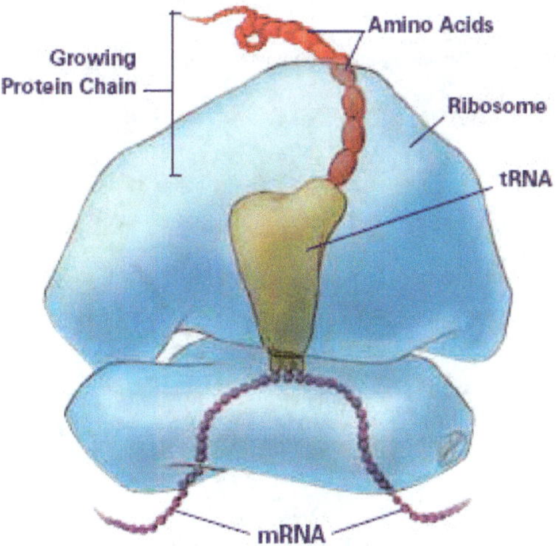


Figure 11.22: Ribosome with mRNA and tRNA

The tRNA molecule's **anticodon loop** aligns with the exposed nucleotides bases of the mRNA molecule located in the groove of the small subunit of the ribosome. Figure 11.23 depicts this alignment. The group of three bases of the mRNA molecule is called a **codon**, and the group of three complementary bases on the tRNA molecules is called the **anticodon**. This explains the name *anticodon loop* for this region of the tRNA molecule. In effect, the anticodon labels each tRNA molecule and associates a particular amino acid with one or more anticodon. For example, in figure 11.23, the anticodon is associated with the amino acid alanine.

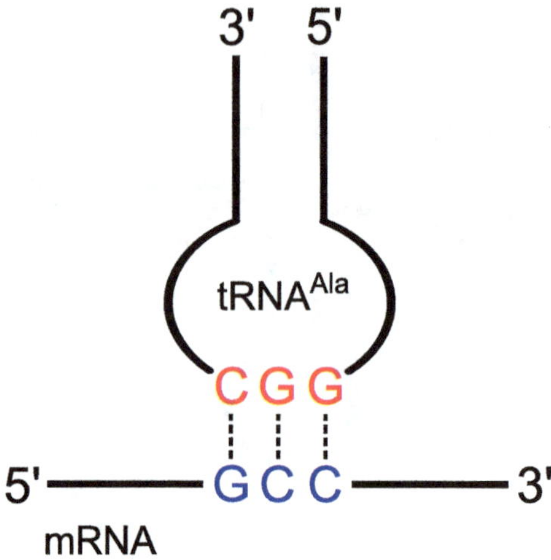


Figure 11.23: Alignment of the Anticodon Loop of tRNA with mRNA

As each tRNA molecule converges on the ribosome with the mRNA molecule located in the groove of the small subunit, a polypeptide consisting of a specific a sequence of amino acids is constructed; the specific sequence of amino acids (the protein's **primary structure**) is controlled by the pattern of the complementary base pairs found in the mRNA and tRNA molecules. This sequential process is summarized in figure 11.24.

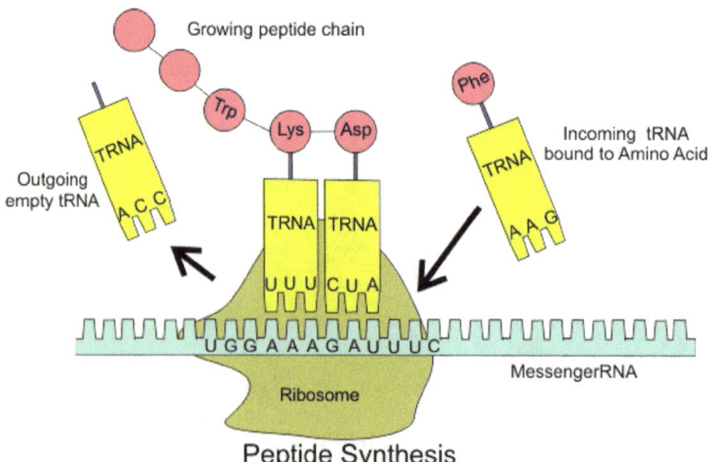


Figure 11.24: Protein Synthesis at the Ribosome

To understand why this process is called **translation**, let's compare this process to the conversion of a document in one language into a document written in another language. In making such conversion, the two different languages are involved; consequently, care must be exercised to choose corresponding words and phrases to be certain that the meaning is preserved in the translation. It is important to notice that while the meaning is preserved, there is a change from one language to another language. Let the tRNA and mRNA molecules correspond to the first language (nucleic acids) and the newly synthesized polypeptide chain correspond to the second language (the primary structure of a protein). The translation process in the cell does change the language; both tRNA and mRNA are nucleic acids, while the polypeptide chain is the primary structure of a new protein! But the complementary base pairing that is used during the translation process (because amino acids correspond to specific codon/anticodon pairs) ensures that the meaning is preserved (the correct primary sequence of a new protein is properly synthesized). The word translation comes from the Latin transfero, meaning "to carry across," and in protein synthesis the meaning is "carried across" from the world of nucleic acids to the world of proteins.

### Termination

Because every protein possesses a finite number of amino acids in its polypeptide sequence, the process described above eventually must stop. This occurs when a **stop codon** is encountered on the mRNA molecule. This encounter is called **termination**; it brings protein synthesis to an end.

## Central Dogma

The three-step process outlined above, **transcription**, **translation**, and **termination**, is often called the **central dogma** of molecular biology. It is the paradigm that has been used since the middle of the twentieth century to understand the link between hereditary information and the synthesis of individual/species specific or species-specific proteins. The ideas embodied in the central dogma are effectively summarized in figure 11.25.

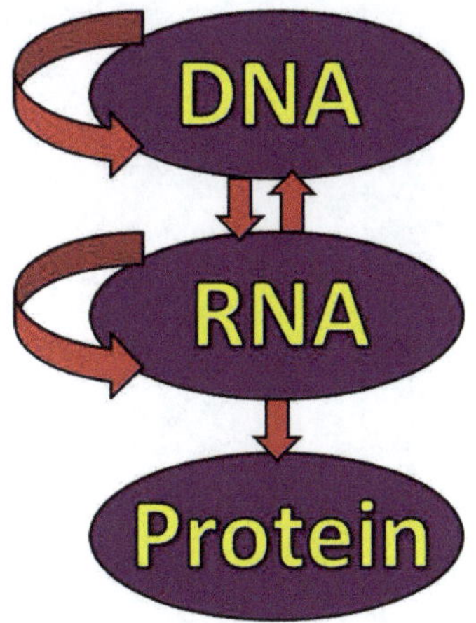


Figure 11.25: Central Dogma of Molecular Biology

## LncRNA

Research done subsequent to the first decade of the twenty-first century has identified subtle aspects of the roles of DNA and RNA that are not encompassed in the standard central dogma paradigm. It appears that the majority of the transcribed RNA molecules do **not** encode information that is used to synthesize proteins. The RNA molecules perform various housekeeping functions in the cell that maintain order and support the synthesis of proteins. Much current research (2014) has focused on long, noncoding RNA molecules (**LncRNA**), nucleic acid molecules that are made up of more than two hundred nucleotides but are **not** utilized to synthesize proteins. A majority of the LncRNA molecules seem to be concentrated in the nucleus and appear to play an important regulatory roles. While the elucidation of the precise contributions of LncRNA molecules awaits further research, it appears clear that the phenomenon of life is not solely the consequence of complex proteins. It appears that RNA molecules play critical roles beyond those required by the central dogma paradigm.

## Epigenetics

The term **epigenetics** refers to the study of changes in the activity of **genes** (the segments of a cell's DNA that are used to synthesize specific polypeptides) that are heritable but are **not** the result of changes in the cell's DNA sequence. In effect, epigenetics is focused on changes in **gene expression** (that is, the production of polypeptides that affect an organism's phenotype) that are not the result of changes in a cell's DNA. Both DNA methylation (the chemical addition of methyl group to DNA molecules) and chemical modification of the histone proteins of the chromosomes are examples of such epigenetic modifications. Gene expression can also be affected by the synthesis of **repressor proteins** that appear to inhibit the transcription of segments of a cell's DNA. Just like the impact of research on LncRNAs, the factors studied in epigenetics imply that the paradigm of the central dogma is likely to be modified as the twenty-first century unfolds.

# Chapter 11 Exercises

1. For a eukaryotic cell, define the following terms:
    a. Nucleus
    b. Cytosol
    c. Ribosome
    d. Mitochondrion
    e. Cytoskeleton
    f. Cell membrane
2. For the mitochondrion, define the following terms:
    a. Inner membrane
    b. Cristae
    c. Matrix
3. What is the meaning of the term "endosymbiont"?
4. What is the function of the ribosomes (both those in the cell and those in the mitochondrion)?
5. Why are proteins such important molecules? Explain.
6. List three important functions of membranes.
7. What is a phospholipid? Describe the structure of a phospholipid.
8. True or false? Phospholipids are amphipathic molecules.
9. Phospholipids form semipermeable membranes. What does "semipermeable" mean?
10. Explain the fluid mosaic model of membranes.
11. Explain the following transport processes:
    a. Simple diffusion
    b. Facilitated transport
    c. Active transport
12. What is osmosis? What is osmotic pressure?
13. Define the following terms:
    a. Isotonic
    b. Hypertonic
    c. Hypotonic
14. What is a phenotype?
15. What is a nucleoside? What is a nucleotide?
16. What is the name of the bond linking multiple nucleotides together in a polynucleotide?
17. What is the primary structure of hereditary molecules?
18. What is the secondary structure of hereditary molecules?
19. What is the tertiary structure of hereditary molecules?
20. What type of bond links complementary base pairs in the DNA (or RNA) double helix?
21. What is supercoiling?
22. What are histones and chromosomes?
23. What are the four levels of structure that occur in proteins? Explain each level.
24. List the various interactions that can participate in determining the tertiary structure of proteins.

25. What is a denatured protein?
26. What is transcription?
27. What is messenger RNA (mRNA)?
28. Where is mRNA synthesized?
29. What is translation? What is the site of translation?
30. What is transfer RNA (tRNA)? What is tRNA activation?
31. What are the codon and the anticodon?
32. What is ribosomal RNA (rRNA)?
33. What are a stop codon and termination?
34. What is the central dogma of molecular biology?

# CHAPTER TWELVE
# Metabolism

## Metabolic Pathways: The Big Picture

In our brief exploration of thermodynamics in chapter 10, we discovered that the Gibbs free energy (with its *energetic* and *entropic* components) is the key to the occurrence of all the transformations in our universe. This basic tenet of thermodynamics applies to **all** chemistry; in particular, it applies to the biochemistry that sustains life. In the last chapter we examined the critical processes of nucleic acid synthesis and protein synthesis; they require an energy source to be consistent with the principles of thermodynamics. What is this energy source?

The term **metabolism** refers to the sum total of all the chemical reactions involved in maintaining the dynamic state of the cell; this dynamic state is exactly the meaning of the word *life*. Consequently, in addition to the synthesis processes (the synthesis of nucleic acids and proteins outlined in the previous chapter), the myriad of chemical reactions that take place in a cell must also provide the energy necessary to sustain the cell. In fact, we are about to discover that **metabolic processes** are responsible for both the synthesis reactions and energy production at the cellular level. Metabolism can be thought of as a dynamic interplay of two distinct cellular activities: **catabolism** is the process of decomposing molecules to supply energy to the cell; the process begins with large molecules (often **very** large molecules) and produces small molecules with an accompanying liberation of energy. **Anabolism** is the process of synthesizing new molecules; we have seen examples of this process in the construction of nucleic acids and proteins in which small molecules are combined to form gigantic molecules.

As the reader might surmise, while there are common aspects of metabolism that occur throughout both the plant and the animal kingdoms that constitute life on this planet, there are numerous species-specific differences. We choose to examine the metabolic processes of the human species, continuing a tacit choice made when examining the synthesis of nucleic acids and proteins in the previous chapter. As a result, the cellular structure of eukaryotic cells introduced earlier will continue to provide a framework for our discussion. Finally, our goal is **not** to examine the metabolic process in minute detail; such an effort would require a text many times the size of this one in order to describe the amazing complexity and intricate interdependencies that characterize human metabolism. Rather, our discussion of metabolism will focus on the catabolic processes crucial for energy production and will emphasize the interconnectedness of all the chemistry we have surveyed in this text. In fact, the **dehydration syntheses** reactions responsible for making glycosidic, ester, and amide linkages, the **hydrolysis reactions** responsible for breaking those same linkages, and the **redox** processes responsible for the rearrangement of electrons in a chemical species are not simply chemical processes. They are intimately related to maintaining a cell's dynamic state, for maintaining life as we know it.

The catabolic process in the human species occurs in **three** very distinct steps characterized primarily by the *location* in the human organism where the step takes place. The first step is called **digestion**; in this step the large, complex molecules that constitute food for a human being are broken down into much smaller molecules. The key to the decomposition of these large molecules (complex carbohydrates, such as starches and polysaccharides, proteins, and lipids) is the *hydrolysis* reaction we examined in chapter 9. The reader will recall that hydrolysis is the route from complex carbohydrates to monosaccharides, from polypeptides to amino acids, and from triglycerides to fatty acids.

Digestion takes places at several points in a human being: mouth, stomach, and small intestines. It involves numerous enzymes that catalyze the hydrolysis reactions, which produce molecules that can be transported across cell membranes into the **cytoplasm** (this term refers to a cell's cytosol and all the organelles contained inside the cell membrane) of individual cells. Of the carbohydrates, only monosaccharides are small enough to be transported across the cell membranes in the small intestines. Lipid molecules (commonly called fats) are hydrolyzed into individual fatty acids or monoglycerides (one glycerol molecule and only one fatty acid molecule); they are then **emulsified** (large globules of lipids are broken into smaller droplets) by soap-like amphipathic molecules to allow effective transport across the cell walls in the small intestines. Once absorbed by the small intestines, the monoglycerides and fatty acids are reassembled into triglycerides that are transported in the bloodstream for use in energy production, (we will discuss this briefly later) or are used as an effective energy storage method. Proteins are denatured and hydrolyzed into individual amino acids, which can be easily transported across the cell membranes in the small intestines.

The second key catabolic step is **glycolysis**; it takes place in the cell's cytoplasm. The central goal of glycolysis is the decomposition of the six carbon glucose molecules (monosaccharides) into the three carbon **pyruvate** molecule. The pyruvate molecule (a pyruvic acid molecule minus a proton, $H^+$) is shown in figure 12.1.

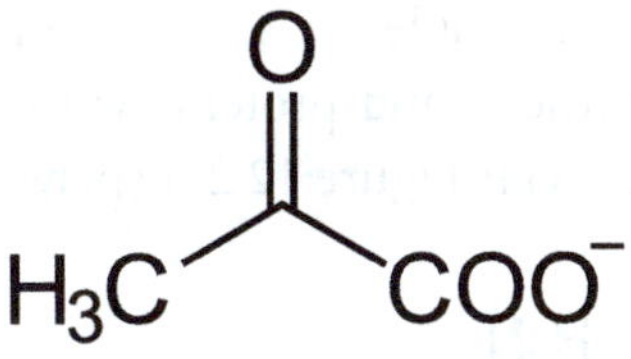

$$H_3C-C(=O)-COO^-$$

Figure 12.1: Pyruvate

The pyruvate molecules generated by glycolysis in the cell cytoplasm can follow two distinct routes. If sufficient oxygen is available, pyruvate molecules enter the cell's mitochondria, where most of the energy that is derived from each starting glucose molecule is produced. This is called the **aerobic path** because it depends on the availability of sufficient oxygen. This **third** step of the catabolic process takes place in two separate stages: (1) the **Krebs cycle** (named for Hans Adolf Krebs, who identified this metabolic pathway; he shared the 1953 Nobel Prize with Fritz Lipmann for this work) and (2) **oxidative phosphorylation.** Each of these stages produces adenosine triphosphate (ATP) molecules, which are the ultimate energy storage molecules, as well as two significant waste byproducts: $CO_2$ and $H_2O$. These two byproducts are the primary waste products (along with urea, as we shall shortly see) of human catabolism. In the event that oxygen is **not** plentiful (we'll examine these circumstances and their implications for human beings), pyruvate can follow another path, known as the **anaerobic path,** to produce energy for the cell.

The **three catabolic steps** produce the energy (stored in the ATP molecules that are synthesized during catabolism) required by the cell to sustain life. During this process, the waste products $CO_2$, $H_2O$, and urea are generated. In addition to providing the energy necessary to maintain the dynamic state of each cell, the catabolic process provides the amino acids necessary for the construction of proteins. While some amino acids are also synthesized by human cells, several amino acids (called **essential** amino acids) **must** be supplied by dietary sources. (The amino acids phenylalanine, valine, threonine, tryptophan, methionine, leucine, isoleucine, lysine, and histidine are regarded as *essential* for human beings.) Now, let's take a closer look at the biochemistry of steps **two** and **three** of the catabolic process, focusing on the production of cellular energy.

## Key Actors in Metabolism: Important Nucleotides

The members of a small group of eight molecules, all of which include the nucleotide *adenosine*, play a central role in catabolism. In fact, the eight molecules are actually four pairs of molecules; one is a **high**-energy molecular form, while the other member of the pair is a **low**-energy form. When a high-energy form is synthesized during catabolism, chemical energy is stored for use by the cell. Let's look at the molecular structures of these **high/low**-energy pairs.

### Adenosine Triphosphate (ATP) and Adenosine Diphosphate (ADP)

The ATP molecule is the basic energy currency of the cell; the synthesis of ATP (high-energy) during catabolism stores energy in a phosphate bond; the *hydrolysis* of ATP to form ADP (low-energy) and a free phosphate group makes energy available for a variety of cellular processes. In particular, the anabolic processes of nucleic acid synthesis and protein synthesis depend on the availability of the energy resulting from the hydrolysis of ATP. Figure 12.2 displays the structures of ATP and ADP.

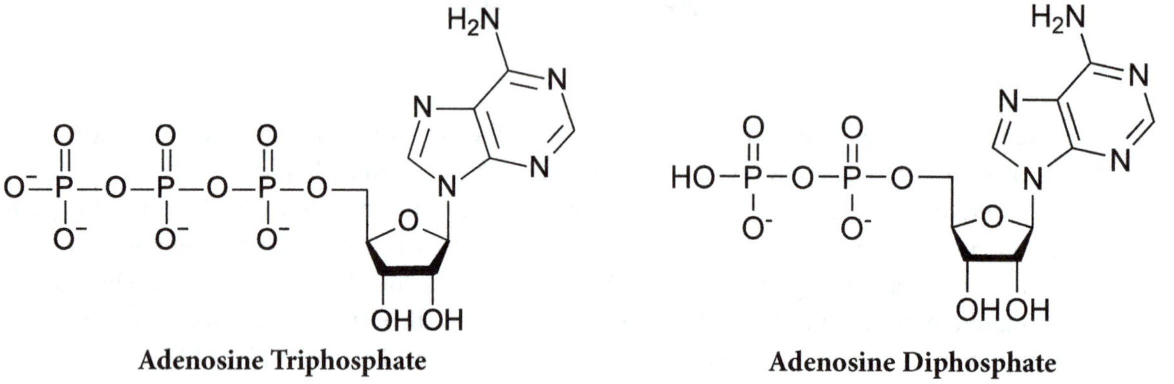

**Adenosine Triphosphate** **Adenosine Diphosphate**

Figure 12.2: ATP and ADP

The relationship between ATP and ADP is summarized in figure 12.3. Note in particular the sign of the Gibbs free energy change displayed in the diagram.

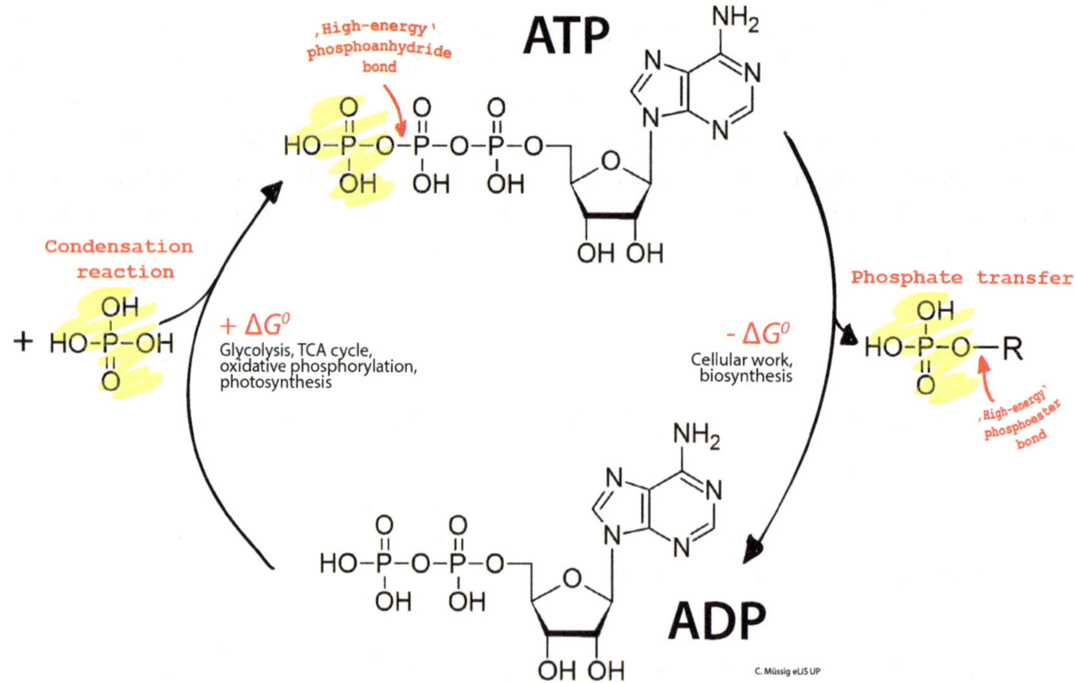


Figure 12.3: ATP/ADP Synthesis and Decomposition

## Nicotinamide Adenine Dinucleotide (NADH and $NAD^+$)

The next pair of molecules are related to one another through the oxidation-reduction process. NADH is the *reduced* (high-energy) form of the molecule, while $NAD^+$ is the oxidized (low-energy) form of the dinucleotide. The molecular structures of NADH and $NAD^+$ are displayed in figure 12.4.

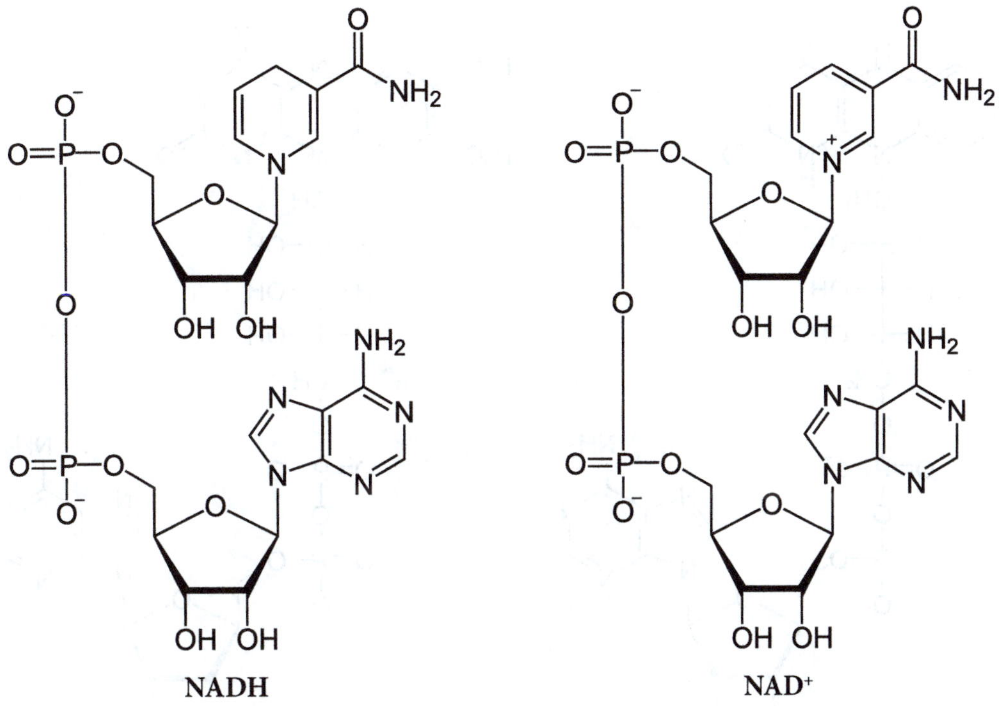

Figure 12.4: Nicotinamide Adenine Dinucleotide

The molecules are called **dinucleotides** because they are made up of two separate nucleotides linked together by the shared oxygen atom of the two phosphate groups. The base that contains the $NH_2$ group is named *nicotinamide* (vitamin $B_3$) and is shown in figure 12.5; the second base contained in the molecules is the purine *adenine*. Note that the nitrogen atom in the nicotinamide ring of $NAD^+$ possesses a formal (+) charge.

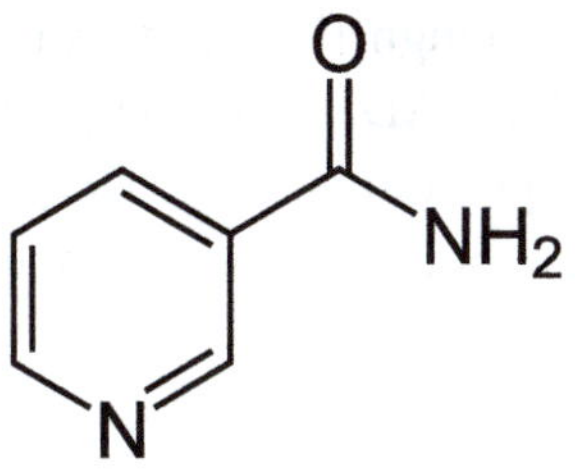

Figure 12.5: Nicotinamide (Vitamin $B_3$)

## Flavin Adenine Dinucleotide ($FADH_2$ and FAD)

The third pair of molecules are also related to one another through the oxidation-reduction process. $FADH_2$ is the *reduced* (high-energy) form of the molecule, while FAD is the oxidized (low-energy) form of the dinucleotide. The molecular structures of $FADH_2$ and FAD are displayed in figure 12.6.

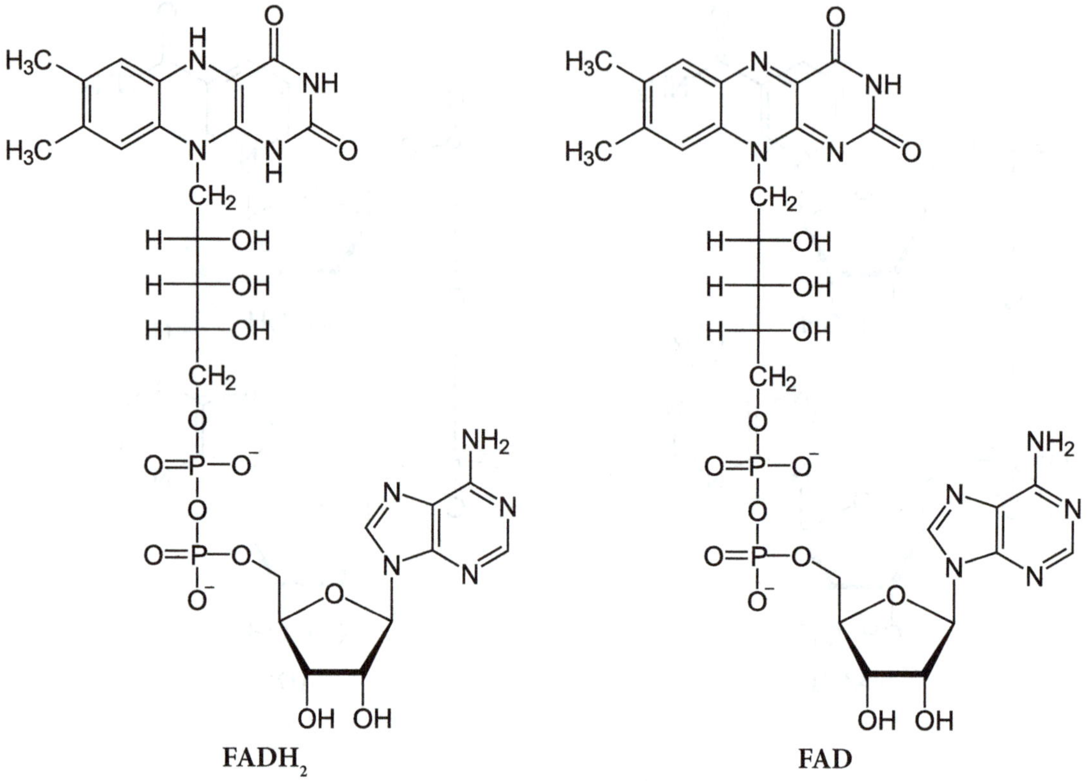

$FADH_2$ FAD

Figure 12.6: Flavin Adenine Dinucleotide

While the molecules are called **dinucleotides**, this is a misnomer because the link between the nitrogen atom in the triple-ring system and the carbon atom of the $CH_2$ group (part of a five-carbon alcohol derived from the ribose called *ribotol*) is **not** a bond between a nitrogen atom and the anomeric carbon atom of a sugar that characterizes a nucleotide. (We noted in chapter 7 that this type of bond is often called an N-glycosidic bond, although IUPAC discourages the use of the term *N-glycoside.*) Consequently, unlike the case of NADH/$NAD^+$, only the portion of the molecule that includes the purine *adenine* is a true nucleotide. The portion of the $FADH_2$/FAD molecules that includes the triple-ring structure linked to the five carbon atoms of ribotol is called *riboflavin* (vitamin $B_3$) and is shown in figure 12.7.

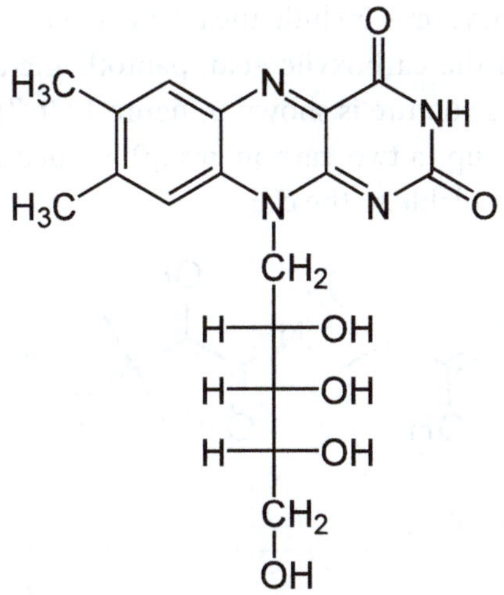

Figure 12.7: Riboflavin (Vitamin $B_2$)

## Acetyl Coenzyme A and Coenzyme A

The final pair of **high/low**-energy molecules are acetyl coenzyme A (high-energy) and coenzyme A (low-energy), both of which include the *adenosine* nucleotide. Their structures are depicted in figure 12.8.

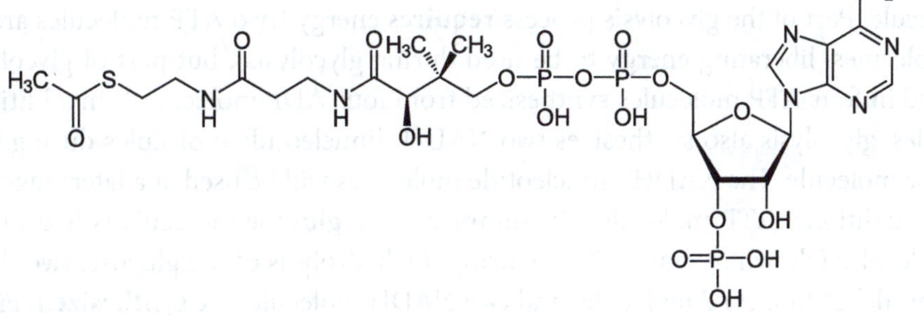

**Acetyl Coenzyme A**

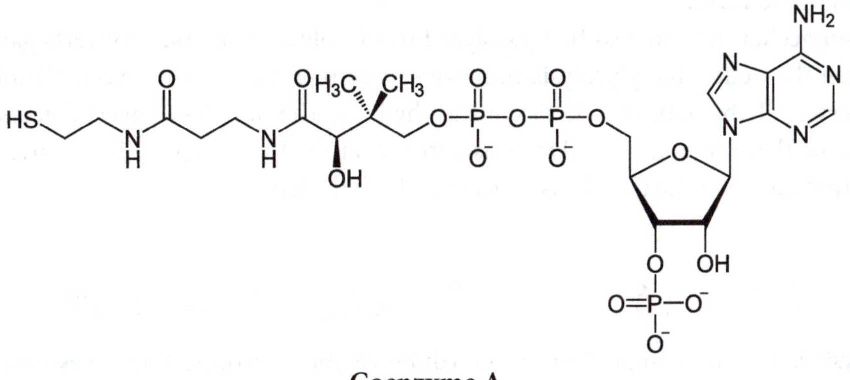

**Coenzyme A**

Figure 12.8: Acetyl Coenzyme A and Coenzyme A

Both acetyl coenzyme A and coenzyme A include the adenosine nucleotide, while the central portion of the molecules is derived from the carboxylic acid, pantothenic acid (vitamin $B_5$); the molecular structure of the pantothenic acid structure is shown in figure 12.9. The significant difference between the two molecules is the acetyl group (a two-carbon group) bonded to the sulfur atom; when the C-S bond is severed, energy is made available to the cell.

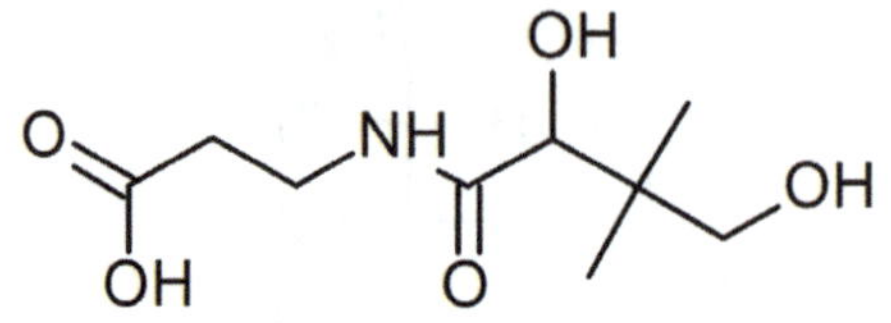

OH O NH OH OH O

Figure 12.9: Pantothenic Acid (Vitamin $B_5$)

## Glycolysis

Glucose is the primary energy source for cells in the human body, and, as we noted earlier, the **second** major catabolic step is called *glycolysis*. This step occurs in the cytoplasm of each cell, and it results in the hydrolysis of the six carbon glucose molecules into two three-carbon molecules called **pyruvates**. For a single glucose molecule, the entire process involves ten distinct molecular reactions, each catalyzed by a specific enzyme, and produces two three-carbon pyruvate molecules from the starting glucose molecule. Part of the glycolysis process **requires** energy (two ATP molecules are converted to two ADP molecules, liberating energy to be used during glycolysis), but part of glycolysis **produces** energy (stored in four ATP molecules synthesized from four ADP molecules). In addition to the four ATP molecules, glycolysis also synthesizes two NADH dinucleotide molecules during the hydrolysis of one glucose molecule. The NADH dinucleotide molecules will be used at a later stage in catabolism to synthesize additional ATP molecules. In summary, **one glucose** molecule is hydrolyzed into **two pyruvate** molecules (shown in figure 12.1). During the hydrolysis of the glucose, **two** ATP molecules are **decomposed**, but **four** ATP molecules and **two** NADH molecules are **synthesized**. Finally, for each glucose molecule **two** water molecules are produced as waste byproducts. The glycolysis pathway is summarized in figure 12.10.

Other monosaccharides can also be hydrolyzed by glycolysis. The body converts galactose to glucose, which can then enter the glycolysis processes directly. Fructose, the sweetest monosaccharide, is absorbed very easily by both muscle tissue and the liver. The muscles convert fructose to an intermediate molecule that can enter the glycolysis pathway at the third step, while the liver produces an intermediate that can enter the glycolysis process at the fifth step.

## Additional Aspects of Carbohydrate Metabolism

While glycolysis is the most important carbohydrate catabolic process that takes place in the cell's cytoplasm, there in fact are several additional biochemical processes (both catabolic and anabolic) that interact with the glycolysis pathway. If there is a shortage of glucose, the human body is capable

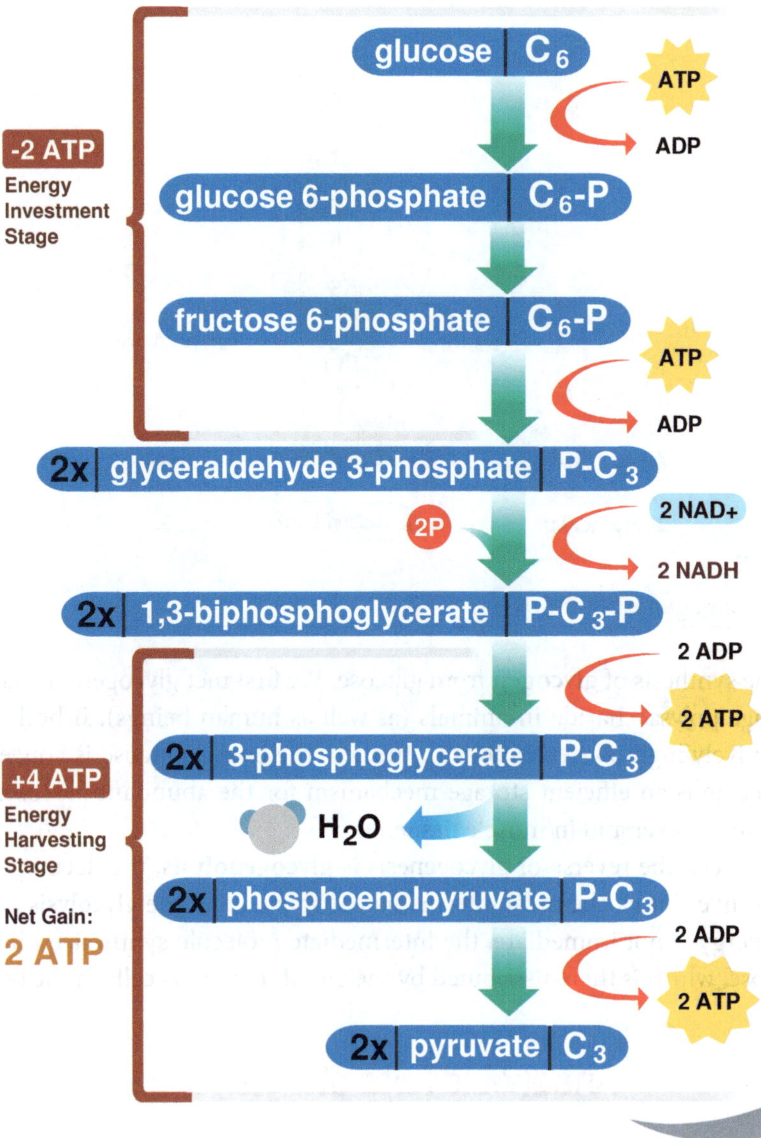


Figure 12.10: Glycolysis Pathway

of synthesizing glucose (consequently, this is an anabolic process) from noncarbohydrate sources. This process is called **gluconeogenesis**, and it can utilize a number of different starting materials, including lactate, amino acids from proteins, fatty acids, and glycerol from lipids. Figure 12.11 sketches a summary of the process. Some of the steps in gluconeogenesis are exactly the reverse of steps in glycolysis; several of the steps are entirely different. The liver is the major site for gluconeogenesis.

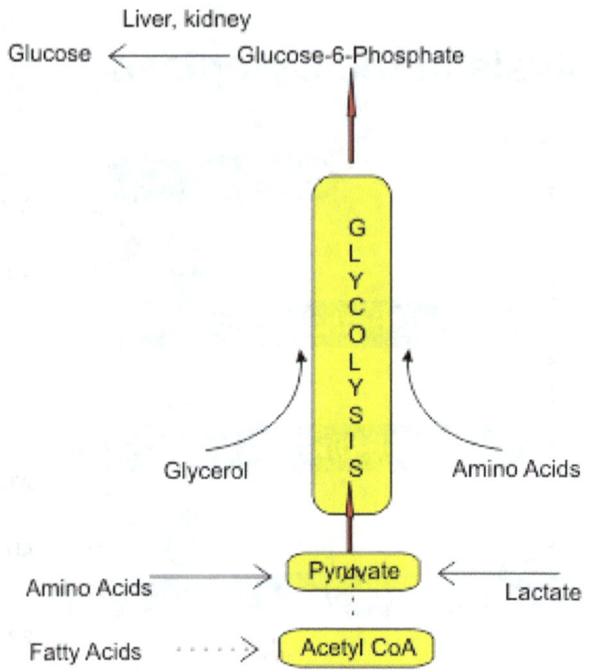


Figure 12.11: Gluconeogenesis Pathway

**Glycogenesis** is the synthesis of **glycogen** from glucose. We first met glycogen in chapter 7 and identified it as the storage polysaccharide in animals (as well as human beings). If both glucose and ATP are present in relatively high concentrations in the cell, the excess glucose is converted to glycogen; the glycogen molecule is an efficient storage mechanism for the abundant glucose. The synthesized glycogen is stored in the liver and in muscle tissue.

As one might expect, the reverse of glycogenesis is **glycogenolysis**, the decomposition of the glycogen stored in the liver into a chemical intermediate that can enter the glycolysis pathway directly. If the demand for energy is not immediate, the intermediate molecule synthesized in glycogenolysis is converted to glucose, which is then distributed by the blood to various cells in the body.

## Anaerobic Fate of Pyruvate

As a result of the glycolysis process, a single glucose molecule is hydrolyzed to two pyruvate molecules. We have already noted that the pyruvate molecules can follow two different paths: one path depends on a sufficient supply of oxygen and is called the **aerobic** path; the second path takes place under conditions in which there is a shortage of oxygen and is called the **anaerobic** path. Why should two paths exist? The answer lies in the fact that the respiratory and circulatory systems supply oxygen to the cells. However, during strenuous exercise or under conditions of extreme danger requiring immediate escape, the supply of oxygen to the muscles may be insufficient to meet the demands of the muscles. Consequently, in the cell's cytoplasm, one of the carbonyl groups in pyruvate is converted to a hydroxyl group, producing the lactate molecule (a lactic acid molecule minus a proton, $H^+$) shown in figure 12.12 and a $NAD^+$ molecule. The $NAD^+$ dinucleotide can then enter the glycolysis pathway, allowing the synthesis of an ATP molecule. The small amount of ATP produced by this mechanism provides an additional source of cellular energy.

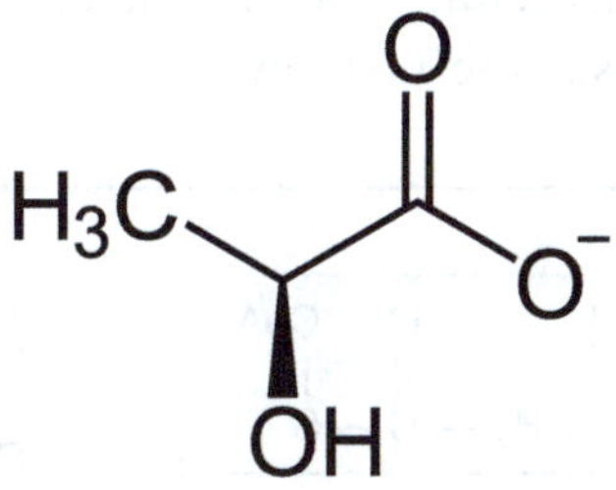

Figure 12.12: Lactate

The increased concentration of lactic acid (formed by adding an $H^+$ to lactate) is responsible for the "muscle ache" that often accompanies extreme exercise.

## Aerobic Fate of Pyruvate

### The Krebs Cycle

The aerobic path of pyruvate now leads us to the **third** step of catabolism that takes place in the mitochondria of the cell. To begin this third step, the pyruvate molecule is decomposed in the matrix of the mitochondrion by removing the carboxylate group ($COO^-$), which is discarded as $CO_2$, and then by linking the remaining two carbon fragments (an acetyl group, $CH_3CO^-$) to coenzyme A by a C-S bond to form acetyl coenzyme A (figure 12.8). Figure 12.13 summarizes the synthesis of acetyl coenzyme A, explicitly showing the elimination of $CO_2$.

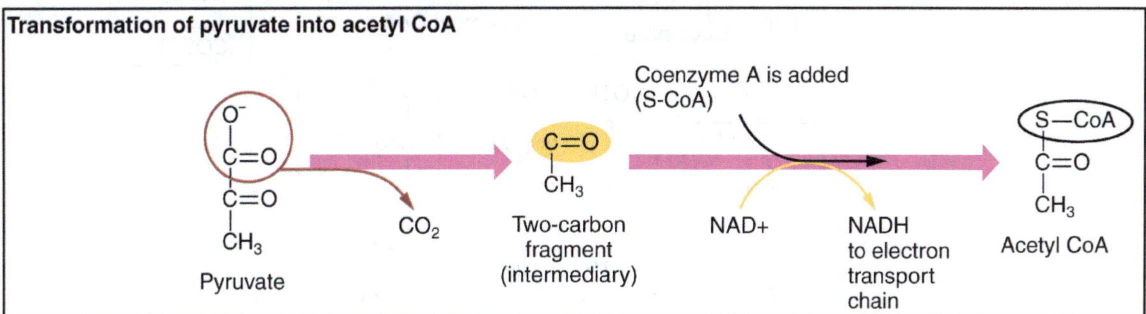


Figure 12.13: Preparation of Acetyl Coenzyme A

The newly created acetyl coenzyme A molecule now enters the Krebs cycle (sometimes called the citric acid cycle, or the tricarboxylic acid [TCA] cycle), which is the **first stage** of the **third** step of the catabolic process. The Krebs cycle is a series of eight reactions that take place in the mitochondrion matrix; each reaction is catalyzed by a specific enzyme. Every glucose molecule produces two acetyl coenzyme A molecules; these two molecules, in turn, produce **two** ATP molecules, **six** NADH molecules, **two** $FADH_2$ molecules, and **four** $CO_2$ molecules. As before, NADH and $FADH_2$ are high-energy molecules that will be used to synthesize additional ATP molecules during the final step of catabolism. Figure 12.14 summarizes the reactions of the Krebs cycle.

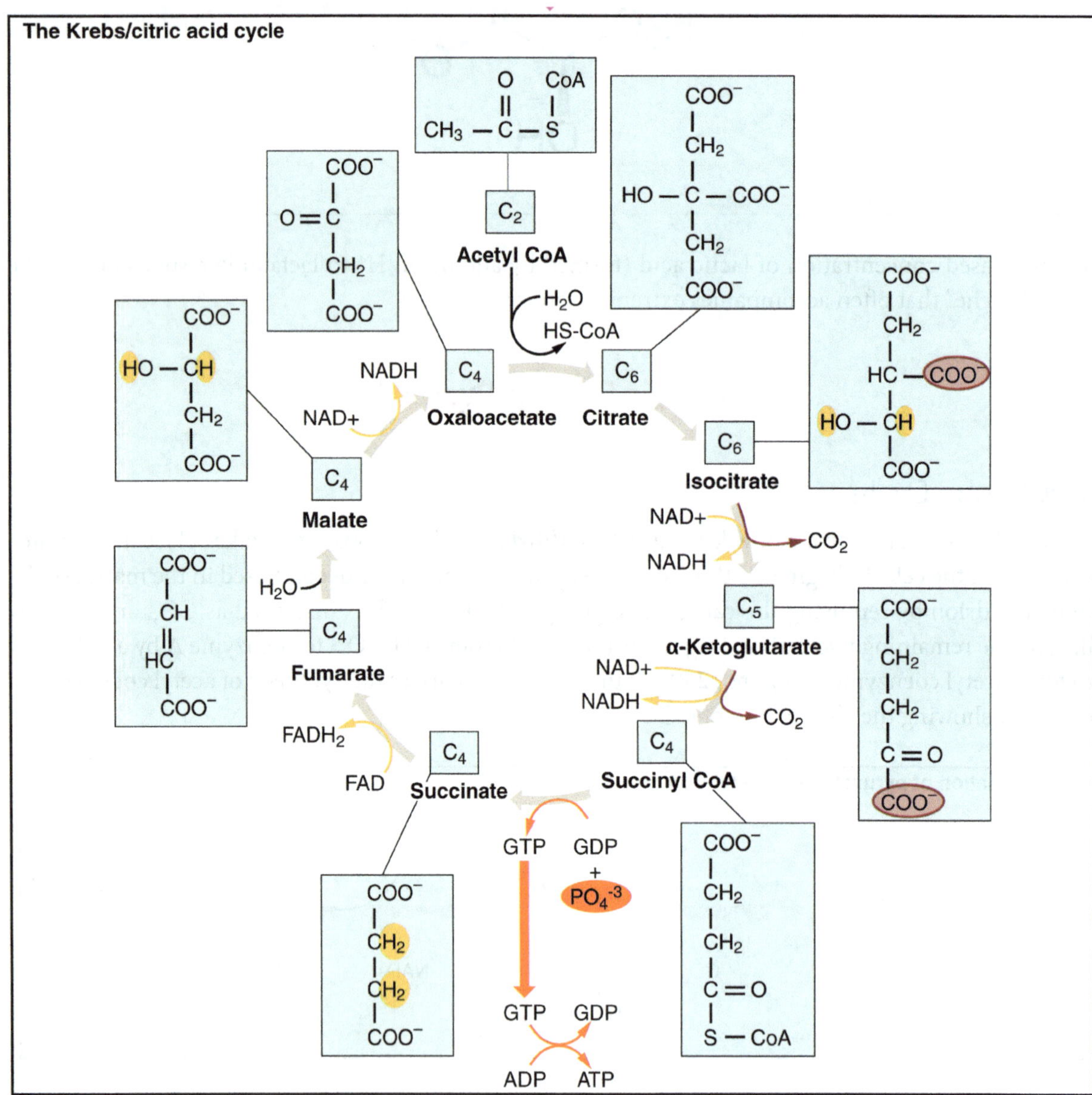


Figure 12.14: The Krebs Cycle

## Electron Transport and Oxidative Phosphorylation

The **second stage** of the **third** step of the catabolic process also occurs in the matrix and in the inner membrane of the mitochondrion. The reduced molecules, NADH and $FADH_2$, (remember that the reduced molecules are the **high-energy** forms of the molecules) are now oxidized in the matrix to produce $-1e$ charges (electrons) and $+1e$ charges (protons, $H^+$) and the corresponding low-energy molecules, $NAD^+$ and FAD, that are reused in the Krebs cycle. Embedded in the inner membrane are **four protein complexes**, conventionally labeled **I, I, III,** and **IV**. Additionally, there are two mobile carriers of the negative charges (the electrons), one named **coenzyme Q** and the other named **cytochrome c**; neither of these carriers is firmly attached to any single complex, and both function to move negative charges between adjacent complexes. The process of moving negative charges between the four protein complexes is called **electron transport**.

Figure 12.15 provides a schematic summary of the mitochondrial structures and the chemical key chemical processes that take place during electron transport. From figure 12.15, the reader should note carefully that the concentration of positive charges (protons, $H^+$) is **greater** in the solution in the inter membrane space than in the matrix solution. Thus, as indicated in the figure, the electrochemical proton gradient **increases** in the direction of the inter membrane space. This also means that the pH of the solution in the inter membrane space is **lower** than the pH in the matrix solution. (Remember that **larger** $H^+$ [$H_3O^+$] concentrations mean **smaller** pH values; see chapter 5.) Finally, the reader should note that at complex **IV** the positive charges (protons, $H^+$) combine with $O_2$ molecules to form $H_2O$, the second waste byproduct of the catabolic process.

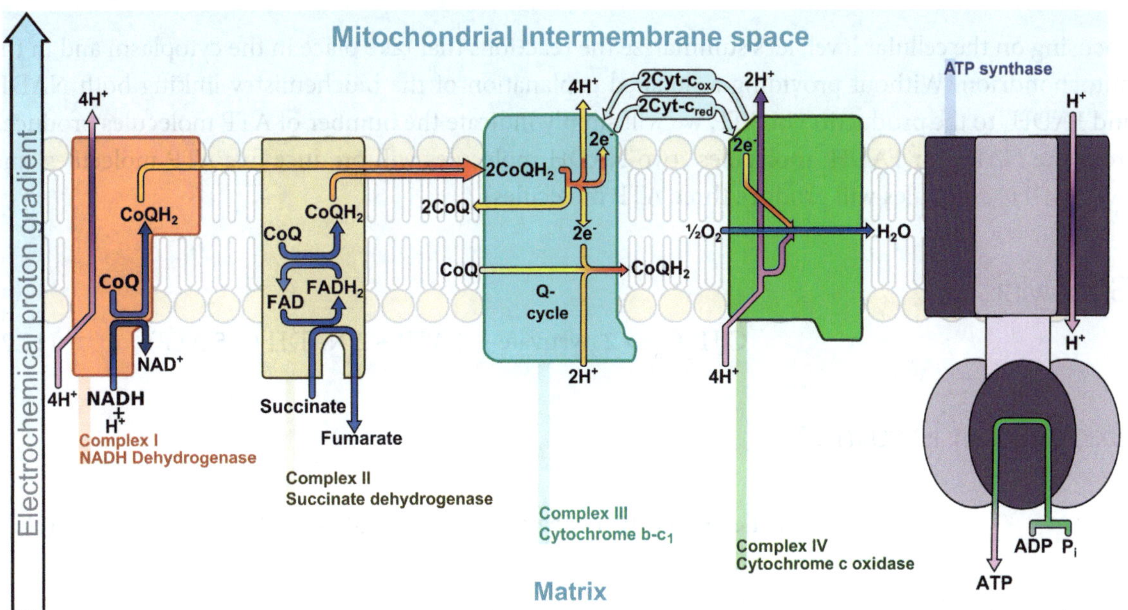


Figure 12.15: Electron Transport and Oxidative Phosphorylation

In the early 1960s, the biochemist Peter Mitchell proposed that most of the ATP synthesized in the cell to store the energy liberated during the catabolic process is the result of the electrochemical proton gradient created by electron transport. Under the regulation of the enzyme ATP synthase, the final step is the synthesis of ATP from ADP, a process called **oxidative phosphorylation** (because a phosphate group is added to ADP, and oxygen is consumed to produce water as a byproduct). As a result of the **third** step of the catabolic process (occurring inside the mitochondrion), one glucose molecule produces a total of **twenty-five** ATP molecules. The far right portion of figure 12.15 summarizes this ATP synthesis. The model of ATP synthesis is called the **chemiosmotic model**. Initially, this proposal by Mitchell was not favorably received; eventually a preponderance of the experimental evidence supported the chemiosmotic model. Peter Mitchell was awarded the 1978 Nobel Prize in Chemistry for his work.

## Summary of ATP Production in Catabolism

Catabolism in the human species occurs in **three** principal steps: **digestion**, which takes place in the mouth, stomach, and small intestines; **glycolysis**, which occurs in the cell cytoplasm; and the two stages of the **Krebs cycle** and **oxidative phosphorylation**, which occur in the mitochondrion. The balanced equation summarizing this chemical reaction for a single glucose molecule is as follows:

$$C_6H_{12}O_6 + 6O_2 \rightarrow 6CO_2 + 6H_2O. \qquad (12.1)$$

Focusing on the cellular level, let's summarize the reactions that take place in the cytoplasm and in the mitochondrion. Without providing a detailed explanation of the biochemistry linking both NADH and $FADH_2$ to the production of ATP, we will simply indicate the number of ATP molecules produced from the NADH or $FADH_2$ molecules (two NADH molecules will produce five ATP molecules, and two $FADH_2$ molecules will produce three ATP molecules).

### Glycolyisis

$$C_6H_{12}O_6 \rightarrow 2\text{ pyruvate} + 2\text{ ATP} + 2\text{ NADH } (= 5\text{ ATP}) \qquad (12.2)$$

### Oxidation of Pyruvate

$$2\text{ pyruvate} \rightarrow 2\text{ acetyl coenzyme A} + 2\text{ NADH } (= 5\text{ ATP}) + 2CO_2 \qquad (12.3)$$

### Krebs Cycle and Oxidative Phosphorylation

$$2\text{ acetyl coenzyme A} \rightarrow$$
$$2\text{ ATP} + 6\text{ NADH } (= 15\text{ ATP}) + 2\text{ FADH}_2\ (= 3\text{ ATP}) + 4CO_2 + 6H_2O \qquad (12.4)$$

Consequently, reacting a **single glucose** molecule with **six** $O_2$ molecules yields **thirty-two** ATP molecules, along with **six** $CO_2$ molecules and **six** $H_2O$ molecules.

## Lipids as an Energy Source

While glucose is the molecule that is **primarily** used by the human body as a source of energy, it is **not** the only molecule that can be used to liberate the energy necessary to sustain life. In fact, two other classes of biomolecules, lipids and proteins, can be used to liberate energy. If sufficient glucose is not available, cells are able to utilize **fatty acids** as an energy source. Free fatty acids are generated by the hydrolysis of lipids (breaking the ester bonds in triglycerides); the fatty acids can be oxidized to form acetyl coenzyme A through a decomposition pathway called **β oxidation** (beta oxidation). Recall that fatty acids are carboxylic acids (hence, there is a carboxyl functional group at one end of the molecule) that also have a long hydrocarbon chain. Through a series of four reactions, the fatty acid is bonded to coenzyme A (this is called an **activated fatty acid**, or **fatty CoA**), a new carbonyl group is created at the original beta carbon (two carbon atoms removed from the carbonyl carbon atom in the original fatty acid), and finally, a two-carbon fragment bonded to coenzyme A is produced that is ready to enter the Krebs cycle. Figure 12.16 provides a graphic summary of these four steps; note particularly the new carbonyl group that is present at the location of the beta carbon atom of the original fatty acid molecule (in the third step). This process can be repeated, removing successive two-carbon fragments until an entire fatty acid is decomposed.

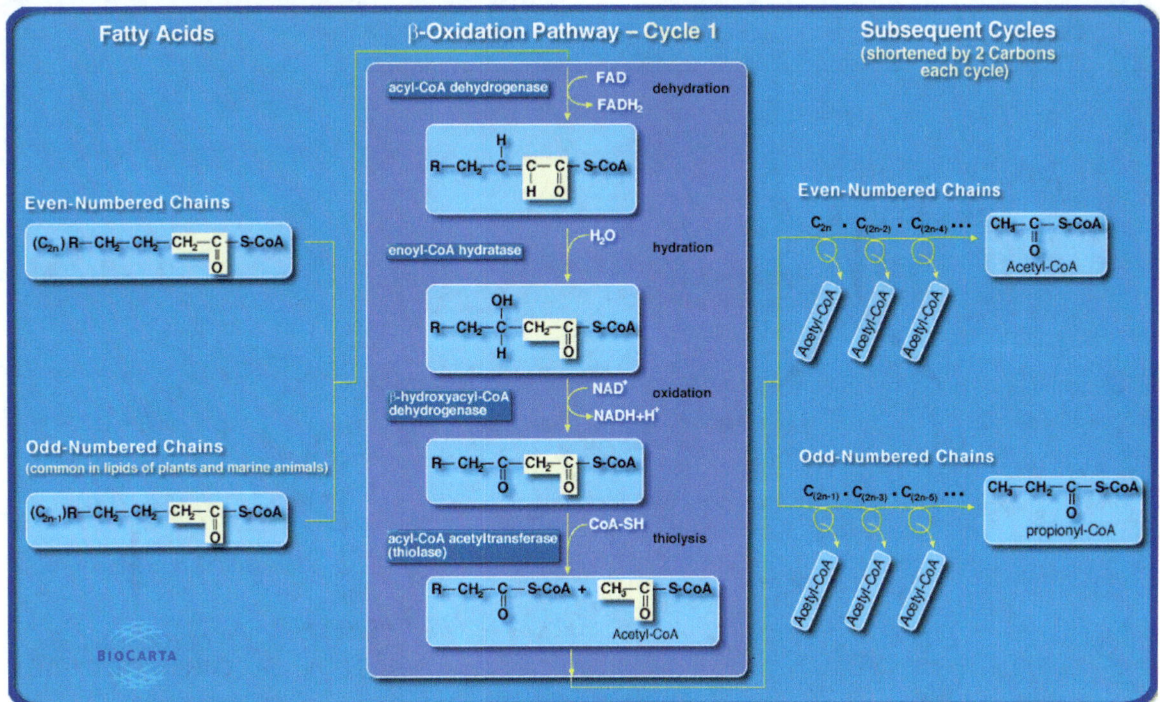


Figure 12.16: β oxidation

## Proteins as an Energy Source

There are three possible fates for ingested proteins: (1) they can be hydrolyzed to amino acids, which are then used in the synthesis of new proteins; (2) if excess protein is present, the amino acid molecules are decomposed and can supply approximately 10% of the human body's energy requirements under normal conditions; and (3) if other sources of energy are not available, proteins can be catabolized to liberate energy. Steps 2 and 3 require the removal of the amino group through a process called **transamination**. The products of the transamination process are then suitable to enter the Krebs cycle at various points (depending on the number of carbon atoms present in the amino acid) and are utilized to synthesize ATP molecules. The removal of the amino group does pose one additional problem for the cell. The $NH_4^+$ ion cannot be converted to ammonia for excretion because ammonia is toxic. The cell converts the $NH_4^+$ ion to **urea** in the **urea cycle**, and it is urea that is excreted in urine. Figure 12.17 shows the urea molecule, and figure 12.18 provides an overview of protein catabolism.

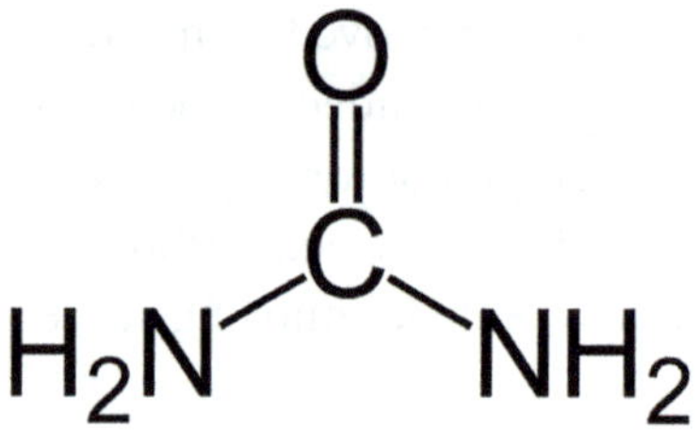

Figure 12.17: Urea

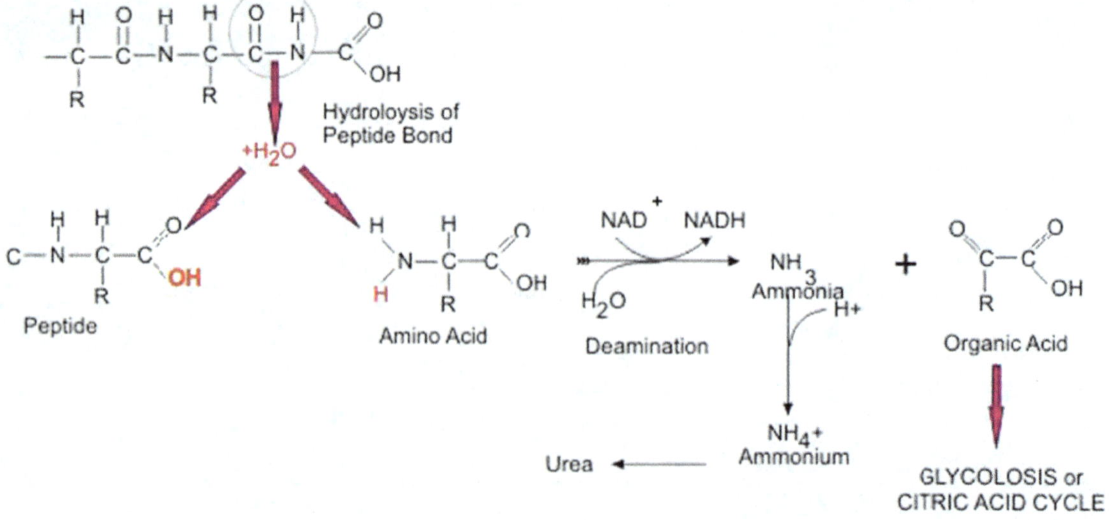

Figure 12.18: Protein Catabolism

# Overview of Catabolic Pathways

The story of catabolism has identified a large number of distinct biochemical steps that rely on principles from diverse segments of the chemical science. Our study of chemical structures, of the reactivity associated with specific structures, of the energetic and entropic constraints on reactivity, and of the critical role played by chemical kinetics (managed by the vast array of enzymes present in a cell) has revealed the amazingly intricate interactions that constitute catabolism in a human being. This overview of catabolism, when combined with our introduction to the synthesis (anabolism) of lipids, carboxylic acids, proteins, and nucleic acids (chapters 9 and 11), provides both parts of the story that is **metabolism**.

Let's step back for a moment at look at the entire catabolic process. Figure 12.19 offers an overview of the catabolic process.

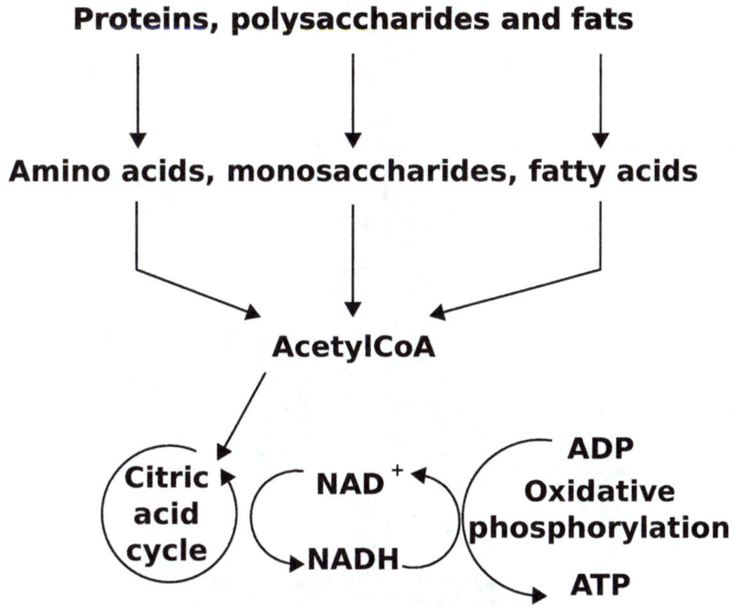


Figure 12.19: Simple Overview of Catabolism

In an attempt to highlight some of the key molecules produced during catabolism, figure 12.20 provides a graphic summary of all the important chemical compounds critical to catabolism, highlighting the production of ATP, $CO_2$ and $H_2O$.

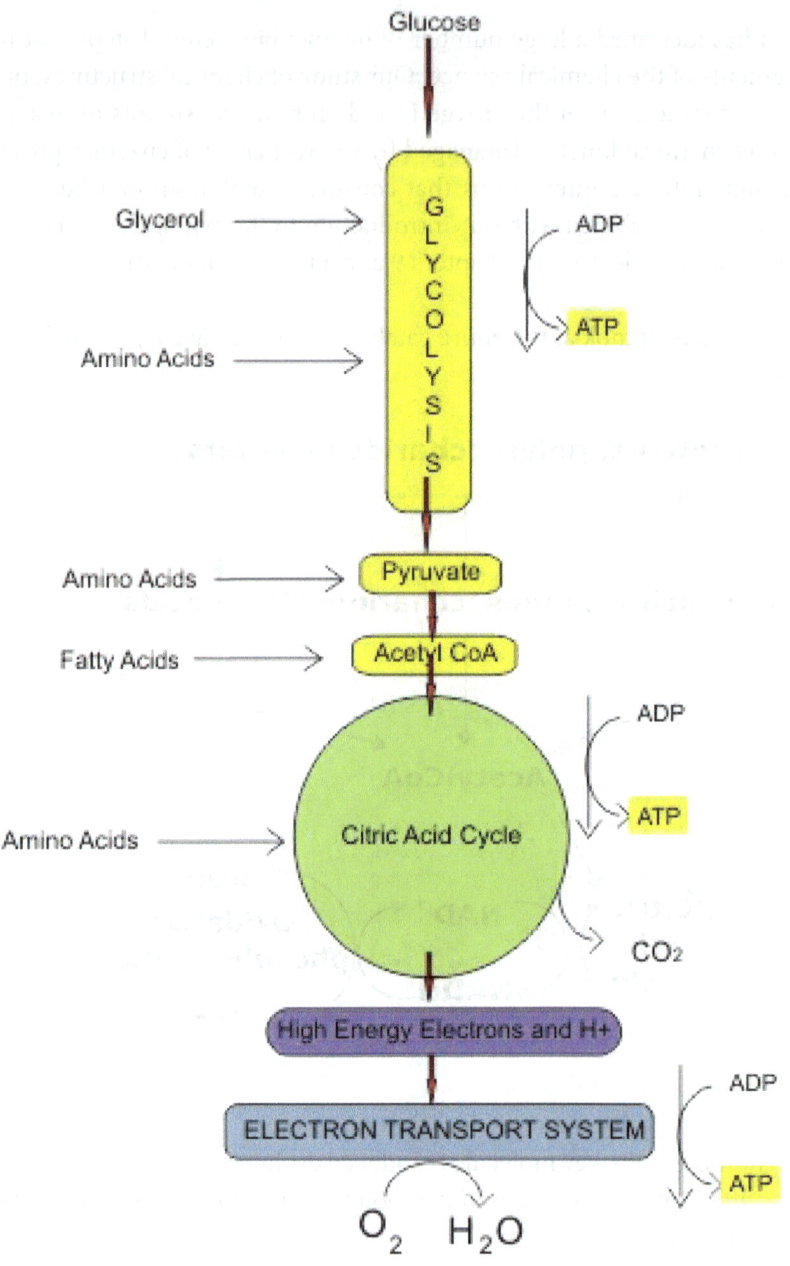


Figure 12.20: Summary of Catabolism

# Chapter 12 Exercises

1. What is metabolism?
2. Distinguish between the terms "catabolism" and "anabolism."
3. What is the first step of catabolism? Where does it occur?
4. What is the second catabolic step, and where does it occur?
5. What is pyruvate?
6. Where does the third step in catabolism take place? What are the two stages that make up this third step?
7. What are the three waste products produced in the third step of catabolism?
8. Distinguish between the aerobic and anaerobic paths.
9. What are essential amino acids?
10. List the amino acids that are essential for human beings.
11. What are ATP and ADP?
12. What are NADH and $NAD^+$?
13. What are $FADH_2$ and FAD?
14. What are acetyl coenzyme A and coenzyme A?
15. What is glycolysis? Explain.
16. How many ATP molecules are decomposed when one glucose molecule is hydrolyzed in glycolysis?
17. How many ATP and NADH molecules are synthesized when one glucose molecule is hydrolyzed in glycolysis?
18. What other monosaccharide molecules can be hydrolyzed by glycolysis?
19. What is gluconeogenesis? In the human body, what is the principal site of gluconeogenesis?
20. What is glycogenesis?
21. What is glycogenolysis?
22. What are two other names for the Krebs cycle?
23. Where does the Krebs cycle take place?
24. Where does the second stage of the third catabolic step take place?
25. Explain the electron transport process.
26. What is oxidative phosphorylation?
27. What is the chemiosmotic model?
28. How many ATP molecules are produced from the catabolism of a single glucose molecule? Explain the key steps.
29. In addition to the monosaccharides, molecules from two other classes of molecules can be used by the human body to produce energy. What are these two additional molecular classes?
30. What is β oxidation?
31. What is transamination?
32. What is the urea cycle?

# APPENDIX

## Chapter Exercises' Answer Key

### Chapter 0 Exercises

1. The two activities were **observation** and **innovative application.**
2. The two people were Galileo Galilei and Isaac Newton.
3. Chemistry is a **body of knowledge** and a **process of discovery**.
4. The five sub-disciplines of chemistry are: Analytical Chemistry, Biochemistry, Inorganic Chemistry, Organic Chemistry, and Physical Chemistry.
5. The three central themes are structures, reactivity, and energetics.
6. In the nineteenth century, the theme that became central to chemistry was the theme of chemical structures.
7. Mathematics is essential for a science to progress from **qualitative** descriptions and predictions to **quantitative** descriptions and predictions.
8. No; there is an inevitable progression from a qualitative explanation to a quantitative explanation.
9. Science is a model-building process.
10. The science of Aristotle tried to answer the question "Why?"
11. The science of Galileo and Newton tried to answer the question "How?"
12. Newton refused to propose "hypotheses"; that is, he refused to answer the question "why".
13. Modern quantum mechanics asks the question "What?"
14. Modern science is both **explanatory** and **predictive**.
15. No; scientific theories are simply **verified**.
16. Groups of atoms exhibiting characteristic reactivity are called **Functional groups**.

### Chapter 1 Exercises

1. The next two steps are: (1.) Formulate a hypothesis and (2.) Design and conduct experiments to test the hypothesis.
2. The models constructed by science are both **explanatory** and **predictive**.
3. Science assumes that an external reality exists and that it is commonly shared.
4. The first step in the scientific method is to ask a question.
5. A hypothesis is a tentative explanation for a collection of observations.

6. The two have a symbiotic relationship. Experimental observations either verify or falsify a theory. Theoretical analyses can suggest additional experiments or can demonstrate that existing experimental data either verify or falsify a particular model.
7. The development of powerful computer resources permits **computational simulations** to complement experiments and theoretical analyses.
8. A law is a concise verbal or mathematical statement of a relationship among phenomena that are always identical when observed under the same conditions.
9. A theory is a unifying principle that explains a body of observations and the "laws" that are based on the observations.
10. A scientific theory is never proven. It can only be verified by repeated experiment. To say that a theory is "falsifiable" means that the theory can be disproven by experimental tests.
11. No; the scientific method is a highly **nonlinear** process.
12. Space: We assume the existence of the concept of **extension**. That is, our physical universe is not a mathematical point having zero dimension. This assumption is consistent with our **experience** (recall that the first step of the scientific method is observation). Further, our experience is consistent with three **mutually perpendicular** (i.e., orthogonal) extensions, which we call **directions** or **dimensions**. The sum total of these directions we call **space** (sometimes labeled as three-space).

    Time: We **experience** (i.e., observe) events **sequentially** and call this physical but non-spatial separation between events our psychological experience of **time**.

    Velocity: We can now use our concepts of space and time to define the **velocity, v**, which is also a **vector** (indicated by the use of boldfaced type) because it possesses both magnitude and direction. The units of velocity are (change in distance) / (change in time).

    Acceleration: Using the idea of velocity, we can now define the concept of **acceleration**, **a**, also a **vector** quantity. The units of acceleration are (change in velocity) / (change in time).
13. No. "Mass" and "matter" are distinct concepts.
14. Another fundamental physical quantity is ($m \times \mathbf{v}$). We give it a special name, defining this quantity as $\mathbf{p} = m \times \mathbf{v}$ where **p** is a **vector** quantity (possessing both magnitude and direction). It is called **momentum** (more properly, **linear momentum**).
15. Energy is a **conserved** quantity.
16. **Heat** energy

    **Electrical** energy

    **Mechanical** energy

    **Radiant** energy
17. The two types of energy are **Kinetic** energy and **Potential** energy.
18. The kinetic energy is $1.1 \times 10^4$ J.
19. This velocity is 9.07 m $s^{-2}$.
20. The percent of balls which are red is 15%.
21. There are 175 silver coins in the collection.

22. (a) $\sqrt[2]{625} = 25.0$
    (b) $\sqrt[2]{81} = 9.0$
    (c) $\sqrt[2]{144} = 12.0$
    (d) $\sqrt[2]{265} = 16.3$
23. (a) $\sqrt[3]{27} = 3.0$
    (b) $\sqrt[3]{8} = 2$
    (c) $\sqrt[3]{-729} = -9.00$
    (d) $\sqrt[3]{1728} = 12.00$
24. (a) $\ln(2.874) = 1.0557$
    (b) $\log(3.589) = 0.5550$
25. (a) $\log(A \times B) = 1.6527$
    (b) $\log(A \div B) -0.2905$
26.

| | | | |
|---|---|---|---|
| 1. | Length | Meter | m |
| 2. | Mass | Kilogram | kg |
| 3. | Time | Second | s |
| 4. | Temperature | Kelvin | K |
| 5. | Amount of substance | Mole | mol |
| 6. | Electric current | Ampere | A |
| 7. | Luminous intensity | Candella | cd |

27. (a) 5 significant figures
    (b) 4 significant figures
    (c) 7 significant figures
    (d) 5 significant figures
28. (a) $2006 \times 375 = 7.52 \times 10^5$
    (b) $1.567 \div 3.0675 = 0.5108$
    (c) $0.617 + 67.3 = 67.9$
    (d) $89.03 + 43 = 132$
29. (a) $2206 \times 375 \div 32.1 = 2.58 \times 10^4$
    (b) $(1.867 \div 5.0675) + 15.1 = 15.5$
    (c) $(0.697 + 17.3) \times 2.5 = 45$
    (d) $(89.451 + 43) \times 1.478 = 196$
30. (a) 34.68
    (b) 5.34
    (c) 79.00
    (d) 13.62

31. (a) $5.673 \times 10^{-1}$ m (or 0.5673 m)
    (b) $6.987 \times 10^{3}$ g (or 6987 g)
    (c) $1.3907 \times 10^{-1}$ m (or 0.13907 m)
    (d) $1.6793 \times 10^{-3}$ L (or 0.0016793 L)
    (e) 4.65 $dm^3$
32. (a) 7.500 °C
    (b) 98.60 °F
    (c) 298.15 K
33. The new volume of the balloon is 11.7 L.
34. The pressure in the pump's cylinder is 2.3 atm.
35. The new pressure of the gas is 5.8 atm.

## Chapter 2 Exercises

1. The cosmological mode is called Big Bang Model.
2. They are the result of a synthesis process that took place in the first generation of stars.
3. As the twenty-first century began, it now appeared that our universe was made up of three distinct components: **ordinary mass-energy** (using Einstein's 1905 statement of the equivalence of energy and mass, $E = mc^2$), **dark matter**, and **dark energy**.
4. Near the end of the first decade of the nineteenth century, the Englishman John Dalton resurrected the atomic concept pioneered by Democritus many centuries earlier. However, the atomic hypothesis was now viewed within the context of a materialist and mechanistic perspective that had come to characterize Western science since the work of Isaac Newton. Dalton articulated four hypotheses that influenced chemical thinking throughout the nineteenth century.
5. 1. All matter consists of indivisible atoms.
   2. The atoms of any single element are **identical** in mass and other properties. They possess the same **size**, the same **mass**, and the same **chemical properties**. The atoms of a single element are **different** from the atoms of any other element. (Note: This statement indicates that Dalton had no knowledge of *chemical isotopes*, which we will soon meet.)
   3. A **chemical** reaction involves **only** the separation, combination, or rearrangement of atoms. It does **not** result in either the **creation** or **destruction** of atoms. This is a statement of the observed conservation of mass. In particular, atoms of one element **cannot** be converted into atoms of another element by a **chemical reaction.**
   4. Compounds are composed of **numbers of atoms of more than one element**. The numbers of atoms of each element present in a compound occur in a **fixed, specific ratio**. These **ratios** are either an **integers** or **simple fractions** (i.e., ratios of small whole numbers).
6. The law of **definite proportions** simply asserted that compounds always contained the same proportions of their constituent substances by mass. The law of **multiple proportions** provided a theoretical basis for understanding the formation of two **different** compounds by the same elements. Articulated by Dalton in 1808, the law of multiple proportions provides an explanation for the formation of **two different compounds by two or more elements.** Choose a constant mass of one of the elements and allow it to form the two compounds by reacting it with two **different** masses of the second element. The **ratios** of the masses of

the second element that completely react with the constant mass of the first element to form two distinct compounds will be **ratios of small whole numbers.**

7. Atoms are **neither** indivisible **nor** eternal.
8. Thomson's critical insight was the recognition that the cathode rays were, in fact, subatomic pieces of matter. This insight led to the recognition of the connection between matter and electricity.
9. The experiments suggested that in the gold foil there were regions of extremely dense mass interspersed among regions containing almost **no** mass. In fact, to explain the nearly backward scattering of some of the $\alpha$ particles required that almost **all** of the gold mass must be concentrated in extremely small volumes of space. In effect, the gold foil, despite its solid appearance, must be mostly **empty space**.
10. The dense central core of an atom is called the **nucleus**.
11. The charge of a proton is **positive**.
12. The charge of a neutron is **neutral**. That is, it is **uncharged**.
13. The charge of an electron is **negative**.
14. The electron exhibits the smallest mass of the three.
15. The **atomic number** of an element is the **number of protons** in the nucleus of each atom of an element. The letter Z is used as the symbol for the atomic number.
16. The **atomic mass number** of an atom of a given element is the **total** number of **neutrons** and **protons** in the nucleus of an atom of that element. The symbol for the atomic mass number is the letter A.
17. (a) 6 protons, 6 neutrons, 6 electrons
    (b) 26 protons, 30 neutrons, 26 electrons
    (c) 16 protons, 16 neutrons, 16 electrons
    (d) 19 protons, 20 neutrons, 19 electrons
18. Atoms that have the **same atomic number** (meaning all the same element because the number of protons is the same) but **different mass numbers** (differing numbers of neutrons) are called **isotopes** (from two Greek words: *iso*, meaning "equal," and *topos*, meaning "place").
19. Since 1961, a single isotope of carbon has been used to define the **unified atomic mass unit**, which is symbolized by u. We define the following:

$$\mathbf{12\ u = Mass\ of\ one\ atom\ of\ {}^{12}_{6}C}$$

Another way stating **exactly the same idea** is

$$\mathbf{1u = One\text{-}twelfth\ the\ mass\ of\ one\ atom\ of\ {}^{12}_{6}C}$$

It is important to note that the definition of the unified atomic mass unit specifies a single isotope of carbon, the carbon-12 isotope.

There are two other terms that often occur in discussions relative atomic masses. First, one often encounters the symbol **amu** for the *atomic mass unit*. This was the official designation prior to the adoption of the unified atomic mass unit standard in 1961. It is no longer appropriate to use the symbol **amu**; however, old traditions are not easily replaced. The second term is the **dalton**, symbolized by Da, which is also a non-SI unit used frequently in biochemistry and molecular biology, although it has not been approved by the Conférence Général des Poids et Mesures (CGPM, the General Conference on Weights and Measures). **The dalton is simply an alternate name for the unified atomic mass unit**; it currently has exactly the same definition as the unified atomic mass unit.

20. The **average atomic mass of** an atom of any element is computed as the weighted average of the atomic masses of the naturally occurring isotopes.

21. A **mole** is defined as **the amount of substance of a system that contains as many distinct elementary objects as there are atoms in exactly 0.012 kilogram of carbon-12.**
22. To answer this question you require additional data. I have collected the following information from the National Institute of Standards and Technology (NIST)

**Data from NIST**

**http://physics.nist.gov/cgi-bin/Compositions/stand_alone.pl?ele=&all=all&ascii=ascii&isotype=some**

**Accessed on September 12, 2014**

| Isotope | | | Relative Atomic Mass | Isotopic Composition | Standard Atomic Weight | Notes |
|---|---|---|---|---|---|---|
| 6 | C | 12 | 12.0000000(0) | 0.9893(8) | 12.0107(8) | g,r |
| | | 13 | 13.0033548378(10) | 0.0107(8) | | |
| | | 14 | 14.003241989(4) | | | |
| 7 | N | 14 | 14.0030740048(6) | 0.99636(20) | 14.0067(2) | g,r,a,d |
| | | 15 | 15.0001088982(7) | 0.00364(20) | | |
| 15 | P | 31 | 30.97376163(20) | 1.0000 | 30.973762(2) | |
| 26 | Fe | 54 | 53.9396105(7) | 0.05845(35) | 55.845(2) | |
| | | 56 | 55.9349375(7) | 0.91754(36) | | |
| | | 57 | 56.9353940(7) | 0.02119(10) | | |
| | | 58 | 57.9332756(8) | 0.00282(4) | | |
| 74 | W | 180 | 179.946704(4) | 0.0012(1) | 183.84(1) | |
| | | 182 | 181.9482042(9) | 0.2650(16) | | |
| | | 183 | 182.9502230(9) | 0.1431(4) | | |
| | | 184 | 183.9509312(9) | 0.3064(2) | | |
| | | 186 | 185.9543641(19) | 0.2843(19) | | |

**Column Descriptions**

**Isotope**: Atomic number, Symbol, and Mass number

**Relative Atomic Mass** (of the isotope): $A_r(X)$, where X is an isotope [formerly called atomic weight (see Standard Atomic Weight)]

These values are scaled to $A_r(^{12}C) = 12$, where $^{12}C$ is a neutral atom in its nuclear and electronic ground state. Thus, the relative atomic mass of entity X is given by: $A_r(X) = m(X) / [m(^{12}C) / 12]$ . If # is present, the value and error were derived not from purely experimental data, but at least partly from systematic trends. *The 2003 Atomic Mass Evaluation* does not extend beyond an atomic mass of 293.

**Representative Isotopic Composition:** Mole fraction of the various isotopes

In the opinion of the Subcommittee for Isotopic Abundance Measurements (SIAM), these values represent the isotopic composition of the chemicals and/or materials most commonly encountered in the laboratory. They may not, therefore, correspond to the most abundant natural material. The uncertainties listed in parenthesis cover the range of probable variations of the materials as well as experimental errors. These values are consistent with the values published in *Atomic Weights of the Elements, 2001.*

**Standard Atomic Weight** (common usage): $A_r$(X), where X is an element [more appropriately called relative atomic mass of the element]

The relative atomic mass of an element is derived by averaging the relative atomic masses of the isotopes of that element. These values are scaled to $A_r(^{12}C) = 12$, where $^{12}C$ is a neutral atom in its nuclear and electronic ground state. These values are dependent on the origin and treatment of the material. The uncertainties are listed in parenthesis. Brackets [ ] indicate the mass number of the most stable isotope.

**Notes:** Notes for Representative Isotopic Composition and Standard Atomic Weight

**a** Air reference material used for the best measurement.

**b** Tank hydrogen has reported $^{2}H$ mole fractions as low as 0.000032.

**c** Materials depleted in $^{6}Li$ and $^{235}U$ are commercial sources of laboratory shelf reagents. In the case of Li, such samples are known to have $^{6}Li$ mole fractions in the range of 0.02007 – 0.07672, with natural materials at the higher end of this range. In the case of U, the $^{235}U$ mole fractions are reported to range from 0.0021 to 0.007207, far removed from the natural value.

**d** The Commission on Atomic Weights and Isotopic Abundances recommends that the value of 272 be employed for $^{14}N/^{15}N$ of $N_2$ in air for the calculation of $^{15}N$ mole fractions from measured $\delta^{15}N$ values.

**e** The original data for Sn were adjusted to account for possible errors due to $^{115}In$ contamination, and an error in the $^{114}Sn$ abundance.

**f** The abundance of this radiogenic isotope may vary substantially.

**g** Geological specimens are known in which the element has an isotopic composition outside the limits for normal material. The difference between the atomic weight of the element in such specimens and that given in the table may exceed the stated uncertainty.

**h** An electron multiplier was used for the Te measurements and the measured abundances were adjusted using a "square root of the masses" correction factor.

i Commercially available Li materials have atomic weights that range between 6.939 and 6.996; if a more accurate value is required, it must be determined for the specific material. Note: The range given in *Atomic Weights of the Elements 1995* was 6.94 to 6.99.

**m** Modified isotopic compositions may be found in commercially available material because it has been subject to an undisclosed or inadvertent isotopic fractionation. Substantial deviations in the atomic weight and isotopic composition of the element from that given in the table can occur.

**r** Range in isotopic composition of normal terrestrial material prevents a more precise value being given; the tabulated value should be applicable to any normal material.

**w** Fresh water reference material used for the best measurement.

(a) 90.08 g
(b) 1378.80 g
(c) 232.30 g
(d) 105.05 g
(e) 418.84 g

23. Yes. The actual number of distinct elementary objects in a mole is an **experimentally determined number** because the actual number of atoms in exactly 12 grams of carbon-12 is also an **experimentally determined number**. This number depends on the ability of humankind, at any point in time, to count individual atoms; this ability to count changes over time. The 2010 recommendation from the Committee on Data for Science

and Technology (CODATA) for the value of this number is 6.022 141 29 × $10^{23}$ $mol^{-1}$. It has been given the name **Avogadro's number.** It is important to note that **Avogadro's number will change**; it is **not** a constant because, as noted earlier, the number of distinct elementary objects in a mole is an **experimentally determined number**. In this text we will use 6.022 × $10^{23}$ $mol^{-1}$ as our working value of Avogadro's number.

24. A sample in which the number of grams in the sample is **numerically** equal to the element's average atomic mass constitutes one mole of the element. This mass is called the **molar mass** of the element.
25. The standard model identifies two classes of particles, the **leptons** and the **hadrons.**
26. The electron is a **lepton.**
27. No; the six hadrons carry **fractional electric charges**: +⅔*e* (up, charm, top) and −⅓*e* (down, strange, bottom).
28. In the least complex view, a proton is composed of three quarks, **two** *up* quarks and **one** *down* quark, and the proton carries a **positive** charge (notice that the charges still add to +1*e*, as in the earlier model). Similarly, the neutron is also understood to be a combination of three quarks, **two** *down* quarks and **one** *up* quark. As before, the neutron is **uncharged.**
29. All the massive particles of the standard model acquire their mass by interacting with a quantum field through the exchange of a new particle, which, for historical reasons, is called the **Higgs boson.** This proposal launched a nearly forty-year search, which culminated with an announcement in July 2012 at the Large Hadron Collider (LHC) that the Higgs particle had been experimentally identified.
30. By 1900, Rutherford and Villard had shown that radioactive materials emitted three distinct types of radiation:
    1. α **radiation** carries a +2*e* charge and was later shown to be made up of helium atoms that had lost two negative charges (associated with the electrons of the three-particle model of the atom). In effect, α radiation possesses nonzero mass and is composed of helium nuclei.
    2. β **radiation** carries a −1*e* charge and was later shown to be identical to the corpuscles observed by Thomson in his cathode ray experiments (ultimately associated with the electron of the three-particle model of the atom). β radiation also possesses nonzero mass.
    3. γ **radiation** carries no charge, is unaffected by electric and magnetic fields, and is a purely electromagnetic phenomenon (like light but possessing much higher energies, similar to the energies associated with X-rays) with which there is **no** associated mass.

    As radioactivity continued to be investigated and the understanding of the three-particle model of the atom matured during the first third of the twentieth century, two additional types of radiation emanating from radioactive materials were identified:

    1. **Positrons** carry a +1*e* charge and have the **same** mass as the particles of **β radiation.**
    2. **Neutrons** are nearly equal in mass to protons, but carry **no** electrical charge.
31. **Radioactivity**, a process by which atoms **spontaneously** emit radiation (first observed by Becquerel in 1896 and named by the Curies two years later) is **not** initiated by an investigator. This process of spontaneously emitting radiation is called **disintegration** because the radiation is emitted from the atomic nucleus and the nucleus, in turn, is changed.
32. The **curie** (Ci) is a non-SI unit originally defined in 1910 as a specific mixture of the radioactive elements radon (a radioactive gas) and radium in which thirty-seven billion atoms disintegrate per second. However, in 1964, the twelfth CGPM agreed to accept the curie as a measure of **activity** and defined the curie as thirty-seven billion disintegrations per second. The SI unit of radioactivity is called the **becquerel** (Bq) and is defined as one disintegration per second. The fifteenth CGPM in 1975 formally adopted the **becquerel** for the SI unit of activity. Consequently, 1 Ci = 3.7 × $10^{10}$ Bq.

33. Two additional categories of units are necessary; one is used to measure the energy transferred per unit mass, the second to measure the average radiation absorbed by biological tissue, taking into account the fact that the different types of radiation cause differing amount of damage to biological systems. As before, in each category there is a SI unit and a non-SI unit. The first of these categories is called the **absorbed dose** (also known as the **total ionizing dose**). The SI unit of the absorbed dose is the **gray** (Gy), which is defined as 1 Gy = 1 J $kg^{-1}$. The older, non-SI unit is the **rad**, which is defined as 1 rad = $10^{-2}$ J $kg^{-1}$. Hence, 1 Gy = 100 rad. Note that this category of units simply indicates how much energy (here, measured in joules) is deposited in every kilogram.
34. The second of the two categories is called the **equivalent absorbed radiation dose** (often shortened to either **equivalent dose** or **dose equivalent**). The SI unit of equivalent absorbed radiation dose is the **sievert** (Sv). For alpha ($\alpha$) radiation, 1 Sv = $5.0 \times 10^{-2}$ Gy, while for beta ($\beta$) and gamma ($\gamma$) radiation, 1 Sv = 1 Gy. The older, non-SI unit of equivalent absorbed radiation dose is called the **rem**. Measuring the effects of radiation with the non-SI units, we find that for alpha ($\alpha$) radiation, 1 rem = $5.0 \times 10^{-2}$ rad, while for beta ($\beta$) and gamma ($\gamma$) radiation, 1 rem = 1 rad. Consequently, 1 Sv = 100 rem.
35. The half-life of a radioactive substance is most simply defined as the amount of time required for one-half of the nuclei in a specified sample of the material to disintegrate. (In figure 2.6, the half-life is represented by the symbol $T_{½}$.) However, this conceptually simple definition becomes meaningless if the sample has only one atom. Clearly, at the end of one half-life, there is either a single atom that **has disintegrated** or a single atom that **has not disintegrated**; it is impossible to have a single atom that has half-disintegrated! Consequently, a better definition of a radioactive substance's half-life can be given in terms of *probability*. **We define a radioactive substance's half-life as the amount of time needed to guarantee that the average probability that a nucleus has disintegrated is 0.5.**
36. The half-life of cesium-131 is 9.68900 days (24 hour days)
37. After 1824.48 hours, 7.81 x $10^1$ g of phosphorus-33 remain.
38. In 1938 it was realized (by the Austrian physicist Lise Meitner and her nephew, Otto Frisch) that the uranium nucleus had been fragmented (colloquially "split") into two less massive nuclei accompanied by the release of energy and several neutrons. Because the nucleus was fragmented, this process is called **nuclear fission**.
39. In the **fusion process** a new, more massive nucleus is produced by combining less massive, lighter nuclei.
40. (a) Binding energy: In fact, the mass of the new nucleus is **less than** the sum of the masses of the individual nucleons. This difference is called the **mass deficit**, or, as a result of Einstein's equation $E = mc^2$ equating mass and energy, the corresponding energy is called the **nuclear binding energy**.

    (b) Critical mass: This minimum amount of radioactive material to sustain a chain reaction varies, depending on the individual radioisotope, and is called the **critical mass**.

    (c) Chain reaction: However, because the fission process often produces two or three additional free neutrons, these free neutrons can initiate another fission cycle. Such a repetitive process is called a **chain reaction**.

    (d) Mass deficit: In fact, the mass of the new nucleus is **less than** the sum of the masses of the individual nucleons. This difference is called the **mass deficit**
41. All of the following are machines used to accelerate subatomic particles: Van de Graaff electrostatic generator, CW generator, Cyclotron, Linear Accelerator, Synchrotron, Collider
42. A "black-body" is a hypothetical object that is both a perfect emitter and a perfect absorber of radiation.

43. The **very *specific amounts*** energy that can be exchanged among the oscillators (or transferred between the black-body and the surrounding environment) or **definite units** are called **quanta.**
44. Like Planck, Einstein used the term "quanta" for the units of electromagnetic energy. In 1926 the American physical chemist G. N. Lewis coined the term "photons" to describe the quanta of electromagnetic radiation.
45. False. The energy depends on the frequency (or equivalently, the wavelength).
46. The conclusions reached by Einstein were significant because during the last half of the nineteenth century, experimental studies had observed that certain metals, when irradiated with visible or ultraviolet radiation, would produce an electrical current (see figure 2.17). This behavior was called the **photoelectric effect**, and the electric current was interpreted as stream of cathode rays (in more modern language, a stream of *photoelectrons*).
47. The **Bohr atom** was both **electromagnetically stable** and successfully **explained a wide range of spectral data.**
48. The principle of **wave-particle duality** suggested that all microscopic transformations of the universe exhibit both particle-like and wave-like characteristics.
49. Statement of the Heisenberg **indeterminacy relation**: it is impossible to know **simultaneously** both the momentum and the position of a particle with **unlimited** accuracy. Mathematically: $(\Delta x) \times (\Delta p_x) \geq h \div (4\pi)$.
50. Pauli formulated a concise statement about the behavior of electrons in any atom that has come to be called the **Pauli exclusion principle**: no two electrons in an atom can have the **same** values of the four quantum numbers (**n**, **1**, $\mathbf{m_l}$, and $\mathbf{m_s}$). In effect, the four quantum numbers act like the "house address" of each electron in an atom; every electron has a **unique** "house address."
51. Hund's rule can stated as follows: the electron configuration with the **maximum** multiplicity is the **most** stable (**lowest-energy**) configuration.
52. (a) Co: $1s^22s^22p^63s^23p^64s^23d^7$ 27 electrons
    (b) Cl: $1s^22s^22p^63s^23p^5$ 17 electrons
    (c) N: $1s^22s^22p^3$ 7 electrons
    (d) S: $1s^22s^22p^63s^23p^4$ 16 electrons

## Chapter 3 Exercises

1. A **pure substance** is a form of matter that has a **definite** or **constant composition** and **distinct properties.**
2. A **mixture** is a combination of **two or more substances** in which the substances **retain their distinct properties.**
3. A **physical process** is any transformation of matter that **does not** change the **chemical identity** of a substance. A **chemical process** is any transformation of matter that **does change** the **chemical identity** of a substance.
4. Physical process
   Chemical process
   Physical process
   Chemical process
   Physical process
   Chemical process

5. Use the data from #22 in the Chapter 2 Exercises key
   44.01 g
   16.04 g
   17.03 g
   32.04 g
6. (a) HO
   (b) $NO_2$
   (c) HCl
   (d) $NH_3$
   (e) $CH_2O$
7. The three elementary states of matter are: solid, liquid, and gas
8. Gases readily fill a container of any shape or volume subject only to the constraints of temperature and pressure. In contrast, liquids possess a definite volume.
9. A **chemical reaction** is any transformation that changes the chemical identity of one or more substances participating in the transformation.
10. A chemist uses a convenient symbolic representation (much like an algebraic equation) called a **chemical equation** to show what happens during a chemical reaction.
11. The numbers written to the left of each molecular formula in a chemical equation are called the **stoichiometric coefficients** (from the Greek words *stoicheion*, meaning "element," and *metron*, meaning "measure"); they represent the relative quantities (or numbers) of reactant and product atoms/molecules that participate in the specific chemical reaction.
12. The stoichiometric coefficients can be readily interpreted two ways. First, from an atomic/molecular point of view, they specify the **exact numbers** of reactant and product atoms/molecules involved in a chemical reaction. In the equation $C_6H_{12}O_6 + 6O_2 \rightarrow 6CO_2 + 6H_2O$ one **molecule** of $C_6H_{12}O_6$ reacts with six **molecules** of $O_2$ to yield six **molecules** of $CO_2$ and six **molecules** of $H_2O$. This is a **microscopic** interpretation. Second, as a result of the definition of the mole given in chapter 2, the equation can also be interpreted as one **mole** of $C_6H_{12}O_6$ reacts with six **moles** of $O_2$ to yield six **moles** of $CO_2$ and six **moles** of $H_2O$.
13. A **solution** is a homogeneous mixture of two or more substances.
14. Because a solution is a mixture, it is composed of two or more distinct substances; the substance present in the largest amount is called the **solvent**, while the substances present in the smaller amounts are called the **solutes**.
15. a. **Bronze** is a *solid* solution of copper and tin.
    **Brass** is a *solid* solution of copper and zinc.
    b. The earth's atmosphere is a *gaseous* solution.
    c. Wine is a solution made up of two liquids (water and alcohol)
16. A **colloid** is a mixture in which **large** particles ("large" when compared to molecular dimensions, ranging in size from approximately 1 nanometer to 1,000 nanometers in diameter) are distributed throughout the solvent. However, a colloid exhibits the qualitative light-scattering effect known as the Tyndall effect. Solutions **do not** exhibit the Tyndall effect.

    On the other hand, while a **suspension** is also a mixture, it is a **heterogeneous mixture**; both solutions and colloids are homogeneous mixtures. Because a suspension is a heterogeneous mixture of relatively large particles that are visible with a common optical microscope (the **minimum** size is approximately 1 micrometer in diameter, which is equal to the **largest** particle present in a colloid), the solvent and the particles of a suspension will separate into two distinct regions over time. Such a separation does **not** occur in either a solution or a colloid.

17. **Solvation** can be understood as the process in which a solute particle is surrounded by solvent molecules arranged in a specific manner.
18. **Solubility** is defined as the maximum number of **grams** of a **solute** present in a given amount of a solvent at a specified temperature.
19. The maximum amount of sucrose (measured in grams) that will dissolve in 335.0 mL of water is $6.700 \times 10^2$ g.
20. The volume of water is 351.3 mL.
21. The term **concentration of a solution** is defined as the amount of solute dissolved in a solvent.
22. Molarity = (Moles of solute) ÷ (Liters of solution)
23. The molarity of the KCl solution is 4.70 M.
24. **Dilution** is a procedure for preparing a **less concentrated** solution from a **more concentrated** solution.
25. The volume of the new solution is 452.9 mL.
26. The molarity of the $NH_4Cl$ solution is 3.57 M.
27. The mass of water in the solution is $3.49 \times 10^3$ g. [assume the density of water is 1.000 g $mL^{-1}$]
28. The volume of acetic acid is the solution is $9. \times 10^1$ mL.
29. a. A **saturated** solution is a solution containing the maximum amount of solute in a given solvent at a specific temperature.
    b. An **unsaturated** solution is a solution containing **less than the maximum** amount of solute at a specific temperature.
    c. A **supersaturated solution** is an unstable solution that contains **more solute than is present in a saturated solution.**
30. A chemical reaction is called **reversible** if it proceeds in the forward direction, from reactants to products **and** in the reverse direction, from products back to reactants.
31. A state of chemical equilibrium exists if the concentrations of reactants and products **do not change** with time. It is a **dynamic** state.

## Chapter 4 Exercises

1. The horizontal rows of the periodic table are called **periods.**
2. The vertical columns are identified as **groups** or **families.**
3. Both Lothar Meyer and Mendeleev, already established professors of chemistry in Germany and Russia, respectively, ordered the elements by **increasing atomic mass**, the key physical parameter utilized in the earlier attempts to organize the elements.
4. The modern periodic table is arranged by **atomic number**, beginning with hydrogen (atomic number = 1) and monotonically increasing in steps of one unit to ununoctium (atomic number = 118).
5. The elements can be segregated into three general categories: (1) the **metals**, which are good conductors of both heat and electricity; (2) the **nonmetals**, which are poor conductors of both heat and electricity; and (3) the **metalloids**, which exhibit properties that are intermediate between the metals and the nonmetals.
6. The element beryllium does not belong. It is in period 2; the other three elements are in period 1.
7. The element oxygen does not belong; fluorine, chlorine, and iodine are halogens while oxygen is not.

8. The field of **electrochemistry** is a branch of chemistry which studies the role of electrical phenomena in chemical reactions.
9. An **electrolyte** is any substance that, when dissolved in a solvent, yields a solution that **conducts electricity**. In contrast, a **nonelectrolyte** is any substance that, when dissolved in a solvent, yields a solution that does **not conduct electricity**.
10. The two types of charge found in our universe are *positive* charge and *negative* charge.
11. A **cation** is positively charged; an **anion** is negatively charged.
12. A **strong electrolyte** is a substance that, when dissolved in water, produces an aqueous solution that is a **very good conductor** of an electrical current.
13. A **weak electrolyte** is a substance that, when dissolved in water, produces an aqueous solution that is a **poor conductor of electricity.**
14. (a) chloride
    (b) sodium ion
    (c) calcium ion
    (d) sulfide
    (e) nitride
15. (a) nitrate
    (b) carbonate
    (c) ammonium
    (d) acetate
    (e) phosphate
16. The outermost electrons of an atom are often called **valence electrons.**
17. Lewis argued that an atom is composed of an *inner kernel* surrounded by a cubical *outer atom*. The eight corners of the outer cube coincided with the observation of the cyclic recurrence of chemical properties for groups of eight elements. Utilizing our idealized aufbau principle (see chapter 2), as each proton and electron are added to build electrically neutral atoms, each added electron occupies one of the eight cubical corners. After all eight cubical corners were filled with electrons (corresponding to the configuration of the noble gases in the periodic table), a new outer cube would form with the next added electron. This new outer cube, which contains a single electron, appears identical to the outer cube of the element whose atomic number is *eight units less*. Because the outermost electrons (often called **valence electrons**) are responsible for chemical reactivity and chemical properties, this model accounts for the periodic recurrence of chemical properties after each group of *eight* elements. Further, as noted earlier, because the filling of all eight cubical corners by electrons corresponds to the configuration of the noble gases, the unusual stability of these gases is attributed to the stability arising from the completely filled cubical arrangement. Lewis used his model of the outer cube of the atom to suggest that electrons can be either **added** to or **lost** from an atom to leave the atom with a **completely filled outer cube containing exactly eight electrons**. This is the highly stable configuration of the noble gases and came to be known as the **octet rule.**
18. Carbon can form a **maximum** of 4 bonds. Carbon is said to be **tetravalent**.
19. But, even more remarkably, the bonds linking carbon atoms and carbon to other atoms were **not** experimentally identical; in fact, it was possible to identify empirically three distinct types of carbon-carbon bonds as well as three distinct types of bonds linking carbon to other atoms.

20. The bond types formed by the carbon atom came to called **single**, **double**, and **triple** bonds, with the double and triples bonds known collectively as **multiple bonds**.
21. In the 1916 paper that introduced his model of chemical bonding, Lewis employed a simple two-dimensional representation of his ideas that came to be known as a **Lewis structure** (also known as a **Lewis dot diagram**, **electron dot diagram**, or **Lewis dot diagram**). Lewis structures use either a *pair of dots* or a simple *straight line* to symbolize the electron pair constituting the chemical bond between two atoms in a molecule.
22. Historically, the Lewis model of bonding in the case of electrolytes came to be called **ionic bonding**, while the bonding characterized by a shared pair of electrons (nonelectrolytes and the *internal* structure of the polyatomic ions) came to be called **covalent bonding**. Chemists associated with these bonding modalities the concepts of an **ionic bond** and a **covalent bond**, as if these two types of bonds are, in fact, constituents of the physical universe. However, they are better understood to be **convenient fictions**, or, from another perspective, **limiting idealizations** that provide only an efficient *model* summarizing a collection of experimental observations.
23. In an attempt to account for molecules such as $SO_2$, the chemists of the early twentieth century invented the concept of **resonance**. Each structure in figure 4.5 is called a **resonance structure** or a **canonical form**, and the $SO_2$ molecule that is part of the physical universe is considered to be an approximate intermediate form with *simultaneous* contributions from *all* the resonance structures. The approximate intermediate is more stable (lower in energy) than all the contributing canonical forms and is called a **resonance hybrid**.
24. (a) hydrogen bromide
    (b) potassium chloride
    (c) magnesium sulfate
    (d) calcium chloride
    (e) magnesium oxide
25. (a) calcium carbonate
    (b) ammonium chloride
    (c) potassium phosphate
    (d) sodium cyanide
    (e) potassium hydroxide
26. (a) lead(IV) oxide
    (b) iron(III) oxide
    (c) copper(II) chloride
    (d) chromium(II) oxide
    (e) gold(III) chloride
27. In 1932, the American chemist Linus Pauling proposed a new property called **electronegativity** that describes the behavior of atoms *linked in a chemical bond*. To define this new property, Pauling combined the convenient fictions of the *ionic bond* and the *covalent bond* and compared the bond energies of homonuclear diatomic molecules with those of a large group of heteronuclear diatomic molecules (for example, $H_2$, $Cl_2$, $F_2$, and $Br_2$ compared with HCl, HF, and HBr). Pauling noted that the bonds formed between elements from opposite sides of the periodic table were more stable (that is, required more energy to rupture them) than the bonds formed either between identical elements or between elements located in close proximity to one another. He suggested that the increased stability was the result of an *ionic* contribution to the bond. He argued that if two dissimilar elements (element A and element B) were linked in a molecule AB, an estimate of the *covalent* contribution to the bond could be made using the bond strengths of the homonuclear diatomic

molecules $A_2$ and $B_2$. On the other hand, there was an *ionic* contribution to the A-B bond because of some *transfer of charge between A and B*. This *ionic* contribution to the bond should arise from a *difference in the electronegativity of element A and element B in the A-B bond*. Consequently, **Pauling defined an element's electronegativity as the tendency for an atom involved in a chemical bond to draw charge (interpreted to mean "electrons") to itself.**

28. The central point is the **trends in electronegativity** exhibited by the elements: electronegativity *tends* to **increase** in a period from left to right and to **decrease** in a group from top to bottom.
29. a. Chlorine is more electronegative.
    b. Chlorine is more electronegative.
    c. Oxygen is more electronegative.
    d. Chlorine is more electronegative.
30. A molecule that exhibits such a **charge separation** is called **polar**. Consequently, the term **polarity** refers to this separation of charge.
31. a. Polar
    b. Nonpolar
    c. Nonpolar [This is difficult!]

## Chapter 5 Exercises

1. The extensive empirical data accumulated particularly during the second half of the nineteenth century when combined with the evolution of the new quantum mechanics after 1925 have strongly demonstrated that the unique chemical properties of an element depend on two key parameters:
   (1) the **number of protons** in the atomic nucleus of an element and
   (2) the **configuration of the electrons** surrounding the nucleus.
2. In order to describe the rearrangement of electrons that takes place **whenever** a molecule is formed (either an electrolyte or a nonelectrolyte), we can assign a *fictitious number*, called an **oxidation number**, to each atom in a molecule. The values of these fictitious numbers range from –4 to +8 usually in *integer* steps. (There are examples of fractional oxidation numbers. These cases will not be considered here.) The value of the fictitious number assigned to each atom in a molecule identifies the **oxidation state** of the atom; an atom assigned an oxidation number of +3 is then said to be in a +3 oxidation states.
3. **Oxidation** is defined as a **chemical process in which the value of an atom's oxidation number becomes *more positive*.** Because this is viewed as the *loss of negative values*, Oxidation is thought of as the *loss of electrons*.
4. Correspondingly, **reduction** is defined as a **chemical process in which the value of an atom's oxidation number becomes *more negative*.** Because this is viewed as the *gain of negative values*, reduction is thought of as the *gain of electrons*.
5. The **Reducing agent** is the species that causes the value of the oxidation number of another atom to become *more negative*. In terms of the above interpretations, the reducing agent *transfers electrons to the atom whose oxidation number value becomes more negative.*
6. The **Oxidizing agent** is the species that causes the value of the oxidation number of another atom to become *more positive*. In terms of the above interpretations, the oxidizing agent *removes electrons from the atom whose oxidation number value becomes more positive.*

7. Arrhenius defined an **acid** as a substance that, when dissolved in water, increases the concentration of **$H^+$ ions**. A **base** is a substance that, when dissolved in water, increases the concentration of **$OH^-$ ions**. (Note that Arrhenius used the term "ions" and did not call the $H^+$ ions *protons.*)
8. A **Brønsted-Lowry acid** (or, more simply, a **Brønsted acid)** is defined as a substance that is able to **lose** (that is, "**donate**" or "give up") a proton. Hence, a Brønsted-Lowry acid is often called a **proton donor.**
9. A **Brønsted-Lowry base** (**Brønsted base)** is defined as a substance that is able to **gain** (that is, "accept") a proton. Hence, a Brønsted-Lowry base is called a **proton acceptor.**
10. The **conjugate base of a Brønsted acid** is the species that remains after a proton is donated by the Brønsted acid.
11. The **conjugate acid of a Brønsted base** is the species that is formed by the addition of a proton to a Brønsted base.
12. We have met the nineteenth-century concepts of a **strong electrolyte** and a **weak electrolyte**; the distinction between the two ideas was defined in terms of the ability of the resulting solution to conduct electricity (chapter 4). We further noted that an aqueous solution that is a very good conductor of electricity must have a large concentration of mobile charges, while an aqueous solution that is a poor conductor of electricity **cannot** have a large concentration of mobile charges.
    Because a **strong** electrolyte produces a solution with a large concentration of ions, this means that, in an aqueous solution, virtually **every** acid molecule has **donated** a proton to form an $H_3O^+$ ion. Such an acid is called a **strong** acid. In the case of a base, because virtually **every** base molecule has **accepted** a proton, resulting in a large excess of $OH^-$ ions in solution, we call the base a **strong** base.
13. As noted in question 12, the concepts of a **strong electrolyte** and a **weak electrolyte date from the 19th century**; the distinction between the two ideas was defined in terms of the ability of the resulting solution to conduct electricity (chapter 4). We further noted that an aqueous solution that is a **very good conductor** of electricity must have a **large concentration** of mobile charges, while an aqueous solution that is a **poor conductor** of electricity **cannot have a large concentration** of mobile charges. If only a small number of acid molecules **donate** a proton, or only a small number of base molecules **accept** a proton, the acid or base is called **weak.**
14. We define the **state of chemical equilibrium** to be the state of a chemical system involved in a reaction when the **concentrations of reactants and products *no longer change with time***. However, it is important to understand that at equilibrium, even though the concentrations of reactants and products no longer change with time, reactants are still **forming** products, and products are still **reforming** reactants. Chemical equilibrium is a **dynamic** state!
15. The **state of chemical equilibrium** to be the state of a chemical system ***involved in a reaction*** when the **concentrations of reactants and products *no longer change with time***. However, it is important to understand that at equilibrium, even though the concentrations of reactants and products no longer change with time, reactants are still **forming** products, and products are still **reforming** reactants. Chemical equilibrium is a **dynamic** state! The reader should note that there is a difference between chemical equilibrium and a much simpler **physical equilibrium**. ***A physical equilibrium involves no chemistry***; that is, the chemical characteristics of a substance do not change.

16. The equilibrium constant is:

$$K_{eq} = \frac{[Cl^-]_{eq}^2[ClO_3^-]_{eq}}{[ClO^-]_{eq}^3}$$

17. Use the following expression for the equilibrium constant:

$$K_{eq} = \frac{[Cl^-]_{eq}^2[ClO_3^-]_{eq}}{[ClO^-]_{eq}^3}$$

The value of the equilibrium constant is:

$$K_{eq} = 0.593$$

18. For the general equation

$$aA + bB \quad cC + dD$$

**Define** a new quantity called the **reaction quotient, *Q***:

$$Q = \frac{[C]^c[D]^d}{[A]^a[B]^b}$$

Note that the *reaction quotient* looks exactly like the *equilibrium constant* **except** that the *reaction quotient* does **not** use the equilibrium concentrations of the reactants and products. This means that the reaction quotient can be calculated at **any time** during the course of a chemical reaction; there is **no requirement** that the chemical reaction must attain equilibrium before the reaction quotient is calculated. Consequently, unlike the *equilibrium constant*, which has a **unique** value determined by the *temperature* and *equilibrium concentrations* of the reactants and the products, the *reaction quotient* has a **different** value at each instant of time during the course of a chemical reaction.

19. Symbolize a monoprotic weak acid by HA (the "A" symbolizes the conjugate base of the acid) and examine its behavior in an aqueous solution:

$$HA(aq) + H_2O(l) \quad H_3O^+(aq) + A^-(aq)$$

Because the behavior of HA in an aqueous solution is an **equilibrium process**, we can write the equilibrium constant as follows:

$$K_{eq} = \frac{[H_3O^+]_{eq}[A^-]_{eq}}{[HA]_{eq}[H_2O]_{eq}}$$

However, because HA is a **weak acid** and only a vanishingly small number of the HA molecules dissociate in the solution, the $[H_2O]_{eq}$ is essentially the same as the original $[H_2O]$. That is, because $[H_2O]_{eq} = [H_2O]$, the $[H_2O]$ is effectively a **constant**. Because the $[H_2O]$ is effectively a constant, it is conventional to make the following **definition**:

$$K_a = K_{eq} \times [H_2O]_{eq}$$

**The newly defined quantity, $K_a$, is called the acid ionization constant**, and it possesses all the characteristics of any other equilibrium constant. In particular, it is **temperature-dependent**.
**The acid ionization constant** can be rewritten as follows:

$$K_a = \frac{[H_3O^+]_{eq}[A^-]_{eq}}{[HA]_{eq}}$$

Now consider the behavior of a weak base. We will symbolize our prototypical weak base by B and examine the following reaction:

$$B(aq) + H_2O(l) \quad HB^+(aq) + OH^-(aq)$$

Because the behavior of the base, B, in an aqueous solution is an **equilibrium process**, we can write the equilibrium constant as follows:

$$K_{eq} = \frac{[HB^+]_{eq}[OH^-]_{eq}}{[B]_{eq}[H_2O]_{eq}}$$

Because B is a **weak base**, only a vanishingly small number of the B molecules accept a $H^+$ ion from a $H_2O$ molecule, making the $[H_2O]_{eq}$ essentially the same as the original $[H_2O]$. Just as in the case of a weak acid, because $[H_2O]_{eq} = [H_2O]$, the $[H_2O]$ is effectively a **constant**. Because the $[H_2O]$ is effectively a constant, it is conventional to make the following **definition**:

$$K_b = K_{eq} \times [H_2O]_{eq}$$

**The newly defined quantity, $K_b$, is called the base ionization constant**, and it possesses all the characteristics of any other equilibrium constant. In particular, it is **temperature-dependent**.
**The base ionization constant** can be rewritten as follows:

$$K_b = \frac{[HB^+]_{eq}[OH^-]_{eq}}{[B]_{eq}}$$

20. Consider the following reaction:

$$H_2O(l) + H_2O(l) \quad H_3O^+(aq) + OH^-(aq)$$

Note that one of the water molecules acts as a **proton donor** (an acid), while the other water molecules acts as a **proton acceptor** (a base). The process described by equation 5.14 is often called the **autoionization** of water because two charged species, one positively charged and one negatively charged (represented by the symbols $H_3O^+$ and $OH^-$), are produced. Significantly, water is a **weak electrolyte** and, consequently, behaves as both a **weak acid** and a **weak base**. We can write an expression for the water equilibrium constant:

$$K_{eq} = \frac{[H_3O^+]_{eq}[OH^-]_{eq}}{[H_2O]_{eq}^2}$$

However, because water is a **weak acid (base)** and, as a result, only a vanishingly small number of the water molecules either accept or donate a proton, the $[H_2O]_{eq}$ is essentially the same as the original $[H_2O]$. That is, because $[H_2O]_{eq} = [H_2O]$, the $[H_2O]$ is effectively a **constant**. Because the $[H_2O]$ is effectively a constant, it is conventional to make the following **definition**:

$$K_w = K_{eq} \times [H_2O]_{eq}^{\ 2}$$

**The newly defined quantity, $K_w$, is called the ion product constant of water**, and it possesses all the characteristics of any other equilibrium constant. In particular, it is **temperature-dependent.**

**The ion product constant of water** can be rewritten as follows:

$$K_w = [H_3O^+]_{eq}[OH^-]_{eq}$$

21. $pH = -\log[H_3O^+]$
    $pOH = -\log[OH^-]$
22. a. 0.340
    b. 9.339
    c. 13.879
23. a. $[H_3O^+]_{eq} = 7.1 \times 10^{-12}$
    b. $[H_3O^+]_{eq} = 4.5 \times 10^{-3}$
    c. $[H_3O^+]_{eq} = 1.41 \times 10^{-4}$
24. a. $pOH = 4.55$
    b. $pOH = 10.71$
    c. $pOH = 6.90$
25. The most elementary chemical reaction that occurs when an acid and a base are mixed is called a **neutralization** reaction. That is,

$$\text{Acid} + \text{Base} \longrightarrow \text{Salt} + \text{Water}.$$

26. A **salt** is defined as a compound composed of a **cation** (other than $H^+$) and an **anion** (other than $OH^-$ or $O^{2-}$).
27. A **buffer** is defined to be a solution (here we continue to restrict our focus to aqueous solutions) that **resists** changes to its pH when **small** amounts of an acid or a base are added to the solution.
28. A **buffer (buffer solution)** is a solution composed of *two* parts; both components **must** be present in the solution.
    (1.) A weak conjugate acid-base pair
    (2.) A salt derived from the weak acid or the weak base

## Chapter 6 Exercises

1. VSEPR stands for *Valence Shell Electron Pair Repulsion.*
2. The geometric arrangement of these regions around the central atom of the molecule is determined by **minimizing the electrostatic repulsion** among the regions of concentrated negative charge.
3. Note that this is the origin of the VSEPR name: the regions of negative charge arise from the pairs of **valence electrons**, and the geometric arrangement of these regions of negative charge depends on **minimizing the repulsion** among them, hence the name **valence shell electron pair repulsion**. This simple idea gives rise

to only **six geometric arrangements** of the negatively charged regions around the central atom: (1) linear, (2) trigonal planar, (3) tetrahedral, (4) trigonal bipyramidal, (5) octahedral, and (6) pentagonal bipyramidal.

4. The fact that the VSEPR theory can successfully predict the ***geometry*** of molecules is a very important attribute of this theory of molecular structure. The reader will recall that the very early effort by Lewis (1916) provided **no** *geometric* information about a molecule's structure.
5. The VSEPR model distinguishes two distinct geometries: the **electronic geometry**, which is determined by the spatial arrangement of the negative charge concentrations, and the **molecular geometry**, which is determined by the spatial configuration of the atomic nuclei. The two geometries are **identical** if there are **no nonbonding pairs (lone pairs)** of electrons arranged around the central atom.
6. The VSEPR model distinguishes two distinct geometries: the **electronic geometry**, which is determined by the spatial arrangement of the negative charge concentrations, and the **molecular geometry**, which is determined by the spatial configuration of the atomic nuclei. The two geometries are **identical** if there are **no nonbonding pairs (lone pairs)** of electrons arranged around the central atom.
7. The simplest approach to answering this question is to draw simple Lewis diagrams and use Figure 6.2. In the Figure, each "E" corresponds to a **nonbonding pair** of electrons in the Figure.

| | | | |
|---|---|---|---|
| (a) Molecular: | Linear | Electronic: | Tetrahedral |
| (b) Molecular: | Bent (Angular) | Electronic: | Tetrahedral |
| (c) Molecular: | Tetrahedral | Electronic: | Tetrahedral |
| (d) Molecular: | Bent (Angular) | Electronic: | Trigonal Planar |

8. The developments of quantum mechanics throughout the twentieth century directly challenge the perspective of the VSEPR model. Even though VSEPR reached its mature development in the mid-1950s, and the developers of the model were well aware of the nearly six decades of work on the quantum model, the model **does not** utilize the quantum perspective. The emergence of the ***dynamic*** electron in the physics of the first third of the twentieth century, accompanied by the probability interpretation of the wave function, Bohr's complementarity, and Heisenberg's indeterminacy principle (chapter 2), all demand an alternative to the ***static structural*** approach that dominated chemical thought from the middle of the nineteenth century into the beginning of the twentieth century.
9. True. VB Theory utilizes the concept of a hybrid atomic orbital.
10. True. In VB Theory, the hybrid orbitals are hybrid **atomic** orbitals; that is, atom–centered orbitals.
11. In VB Theory, a chemical bond is defined as the **overlap of atomic orbitals**. But what does this mean? The phrase *overlap of atomic orbitals* means that orbitals associated with electrons from two distinct atoms (hence, they are atomic orbitals) delineate a probability distribution for a common region of space; this probability distribution specifies the probability of locating two electrons in some small but essentially finite region of space proximate to the two nuclei.
12. From the VB perspective, an electron is **not** added to one atom or lost from another atom (the Heisenberg indeterminacy principle prohibits the use of such macroscopically inspired terms); rather, a pair of electron shares a spatial probability distribution whose experimental consequence is the observation of *charge carriers*. So, the key to understanding the chemical bonding again rests on the idea of a shared pair of electrons. VB theory predicts *bond energies* quantitatively, and further, demonstrates that stable molecules form from reacting atoms or molecules when the energy characterizing the system has **decreased to a minimum**. For the first time, a theory of molecular bonding tells us why molecules exist: **minimization of a system's characteristic energy**. Finally, VB theory is capable of explaining the quantitative difference in the *bond lengths* of

the two molecules by noting that the common spatial probability distribution defined by the orbital overlap involves **different atomic orbitals** in $H_2$ compared to those used in $F_2$.

13. By utilizing the resonance property of quantum mechanics, Pauling devised a procedure to *couple* together **nonequivalent atomic orbitals** (that is, orbitals characterized by **different** values of the *angular momentum quantum number*, see chapter 2) to form a new set of atomic orbitals. These new orbitals are called **hybrid atomic orbitals.**
14. The procedure that forms hybrid atomic orbitals is called **hybridization.**
15. Yes, the hybrid orbitals are atomic orbitals. They are centered on individual atoms.
16. The hybridization procedure **conserves the *number* of orbitals.** That is, the ***number* of hybrid atomic orbitals** produce by the hybridization procedure ***exactly equals the number* of pure atomic orbitals** coupled together by the phenomenon of quantum mechanical resonance.
17. The hybridization procedure does require energy. However, because the resulting set of hybrid atomic orbitals identifies quantum states that are energetically **more stable** than the quantum states associated with the pure atomic orbitals used by the hybridization procedure, the hybridization procedure is *advantageous*; the achievement of an energetically more stable state (identified by a hybrid atomic orbital) by a molecule compensates for the additional energy required by the hybridization procedure.
18. (a) **One** *s orbital* and **one** *p orbital*; **two** hybrid orbitals; **linear** geometry.
    (b) **One** *s orbital* and **two** *p orbitals*; **three** hybrid orbitals; **trigonal planar** geometry.
    (c) **One** *s orbital* and **three** *p orbitals*; **four** hybrid orbitals; **tetrahedral** geometry.
19. The VB model (again, like VSEPR) distinguishes two distinct geometries: the electronic geometry, which is determined by the spatial arrangement of the hybrid atomic orbitals, and the molecular geometry, which is determined by the spatial configuration of the atomic nuclei. The two geometries are **identical** if the hybrid atomic orbitals of VB theory are **not** associated with any **nonbonding pairs (lone pairs)** of electrons around the central atom. Consequently, as in the VSEPR model, the electronic geometry and the molecular geometry are distinct.
20. A **sigma bond** is a chemical bond that is a result of the end-to-end *overlap of atomic orbitals*. In the case of a sigma bond, this common region of space is located geometrically **between** the nuclei of two atoms participating in the bond. It is a region of space that lies **symmetrically along a line connecting the two nuclei.**
21. In contrast to a sigma bond, a **pi bond** is a chemical bond that is formed by atomic orbitals *overlapping **above** and **below** the molecular plane* either of a planar molecule or of a planar segment of a larger molecule. A pi bond (or, more generally, a pi bond system, as in benzene) involves the overlap of *two or more* unhybridized ***p** atomic orbitals*. In the case of a pi bond, this common region of space is located geometrically **above** and **below** the molecular plane either of a planar molecule or of a planar segment of a larger molecule.
22. As the name implies, a *double* bond represents *two* linkages between a pair of atoms, or, in terms Lewis would understand, a bond that requires *two* electron pairs to describe it. Using the terminology from VB theory, a double bond means that *both* a **sigma bond** and a **pi bond** determine the chemical link between two atoms. The presence of a *double* bond is understood as the formation of *one* **sigma bond** and *one* **pi bond.**
23. A **triple bond** requires *one* **sigma bond** and *two* **pi bonds** linking two atoms. In $C_2H_2$ the two carbon nuclei are first linked (that is, *bonded*) by the end-to-end overlap of ***sp** hybrid atomic orbitals* (a **sigma bond**). The unhybridized *s atomic orbital* of the two hydrogen atoms then form sigma bonds with the remaining ***sp** hybrid atomic orbitals* on each carbon atom. The unhybridized *p orbitals* (two on each carbon nucleus) are **perpendicular** to both the ***sp** hybrid atomic orbitals* and one another. The experimental data support the

interpretation that $C_2H_2$ is a linear molecule. This can be understood within the framework of VB theory by allowing the remaining two *p orbitals* centered at each carbon nucleus to overlap and form *four* probability distributions. These probability distributions specify the probability of locating two electrons in a region of space possessing a characteristic geometry and located symmetrically *around* the line joining the two nuclei (this is the spatial region defined by the sigma bond) **but which do not share the *same* space as the sigma bond**. This is nothing more than the formation of two **pi bonds**! Consequently, the presence of a *triple* bond is understood as the formation of *one* **sigma bond** and *two* **pi bonds.**

24. Developed in the 1930s following the initial introduction of VB theory, MO theory is also an **explicitly quantum mechanical model of chemical bonding.**
25. False. MO theory is **molecule-centered**, making the fundamental assumption that the orbitals (the one-electron solutions of the Schrödinger equation that we met in chapter 2) associated with each electron are properties of the **entire molecule**. The molecular orbitals of MO theory are the result of the interactions of the atomic orbitals of the bonded atoms, but the molecular orbitals are **not** localized; they are associated with the **entire** molecule.
26. Within the conceptual framework of MO theory, the interaction of atomic orbitals (again, a quantum mechanical resonance phenomenon) produces two types of molecular orbitals, **bonding molecular orbitals** and **antibonding molecular orbitals**. Bonding molecular orbitals are **lower** in energy and exhibit **more stability** than the starting atomic orbitals, while antibonding molecular orbitals are **higher** in energy and exhibit **less stability** than the starting atomic orbitals.
27. Just like VB theory, however, the quantum mechanical procedure for forming bonding and antibonding orbitals **conserves orbital number**. This means that the number of bonding and antibonding orbitals that form as a result of the quantum mechanical resonance is **exactly equal** to the number of interacting atomic orbitals.
28. In the context of MO theory, the orbitals formed by the interaction of atomic orbitals may be either **sigma orbitals** or **pi orbitals**. A sigma molecular orbital defines a probability distribution that identifies a common region of space that is located geometrically **between** the nuclei of two atoms participating in the bond; it is a region of space that lies **symmetrically along a line connecting the two nuclei**. In contrast, a pi molecular orbital defines a probability distribution that identifies a common region of space that is located geometrically **above** and **below** the molecular plane either of a planar molecule or of a planar segment of a larger molecule. This spatial region is always located symmetrically *around* the line joining the two nuclei participating in a bond, **but it is *spatially disjoint* (sharing no common spatial coordinates) from any specified sigma orbitals**. Because the designations *sigma* and *pi* identify the geometric symmetry of the molecular orbitals, there are, in fact, **four** possibilities: **sigma bonding molecular orbitals** (symbolized by $\sigma$), **sigma antibonding molecular orbitals** (symbolized by $\sigma^*$), **pi bonding molecular orbitals** (symbolized by $\pi$), and **pi antibonding molecular orbitals** (symbolized by $\pi^*$).
29. Bond order = (½) × [ (Bonding valence electrons) – (Antibonding valence electrons ) ]
    The bond order of $O_2$ is 2.
30. See Figure 6.16. Note that there are **two unpaired electrons**.

## Chapter 7 Exercises

1. The element carbon is the focus of organic chemistry.
2. The carbon atom forms a maximum of 4 bonds. In stable compounds, the carbon atom appears to form **four chemical bonds.** We say that carbon is **tetravalent.**
3. The carbon atom can form **multiple bonds** (both double bond and triple bonds).
4. 

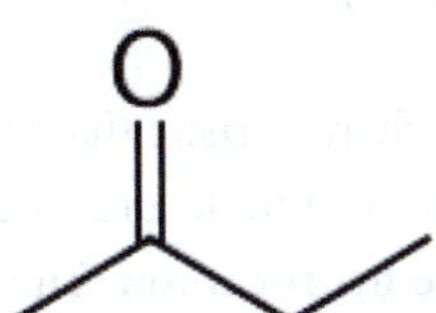


**Butanone Skeletal Structure**

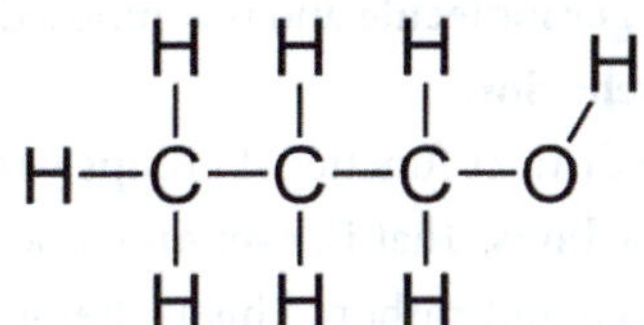


**1-Butanol Structural Formula**

**$CH_3CH_2CH_2CH_3$**

**Condensed Structure**

5. a. A model of an organic molecule that depicts a three-dimensional structure is often called a *ball-and-stick* model of the organic molecule.
   b. There is an alternate approach to representing the three-dimensionality of actual molecules. The image in Figure 7.4 is often called a *wedge-and-dashed model.* In this model the *wedge* bond is used to indicate that the hydrogen atom is projecting *above* the plane of the textbook page, while the *dashed* bond indicates that the hydrogen atom projects *below* the plane of the textbook page.
6. a. A **primary** carbon atom is a carbon atom with only one other carbon atom attached to it.
   b. A **secondary** carbon atom is a carbon atom with two other carbon atoms attached to it.
   c. A **tertiary** carbon atom is a carbon atom with three other carbon atoms attached to it.
   d. A **quaternary** carbon atom is a carbon atom with four other carbon atoms attached to it.
7. An organic (and, of course, biochemical) compound is called **saturated** if it contains ***only*** single bonds; that is, all the atoms in the molecule are linked by single bonds. In contrast, a molecule in which ***one or more*** of the links between atoms in the molecule are ***multiple bonds*** is called **unsaturated.**
8. The atom can form four single bonds. The most elementary identification is between the set of *four* ***sp3*** *hybrid atomic orbitals* and a tetravalent carbon atom exhibiting *four* single bonds. It is particularly important to note that such a carbon atom not only exhibits *four* bonds but also possesses an experimental geometry that is consistent with the *tetrahedral* geometry of $\mathbf{sp^3}$ hybridization.
9. In the case of the $\mathbf{sp^2}$ hybridization of carbon, the *three* ***sp2*** *hybrid atomic orbitals* that are centered on the carbon atom are *coplanar,* leaving a single unhybridized ***p*** *atomic orbital* centered on the same carbon atom and perpendicular to the plane defined by the *three* ***sp2*** *hybrid atomic orbitals.* The end-to-end *resonance interaction* of *two* ***sp2*** *hybrid atomic orbitals* (or of one ***sp2*** *hybrid atomic orbital* with one unhybridized atomic orbital) to define a *single* bond as well as the *resonance interaction* of the unhybridized ***p*** *atomic orbitals* centered on **adjacent** atoms (carbon or noncarbon) directly correlate with the experimental observation of a *double* bond. Importantly, the experimental data are consistent with the planar geometry required by the *double* bond of VB theory and the necessary *four* bonds associated with each $\mathbf{sp^2}$ hybridized carbon atom.
10. The **sp** hybridization of the carbon atom produces a set of *two* ***sp*** *hybrid atomic orbitals* that exhibit a *linear* geometry while leaving *two* unhybridized ***p*** *atomic orbitals,* all centered on the same carbon atom. The unhybridized ***p*** *atomic orbitals* are *mutually perpendicular* as well as *perpendicular* to the imaginary line passing through the *two* ***sp*** *hybrid atomic orbitals* (because their geometry is *linear*). The end-to-end *resonance*

*interaction* of *two* **sp** *hybrid atomic orbitals* (or of one **sp** *hybrid atomic orbital* with one unhybridized atomic orbital) to define a *single* bond as well as the *resonance interaction* of the two unhybridized ***p*** *atomic orbitals* centered on **adjacent** atoms (carbon or noncarbon) directly correlate with the experimental observation of a *triple* bond. Again, the linear geometry required by the *triple* bond of VB theory and the *four* bonds associated with each **sp** hybridized carbon atom are consistent with the experimental data.

11. A **functional group** is a **part of a larger molecule** and is composed of an **atom**, a **group of atoms**, or a **bond** that has a **characteristic chemical behavior**.
12. The first family is composed of *four* distinct functional groups with the common characteristic that all the molecules of this family are **hydrocarbons**. That is, every member of this first family of molecules is composed of only **two** elements: hydrogen and carbon. This is the origin of the name **hydrocarbon**. The four functional groups that belong to the hydrocarbon family are the **alkanes**, the **alkenes**, the **alkynes**, and the **aromatic hydrocarbons**.
13. All the molecules of the first family of functional groups family are **hydrocarbons**. That is, every member is composed of only **two** elements: hydrogen and carbon. This is the origin of the name **hydrocarbon**.
14. a. In the **alkanes, all** the carbon-carbon bonds are **single bonds.**
    b. In the **alkenes, at least one** of the carbon-carbon bonds **must be a double bond.**
    c. In the **alkynes, at least one** of the carbon-carbon bonds **must be a triple bond.**
15. The molecules constituting the **aromatic hydrocarbon** functional group contain planar, benzene-like ring structures, and all the aromatic hydrocarbons exhibit the phenomenon of **delocalized electrons** that is characteristic of benzene.
16. The second family of functional groups consists of molecules in which a **single** bond links a carbon atom to an **electronegative** atom. (These are atoms that are assigned *large* values for their electronegativity; see chapter 4.) This second family is significant because it marks the appearance in organic molecules of several crucially important electronegative atoms. (These atoms are members of a class of atoms called **heteroatoms** by organic chemists. In general, a heteroatom is *any* atom that is ***not*** either hydrogen or carbon.) In particular, we see for the first time the inclusion of oxygen atoms, nitrogen atoms, and sulfur atoms in organic molecules. There are again four principal members of this family of functional groups: the **alcohols**, the **ethers**, the **sulfur-containing compounds**, and the **amines**.
17. The second family of functional groups consists of molecules in which a **single** bond links a carbon atom to an **electronegative** atom.
18. The *OH* atom combination is the characteristic molecular fragment of **all** alcohols. The *OH* group of atoms is the **hydroxyl functional group**, and often the adjective **hydroxy** is used to describe it.
19. The **ethers** are a collection of molecules in which a single oxygen atom is linked to *two different* carbon atoms by **single** bonds.
20. The **thiols** are the sulfur analogs of alcohols in which the oxygen atom has been replaced by a sulfur atom.
21. The **thioethers** are the sulfur analogs of ethers in which the oxygen atom has been replaced by a sulfur atom.
22. a. A primary amine has one carbon atom bonded to the nitrogen atom in an amine.
    b. A secondary amine has two carbon atoms bonded to the nitrogen atom in an amine.
    c. A tertiary amine has three carbon atoms bonded to the nitrogen atom in an amine.
    d. A quaternary amine has four carbon atoms bonded to the nitrogen atom in an amine.

23. The third family of functional groups consists of five classes of molecules, **all** of which contain a carbon-oxygen **double** bond: the **aldehydes**, the **ketones**, the **carboxylic acids**, the **esters**, and the **amides**.

24.

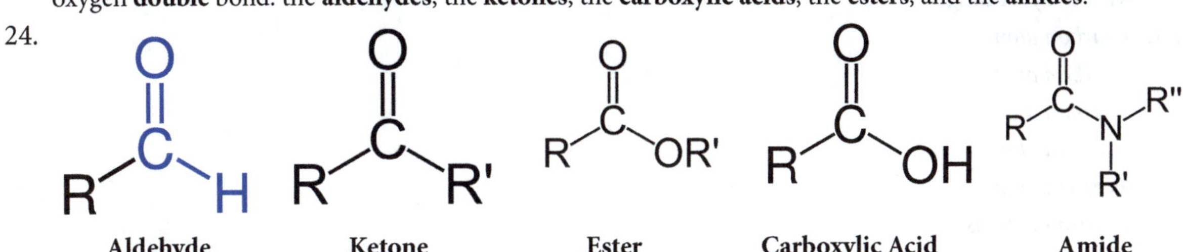


25.

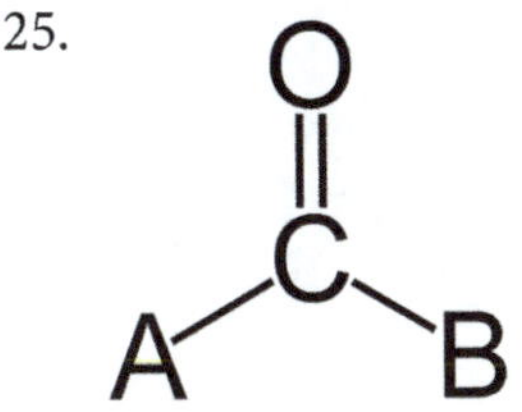


26. The aliphatics occur as **open-chain** molecules and as **cyclic molecules** (molecules with ring structures).
27. We summarize here some of the rather unremarkable properties that characterize the alkane functional group:
    (1.) They are either odorless or possess a mild odor; they are colorless and tasteless.
    (2.) The alkanes are nonpolar (see chapter 4) molecules; they are insoluble in water but soluble in nonpolar organic solvents.
    (3.) The alkanes are less dense than water.
    (4.) The alkanes **are** flammable, reacting vigorously with oxygen.
    (5.) The alkanes exhibit a monotonic increase in both melting and boiling points as a function of their molecular weight; alkanes containing four or fewer carbon atoms are gases at 298.15 K and a pressure of 1 atmosphere.
    (6.) With the exception of their flammability, the alkanes are not very reactive.
28. False. The alkenes are unsaturated hydrocarbons and, like the alkane family, occur as both open-chain molecules and cyclic molecules.
29. Each organic name consists of three parts:
    (1.) A prefix
    (2.) A root
    (3.) A suffix
30. In order to understand the meaning of these three principal parts of an organic name, we *imagine* that every organic molecule can be constructed by starting with either an alkane (open-chain or ring) or an aromatic hydrocarbon. We call this imaginary molecule the **parent compound**.
31. A substituent is either a single atom or a group of atoms (other than one of the functional groups noted above) that replaces a hydrogen atom in the molecule's **parent compound**.
32. The *suffix* identifies the **functional group** to which the molecule belongs.
33. One of the most significant groups of atoms that plays the role of a substituent is the **alkyl group**. An alkyl group is simply an alkane molecule with one hydrogen atom removed. The removal of the hydrogen atom during the formation of an alkyl group provides an open bonding site, allowing the alkyl group to form a

link with a parent compound. The name **alkyl** is derived from the word **alkane** by replacing the *ane* ending with the *yl* ending.

34. *1 carbon atom*
    *4 carbon atoms*
    *3 carbon atoms*
    *8 carbon atoms*
    *9 carbon atoms*
    *10 carbon atoms*
    *3 carbon atoms*
35. a. 1-chloropentane
    b. 1-bromo-2-ethyl-4-methylpentane
    c. 1-bromo-2-isobutylhexane
36. a. 1-chloro-1-pentene or 1-chloropent-1-ene
    b. 6-bromo-3,5-dimethyl-1-hexene or 6-bromo-3,5-dimethylhex-1-ene
    c. 8-bromo-5,7-dimethyl-2-octene or 6-bromo-5,7-dimethyloct-2-ene
37. a. $CH_3CH\text{-}CHCH_2CH_2CH_2CH_2CH_3$
    | |
    $CH_3$ $CHCH_2CH_2CH_3$
    |
    $CH_3$
    b. $CH_2{=}CCH_2CHCH_2CH_3$
    | |
    Br $CH_2CH_3$
38. a. The 1,2 arrangement of substituents (adjacent carbon atoms) on a benzene ring is called the **ortho** relationship.
    b. The 1,3 arrangement of substituents (*one* carbon atom *between* the carbon atoms to which the substituents are bonded) on a benzene ring is called the **meta** relationship.
    c. The 1,4 arrangement of substituents (*two* carbon atom *between* the carbon atoms to which the substituents are bonded) on a benzene ring is called the **para** relationship.
39. Carbohydrates are polyhydroxy aldehydes, or polyhydroxy ketones, or substances that yield these compounds when hydrolyzed.
40. The tem **hydrolysis** is built from two Greek words, *hydro* (water) and *lysis* (to separate) and literally means *to cut or separate with water.*
41. a. **Monosaccharides** are carbohydrates that cannot be hydrolyzed into simpler carbohydrate units. The suffix *saccharide* is derived from the Greek *sakkharon*, meaning "sugar" (the Latin word for *sugar* is *saccharum*), and is an indication that the simplest members of the carbohydrate class of molecules are the sugars that are commonly part of our daily experience. The formal definition requires a monosaccharide molecule to contain a minimum of three carbon atoms.
    b. **Disaccharides** are carbohydrates that yield two monosaccharides upon hydrolysis.
    c. **Polysaccharides** are carbohydrates that yield **many** monosaccharide units when hydrolyzed.
    d. **Oligosaccharides** are carbohydrates composed of three to nine monosaccharide units.

42. a. 1-pentanol or pentan-1-ol
    b. 6-bromo-3,5-dimethyl-1-hexanol or 6-bromo-3,5-dimethylhexan-1-ol
    c. 5-ethyl-7-methyl-2-octanol or 5-ethyl-7-methyloctan-2-ol
43. a. 4,6-dimethyl-2-propylheptanal
    b. 4,7-dimethyl-6-propyl-3-nonanone or 4,7-dimethyl-6-propylnonan-3-one

    Or

    4-methyl-6-sec-butyl-3-nonanone or 4-methyl-6-sec-butylnonan-3-one
44. The suffix *ose* is often preceded by either the prefix *aldo* (from the word *aldehyde*) or by the prefix *keto* (from the word *ketone*), giving rise to the terms *aldose* for a monosaccharide containing an aldehyde functional group and *ketose* for a monosaccharide containing a ketone functional group.
45. Glucose, Fructose, Galactose, Ribose, 2-Deoxyribose
46. a. ketose
    b. aldose
    c. aldose
47. The two pentose sugars that play an important structural role in nucleic acids are ribose and 2-deoxyribose.
48. a. An acetal is a functional group in which **two ether-like** structures share a common carbon atom.
    b. A hemiacetal is a functional group in which **one hydroxyl** group is bonded to a carbon atom that is part of an **ether-like** structure.
49. The conventional representation of the cyclic form of the monosaccharides is called a **Haworth projection.**
50. Carbon atom 1 in Figure 7.23 is called the **anomeric** carbon; note that it is the key carbon atom of a hemiacetal functional group. The anomeric carbon along with the oxygen atom and carbon atom 5 constitute the **ether-like** structure of the hemiacetal, while the hydroxyl group bonded to the anomeric carbon is the other constituent of the hemiacetal.
51. The formation of cyclic structures by the monosaccharides actually produces **two** interconvertible forms of the ring molecule that are in equilibrium with one another. These two interconvertible forms are called **anomers** (this is the origin of the name for carbon atom 1 in figure 7.23) and are designated as the **alpha anomer** ($\alpha$) and the **beta anomer** ($\beta$). Figure 7.24 depicts the equilibrium between the two anomers.
52. The six-member rings are given the generic name **pyranose**, where the reader will recognize the *ose* suffix that is indicative of a sugar. In contrast, the five-member rings are given the name **furanose**.
53. Disaccharides, as the name implies, are constructed from two monosaccharides through the formation of a **glycosidic bond**. Figure 7.28 is a schematic highlighting the key features of a glycosidic bond between two glucose molecules. By examining the figure, let's carefully note the important characteristics of the glycosidic bond. Beginning with the upper portion of the figure, note that the two leftmost glucose molecules are designated as a anomers, meaning that the hydroxyl groups bonded to the anomeric carbon of each molecule (carbon atom 1 in the figure) lie *below* the molecular rings. The glycosidic bond forms as a result of a reaction between the hydroxyl group bonded to carbon atom 1 of the first glucose molecule and the hydroxyl group bonded to carbon atom 4 of the second glucose molecule (both are encircled by the dashed region in the figure). The result of this reaction leaves an oxygen atom bonded both to carbon atom 1 (the anomeric carbon) of the first glucose molecule and to carbon atom 4 of the second glucose molecule while producing a disaccharide molecule and a water molecule as the products of the reaction (right half of the upper portion of the figure). The disaccharide (also called a **glycoside** because it is a molecule made up of two simpler components joined by a **glycosidic bond**) also exhibits one of our two new functional groups:

the acetal. To identify the acetal structure, examine the disaccharide molecule and note that the anomeric carbon (labeled 1) of the left molecular ring in the disaccharide is *shared by* ***two ether-like*** *structures*. This is the defining signature of the acetal functional group. Because the glycosidic bond formed in the upper half of the figure links the anomeric carbon of the first α anomer with carbon atom 4 of a second molecule, the bond is given the formal name **α(1→4) glycosidic bond.**

54. The disaccharide that is produced when a glycosidic bond forms is also called a **glycoside**.
55. The glycosidic bonb is givn the name: α, **β(1→2) glycosidic bond**.
56. As we noted above, the α, **β(1→2) glycosidic bond** links together the anomeric carbon atoms from the glucose and fructose monosaccharide components of sucrose. Because both anomeric carbon atoms are part of the glycosidic bond and are **not** available to participate in other reactions, sucrose **cannot** participate in oxidation-reduction reactions (see chapter 5). In contrast, the monosaccharide glucose **can** participate in oxidation-reduction reactions and is called a **reducing** sugar. **Sucrose is not a reducing sugar.**
57. By the successive addition of glucose units, a **polysaccharide** molecule (that is, a molecule containing *many* monosaccharide units) is produced. In effect, polysaccharides are highly effective structures used for **storing** glucose. The polysaccharide known as **starch** is the glucose storage mechanism used by plants.
58. The storage polysaccharide used by animals is called **glycogen.**
59. **Cellulose** is a polysaccharide in which large numbers of glucose units are linked together in very long linear chains by β(1→4) glycosidic bonds (figure 7.34). Cellulose is a major component of the cell walls of green plants, and its linear alignment allows it to form strong, rigid structures. Cellulose constitutes more than 40% of the wood in trees and is a major component of plant fibers: cotton (~90%), flax (~65%), and hemp (~67%).
60. **Chitin** is a modified polysaccharide in which one of the hydroxyl groups of glucose is replaced by an acetyl amine group and the glucose units are linked together by β(1→4) glycosidic bonds (figure 7.35). Chitin is the major component of the cell walls of fungi, the exoskeletons of arthropods, the radulas of mollusks, and the beaks of cephalopods. Interestingly, because of its strength, flexibility, and biodegradability, chitin has been used as surgical thread.
61. In fact, two of the categories of biological acids, the **fatty acids** and the **amino acids**, are members of the more general category of organic weak acids known as the **carboxylic acids** (one of the organic functional groups introduced early in this chapter).
62. Because the long, hydrocarbon chain is *nonpolar*, while the carboxylic acid functional group is *polar*, the fatty acids are called **amphipathic** or **amphiphilic**. (The origin of the terms is the Greek *amphis*, meaning "both," and *philia*, meaning "love"; the molecule is attracted to **both** *polar* and *nonpolar* chemical environments.)
63. Four molecules react to form a triglyceride: one glycerol molecule and three fatty acid molecules
64. Triglycerides are lipids.
65. Like all other members of the steroid family, cholesterol is built around **a core of four fused rings of carbon atoms** (a total of seventeen carbon atoms often identified as ***four fused carbocyclic rings***).
66. Unsaturated fatty acids with double bonds are often identified using a numbering system that originated with the biochemistry community. The carbon atom that is located **farthest** from the carboxylic acid functional group is designated as the ω (**omega**) carbon atom. The carbon atoms of the long hydrocarbon chain are then numbered beginning with the **ω carbon atom.** The ***number*** *of the carbon atom at which the first double bond occurs,* along with *the* ***total number*** *of double bonds*, is used to generate a name for the fatty acid. Figure 7.43 contains examples of this naming convention (with the omega numbering in red) for unsaturated fatty acids.

67. From figure 7.42 it is clear that both the number and location of the multiple bonds in an unsaturated fatty acid dramatically affect the spatial arrangement of the molecule. The differences in the spatial arrangements of the saturated and unsaturated fatty acids have far-reaching effects on the resulting triglycerides. Figure 7.44 show a triglyceride formed from three *saturated* fatty acids. Note that the skeletal structure implies that the long hydrocarbon portions of the molecule lie **parallel** and **in close proximity** to one another. Compare this spatial arrangement to that of figure 7.44, which shows the skeletal structure of a triglyceride formed from three *unsaturated fatty acids.*

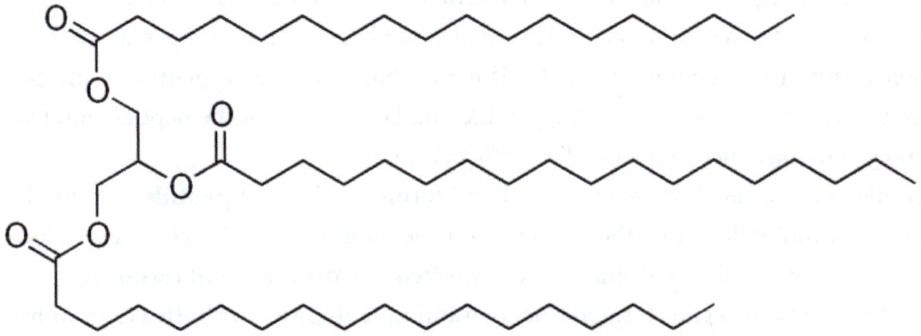

**Figure 7.44: Triglyceride with Saturated Fatty Acids**

In figure 7.45 the skeletal structure suggests that the three long hydrocarbon portions of the molecule are **neither** parallel **nor** in close proximity to one another. This difference has the important observational consequence that triglycerides formed from saturated fatty acids tend to be **solids** at room temperature and normal atmospheric pressure. The approximately parallel hydrocarbon chains interact more strongly with one another (via weak intermolecular and intramolecular forces), resulting in a more closely packed and more dense (solid or semisolid) configuration. In contrast, triglycerides formed from unsaturated fatty acids tend to remain **liquid** at room temperature and normal atmospheric pressure.

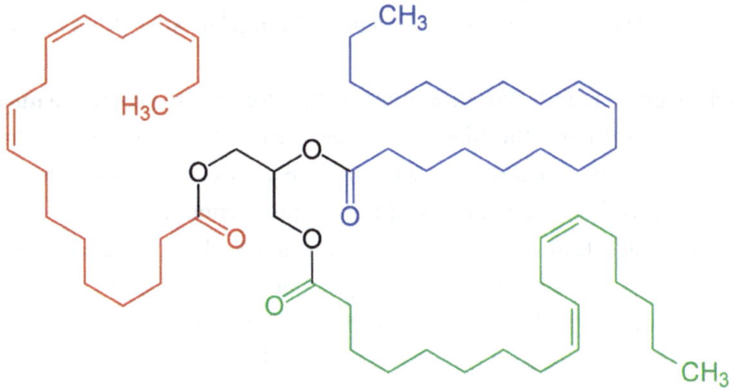

**Figure 7.45: Triglyceride with Unsaturated Fatty Acids**

68. Like the fatty acids, the **amino acids** are a subset of the carboxylic acid family of molecules and possess the carboxylic functional group (See figure 7.11 and figure 7.36). In addition, the amino acids possess a second

functional group, the **amino functional group** (see figure 7.10), which we also met much earlier in this chapter.

69. The doubly charged species (one end positively charged and one end negatively negative charged, with a **net** charge of **zero**) in Figure 7.48 is called a **zwitterion.**
70. The concentration of the zwitterion is greater than the concentrations of the other two species (we say that it *dominates* the amino acid equilibrium) at a **unique** pH value called the **isoelectric point.**
71. The carboxylic acid functional group is frequently called the **C-terminus** of an amino acid molecule while the amino functional group is called the **N-terminus** of an amino acid molecule.
72. The consequence of the reaction establishes a new bond between the nitrogen atom of amino acid (2) with the carbonyl carbon of the amino acid (1). This new bond is called a **peptide bond** by the biochemistry community. A careful examination of the peptide bond discloses that the peptide bond is part of the **amide functional group** introduced much earlier in this chapter.
73. The molecule that is created when a peptide bond forms is called a **dipeptide**, because it was formed from **two** amino acid molecules (note the parallel with the naming of a *disaccharide* produced by linking **two** *monosaccharides*). The individual amino acids are often called **amino acid residues.**
74. Of course, because the *dipeptide* formed from two amino acids possesses **both** a C-terminus and an N-terminus, additional peptide bonds can be created at either end of the molecule, producing a **polypeptide** molecule (exactly analogous to the polysaccharide molecules we observed earlier). Consequently, a **polypeptide** is a very large molecule constructed from a very large number of amino acid molecules.
75. While the **nucleic acids** were initially identified as constituents of the cell nucleus (thereby associating them with eukaryotic life forms) and exhibited the acidic properties associated with the phosphate group, this simple characterization was subsequently replaced over a period of some seventy-five years by a new understanding that was to place these molecules at the very center of all life on this planet. Further investigations of the nucleic acids revealed both a *range of reactions* and *chemical structures* that rapidly overshadowed the early observation of their acidic properties. Like the polysaccharides and polypeptides, the nucleic acids are *large* molecules composed of many thousands of subunits known as **nucleotides.**
76. The two classes of nucleic acids found on this planet: **ribonucleic acid (RNA)** and **deoxyribonucleic acid (DNA)**.
77. Finally, a phosphate group is added to the *nucleoside*, thus forming the complete **nucleotide** subunit.
78. The process begins by bonding either the ribofuranose or the deoxyribofuranose to one of the five nitrogenous bases to form a structure that is called a **nucleoside.** Consequently, a nucleoside is produced from **two** simpler molecules: a 5-carbon sugar and one of the five nitrogenous bases.
79. The two general molecular classes to which the nitrogenous bases belong are the purines and the pyrimidines.
80. The nitrogenous bases found in DNA are adenine, guanine, cytosine, and thymine.
81. The nitrogenous bases found in RNA are adenine, guanine, cytosine, and uracil.
82. A careful examination of the bond between the sugar and the phosphate group reveals that it appears to be almost identical to an *ester functional group*, with the phosphorus atom replacing the carbon atom of the ordinary ester. Consequently, the linkage is called a **phosphoester**. The last step in the formation of a complete nucleic acid is to link together many thousands of the nucleotide subunits (remember, nucleic acids are *very large* molecules). This final step requires the formation of a **second phosphoester** bond between a second pentose sugar and the phosphate group.

83. Because there are **two phosphoester** linkages that together link together two subunits in a nucleic acid, it is conventional to name the total linkage between the two subunits a **phosphodiester bond.**
84. Note in the RNA panel that the indication "5′ to 3′ direction" indicates that the upper pentose is bonded to a phosphate group at the 5′ carbon atom, while the lower pentose sugar is bonded to a phosphate group at the 3′ carbon atom. This means that additional nucleotide subunits can be added **at either end of the molecule.** Hence, just like the *polysaccharides* and the *polypeptides*, it is possible to form **polynucleotides** by continuing the steps used to link two subunits together. As we shall see later, this capability is crucial to life as we understand it.

## Chapter 8 Exercises

1. The word **isomer** comes from two Greek words, *isos*, meaning "equal," and *méros*, meaning "part," and identifies a compound possessing a **single** *molecular formula* (see chapter 3) but whose atoms are **arranged differently in space.** As we shall soon see, the phrase *"arranged differently in space"* conceals an amazingly rich range of possibilities that makes *isomerism* one of the most important characteristics determining the chemistry of our universe.
2. The **constitutional (structural) isomers** are molecules that have the **same** molecular formulas but **different** structural formulas; the atoms constituting these isomeric molecules are **arranged differently in space** by simply being *bonded differently to one another.*
3. There are 9 possible structural isomers of heptane

$H_3C$ $CH_3$

(1)

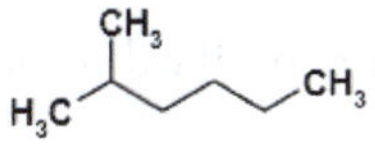

$CH_3$ $H_3C$ $CH_3$

(2)

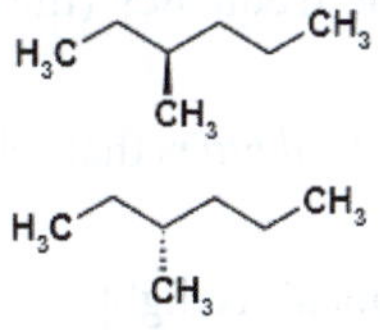

$H_3C$ $CH_3$ $CH_3$ $H_3C$ $CH_3$ $CH_3$

(3a,b)

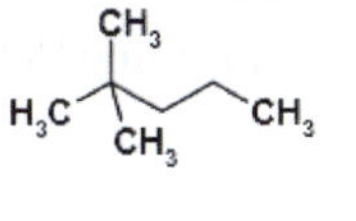

$CH_3$ $H_3C$ $CH_3$ $CH_3$

(4)

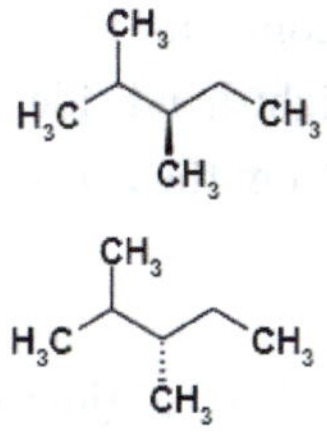

$CH_3$ $H_3C$ $CH_3$ $CH_3$ $CH_3$ $H_3C$ $CH_3$ $CH_3$

(5a,b)

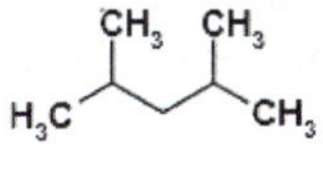

$CH_3$ $CH_3$ $H_3C$ $CH_3$

(6)

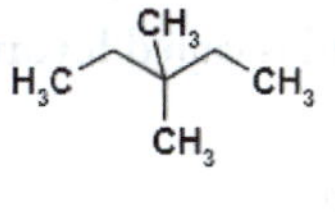

$CH_3$ $H_3C$ $CH_3$ $CH_3$

(7)

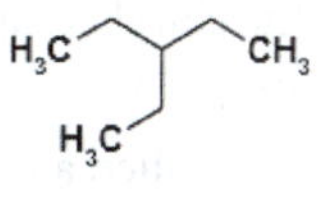

$H_3C$ $CH_3$ $H_3C$

(8)

$CH_3$ $H_3C$ $CH_3$ $H_3C$ $CH_3$

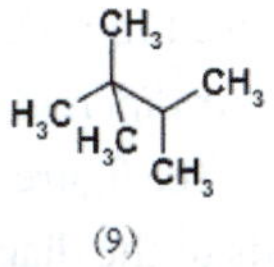

(9)

4. Stereoisomers are molecules that have the **same** molecular formula, have the **same** structural formula, but whose atoms are **arranged differently in space.**
5. In discussing the members of the stereoisomer class of molecules, we shall encounter a number of new terms: **enantiomer**, **diastereomer**, **cis/trans isomer**, **conformer**, and **rotamer** (see figure 8.1). It is important for the reader to note that these new terms are *relational* terms; they draw connections between and among the members of the stereoisomer class of molecules. These terms are not simply categories into which *individual* molecules are placed; this is the role of the two general groups: *constitutional isomers* and *stereoisomers*. In contrast, the terms *enantiomer, diastereomer, cis/trans isomer, conformer, and rotamer* apply to at least **two** and often **more than two** molecules *simultaneously*, establishing important *relationships* between and among stereoisomers.
6. The left hand is **not superposable** on the right hand. This observation suggests that there is an **essential** or **inherent** difference between a left hand and a right hand. We call this difference **handedness.** The left hand **is identical** to the mirror image of the right hand. This means that the property that we have called *handedness* implies that two objects are **handed if they are nonsuperposable mirror images of one another!**
7. A molecule that possesses a **nonsuperposable mirror image** is called **chiral**. If a molecule is **not** *chiral*, we say that it is **achiral.**
8. If **four** different groups of atoms are bonded to the carbon atom, it is called an **asymmetric carbon atom**, and the molecule itself is chiral.
9. It was Maxwell who united the ideas of **magnetism** and **electricity** (connected to our earlier concept of *charge* and represented by the electric field in the figure) to explain a *light wave* as a three-dimensional disturbance in space.
10. One complete cycle of either the electric or magnetic properties is called the **wavelength** (symbolized by the Greek letter *lambda* [$\lambda$]), while **c** is the **velocity** of light (the distance light travels per unit time; usually the time unit is *seconds*), and the quantity ($\mathbf{c} \div \lambda$) is called the **frequency** (the number of complete cycles completed in one second).
11. *Polarizers* are materials (labeled 3 in figure 8.6) which act as *optical filters* that select one specific direction of the electric field (shown as item 4 in figure 8.6).
12. Every chiral compound will **rotate** the electric field of plane-polarized light.
13. Every chiral compound will rotate the electric field of plane-polarized light. Any material that interacts in this way with light is called **optically active**.
14. If the electric field is rotated in a *clockwise* direction, the chiral compound is called **dextrorotatory** or **dextrorotary** (from the Latin *dexter*, meaning "right," or "on the right-hand side"); for a *counterclockwise* rotation the chiral compound is called **levorotatory** or **levorotary** (from the Latin *laevus*, meaning "left" or "on the left side").
15. The device is called a polarimeter.
16. Substances that are optically active are often given the designation (+) or "d" (lowercase *d*) if they are dextrorotatory or (-) or "l" (lowercase *l*) if they are levorotatory. Another system is used in chemistry (known as the **D/L system**) that is based on the molecule glyceraldehyde and a representation method called a **Fischer projection**. The cartoon presented in figure 8.9 will help us understand the conventions used in a *Fischer projection*. In the figure, the simple molecule is a single carbon atom to which are bound four *different* substituents. Consequently, the carbon atom is an *asymmetric carbon* and the molecule is *chiral* (see figure 8.4). The observer understands that the three-dimensional structure of the molecule corresponds to the diagram

shown in the upper right portion of the figure: groups A and B lie **below** the plane of the page (indicated by the dashed lines), while groups C and D lie **above** the plane of the page (indicated by the wedges). This is translated to the *Fischer projection* in the lower right portion of the figure. The intersection point of the two lines (crossing at 90°) is the location of the asymmetric carbon (note that the symbol for carbon is **not** written). The **horizontal** lines passing through the asymmetric carbon correspond to the *wedges* of the upper right portion of the figure, meaning that C and D in the *Fischer projection* lie **above** the plane of the page. The **vertical** lines passing through the asymmetric carbon correspond to the *dashes* of the upper right portion of the figure, meaning that A and B in the *Fischer projection* lie **below** the plane of the page.

17. Using Fischer projections, chemists in the late nineteenth century defined the D/L nomenclature system. In the Fischer projection of a stereoisomer, first locate the **chiral** carbon atom that is **farthest from the carbonyl group**. If the hydroxyl group (*OH*) that is bonded to that carbon atom lies on the **right side of the vertical line in the Fischer projection**, the stereoisomer is designated as "D." If the hydroxyl group that is bonded to that carbon atom lies on the **left side of the vertical line in the Fischer projection**, the stereoisomer is designated as "L."
18. False. Polypeptides are chiral.
19. Glycine is achiral.
20. Two *chiral* molecules that are nonsuperposable mirror images of one another are called **enantiomers.**
21. Either isomer of thalidomide will almost immediately **racemize.** That is, any sample containing **only one** of the isomers will very rapidly convert into a 50:50 mixture of **both** isomers (a *racemic mixture*). Hence, a *racemic mixture* is a 50:50 mixture of **both** isomers.
22. A compound that causes severe birth defects is called **teratogenic.**
23. Yes, thalidomide is chiral.
24. A ***meso*** compound is a molecule that **is superposable on its mirror image.**
25. Stereoisomers that are **not** *enantiomers* are called **diastereomers.**
26. The conformational isomers (*conformers*) and *rotamers* are molecules related as diastereomers that differ from one another only as the result of a **simple rotation** around one or more formal carbon-carbon single bonds. Such a rotation around a single bond is energy dependent, and, consequently, the stability of *conformational isomers* and *rotamers* depends greatly on the energy available to the molecule. Many *conformational isomers* and *rotamers* rapidly interconvert between various diastereomers and are not considered to be **stable** isomers.
27. The three rotational isomers of butane are identified by the terms *gauche, anti,* and *eclipsed.*
28. While a large fraction of the molecules related as diastereomers either are stable and optically active or are capable of isomeric interconversion as conformers or rotamers, there is a third group of molecules whose members are neither optically active nor conformers or rotamers. This third group of stable achiral isomers is distinguished by characteristics of their molecular structure that prevent even the **hindered rotation** associated with conformers and rotamers from taking place. (The term **hindered rotation** is rotational motion requiring energy but **not** the breaking of chemical bonds.) In effect, members of this third group **cannot** rotationally interconvert specifically because their molecular structure imposes the constraint of **restricted rotation**. Such isomers were initially called **cis/trans** isomers. While the variety of molecules belonging to this third group is very large, our focus will be on two major groups of molecules: the substituted cycloalkanes and the open-chain alkenes. The focus on the substituted cycloalkanes excludes molecules such as cyclohexane, which, as we have already seen, exhibits conformational isomerism.

## Chapter 9 Exercises

1. A **condensation reaction** is a chemical *synthesis* in which two or more chemical species (most often, complete molecules) are joined to form a new chemical compound while at the same time producing a small molecule as a byproduct of the reaction. Its name is derived from the Latin *condensare,* meaning "press close together."
2. Among the common small molecules generated as byproducts of a *condensation* reaction are water, ammonia, methanol, hydrogen chloride, and even acetic acid.
3. In a biological environment the most common byproduct is water. In this context, *condensation* reactions are given the name **dehydration synthesis reactions** because water is, in effect, "squeezed out" (the reactants are dehydrated).
4. Even though the central focus of our story will remain on the dehydration synthesis reactions and their critical role in the biochemistry of life, it is worthwhile to observe that **not all *dehydration* reactions are necessarily *condensations*. That is, there is a class of dehydration** reactions that involve only a **single** molecule (hence, they are **not** *condensations*), and they are called **intramolecular dehydration reactions**.
5. Zaitsev recognized that the process of eliminating water could produce two possible products, either 2-butene or 1-butene. His experimental observations confirmed that the two products were **not** produced in equal numbers; one isomer (called the **major product**) was favored in that it was produced in larger numbers. The second isomer (called the **minor product**) was produced in smaller numbers.
6. Zaitsev formulated the **empirical rule** now known as **Zaitsev's rule**: "The alkene that is formed in the largest amount (the major product) is the one that corresponds to the removal of a hydrogen atom from the β-carbon (the carbon atom **adjacent to** the carbon atom to which the *OH* group is bonded), which has the fewest number of hydrogen atom substituents."
7. a. $CH_3CH{=}CCHCH_2CH_3 + H_2O$ [Major]

   (with $CH_3$ attached below the third carbon)

   $CH_2{=}CHCH_2CHCH_2CH_3 + H_2O$ [Minor]

   (with $CH_3$ attached below the fourth carbon)

   b. $CH_3CH_2CHCH{=}CCH_3 + H_2O$ [Major]

   (with $CH_3$ attached above CH and $CH_3$ attached below C)

   $CH_3CH_2CHCH_2C{=}CH_2 + H_2O$ [Minor]

   (with $CH_3$ attached above CH and $CH_3$ attached below C)

c. $CH_3C{=}CHCH_2CH_2CHCH_3 + H_2O$ [Major]

(with $CH_3$ branches on C-2 and C-6)

$CH_3CHCH{=}CHCH_2CHCH_3 + H_2O$ [Minor]

(with $CH_3$ branches on C-2 and C-6)

8. An **esterification reaction** is certainly a *condensation* reaction, but, more importantly, it is a **dehydration synthesis**! In an esterification, esters are formed as the product of a *condensation* reaction between a **carboxylic acid** and an **alcohol** in which a **water molecule** is the small molecule byproduct.
9. It is perfectly reasonable to expect to find a group of reactions that are the **reverse** of *condensation* reactions. Indeed, such a collection of reactions **does** exist and **does** play a critical role in life processes. These reactions are the **hydrolysis reactions**; the name originates with the Greek word *hydro*, meaning "water," and the Greek word *lysis*, meaning "a loosening, dissolution, decomposition, or untying," and describes chemical reactions that cleave bonds by **adding water molecules.**
10. The water molecule participates in every hydrolysis reaction.
11. **Markovnikov's rule** states: "When an *unsymmetrical* molecule such as HX, where the 'X' can be a single atom, such as *Br*, or a group of atoms, such as *OH*, is inserted at the site of a carbon-carbon double bond, the hydrogen atom from HX becomes bonded to the carbon atom that has the greater number of hydrogen atoms, and the X is bonded to the carbon with the fewer number of hydrogen atoms."
12. True. Both Zaitsev's rule and Markovnikov's rule are empirical rules.
13. ***Hydration* reactions** are reactions that add $H_2O$ to the starting alkene. In general, hydration reactions yield a product by adding water.
14. a.

$CH_3CH_2CH_2CCH_2CH_3$ [Major] (OH above and $CH_3$ below the C)

$CH_3CH_2CH_2CHCHCH_3$ [Minor] (OH on the second CH, $CH_3$ on the first CH)

b.

$CH_3CH_2CHCHCHCH_3$ [Major] ($CH_3$ on the first CH, OH on the second CH, $CH_3$ on the third CH)

$CH_3CH_2CHCHCH_2CH_2$ [Minor] ($CH_3$ above the first CH, OH above the $CH_2$, $CH_3$ below the second CH)

15. The critical significance of water becomes apparent in *condensation* reactions (a chemical *synthesis* in which two or more chemical species are joined to form a new chemical compound while at the same time **producing** water as a byproduct) and *hydrolysis* reactions (chemical reactions that cleave bonds by **adding** water molecules). Note the symmetric role of water: in a *condensation*, water is **produced**; in a *hydrolysis*, water is **consumed**. In effect, these two classes of reactions are symmetrically responsible for the central biochemical processes that characterize life: the **production** of complex biomolecules meeting the structural, functional, and replication requirements of livening organisms and the **breakdown** of complex biomolecules to produce energy and the building blocks of new biomolecules.
16. A *hydrolysis* reaction will cleave the glycosidic bond, **add** water, and produce two monosaccharide molecules from a single disaccharide molecule. A simple extension of this process makes it clear that a polysaccharide molecule can be reduced to its monosaccharide constituents by repeated hydrolysis reactions. The addition of each water molecules results in the cleavage of one glycosidic bond.
17. The building of a dipeptide from two amino acid residues is a *condensation* reaction whose small molecule byproduct is water! The repeated addition of amino acids either at the C-terminus or at the N-terminus via condensation reactions produces a polypeptide; each condensation reaction eliminates a water molecule as a byproduct.
18. A *hydrolysis* is the chemical reaction that provides the reverse route, cleaving the peptide bonds in a polypeptide, adding one water molecule for each peptide bond that is broken, and reducing a polypeptide to its constituent amino acids.
19. The formation of a nucleoside, composed of a ribose (or deoxyribose) sugar linked to either a purine or pyrimidine molecule, requires the creation of a new bond between a nitrogen atom and the anomeric carbon atom of the sugar. We now recognize that this bond formation process is a *condensation* reaction eliminating water as a byproduct.
20. The formation of a nucleoside is only one step along the construction path of a complete nucleic acid molecule. Because a nucleic acid is a truly gigantic molecule composed of thousands of *nucleotide* components linked together, the next essential step is to build the *nucleotide* components of the molecule. This is accomplished by linking a phosphate group to a *nucleoside* through the creation of a *phosphoester* bond (an analogue to the ester functional group in which a phosphorus atom has replaced the usual carbon atom). Figure 9.8 summarizes the process for the cytidine nucleoside. Note carefully that the figure depicts another *condensation* reaction! Notice that an OH group is lost by the sugar, and a hydrogen atom is lost from the phosphate group. The phosphate is bonded to the nucleoside, and the hydrogen atom combines with the OH group to form a water molecule as a byproduct. Note that the bond that is formed is called a ***phosphoester* bond**.
21. The final step in the formation of a complete nucleic acid is the linking together of many thousands of the nucleotide subunits through the formation of a **second phosphoester** bond between a second pentose sugar and the phosphate group.

## Chapter 10 Exercises

1. The central role played by *energy* is reflected in the identification of energy conservation (see chapter 1) as the **first law of thermodynamics**; this conservation principle understood a thermodynamic system's energy to be the sum total of all the *heat* added to or removed from the system and the *work* done by or to the system.
2. A thermodynamic system's energy to be the sum total of all the ***heat*** added to or removed from the system and the ***work*** done by or to the system.
3. False. *Heat* has been recognized as an **energy transfer process**. Specifically, when the *energy* of a thermodynamic system **changes as a consequence of a temperature difference between the system and the surroundings**, we say that *energy* has been transferred as **heat**.
4. False. When the *energy* of a thermodynamic system **changes as a consequence of a temperature difference between the system and the surroundings**, we say that *energy* has been transferred as **heat**. There are two key points to remember here: (1) *heat* is an energy **transfer process**, and (2) there **must** be a **temperature difference** between the thermodynamic system and everything else in the universe that is **not part** of the system.
5. The kinetic molecular theory of gases identifies *temperature* as a **quantitative measure** of the *average kinetic energy* of molecules.
6. A **change** in the thermodynamic variable (more precisely, the mathematical function) called the **enthalpy** (symbolized by an uppercase **H**) provides a quantitative measure of the *heat* associated with a **constant** pressure physical process. Because the enthalpy change is a quantitative measure of the *heat* associated with a constant pressure physical process, and *heat* itself is an energy transfer process, an *enthalpy change* is measured in energy units.
7. True. Just as with mechanical energy, we emphasize that it is **only** possible to measure **differences** in the energy of two different states of a thermodynamic system.
8. When an energy transfer occurs between the reacting system and its surroundings, the experimental observable associated with this energy transfer is a measured **temperature difference**.
9. $\Delta_R H = H_{\text{products}} - H_{\text{reactants,}}$ a reaction for which $\Delta_R H > 0$ is called **endothermic (endoergic)**.

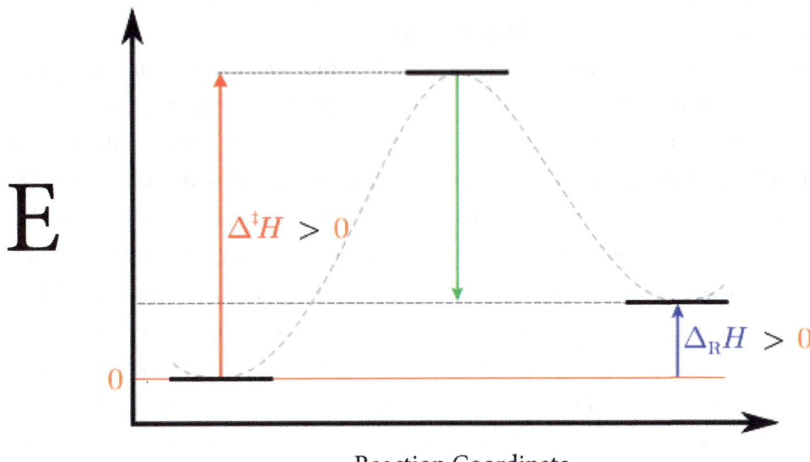

10. $\Delta_R H = H_{\text{products}} - H_{\text{reactants,}}$ a reaction for which $\Delta_R H < 0$ is called **exothermic** (more generally, **exoergic**)

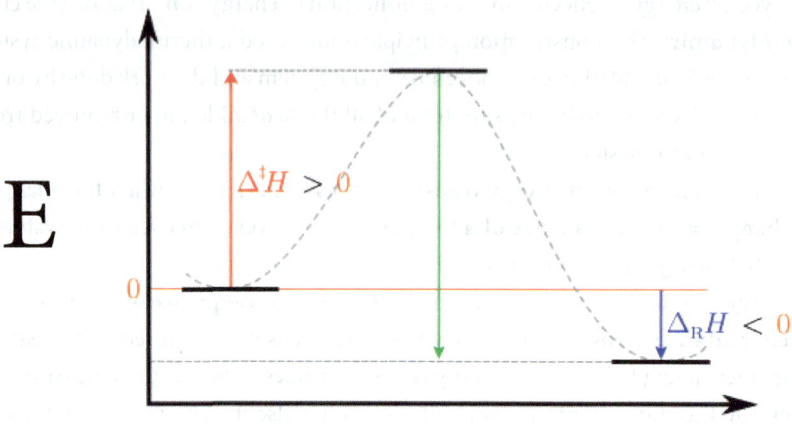


Reaction Coordinate

11. If $\Delta_R H = 0$, the reaction is called **thermal neutral.**
12. Formally, a **spontaneous process** is a reaction that simply **occurs** or **takes place** under a given set of conditions (*i.e.*, without *useful* work being done, where *useful* work means work for which the forces acting over a distance are **not** the result of pressure and volume changes in the system). The most crucial aspect of a spontaneous process is that it has a *preferred* direction. If, under a set of specified conditions, a reaction **does not occur**, the reaction is said to be **nonspontaneous.**
13. The **entropy** of a thermodynamic system (represented by an uppercase **S**) is a **direct measure of the randomness or disorder of a system**; the **greater** the **disorder**, the **greater** the **entropy**. From another point of view, entropy is a description of the number of possible distinct arrangements of the constituents of a system, and, in this sense, it is related to the concept of probability. (Ludwig Boltzmann made this connection explicit with his formulation of the kinetic molecular theory of gases.) The entropy of a system is related to the frequency with which a particular arrangement of the constituents of the system occurs. An ordered system has a **low probability** of occurring and, hence, has a **small entropy value**; a disordered system has a **high probability** of occurring and, hence, has a **large entropy value**.
14. Because the energy conservation principle is at the heart of thermodynamics, it is tempting to use energy as a means to explain spontaneity. For example, one is tempted to argue that processes occur spontaneously because a thermodynamic system seeks the state of *greatest stability*, that is, the state of *lowest* energy. However, this line of reasoning does not work! If the ambient temperature is greater than 273.15 K (and the pressure is equal to 1 atm), an ice cube spontaneously melts; however, the resulting puddle of water is in a *higher* energy state than the ice cube. That is, the melting process (a simple physical phase transition from a solid to a liquid; we'll look at **phase changes** in more detail shortly) is endothermic, but it is also spontaneous. The examples in equations 10.3 and 10.4 seem to indicate that while exothermicity (the transition to a *lower* energy state) *favors* the spontaneity of a reaction, it is *not* the sole factor causing spontaneous reactions. Hence, the concept of energy (and the principle of energy conservation) is **not** sufficient to explain all the observed characteristics of the universe. In particular, the central concept of energy conservation leaves the observation of spontaneity unexplained.
15. An ordered system has a **low probability** of occurring and, hence, has a **small entropy value**; a disordered system has a **high probability** of occurring and, hence, has a **large entropy value**. This linking of entropy

to the degree of disorder has a very simple and direct interpretation when observing the physical universe. If we select a fixed molar amount of a substance, we can make a very general statement about the entropy associated with the sample, as follows: $S_{solid} < S_{liquid} << S_{gas}$.

16. The consequence of introducing the new variable, **S**, denoting entropy (it is also a mathematical function, just like enthalpy, ***H***, is a function) was the articulation of the **second law of thermodynamics**: the **total entropy** of an **isolated** system *increases* in the course of a **spontaneous** change. Hence, the second law connects the concept of a **spontaneous change** with the concept of **entropy**.
17. The **Gibbs free energy** (often simply called the **free energy**) is defined as follows:

$$G = H - TS$$

where the symbol $TS$ means the product of $T \times S$. The first important observation is that all the variables in equation 10.7 apply **only** to the thermodynamic system. Second, just as we saw with enthalpy, by restricting ourselves to idealized cases in which the pressure of the thermodynamic system remains **constant**, millions of chemical reactions on this planet can be analyzed with equation 10.7. Finally, we add one additional restriction: we require that the temperature of the thermodynamic system remain **constant**. Again, this is a commonly occurring feature of an immensely large number of chemical reactions on Earth.
18. An analysis of the change in the Gibbs free energy for a reaction ($\Delta_R G$) under the **dual** conditions of **constant temperature** and **constant pressure** reveals an amazing conclusion: a chemical reaction taking place under the dual conditions of **constant temperature** and **constant pressure** is spontaneous if and only if $\boldsymbol{\Delta_R G < 0}$. This statement is a shorthand method of stating two separate ideas: (1) if $\boldsymbol{\Delta_R G < 0}$ under the conditions of **constant temp erature** and **constant pressure**, a reaction is *spontaneous*; and (2) if a reaction is spontaneous under the conditions of **constant temperature** and **constant pressure**, $\boldsymbol{\Delta_R G < 0}$!
19. Use Figure 10.5 to determine the answers.
    a. $\boldsymbol{\Delta_R G < 0}$ **for large** $T$ — spontaneous
    b. $\boldsymbol{\Delta_R G > 0}$ **for ALL** $T$ — nonspontaneous
    c. $\boldsymbol{\Delta_R G < 0}$ **for ALL** $T$ — spontaneous
    d. $\boldsymbol{\Delta_R G < 0}$ **for small** $T$ — spontaneous
20. a. Fusion is the processes of **melting and freezing.**
    b. Friction is the force that resists relative motion between two bodies when they are in contact.
    c. Adhesion is an attraction originating at the molecular level and exerted between the macroscopic surfaces of objects that are in contact.
    d. Cohesion is a molecular attraction by which the individual particles of a body are bound together in a macroscopic mass.
    e. Viscosity is the property which measures the tendency of a fluid or semifluid to resist flow.
    f. Surface tension is an attractive force between the molecules at the surface of a fluid and the molecules beneath the surface, which deforms the macroscopic surface to achieve the smallest possible surface area.
21. The processes and properties that depend on weak interactions either **between** atoms and molecules or **between** entire molecules are called **intermolecular forces.** Be careful to make a distinction between the terms *intermolecular* and *intramolecular*. The term *intramolecular* refers to properties **solely inside** or **belonging to** an individual molecule. The parameters that characterize chemical bonds between the atoms of a molecule, for example, are *intramolecular* properties.

22. The spin quantum number of all bosons exhibits **only** integer values. In contrast, all objects whose spin quantum number is a half-integer (±½, for example) are called **fermions**; in particular electrons are members of the fermion class.
23. Van der Waals forces are the interactions of complete, neutral molecules either with one another or with neutral atoms. The van der Waals class of intermolecular forces is divided into two groups; one is called the **dispersion forces** (or **London forces**, in honor of Fritz London, who studied these forces using quantum theory), while the second group is named the **dipole-dipole forces**.
24. A **hydrogen bond** is formally defined as a special dipole-dipole interaction between a hydrogen atom that is part of a polar bond (e.g., bonds of the type N-H, O-H, and F-H) and an electronegative atom (N, O, and F).
25. A very simple but very powerful generalization that is based on early macroscopic observations is called the **golden rule** and is easily stated as follows: ***like dissolves like***. From the perspective of intermolecular forces and the solvation process, this means that solvent molecules can effectively surround a solute particle if **the same intermolecular forces** are shared by the solvent molecules and the solute particle. Consequently, polar solvents easily solvate polar solutes while nonpolar solvents solvate nonpolar solutes. However, polar and nonpolar substances **do not** easily form a solution: note that oil and water do **not** mix!
26. For our purposes, the three classical states of matter will be treated as only three distinct phases, and the phrase ***phase transition*** has exactly the same meaning as the term ***change of state***. The reader should be aware, however, that the terms *state* and *phase* are not identical.
27. The **vapor pressure** is the pressure exerted by the gaseous molecules that have escaped from the surface of a liquid.
28. The **boiling point** is the temperature at which the vapor pressure of a liquid is **equal** to the external pressure exerted on the liquid. The **normal boiling point** is the temperature when the external pressure is exactly to 1 atm.
29. Finally, let's examine the behavior of $H_2O$ as the compound undergoes **several** phase transitions, the first phase transition (melting/freezing) from the solid phase to the liquid phase and a second phase transition (vaporization/condensation) from the liquid phase to the vapor phase. Figure 10.11 displays the phase transitions as a function of increasing energy (plotted along the horizontal axis from left to right), while the vertical axis plots the increasing temperature. Note that during the two phase transitions, the plotted curve is **horizontal**, meaning that the **temperature is constant during the phase transition**.
30. The rate of a chemical reaction analyzes a chemical process as a function of **time**. The **rate of reaction** is determined by measuring either the amount of a product generated or the amount of a reactant consumed by a chemical reaction during a specified time period.
31. A **catalyst** is a substance that **increases the rate of reaction**, even though it is **not consumed** during the course of the reaction; while it participates in the reaction, it is **not** "used up," degraded, or changed. The central effect of a catalyst is to **lower** activation energy barrier (and perhaps changing the overall shape of the barrier as shown in the figure, implying that there may be additional paths from reactants to products as a result of the action of a catalyst.) and thereby **increasing the rate of reaction**.
32. Note that a catalyst is a fourth way to influence the rate of reaction (as noted previously, the other three are temperature, reactant concentrations, and pH); however, unlike the other three factors, a catalyst will **only increase** the rate of reaction. Further, it should be noted that a catalyst **will *not* change** the chemical equilibrium of a reaction, because a lower activation energy barrier **increases both** the forward reaction rate and the reverse reaction rate. Consequently, the point at which the concentrations of the reactants and the products no longer change with time remains **unchanged**.

33. Biochemical catalysts are large polypeptides ranging in size from just over fifty to well over two thousand amino acid residues and are given the name **enzyme**, from the Greek *enzymon*, meaning "yeast" or "leaven."
34. Every enzyme molecule possesses a specialized region with a highly specific three-dimensional geometry that is called the **active site.**
35. The active site is the region of the enzyme that binds to the reactant to form an enzyme-substrate complex (ES-complex). (The reactant is known as the **substrate**; typically, the name of an enzyme is formed by adding the suffix *-ase* to the name of the substrate.) The highly specific geometry of an enzyme's active site limits the number of substrates that will form an ES-complex to only a few and, for many enzymes, to only one. This characteristic is called **substrate specificity**.
36. There are **two** types of complementary components that participate in enzyme activity. **Cofactors** are inorganic substances (some as simple as metallic ions) often located at the active sites of enzymes that are not part of the molecular structure of the enzyme but are held tightly to the enzyme by intermolecular forces. **Coenzymes** are small *organic* molecules that can be either tightly or loosely bound to an enzyme itself.
37. Because the active sites of enzymes exhibit a highly specific three-dimensional geometry, the earliest proposed model was the **lock-in-key model**, which suggests that the geometry of a substrate exactly matches the geometry of the associated enzyme as a key fits into a lock. This model very easily explains the **high specificity** of enzyme catalysis: the geometries of both the substrate and the enzyme must **exactly** match for a reaction to proceed. Figure 10.14 displays a simple representation of the lock-in-key model.
38. In contrast, the **induced-fit** model (proposed in the late 1950s) suggests a more *dynamic* interpretation of the enzyme-substrate interaction. This model proposes that the geometry of the enzyme's active site actually **changes shape** as the enzyme-substrate complex forms. In fact, the geometry of the substrate itself may also change during its interaction with the flexible geometry of the active site. Once the geometric adaptation is complete, the products form and leave the enzyme unaltered, ready to accept another substrate. Figure 10.15 is a schematic representation of the induced-fit model, showing the adaptation of the enzyme's active site geometry (however, the figure does **not** show any change in the geometry of the substrate itself).
39. Finally, as the twentieth century drew to a close, a third model of enzyme activity was proposed; it is called the **conformational selection/population shift model**. This model suggests that the active site of an enzyme exists in multiple conformational states, and the substrate geometry has a high affinity of bonding with one of these states. In effect, a population of active site geometries is present, and one of these geometries exhibits a high probability of matching the substrate's geometry. This maximum probability determines the binding of the substrate to the enzyme. Figure 10.16 is a simple representation of the conformational selection/population shift model.
40. One key role of the enzyme's active site is to maintain an optimal spatial separation among the enzyme and all the interacting substrates that effectively promotes a chemical reaction. The ability of the enzyme's active site to maintain this optimal spatial separation is called the **proximity effect** of an enzyme. The active site of the enzyme, because it possesses a specific geometric shape, imposes a very particular three-dimensional arrangement on the enzymes and the substrate(s). The fact that the active site of an enzyme determines a specific three-dimensional geometric arrangement of the enzyme and the substrate(s) is called the enzyme's **orientation effect**. Finally, the formation of the ES-complex **reduces** the magnitude of the activation energy barrier separating the reactants (the substrate[s]) from the products of the reaction, thereby **increasing** the rate of reaction. In effect, the strength of the chemicals bonds that maintain the chemical structure of the substrate(s) is **reduced**, and these reduced energetics promote the formation of the products. This energetic

change in the intramolecular bonds of the substrate(s) as a result of the enzyme's participation in the reaction is called the **bond energy effect.**

41. An **inhibitor** is any substance that reduces an enzyme's activity; there are several type of enzyme inhibition. **Competitive inhibitors** directly contend with an enzyme's substrate for access to the active site of the enzyme.
42. An enzyme's activity can also be affected by the binding of an **effector molecule** at the enzyme's **allosteric site**, which is a region of the enzyme molecule that is distinct from the active site. Interestingly, the allosteric control of an enzyme's activity can either **enhance** an enzyme's activity or **inhibit** an enzyme's activity. The binding of the effector molecule changes the geometry of the active site, either making it possible for a substrate molecule to bind at the enzyme's active site or changing the geometry so a substrate **cannot** bind at the active site.
43. While the competitive and noncompetitive inhibition processes are often **reversible**, in that neither process permanently deactivates an enzyme, there is a class of inhibitors, called an **irreversible inhibitor**, which permanently changes an enzyme so that it can no longer catalyze a specific chemical reaction. An irreversible inhibitor permanently binds at the active site, structurally modifying the active site so that its geometry and the geometry of the substrate are **no longer compatible**. Consequently, the enzyme is **permanently deactivated**.

## Chapter 11 Exercises

1. a. This is the part of the cell that contains most of the cell's hereditary information, organized in structures called **chromosomes.**
   b. The cytosol is the gel-like intracellular fluid contained within the cell membrane; it is approximately 80% water and is usually transparent. It is the location of multiple processes that occur within a cell: the transmission of molecular signals from the cell's outer membrane to other regions within the cell and the transport of the products of metabolism to other sites within the cell.
   c. The ribosome is the site of **protein synthesis.** Proteins are one of the three complex structures we will examine later in this chapter. They are highly specialized **polypeptides**, our old friends from an earlier chapter.
   d. The mitochondria are cell organelles that generate **most** of the cell's chemical energy. The mitochondria possess **hereditary repositories** that are **independent** of the cell's hereditary molecules.
   e The cytoskeleton is a protein mosaic that forms a skeleton within the cytosol and participates in intracellular transport and cellular division.
   f This is the biological **membrane** that separates a cell's interior from the external environment. It is selectively permeable, and its central function is to protect the cell multiple organelles from the cell's surroundings.
2. a. The inner membrane of the mitochondrion forms the internal compartments known as **cristae** (see below). The inner mitochondrial membrane is also the site at which a key component of metabolism takes place. This metabolic process is known as the **electron transport chain** and **oxidative phosphorylation.**
   b. These are the compartments formed by the folding of the inner mitochondrial membrane. As a result of the folding, the mitochondria have a very large surface area at which metabolic processes occur.

c. This is the actual space bounded by the inner mitochondrial membrane, and it contains the mitochondrion's ribosome and hereditary material (DNA). The fluid of the matrix is more viscous than the relatively aqueous cytosol.

3. The term **endosymbiont** is used to describe the mitochondria. It has been hypothesized that the mitochondria were originally independent **prokaryotic** organisms capable of performing the redox processes that eukaryotic cells cannot perform by themselves. These prokaryotic cells established a **symbiotic relationship**, living inside the eukaryotic cells. This hypothesis has been supported (as recently as 2011) by noting that a group of bacteria appear to share a relatively recent common ancestor with the mitochondria found in eukaryotic cells.
4. The ribosomes both in the eukaryotic cell and in the mitochondria perform the process of **protein synthesis**.
5. The word "protein" comes from the Greek *proteios*, meaning "primary," "in the lead," or "standing in front," which is indicative of the more profound importance of proteins. We will soon find that proteins are essential to almost every facet of life. The structural proteins are the basis of every muscle and are essential to the cytoskeleton of every eukaryotic cell. Proteins are the enzymes that catalyze biochemical reactions and are critical participants in metabolic processes; without the tremendous increase in the rate of reaction achieved by these biochemical catalysts, life processes would occur millions of times more slowly. Proteins play crucial roles in **intercellular signaling**, the **immune response**, and the ability of cells to **adhere** to one another (essential for wound healing, for example). It is not an exaggeration to say that the story of life on this planet is really the story of proteins.
6. The membranes serve several critical functions for both the cell and the various cell organelles:
   (1.) They separate cells and organelles from their external environment, thereby defining a cell and each organelle within a cell.
   (2.) They provide selective transport of nutrients into cells and organelles and transport the waste products generated by cells and organelles to the external environment.
   (3.) They selectively control the transfer of the specialized product molecules generated in each organelle, making them available to other regions in the cell.
7. Phospholipids are members of the class of molecules called *lipids* and are constructed much like our earlier example of a lipid molecule, the triglyceride. As with the triglycerides, the glycerol molecule forms the structural keystone of the phospholipids; however, a phospholipid has only **two fatty acid** molecules linked to the glycerol backbone by ester bonds. In place of the third fatty acid found in triglycerides, a phosphate group is linked to the glycerol by a phosphoester bond.
8. True. Phospholipid molecules are *amphipathic* molecules (we first met this term in chapter 7), possessing **both** a *hydrophobic* ("water-fearing") segment and a *hydrophilic* ("water-loving") segment.
9. The term **semipermeable** means that the membranes **allow** the passage of solvent molecules (principally water molecules) but **prevent** the passage of solute molecules.
10. The **fluid mosaic model** suggests that a number of distinct molecules and structures dynamically "float" in the lipid bilayer of the membrane: carbohydrates, glycoproteins (molecules formed from a sugar and a polypeptide), cholesterol, protein channels, and the cytoskeleton. Each of these molecules and structures contributes to the dynamic functioning of a cell and the organelles within a cell. Cholesterol, for example, can both increase and decrease the fluidity of the membrane, depending on its concentration in a particular membrane region. In a very real sense, the membrane is not simply a passive barrier layer separating different

cellular environments. On the contrary, the membrane is the locus of a complex interplay of molecules, structures, and molecular interactions that make life possible.

11. a. Simple diffusion (from the Latin *diffudere*, meaning "to spread out") is best understood from a particulate point of view and is more precisely termed *molecular diffusion*. It is a microscopic phenomenon that occurs spontaneously in our world and is caused by the **randomly distributed kinetic energy** associated with the particles that are spreading out. Diffusion effectively explains the movement of particles (molecules) from a region of **higher** concentration to a region of **lower** concentration, although a concentration gradient (a difference in the concentration between two regions) is not required by the diffusion process. In the context of a membrane separating regions of two **different** concentrations, simple diffusion describes the observed movement of particles across the membrane.
    b. To make it possible for polar molecules and charged species to traverse a membrane in a more timely fashion, polar protein structures (ionophores and aquaporins are two examples) are integrated directly into the membrane itself and provide a channel through the hydrophobic portion of the membrane. This **facilitated transport** still relies on the **randomly distributed kinetic energy** associated with the molecules crossing the membrane and, like simple diffusion, involves movement from a region of **higher** concentration to a region of **lower** concentration.
    c. Active transport requires the expenditure of cellular energy. This process moves molecules from a region of **lower** concentration to a region of **higher** concentration and must utilize stored energy to complete the process.
12. **Osmosis** is the net movement of solvent molecules through a **semipermeable membrane**, either from a **pure solvent** or from a **dilute solution** into a region with a **higher** solute concentration. At the microscopic level, osmosis is understood as the accelerated movement of a mass (the solvent molecules) across the area of the semipermeable; but the acceleration of a mass across the area of the membrane is simply a force exerted on an area, that is, a **pressure**! In this case the pressure is called the **osmotic pressure**. The osmotic pressure of a solution is measured by determining the opposing pressure required to **stop** the osmosis.
13. a. Two solutions of **equal** concentrations and, hence, exhibiting the **same** osmotic pressure are called isotonic.
    b. For two solutions having **different** concentrations and, hence, **unequal** osmotic pressures, the **more concentrated** one (the one with a higher concentration of solute) is called hypertonic.
    c. For two solutions having **different** concentrations and, hence, **unequal** osmotic pressures, the **more dilute** one (the one with a lower concentration of solute) is called hypotonic.
14. Over a period that extends through thousands of years of human history, humankind has recognized identifiable patterns in the physical characteristics, commonly called *traits*, which are found within the human population. Eye color, hair color, height, skin color, and facial shapes, to name only a few of the most obvious traits, have been recognized to occur in identifiable patterns within families, clans, tribes, and races. The total collection of these traits or characteristics is known as a **phenotype**.
15. The nitrogenous bases are bonded to the sugars by a bond resembling the *glycosidic* bonds of carbohydrates to form a **nucleoside**; consequently, a nucleoside is made up of two parts: a sugar and one of the nitrogenous bases. The nucleosides are bonded to a phosphate group via a phosphoester bond to form a **nucleotide**, a molecule made up of three parts: a sugar, one of the nitrogenous bases, and a phosphate group.
16. Multiple nucleotide units can be linked together by **phosphodiester bonds** to form a **polynucleotide**.
17. The nucleotide sequence is called the **primary structure** of a nucleic acid.

18. The formation of the complementary base pairs, producing the **double helix** configuration of the DNA molecule, is called the **secondary structure** of nucleic acids.
19. The overall shape achieved by a nucleic acid as the result of folding and twisting is called its **tertiary structure**.
20. Because of the different molecular configurations of the purine and pyrimidine molecules (see chapter 7), it is possible to form only adenine-thymine (A-T) and cytosine-guanine (C-G) **complementary pairs** through a **hydrogen bonding interaction**. Remember, the hydrogen bonds between the complementary purine and pyrimidine pairs are **strong enough** to provide the nucleic acid molecule with moderate stability. However, the hydrogen bonds **are not so strong** as to prevent the formation of two **single-stranded** polynucleotide molecules from which, through the complementary base pairing process, two identical molecules are formed.
21. The further twisting of DNA is called **supercoiling**. In supercoiling, the DNA is wrapped around protein molecules called **histones**; once the DNA has wound itself around the histone molecules, it is packed into a structure called the **chromosome**.
22. **Histones** are protein molecules that are central to the supercoiling process. The resulting structures in the cell nucleus are called **chromosomes**.
23. The **primary structure** of a protein is determined by the **number, kind, and sequence** of the amino acid units comprising the polypeptide chain or chains making up the molecule. Because each amino acid is characterized by its side-chain, the primary structure determines the **alignment of side-chains** in a protein, which, in turn, determines the **three-dimensional shape** into which a protein folds. **Hence, the amino acid sequence is the most important factor determining a protein's three-dimensional shape**. One of the initial observations made in the study of proteins was that large segments of the molecules exhibit a particular characteristic regularity. Two large structures were easily identified: the **$\alpha$-helix** and the **$\beta$-pleated-sheet**. These structures can occur multiple times within a single protein molecule and clearly determine a very characteristic three-dimensional spatial arrangement of segments of a protein. These regular three-dimensional spatial arrangements constitute the **secondary structure** of proteins. The **tertiary structure** of a protein is the distinctive and characteristic conformation or shape of a protein molecule. Finally, a fourth type of structure, called the **quaternary structure**, is found in **only** some complex proteins. These complex proteins are composed of **two or more** smaller subunits (these subunits are independent polypeptide chains). In some instances, nonprotein components (called prosthetic groups) may also be present. **The quaternary structure refers to the shape of the entire complex molecule**, and this final shape is determined by the way in which the independent subunits are held together (usually hydrogen bonds and bonding interactions resulting from charge concentration are most important).
24. The tertiary structure of proteins (the overall three-dimensional conformation) is held together by a variety of interactions:
    (1.) hydrogen bonding
    (2.) hydrophobic effects
    (3.) polar interactions
    (4.) bonding interactions resulting from charge concentration
    (5.) disulfide bonding (this is a bonding interaction between two sulfur atoms; the sulfur atom occurs in the side-chains of cysteine and methionine).
25. If the interactions that maintain a protein's secondary, tertiary, and quaternary structure are disrupted by either physical or chemical means, the protein is said to be **denatured**. The denaturation process either completely destroys or significantly diminishes those interactions that maintain a protein's characteristic

three-dimensional arrangement. Because it is this characteristic three-dimensional arrangement that determines the activity of a protein, a **denatured protein** loses its ability to be biologically active. The denaturation process **does not**, however, change a protein's primary structure.

26. The first step in protein synthesis is called **transcription**. In this step, a new molecule called messenger ribonucleic acid (messenger RNA, or often abbreviated as mRNA) is synthesized using the cell's DNA as a template and controlled by the enzyme RNA polymerase.
27. Messenger RNA (mRNA) is a new molecule synthesized during transcription using the cell's DNA as a template. The synthesis is controlled by the enzyme RNA polymerase.
28. It is synthesized in the nucleus of the cell.
29. The second step of protein synthesis is called **translation** and occurs **outside** the nucleus at the **ribosome**. (Recall that ribosomes are specialized structures located on the cell's rough endoplasmic reticulum. Do not confuse these ribosomes with the specialized ribosomes located in the mitochondria.) This step requires a second molecule found in the cell's cytosol: transfer ribonucleic acid (transfer RNA, often abbreviated tRNA).
30. Translation requires a second molecule found in the cell's cytosol: transfer ribonucleic acid (transfer RNA, often abbreviated tRNA). tRNA is a nucleic acid that is responsible for *transferring* or *transporting* "something" to a specific location in the cell. The "something" that is transported is an amino acid. The formation of an ester bond between an amino acid and a tRNA molecule is called **tRNA activation**.
31. The group of three bases of the mRNA molecule is called a **codon**, and the group of three complementary bases on the tRNA molecules is called the **anticodon**.
32. The ribosome is composed of two protein/**ribosomal RNA** (abbreviated rRNA) subunits; the small subunit possesses a grove into which the mRNA fits with its nucleotide bases pointing toward the large subunit. Figure 11.22 depicts the relationship among mRNA, tRNA, and the ribosome.
33. Because every protein possesses a finite number of amino acids in its polypeptide sequence, the process described above eventually must stop. This occurs when a **stop codon** is encountered on the mRNA molecule. This encounter is called **termination**; it brings protein synthesis to an end.
34. The three-step process outlined above, **transcription**, **translation**, and **termination**, is often called the **central dogma** of molecular biology.

## Chapter 12 Exercises

1. The term **metabolism** refers to the sum total of all the chemical reactions involved in maintaining the dynamic state of the cell; this dynamic state is exactly the meaning of the word *life*.
2. Metabolism can be thought of as a dynamic interplay of two distinct cellular activities: **catabolism** is the process of decomposing molecules to supply energy to the cell; the process begins with large molecules (often **very** large molecules) and produces small molecules with an accompanying liberation of energy. **Anabolism** is the process of synthesizing new molecules; we have seen examples of this process in the construction of nucleic acids and proteins in which small molecules are combined to form gigantic molecules.
3. The catabolic process in the human species occurs in **three** very distinct steps characterized primarily by the *location* in the human organism where the step takes place. The first step is called **digestion**; in this step the large, complex molecules that constitute food for a human being are broken down into much smaller molecules. Digestion takes places at several points in a human being: mouth, stomach, and small intestines.

4. The second key catabolic step is **glycolysis**; it takes place in the cell's cytoplasm.
5. Pyruvate is a three carbon molecule. The pyruvate molecule is shown in the following figure.

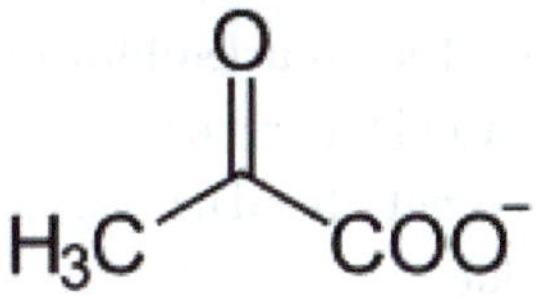

$$H_3C-C(=O)-COO^-$$

6. If sufficient oxygen is available, pyruvate molecules enter the cell's mitochondria, where most of the energy that is derived from each starting glucose molecule is produced. This is called the **aerobic path** because it depends on the availability of sufficient oxygen. This **third** step of the catabolic process takes place in two separate stages: (1.) the **Krebs cycle** (named for Hans Adolf Krebs, who identified this metabolic pathway; he shared the 1953 Nobel Prize with Fritz Lipmann for this work) and (2.) **oxidative phosphorylation.** Each of these stages produces adenosine triphosphate (ATP) molecules, which are the ultimate energy storage molecules, as well as two significant waste byproducts: $CO_2$ and $H_2O$. These two byproducts are the primary waste products (along with urea, as we shall shortly see) of human catabolism. In the event that oxygen is **not** plentiful, pyruvate can follow another path, known as the **anaerobic path**, to produce energy for the cell.
7. The **three catabolic steps** produce the energy (stored in the ATP molecules that are synthesized during catabolism) required by the cell to sustain life. As a result of the third step of catabolism, the **waste products** $CO_2$, $H_2O$, and urea are generated.
8. If sufficient oxygen is available, pyruvate molecules enter the cell's mitochondria, where most of the energy that is derived from each starting glucose molecule is produced. This is called the **aerobic path** because it depends on the availability of sufficient oxygen. In the event that oxygen is **not** plentiful, pyruvate can follow another path, known as the **anaerobic path**, to produce energy for the cell.
9. In addition to providing the energy necessary to maintain the dynamic state of each cell, the catabolic process provides the amino acids necessary for the construction of proteins. While some amino acids are also synthesized by human cells, several amino acids (called **essential** amino acids) **must** be supplied by dietary sources.
10. The amino acids phenylalanine, valine, threonine, tryptophan, methionine, leucine, isoleucine, lysine, and histidine are regarded as *essential* for human beings.
11. The molecule **Adenosine Triphosphate** is known as **ATP and** the molecule **Adenosine Diphosphate** is known as **ADP**. The ATP molecule is the basic energy currency of the cell; the synthesis of ATP (high-energy) during catabolism stores energy in a phosphate bond; the *hydrolysis* of ATP to form ADP (low-energy) and a free phosphate group makes energy available for a variety of cellular processes.
12. The molecule **Nicotinamide Adenine Dinucleotide** is known as **NADH** and **$NAD^+$** (two distinct oxidation states). The next pair of molecules are related to one another through the oxidation-reduction process. NADH is the *reduced* (high-energy) form of the molecule, while $NAD^+$ is the oxidized (low-energy) form of the dinucleotide. The molecules are called **dinucleotides** because they are made up of two separate nucleotides linked together by the shared oxygen atom of the two phosphate groups. The base that contains the $NH_2$ group is named *nicotinamide* (vitamin $B_3$) and is shown in figure 12.5; the second base contained in the molecules is the purine *adenine*. Note that the nitrogen atom in the nicotinamide ring of $NAD^+$ possesses a formal (+) charge.

13. The molecule **Flavin Adenine Dinucleotide** is known as **$FADH_2$** and **FAD** (two distinct oxidation states). The third pair of molecules are also related to one another through the oxidation-reduction process. $FADH_2$ is the *reduced* (high-energy) form of the molecule, while FAD is the oxidized (low-energy) form of the dinucleotide. While the molecules are called **dinucleotides**, this is a misnomer because the link between the nitrogen atom in the triple-ring system and the carbon atom of the $CH_2$ group (part of a five-carbon alcohol derived from the ribose called *ribotol*) is **not** a bond between a nitrogen atom and the anomeric carbon atom of a sugar that characterizes a nucleotide.
14. The final pair of **high/low**-energy molecules are acetyl coenzyme A (high-energy) and coenzyme A (low-energy), both of which include the *adenosine* nucleotide. Both acetyl coenzyme A and coenzyme A include the adenosine nucleotide, while the central portion of the molecules is derived from the carboxylic acid, pantothenic acid (vitamin $B_5$).
15. Glucose is the primary energy source for cells in the human body, and, as we noted earlier, the **second** major catabolic step is called *glycolysis*. This step occurs in the cytoplasm of each cell, and it results in the hydrolysis of the six carbon glucose molecules into two three-carbon molecules called **pyruvates**. For a single glucose molecule, the entire process involves ten distinct molecular reactions, each catalyzed by a specific enzyme, and produces two three-carbon pyruvate molecules from the starting glucose molecule.
16. **One glucose** molecule is hydrolyzed into **two pyruvate** molecules. During the hydrolysis of the glucose, **two** ATP molecules are **decomposed**. See Figure 12.10.
17. When one glucose molecule is hydrolyzed, **four** ATP molecules and **two** NADH molecules are **synthesized**. See Figure 12.10.
18. Other monosaccharides can also be hydrolyzed by glycolysis. The body converts galactose to glucose, which can then enter the glycolysis processes directly. Fructose, the sweetest monosaccharide, is absorbed very easily by both muscle tissue and the liver. The muscles convert fructose to an intermediate molecule that can enter the glycolysis pathway at the third step, while the liver produces an intermediate that can enter the glycolysis process at the fifth step.
19. If there is a shortage of glucose, the human body is capable of synthesizing glucose (consequently, this is an anabolic process) from noncarbohydrate sources. This process is called **gluconeogenesis**, and it can utilize a number of different starting materials, including lactate, amino acids from proteins, fatty acids, and glycerol from lipids. The liver is the major site for gluconeogenesis.
20. **Glycogenesis** is the synthesis of **glycogen** from glucose. We first met glycogen in chapter 7 and identified it as the storage polysaccharide in animals (as well as human beings). If both glucose and ATP are present in relatively high concentrations in the cell, the excess glucose is converted to glycogen; the glycogen molecule is an efficient storage mechanism for the abundant glucose. The synthesized glycogen is stored in the liver and in muscle tissue.
21. The reverse of glycogenesis is **glycogenolysis**, the decomposition of the glycogen stored in the liver into a chemical intermediate that can enter the glycolysis pathway directly. If the demand for energy is not immediate, the intermediate molecule synthesized in glycogenolysis is converted to glucose, which is then distributed by the blood to various cells in the body.
22. The **Krebs cycle** is sometimes called the **citric acid cycle**, or the **tricarboxylic acid [TCA] cycle**.
23. The Krebs cycle (aerobic path of pyruvate) takes place in the mitochondria of the cell. It is a series of eight reactions that take place in the mitochondrion matrix; each reaction is catalyzed by a specific enzyme.

24. The **second stage** of the **third** step of the catabolic process also occurs in the matrix and in the inner membrane of the mitochondrion.
25. The reduced molecules, NADH and $FADH_2$, (remember that the reduced molecules are the **high-energy** forms of the molecules) are now oxidized in the matrix to produce $-1e$ charges (electrons) and $+1e$ charges (protons, $H^+$) and the corresponding low-energy molecules, $NAD^+$ and FAD, that are reused in the Krebs cycle. Embedded in the inner membrane are **four protein complexes**, conventionally labeled **I, I, III,** and **IV**. Additionally, there are two mobile carriers of the negative charges (the electrons), one named **coenzyme Q** and the other named **cytochrome c**; neither of these carriers is firmly attached to any single complex, and both function to move negative charges between adjacent complexes. The process of moving negative charges between the four protein complexes is called **electron transport**.
26. In the early 1960s, the biochemist Peter Mitchell proposed that most of the ATP synthesized in the cell to store the energy liberated during the catabolic process is the result of the electrochemical proton gradient created by electron transport. Under the regulation of the enzyme ATP synthase, the final step is the synthesis of ATP from ADP, a process called **oxidative phosphorylation** (because a phosphate group is added to ADP, and oxygen is consumed to produce water as a byproduct).
27. The model of ATP synthesis is called the **chemiosmotic model**. Initially, this proposal by Mitchell was not favorably received; eventually a preponderance of the experimental evidence supported the chemiosmotic model. Peter Mitchell was awarded the 1978 Nobel Prize in Chemistry for his work.
28. As a result of the **third** step of the catabolic process (occurring inside the mitochondrion), one glucose molecule produces a total of **twenty-five** ATP molecules. Figure 12.15 provides a schematic summary of the mitochondrial structures and the chemical key chemical processes that take place during electron transport. From figure 12.15, the reader should note carefully that the concentration of positive charges (protons, $H^+$) is **greater** in the solution in the intermembrane space than in the matrix solution. Thus, as indicated in the figure, the electrochemical proton gradient **increases** in the direction of the intermembrane space. This also means that the pH of the solution in the intermembrane space is **lower** than the pH in the matrix solution. (Remember that **larger** $H^+$ [$H_3O^+$] concentrations mean **smaller** pH values; see chapter 5.) Finally, the reader should note that at complex **IV** the positive charges (protons, $H^+$) combine with $O_2$ molecules to form $H_2O$, the second waste byproduct of the catabolic process.
29. Two other classes of biomolecules, lipids and proteins, can be used to liberate energy.
30. Free fatty acids are generated by the hydrolysis of lipids (breaking the ester bonds in triglycerides); the fatty acids can be oxidized to form acetyl coenzyme A through a decomposition pathway called **β oxidation** (beta oxidation). See Figure 12.16.
31. There are three possible fates for ingested proteins: (1) they can be hydrolyzed to amino acids, which are then used in the synthesis of new proteins; (2) if excess protein is present, the amino acid molecules are decomposed and can supply approximately 10% of the human body's energy requirements under normal conditions; and (3) if other sources of energy are not available, proteins can be catabolized to liberate energy. Steps 2 and 3 require the removal of the amino group through a process called **transamination**. The products of the transamination process are then suitable to enter the Krebs cycle at various points (depending on the number of carbon atoms present in the amino acid) and are utilized to synthesize ATP molecules.
32. The removal of the amino group does pose one additional problem for the cell. The $NH_4^+$ ion cannot be converted to ammonia for excretion because ammonia is toxic. The cell converts the $NH_4^+$ ion to **urea** in the **urea cycle**, and it is urea that is excreted in urine. See Figure 12.17.

# INDEX

## C

## F



## G

## H

## I

## J



## K

## M

## N

## O

## P

## Q

## U

## Y



## Z

# IMAGE CREDITS

Cover: Copyright © 2013 Depositphotos Inc./2nix.

Figure 1.2: Copyright © Volker Sperlich (CC by 2.0) at http://commons.wikimedia.org/wikiFile:Prinzip_Torricelli.jpg.

Figure 1.3: Copyright © Krishnavedala (CC by 3.0) at http://commons.wikimedia.org/wiki/File:BoylesLaw.svg.

Figure 1.4: Copyright © Kismalac (CC by 3.0) at http://commons.wikimedia.org/wikiFile:CelsiusKelvin.svg.

Figure 1.5: Copyright © Jörg Rittmeister (CC by 3.0) at http://commons.wikimedia.org/wiki/File:Gay-lussac_schema.jpg.

Figure 2.2: Copyright © D-Khru (CC by 2.0) at http://commons.wikimedia.org/wiki/File:Crookes_tube-in_use-lateral_view-standing_cross_prPNr%C2%B011.

Figure 2.4: Copyright © Cush (CC by 3.0) at http://commons.wikimedia.org/wiki/File:Standard_Model_of_Elementary_articles.svg.

Figure 2.8: Copyright © Bobmumgaard (CC by 3.0) at http://commons.wikimedia.org/wiki/File:Alcator_C-Mod_tokamak.jpg.

Figure 2.10: Copyright © Florian Nolz (CC by 3.0) at http://commons.wikimedia.org/wiki/File:Wideroe_linac_en.svg.

Figure 2.11: Copyright © Superborsuk (CC by 3.0) at http://commons.wikimedia.org/wiki/File:Isis_schema.svg.

Figure 2.13: Copyright © Tomasz W. Kozlowski (CC by 2.5) at http://commons.wikimedia.org/wiki/File:LHC.svg.

Figure 2.14: Copyright © Jim.belk (CC by 3.0) at http://commons.wikimedia.org/wiki/File:Coupled_Harmonic_Oscillator.svg.

Figure 2.15: Copyright © Theresa Knott (CC by 3.0) at http://commons.wikimedia.org/wiki/File:Waves_on_an_oscillascope.png.

Figure 2.15: Copyright © Badseed (CC by 3.0) at http://commons.wikimedia.org/wiki/File:Sine_wave_amplitude_wavelength.svg.

Figure 2.17: Copyright © Björn Nordberg (CC by 3.0) at http://commons.wikimedia.org/wiki/File:Fotoelektrisk_effekt2.png.

Figure 2.18: Copyright © Jan Homann (CC by 3.0) at http://commons.wikimedia.org/wiki/File:Visible_spectrum_of_hydrogen.jpg.

Figures 2.19 and 2.20: Copyright © Haade (CC by 3.0) at http://commons.wikimedia.org/wiki/File:Single_electron_orbitals.jpg.

Figure 2.21: Copyright © CK-12 Foundation (CC by 3.0) at http://commons.wikimedia.org/wiki/File:D_orbitals.png.

Figure 2.24: Copyright © Adrignola (CC by 3.0) at http://commons.wikimedia.org/wiki/File:Atomic_orbital_diagonal_rule.svg.

Figure 7.48: Copyright © YassineMrabet (CC by 3.0) at http://commons.wikimedia.org/wiki/File:Amino_acid_zwitterions.svg.

Figure 7.51: Copyright © Protein Chemist (CC by 3.0) at http://commons.wikimedia.org/wiki/File:Peptide-Figure-Revised.png.

Figure 7.54: Copyright © Derekleungszhei (CC by 3.0) at http://commons.wikimedia.org/wiki/File:Adenine_chemical_structure.png; Copyright © Derekleungszhei (CC by 3.0) at http://commons.wikimedia.org/wiki/File:Thymine chemical_structure.png; Copyright © Derekleungszhei (CC by 3.0) at http://commons.wikimedia.org/wiki File:Uracil_chemical_structure.png.

Figure 8.1: Copyright © Vladsinger (CC by 3.0) at http://commons.wikimedia.org/wiki/File:Isomerism.svg.

Figure 8.2: Copyright © Lanzi (CC by 3.0) at http://commons.wikimedia.org/wiki/File:Isomerie_Butan-Pentan-Hexan.png.

Figure 8.3: Copyright © v8rik (CC by 3.0) at http://commons.wikimedia.org/wiki/File:Structural_isomers.png.

Figure 8.5: Copyright © Illusterati (CC by 3.0) at http://commons.wikimedia.org/wiki/File:The_Photon.png.

Figure 8.6: Copyright © Kaidor (CC by 3.0) at http://commons.wikimedia.org/wiki/File:Vertical_polarization.svg.

Figure 8.8: Copyright © Kaidor (CC by 3.0) at http://commons.wikimedia.org/wiki/File:Polarimeter_%28Optical_rotation%29.svg.

Figure 8.9: Copyright © Slashme (CC by 3.0) at http://commons.wikimedia.org/wiki/File:Fischer_Projection2.svg.

Figure 8.11: Copyright © Olaf Matyja (CC by 3.0) at http://commons.wikimedia.org/wiki/File:Aldehyd_D-glicerynowy.svg.

Figure 8.23: Copyright © Christopher King (CC by 3.0) at http://commons.wikimedia.org/wiki/File:DGalactose_Fischer.svg; Copyright © Christopher King (CC by 3.0) at http://commons.wikimedia.org/wiki/File:DTalose_Fischer.svg.

Figure 8.24: Copyright © Cvf-ps (CC by 3.0) at http://commons.wikimedia.org/wiki/File:Cyclohexane_universe.png.

Figure 9.2: Copyright © Ninomy (CC by 3.0) at http://commons.wikimedia.org/wiki/File:Esterification-ja.png.

Figure 9.3: Copyright © OpenStax College (CC by 3.0) at http://commons.wikimedia.org/wiki/File:220_Triglycerides-01.jpg.

Figure 9.7: Copyright © Derekleungszhei (CC by 3.0) at http://commons.wikimedia.org/wiki/File:Adenine_chemical_structure.png; Copyright © Ninomy (CC by 3.0) at http://commons.wikimedia.org/wiki/File:Esterification-ja.png.

Figure 9.8: Copyright © Ninomy (CC by 3.0) at http://commons.wikimedia.org/wiki/File:Esterification-ja.png.

Figure 9.10: Copyright © Reinoutr (CC by 3.0) at http://commons.wikimedia.org/wiki/File:3%27_RNA_degradation.png.

Figure 10.3: Copyright © Provenzano15 (CC by 3.0) at http://commons.wikimedia.org/wiki/File:Exergonic_Reaction.svg.

Figure 10.4: Copyright © Provenzano15 (CC by 3.0) at http://commons.wikimedia.org/wiki/File:Endergonic_Reaction.svg.

Figure 10.7: Copyright © Michal Maňas (CC by 3.0) at http://commons.wikimedia.org/wiki/File:3D_model_hydrogen_bonds_in_water.jpg.

Figure 10.9: Copyright © Jkwchui (CC by 3.0) at http://commons.wikimedia.org/wiki/File:Boiling-points_Chalcogen-Halogen.svg.

Figure 10.11: Copyright © Cawang (CC by 3.0) at http://commons.wikimedia.org/wiki/File:Water_Phase_Change_Diagram.png.

Figure 10.14: Copyright © Webridge (CC by 3.0) at http://commons.wikimedia.org/wiki/File:Enzymatic_mechanism_model.png.

Figure 10.16: Copyright © Webridge (CC by 3.0) at http://commons.wikimedia.org/wiki/File:Enzymatic_mechanism_model.png.

Figure 10.18: Copyright © Mcstrother (CC by 3.0) at http://commons.wikimedia.org/wiki/File:Non-competitive_inhibition.svg.

Figure 10.19: Copyright © Mcstrother (CC by 3.0) at http://commons.wikimedia.org/wiki/File:Allosteric_comp_inhib_2.svg.

Figure 11.4: Copyright © OpenStax College (CC by 3.0) at http://commons.wikimedia.org/wiki/File:0301_Phospholipid_Structure.jpg.

Figure 11.5: Copyright © OpenStax College (CC by 3.0) at http://commons.wikimedia.org/wiki/File:0302_Phospholipid_Bilayer.jpg.

Figure 11.6: Copyright © Dhatfield (CC by 3.0) at http://commons.wikimedia.org/wiki/File:Cell_membrane_detailed_diagram_3.svg.
Figure 11.13: Copyright © OpenStax College (CC by 3.0) at http://commons.wikimedia.org/wiki/File:229_Nucleotides-01.jpg.
Figure 11.14: Copyright © KES47 (CC by 3.0) at http://commons.wikimedia.org/wiki/File:Chromosome_en.svg.
Figure 11.15: Copyright © Madprime (CC by 3.0) at http://commons.wikimedia.org/wiki/File:DNA_replication_split_horizontal.svg.
Figure 11.17: Copyright © Preston Manor School + JFL (CC by 3.0) at http://commons.wikimedia.org/wiki/File:Bsheet.gif.
Figure 11.20: Copyright © Boumphreyfr (CC by 3.0) at http://commons.wikimedia.org/wiki/File:Rna_syn.png.
Figure 11.21: Copyright © Yikrazuul (CC by 3.0) at http://commons.wikimedia.org/wiki/File:TRNA-Phe_yeast_1ehz.png.
Figure 11.23: Copyright © Yikrazuul (CC by 3.0) at http://commons.wikimedia.org/wiki/File:Codon-Anticodon_pairing.svg.
Figure 11.24: Copyright © Boumphreyfr (CC by 3.0) at http://commons.wikimedia.org/wiki/File:Peptide_syn.png.
Figure 12.2: Copyright © Cacycle (CC by 3.0) at http://commons.wikimedia.org/wiki/File:ADP_chemical_structure.png.
Figure 12.3: Copyright © Muessig (CC by 3.0) at http://commons.wikimedia.org/wiki/File:ADP_ATP_cycle.png.
Figure 12.9: Copyright © LHcheM (CC by 3.0) at http://commons.wikimedia.org/wiki/File:Structure_of_Pantothenic_acid.png.
Figure 12.10: Copyright © Regis Frey (CC by 3.0) at http://commons.wikimedia.org/wiki/File:CellRespiration.svg.
Figure 12.11: Copyright © Boumphreyfr (CC by 3.0) at http://commons.wikimedia.org/wiki/File:Gluconeogenesis.png.
Figure 12.12: Copyright © (CC by 3.0) at http://commons.wikimedia.org/wiki/File:L-Lactat-Ion.svg.
Figure 12.13: Copyright © OpenStax College (CC by 3.0) at http://commons.wikimedia.org/wiki/File:2507_The_Krebs_Cycle.jpg.
Figure 12.15: Copyright © Henrik Vestin (CC by 3.0) at http://commons.wikimedia.org/wiki/File:Electron_transport_chain.svg.
Figure 12.16: Copyright © Modre-Osprian, Robert; Osprian, Ingrid; Tilg, Bernhard; Schreier, Günter; Weinberger, Klaus; Graber, Armin, (CC by 3.0) at http://commons.wikimedia.org/wiki/File:BetaOxidationSpiral.tiff.
Figure 12.18: Copyright © Boumphreyfr (CC by 3.0) at http://commons.wikimedia.org/wiki/File:Protein_catabolism.png.
Figure 12.20: Copyright © Boumphreyfr (CC by 3.0) at http://commons.wikimedia.org/wiki/File:ATP_Production_Pathways.jpg.
All other images Copyright in the Public Domain.

CPSIA information can be obtained at www.ICGtesting.com
Printed in the USA
LVOW01s1225190815

450643LV00002B/12/P